M.-E. Michel-Beyerle (Ed.)

# The Reaction Center of Photosynthetic Bacteria

## Structure and Dynamics

Springer

*Berlin*
*Heidelberg*
*New York*
*Barcelona*
*Budapest*
*Hong Kong*
*London*
*Milan*
*Paris*
*Santa Clara*
*Singapore*
*Tokyo*

Proceedings of the workshop on

**The Reaction Center of Photosynthetic Bacteria**

Structure and Dynamics

held from March 1 -4, 1995,
in Feldafing, Bavaria (Germany)

M.-E. Michel-Beyerle (Ed.)

# The Reaction Center of Photosynthetic Bacteria

## Structure and Dynamics

With 167 Figures

Springer

Professor Dr. Maria-Elisabeth Michel-Beyerle
Institut für Physikalische und Theoretische Chemie
Technische Universität München
Lichtenbergstraße 4
D - 85748 Garching, Germany

Cataloging-in-Publication Data applied for

Die Deutsche Bibliothek - CIP-Einheitsaufnahme

The **reaction center of photosynthetic bacteria** : structure and dynamics ; [proceedings of the Workshop on Reaction Center of Photosynthetic Bacteria - Structure and Dynamics, held from March 1 - 4, 1995, in Feldafing, Bavaria (Germany)] / M.-E. Michel-Beyerle (ed.). - Berlin ; Heidelberg ; New York ; Barcelona ; Budapest ; Hong Kong ; London ; Milan ; Paris ; Santa Clara ; Singapore ; Tokyo : Springer, 1996

NE: Michel-Beyerle, Maria E. [Hrsg.]; Workshop on Reaction Center of Photosynthetic Bacteria - Structure and Dynamics <3, 1995, Feldafing>

ISBN-13: 978-3-642-64723-9 e-ISBN-13: 978-3-642-61157-5
DOI: 10.1007/978-3-642-61157-5



**Softcover reprint of the hardcover 1st edition 1996**



Production: PRODUserv Springer Produktions-Gesellschaft, Berlin
Typesetting: Camera-ready by authors
SPIN: 10534506 39/3020 - 5 4 3 2 1 0 - Printed on acid-free paper

# Preface

The workshop on "The Reaction Center of Photosynthetic Bacteria - Structure and Dynamics" held from March 1-4, 1995 at the Hotel "Kaiserin Elisabeth" in Feldafing, Bavaria, constitutes the third Feldafing-Meeting in a series which started in 1985.

The first Feldafing-Meeting (Vol.42, Springer Series in Chemical Physics 1985) had been dominated by the fresh impression of the *three-dimensional structure of the reaction center* from the photosynthetic bacterium *Rhodopseudomonas viridis*. During the following years the riddles in the reaction center function have stimulated a variety of innovative developments of spectroscopic techniques and an explosively fast progress in site-specific modification of the reaction center. The latter has been achieved along two lines, mutagenic replacement of amino acid residues in the protein matrix and thermally induced exchange of cofactors. This expansion of the field has been recorded in "Reaction Centers of Photosynthetic Bacteria" (Vol.6, Springer Series in Biophysics 1990).

Contributions to the '95 Feldafing-Meeting can be considered as the harvest of these novel techniques in spectroscopy, biochemistry and molecular biology. The material is organized in two sections, Part 1 addresses *Crystallographic and Electronic Structure*, Part 2 *Dynamics of Primary Events and the Role of Protein*. One long-standing problem of reaction center function, the mechanism of primary charge separation has finally been solved, in particular, by the most recent generation of femtosecond-time resolved spectroscopy on reaction center preparations with selectively modified energetics. With respect to detailed structure and energy of ionic states an impressive wealth of data has been presented. It is characteristic for the present status of the field that this information pertains to more general aspects of protein function as e.g. accumulative electrostatic interactions, the role of hydrogen bridges in the function of cofactors, dispersive kinetics and energetics or memory effects.

We like to express our thanks to all the participants for their contributions in lectures, in discussions and finally in the papers of this volume. With respect to the organization of the meeting and the editing of the book I am highly indebted to members of my research group, especially to Dipl.Phys. Peter Müller and Dr. Alexander Ogrodnik.

In line with the previous Feldafing-Meetings, also this workshop was organized under the auspices of the two Munich universities, the Technical University and the Ludwig-Maximilians-University, and the Max-Planck-Society. It was supported by "Deutsche Forschungsgemeinschaft" within the special program "Sonderforschungsbereich 143".

Garching, November 1995 M.-E. Michel-Beyerle

# Contents

**PART 2: Dynamics of Primary Events and the Role of Protein**

# PART 1

# Crystallographic and Electronic Structure

# Crystallization and Structure of the Photosynthetic Reaction Centres from *Rhodobacter sphaeroides* - wild type and mutants

Günter Fritzsch, Ulrich Ermler, Michael Merckel and Hartmut Michel

Max-Planck-Institut für Biophysik, Abteilung Molekulare Membranbiophysik, D-60528 Frankfurt a.M., Germany

**Abstract**. Trigonal crystals of the photosynthetic reaction centres from the purple bacterium *Rhodobacter sphaeroides* have been grown with potassium phosphate as precipitant. These crystals diffract to 2.6 Å and are suitable for a detailed structure determination. The cofactor binding is similar to that observed in the reaction centres of *Rhodopseudomonas viridis*. A 23 Å long water chain connects the secondary quinone $Q_B$ with the cytoplasm. The structures of the mutants Thr M222 → Val, Trp M252 → Phe, and Trp M252 → Tyr show minor differences compared with the wild type. In the carotenoidless strain R26 the carotenoid-binding site is occupied by a detergent molecule.

**Keywords.** Photosynthesis, reaction centre, crystallization, structure

## 1. The orthorhombic crystal form

The photosynthetic reaction centre (RC) from the purple bacterium *Rhodobacter (Rb.) sphaeroides* is an integral membrane protein with an $M_r$ = 102,000. In the past, its molecular structure has been studied by analysis of orthorhombic crystals [1,2,3,4,5,6]. When solubilized with N,N-dimethyldodecylamine N-oxide (LDAO) or octyl ß-glucoside orthorhombic crystals of this RC can be grown with poly(ethyleneglycol) ($PEG^{4000}$) as the precipitant in the presence of sodium chloride (300 to 500 mM). Indispensable for the formation of these crystals is the addition of ≥ 3% (w/v) heptanetriol as was found for the crystallization of the RC's from *Rhodopseudomonas (Rps.) viridis* [7].

Applying the vapour diffusion technique orthorhombic crystals appear after a few days or weeks. They are needle-like and may grow to 4 mm long, the thickness is usually a few tenth of a mm, sometimes almost 1 mm (Fig. 1). In spite of these huge dimensions the crystals lack compactness. Generally, considerable parts inside the needles are hollow. When exposed to X-rays the orthorhombic crystals are of low quality. The resolution is about 3 Å in the horizontal direction of the X-ray film. In the vertical direction, however, it is only 4 Å or even worse. This anisotropic diffraction of the reflections is due to fact that the contacts in the crystal lattice are relatively strong in direction to the long crystal axis (z-axis) but rather weak in the directions perpendicular to the z-axis. The observed number of reflections (ca. 22,000 in [4]) is much less than required in order to refine the three spatial coordinates and one isotropic atomic B-factor for each of about 7,000 atoms. The completeness of the data sets (independent reflections) is less than 70% in all cases. For these reasons and for the lack of heavy metal derivatives the structure of the RC from *Rb. sphaeroides* could only be solved with the information from the known RC structure of *Rps. viridis* [8]. Due to the limited resolution the minor differences between both RC's could not be detected by analysis of the orthorhombic crystals.

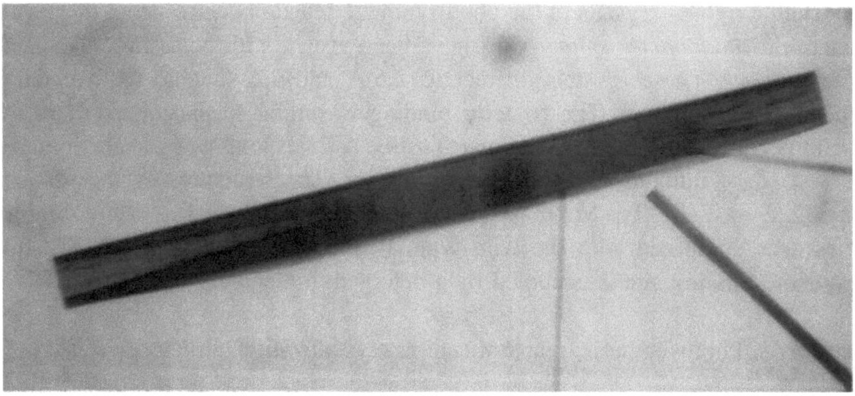

Fig. 1. Orthorhombic crystal grown from $PEG^{4000}$. The length of the crystal is 3.5 mm, the thickness is 0.3 mm. The light, wedge-shaped areas at the right and left part of the crystal indicate low compactness.

## 2. The trigonal crystal form

With the introduction of potassium phosphate as precipitant another crystal form of this protein could be grown [9]. The favorable concentration of RC is 100 μM (10.6 mg/ml). Again LDAO is a proper detergent with an optimal concentration of 0.1 ± 0.05% (v/v). The suitable concentration of potassium phosphate in the mother solution is 0.5 to 1.0 M whereas 1.4 to 1.8 M are optimal for the bottom solution. Heptanetriol is absolutely necessary for this crystallization. Its favourable concentration is 1 to 2% (w/v) with an optimum at about 1.8% (w/v). The addition of 1% to 10% dioxane supports the growth of single trigonal crystals, probably due to a decrease of the dielectric constant. Crystals appear after one or several weeks. They are compact, up to 3 mm long and sometimes more than 1 mm thick (Fig. 2). They belong to the trigonal space group $P3_121$ with unit cell dimensions of 141.4 Å, 141.4 Å, 187.2 Å. The solvent content is 77%.

Fig. 2. Trigonal crystals from RC of *Rb. sphaeroides* grown from potassium phosphate. The length of the biggest crystal is 0.8 mm, its thickness 0.5 mm. The crystal with the hexagonal shape is seen along its z-axis.

With potassium phosphate as precipitating agent the trigonal crystals can coexist with the orthorhombic ones in the same crystallization sample (Fig. 3).

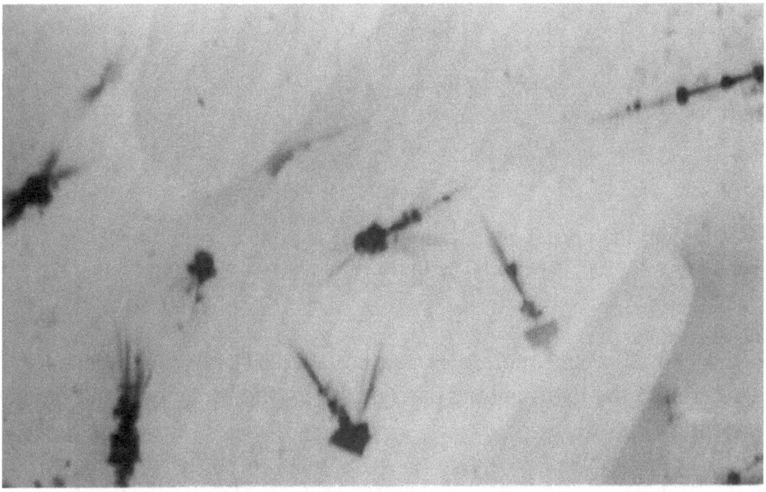

Fig. 3. Coexisting trigonal and orthorhombic crystals. Here, 5 µl of mother solution have been put on a glass slide. The thin, needle-like structures are orthorhombic - the flat structures trigonal crystals.

The needle-like orthorhombic crystals resemble in their low diffraction power the crystals obtained with $PEG^{4000}$. For this reason the conditions in the mother solution must be chosen in such a way that preferentially trigonal crystals grow. This can be achieved 1) by applying concentrations of heptanetriol less than 2.0% and 2) by choosing pH < 6.7. The quality of the trigonal crystals in the X-ray beam, however, is less good at pH < 7.0, but it is much better at pH values around 7.5. So, a pH ≥ 7 should be choosen for the crystallization, even if orthorhombic needles appear and compete with the trigonal crystals. Up to now, the results of the crystallization attempts are not yet reproducible and the physical-chemical background of crystal formation is not understood very well.

The crystal packing density is rather low. In Fig. 4 a view along the crystallographic z-axis is shown where the protein density is represented in black and the solvent space (buffer plus detergent) in white. Fig. 5 shows the crystal packing along the x-axis.

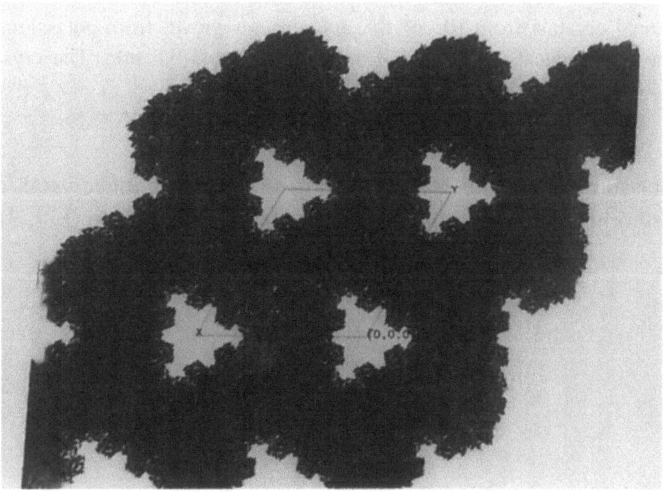

Fig. 4. A view into the crystal along the z-axis. The protein is represented in black, the solvent-containing "holes" in white. The unit cell is marked with thin lines.

More than 10 detergent molecules can be detected crystallographically. They are located around the hydrophobic parts of the L- and M-subunit and replace the lipid molecules (Fig. 6).

Recently, a third crystal form of the RC of *Rb. sphaeroides* has been grown [10]. These crystals have a tetragonal space group and diffract to 2.8 Å resolution. Thus, three different crystal forms of the *Rb. sphaeroides* RC are now available [Table 1].

Fig. 5. A view into the crystal along the x-axis. Further explanations see Fig. 4.

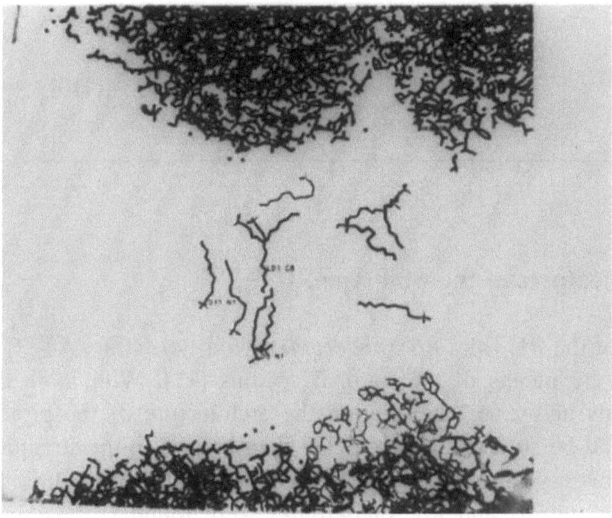

Fig. 6. The crystallographically detectable LDAO molecules. In the upper and lower part of the figure the protein density in a crystal is shown. In the middle part, one RC molecule is omitted (ghost molecule), merely its LDAO molecules are presented. The ghost molecule is envelopped by the LDAO's, its H-subunit touches the residues of the RC molecule shown below. The white regions left and right from the LDAO's are holes filled with buffer (see Fig. 4 and 5).

**Table 1.** Crystal types of RC from *Rb. sphaeroides*

| | Orthorhombic | Trigonal | Tetragonal |
|---|---|---|---|
| Precipitant | PEG | Potassium phosphate | PEG |
| Detergent | LDAO | LDAO | ß-OG |
| Space group | $P2_12_12_1$ | $P3_121$ | $P4_{1,3}2_12_1$ |
| Number of unique reflections | 23,000 | 63,000 | 51,000 |
| Number of molecules in the asymmetric unit | 1 | 1 | 2 |
| Resolution | 3 - 4 Å | 2.6 Å | 2.8 Å |
| References | [1,2,3] | [11, unpubl.] | [10] |

## 3. Structural features of the wild type RC

The structure of the RC from *Rb. sphaeroides*, wild-type strain ATCC 17023, has been analyzed by means of the trigonal crystals [11]. With more than 56,000 reflection points and 2.65 Å resolution the architecture of the protein-cofactor interaction could be studied in more detail than before. In the structure based on the orthorhombic crystal form, some differences in cofactor binding between the RC's from *Rps. viridis* and *Rb. sphaeroides* were found [1,2,3,4,5,6]. Most of these differences could not be confirmed by the analysis of the structure from the trigonal crystals. On the contrary, the protein-cofactor interactions along the electron transfer pathway is rather similar to the *Rps. viridis* RC.

Generally, the distances between the centres of mass of the cofactors along the active branch (A branch) are slightly larger (ca. 0.2 Å) in the *Rb. sphaeroides* than in the *Rps. viridis* RC's. It is unclear to what extent these small deviations are a consequence of the different crystallization conditions like ionic strength or dielectric constant. The major difference in the structures of both RC's is observed at the $Q_B$-binding site. Whereas in the *Rps. viridis* RC the secondary quinone $Q_B$ is located 8 Å away from the non-heme iron it is 12.3 Å in the *Rb. sphaeroides* RC. Further refinement of mutated resp. carotenoidless RC from *Rb. sphaeroides* shows an ambiguous pattern of the $Q_B$ position (Fig. 7). The nearest $Q_B$ is about 10 Å away from the non-heme iron. Detailed studies of these discrepancies are under way.

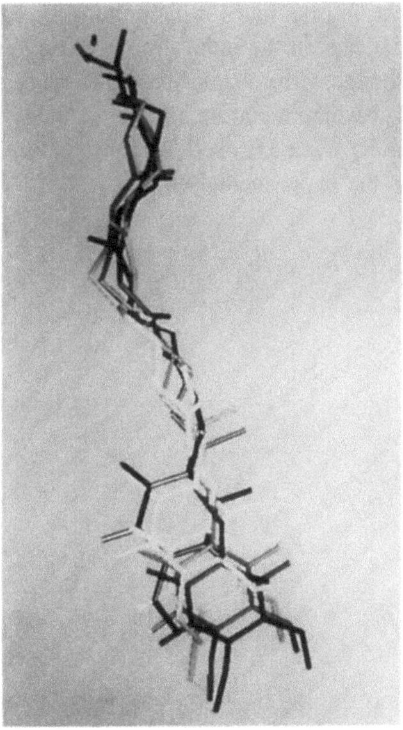

Fig. 7. Superposition of $Q_B$ molecules from five different RC structures (wild type, R26, mutants Trp M252 → Tyr, Ser M222 → Val and Trp M252 → Phe). There is no common site for the five head groups. They are distributed over a distance of about 3 Å. The $Q_B$ head group in the wild type is 12.3 Å away from the non-heme iron (black) whereas the one from Trp M252 → Phe is about 10 Å away.

Between the $Q_B$-binding site and the cytoplasm a water chain of 23 Å length has been found. It consists of 14 fixed water molecules which are close enough to each other to form a linear structure of water molecules connected by hydrogen bonds (Fig. 8). Along this chain mostly acidic and basic amino acid residues are observed which are in contact with the water molecules. This arrangement implies an electrostatically favourable pathway for proton transport from the cytoplasm to the reduced quinone molecule.

## 4. The structures of mutant and carotenoidless RC's

Trigonal crystals have been obtained also from three site-directed mutants [see 12] of amino acids near the $Q_A$-binding site. Compared with the wild type, the mutant Thr M222 → Val does not show significant structural differences beyond the valine residue. The situation is similar for the mutant Trp M252 → Phe. Although the phenyl group is smaller than the indol group of tryptophan, it does not cause remarkable structural changes. More significant differences are observed with the mutant Trp M252 → Tyr where the aromatic part of the tyrosine is shifted almost 1 Å towards the quinone $Q_A$. The $Q_A$ itself is slightly shifted towards the tyrosine, but this effect is within the experimental error.

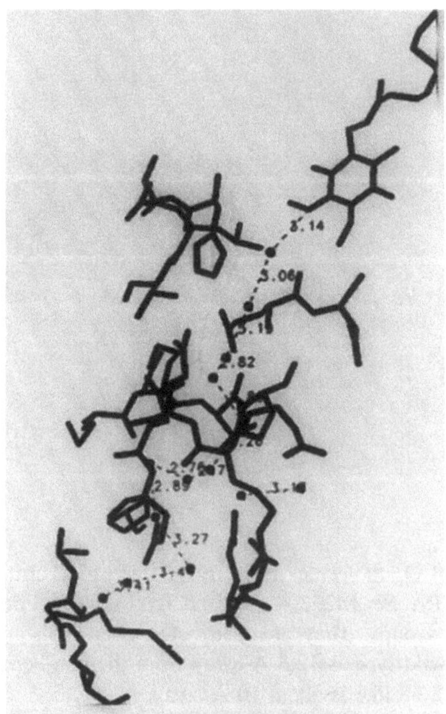

Fig. 8. Image of the water chain and neighbouring residues between the $Q_B$ head group (right above) and the cytoplasm (left below) in the RC from *Rb. sphaeroides*. The black circles represent the water molecules. The numbers show the distances between the individual water molecules in Å. Compared with Fig. 7 the $Q_B$ molecule is shown from a different point of view.

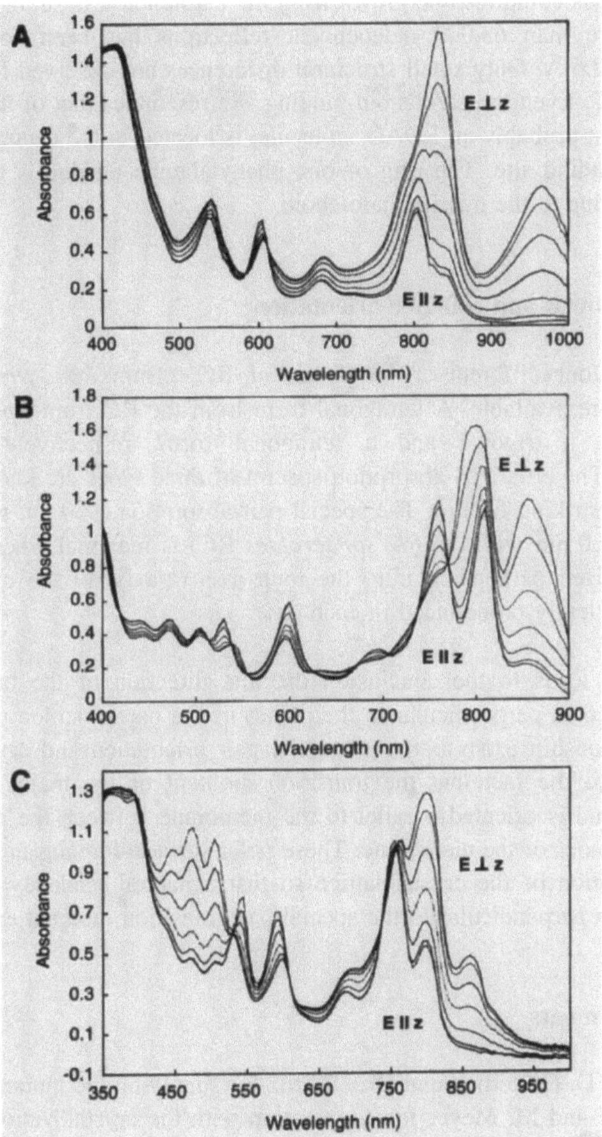


Fig. 9. The polarized absorption spectra of tetragonal crystals from *Rps. viridis* RC (A), of the orthogonal (B) and trigonal (C) crystals from *Rb. sphaeroides* RC. E ∥ z (E ⊥ z): The direction of the polarized light is oriented parallel (perpendicular) to the long axis (z-axis) of the crystal. In each case, a strong dichroism is observed in the region of the special pair absorption (960 nm for the RC from *Rps. viridis* and 860 nm for the RC from *Rb. sphaeroides*). If the light is polarized perpendicular to the z-axis the special pair absorption for each of the three crystal forms is much stronger than for light polarized parallel to the z-axis.

Good crystals could also be grown from the carotenoidless strain R26. A data set with more than 63,000 independent reflections has been collected. The resolution is 2.6 Å. Only small structural differences are observed for the major part of the RC. Even the carotenoid-binding site resembles that of the wild type. A ligand, most probably an LDAO molecule, is located in the central part of the carotenoid binding site. The ring of one phenylalanine residue is tilted by 90° filling a gap due to the missing carotenoid.

## 5. Crystal growth and cofactor orientation

Up to now, four different crystal forms of RC's from *Rps. viridis* and *Rb. sphaeroides* are available: A tetragonal form from the RC from *Rps. viridis*; an orthorhombic, a trigonal and a tetragonal form, respectively, from *Rb. sphaeroides*. The polarized absorption spectra of three types are known (Fig. 9). They share a striking feature: The special pair absorption (960 nm with the *Rps. viridis* RC, 860 nm with the *Rb. sphaeroides* RC) is maximal when the actinic light is polarized perpendicular to the long axis (z-axis) of the crystals. This dichroism is clearly pronounced in each case.

This result leads to the conclusion that the direction of the fastest crystal growth is oriented perpendicular to the planes of the bacteriochlorophyll dimers. Thus, a relationship exists between special pair orientation and crystal growth. This is due to the fact that the transition moment of the major special pair absorption band is oriented parallel to the membrane between the polar surface parts at both sides of the membrane. These polar surface domains are responsible for the formation of the crystal lattice so that a natural tendency of preferred crystal growth perpendicular to the special pair transition moment exists.

## Acknowledgements

We thank Dr. D. Oesterhelt and Dr. U. Stilz for supplying the mutants as well as Dr. H. Scheer and M. Meyer for cooperation with the crystallization of the RC from R26. We acknowledge help in use of the facilities at DESY, Hamburg, by the staff of the EMBL outstation. This work was supported by the Max-Planck-Gesellschaft and the Fonds der Chemischen Industrie.

## References

1. Allen, J.P., Feher, G., Yeates, T.O., Rees, D.C., Deisenhofer, J., Michel, H. and Huber, R. (1986) Proc. Natl. Acad. Sci. USA 83, 8589-8593.
2. Chang, C.-H., Tiede, D., Tang, J., Smith, U., Norris, J. and Schiffer, M. (1986) FEBS Lett. 205, 82-86.

3. Arnoux, B., Ducruix, F., Reiss-Husson, F., Lutz, M., Norris, J., Schiffer, M. and Chang, C.-H. (1989) FEBS Lett. 258, 47-50.
4. Yeates, T.O., Komiya, H., Chirino, A., Rees, D.C., Allen, J.P. and Feher, G. (1988) Proc. Natl. Acad. Sci. USA 85, 7993-7997.
5. Allen, J.P., Feher, G., Yeates, T.O., Komiya, H. and Rees, C.D. (1988) Proc. Natl. Acad. Sci. USA 85, 8487-8491.
6. El-Kabbani, O., Chang, C.-H., Tiede, D., Norris, J. and Schiffer, M. (1991) Biochemistry 30, 5361-5369.
7. Michel, H. (1982) J. Mol. Biol. 158, 657-572.
8. Deisenhofer, J., Epp, O., Miki, R., Huber, R. and Michel, H. (1985) Nature 318, 618-624.
9. Buchanan, S.K., Fritzsch, G., Ermler, U. and Michel, H. (1993) J. Mol. Biol. 230, 1311-1314.
10. Allen, J.P. (1994) Proteins Struct. Funct. Genet. 20, 283-286.
11. Ermler, U., Fritzsch, G., Buchanan, S.K. and Michel, H. (1994) Structure 2, 925-936.
12. Stilz, U., Finkele, U., Holzapfel, W., Lauterwasser, C., Zinth, W. and Oesterhelt, D. (1994) Europ. J. Biochem. 223, 233-242.

# THE CRYSTAL STRUCTURE OF ALLOPHYCOCYANIN AND ITS COMPARISON TO C-PHYCOCYANIN AND b-PHYCOERYTHRIN

Katjusa Brejc, Ralf Ficner[¶], Stefan Steinbacher and Robert Huber

Max-Planck-Institut für Biochemie, Martinsried, Germany
[¶] Present address: EMBL, Heidelberg, Germany

**Abstract.** The phycobiliprotein allophycocyanin (APC[1])from the cyanobacterium *Spirulina platensis* has been isolated and crystallised. The three-dimensional structure of the ($\alpha\beta$) monomer has been solved by multiple isomorphous replacement techniques. The molecular structure of APC is similar to other known phycobiliproteins. In comparison to C-phycocyanin (C-PC) and b-phycoerythrin (b-PE) the major differences arise from deletions and insertions of segments involved in protein-chromophore interactions. In C-PC the absorption spectra of the monomers and the trimers do not differ markedly (near 620 nm). In APC the monomer spectrum is very similar to that of monomeric C-PC, but the spectrum of the trimer is shifted to 650 nm. The exciton interactions between the neighbouring pigments or changed chromophore conformation were proposed as the cause for the 650 nm absorption maximum in APC trimers. The comparison of the chromophore binding regions of both proteins indicates that a significant conformational change occurs in the $\alpha$84 chromophore of APC due to the different protein environment of the $\alpha$ and neighbouring $\beta$-subunit.

**Keywords.** Allophycocyanin, phycobiliproteins, cyanobacteria, X-ray crystallography

## 1 Introduction

The Cyanobacteria and the eukaryotic algae Rhodophyta (red algae) contain phycobilisomes as light-harvesting protein-pigment complexes. These supramolecular aggregates are attached to the outer surface of the thylakoid or photosynthetic membrane [1]. Phycobilisomes absorb light in the range of 450 to 660 nm and transfer the excitation energy to the reaction centres, with an overall quantum efficiency of over 90% [2].

Phycobilisomes consist of two distinct structural parts: the core and the peripheral rods [3, 4]. Both parts contain disc-like structures, composed of phycobiliproteins and linker molecules. The core, which is in direct contact with

[1] Abbreviations used: APC, allophycocyanin; PC, phycocyanin; PE, phycoerythrin; PCB, phycocyanobilin; PCFR, C-phycocyanin from *Fremyella diplosiphon*; PEPC, b-phycoerythrin from *Porphyridium cruentum*.

the photosynthetic membrane, consists predominantly of APC discs ($A_{max}$=650 nm). In the peripheral rods, proximal to the core are phycocyanin discs (PC, $A_{max}$=620 nm), while phycoerythrin discs (PE, $A_{max}$=560 nm) form the tip of the rods.

Phycobiliproteins are brilliantly coloured molecules which unique colours originate from pigment molecules, called phycobilins. Phycobiliproteins are composed of two different subunits, α and β, which occur in equal molar amounts in the protein complex building up an (αβ) heterodimer by convention termed as a monomer. Both subunits carry one or more phycobilins. Phycobilins are open-chain tetrapyrroles covalently bound by thioether linkages to cysteine residues. The spectroscopic variety of the phycobiliproteins is based on chemically distinct pigments, chemically distinct chromophore-protein linkages, specific pigment-protein interactions, and diverse protein environment of the chromophore.

APC is the minor component of the phycobiliprotein fraction in phycobilisomes. The two subunits, α and β, consist of 160 and 161 amino acid residues, respectively [5]. Each subunit has only one pigment, phycocyanobilin (PCB).

APC has been isolated from the cyanobacterium *Spirulina platensis,* [6,7] and crystallised [7]. The crystals belong to space group $P6_322$ with cell constants $a=b=101.9$Å, $c=130.6$Å, $\alpha=\beta=90^\circ$, $\gamma=120^\circ$, with one (αβ) monomer in the asymmetric unit. The three-dimensional structure was solved by multiple isomorphous replacement techniques. The conventional crystallographic *R*-factor of the final structure is 19.6% with data from 8.0 to 2.3 Å resolution [7].

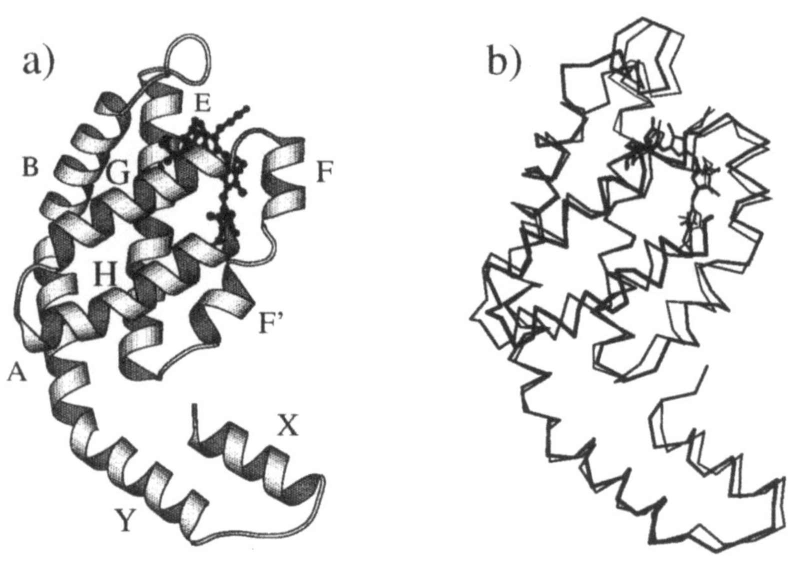


**Fig. 1.** (a) Ribbon representation of the APC α-subunit. Helical segments are denoted X, Y, A, B, E, F', F, G and H. (b) Superposition of the $C^\alpha$ backbone of the α (thin line) and β-subunit (thick line) of the APC structure.

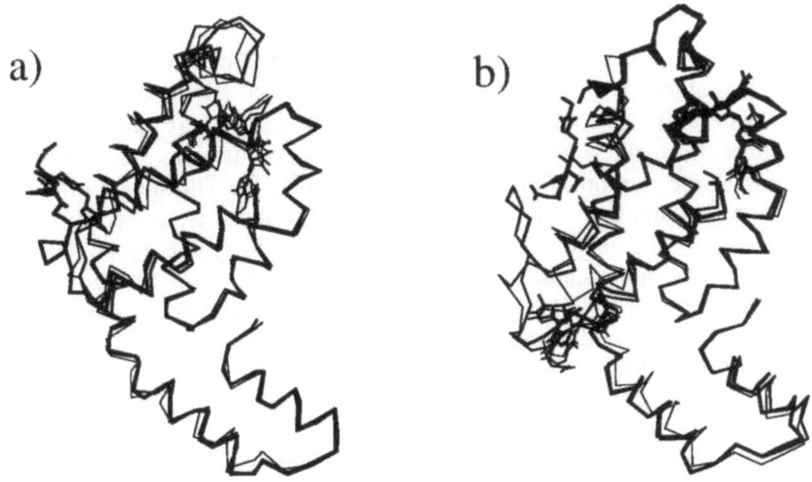


**Fig. 2.** Superposition of the $C^{\alpha}$ backbone of APC (very thin lines), PCFR (thin lines) and PEPC (thick lines) structures: (a) α-subunit, (b) β-subunit.

## 2 Molecular Structure

### 2.1 Comparison of the α and β-subunit of the APC

The α and β-subunits of the APC monomer show the same folding pattern (Fig. 1a). Although the sequence identity between the subunits is rather low (ca. 38%), their tertiary structures are almost identical. There are some minor differences in the spatial arrangement of the secondary structure elements. One of the differences arises from a deletion in the loop connecting helices B and E in the β-subunit (BE loop) as shown in Fig. 1b. The deletion of one residue causes a conformational change at the beginning of the BE loop. In the α-subunit the BE loop is completely exposed to the solvent in contrast to the β-subunit, where the same loop interacts with the chromophore of the neighbouring monomer (Fig. 7). The N-terminus of the α-subunit is two residues shorter than the N-terminus of the β-subunit and is buried within the protein.

### 2.2 Comparison of the APC, PCFR and PEPC structures

The three-dimensional structure of APC confirmes the general principle of phycobiliprotein architecture [8,9,10]. In the phycobiliprotein family the deletions and insertions, involving attachment of additional chromophore molecules, take place only in the BE and GH loops (Fig. 2a and 2b).

In the α-subunit of APC a deletion of two residues in the BE loop changes the loop conformation and results in protein-pigment interactions, which are not present in PCFR and PEPC structures. Another difference occurs between the α-subunits of three phycobiliproteins in the GH loop. Compared to PCFR in APC there is an additional insertion of two residues, and in PEPC the GH loop is enlarged and contains the attachment site for the additional chromophore (Fig. 2a).

In the β-subunits of phycobiliproteins the BE loop is highly conserved. This might be explained by the fact that the residues in the BE loop participate in interactions with two chromophores: β84 chromophore and the 3α84 chromophore from the neighbouring monomer of the trimer. Additionally, it contributes to the monomer-monomer contacts (Fig. 7). The major difference between the β-subunits of the compared phycobiliproteins is located in the GH loop (Fig. 2b). APC is the only phycobiliprotein which does not contain an additional chromophore binding site. Compared to PCFR the GH loop in APC is ten residues shorter. The GH loop in PEPC, compared to the PCFR structure, has five additional amino acid residues which are directed towards the double-linked chromophore (Fig. 2b).

## 3 Structure and environment of the chromophores

APC carries two PCB chromophores at positions α84 and β84. Both sites are topological equivalent to the pigment binding sites in PCFR and PEPC. The stereochemistry of the chiral atoms for the α84 and β84 chromophores is $C_{(2)}$-R, $C_{(3)}$-R and $C_{(31)}$-R. The configuration ($C_{(4)}$-Z, $C_{(10)}$-Z and $C_{(15)}$-Z) and the conformation ($C_{(5)}$-*anti*, $C_{(9)}$-*syn* and $C_{(14)}$-*anti*) are equal for both chromophores. The stereochemistry and configuration of the chromophores in APC are identical with the PCB pigments in PCFR. In all three phycobiliprotein structures a "canonical" residues keeping the chromophore in an extended conformation are conserved: Argα86, Aspα87 and Thrβ77 for the α84 chromophore and Argβ86 and Aspβ87 for the β84 chromophore.

### 3.1 Geometry and the environment of the α84 chromophore

The residues of two different subunits modulate the spectral properties of the α84

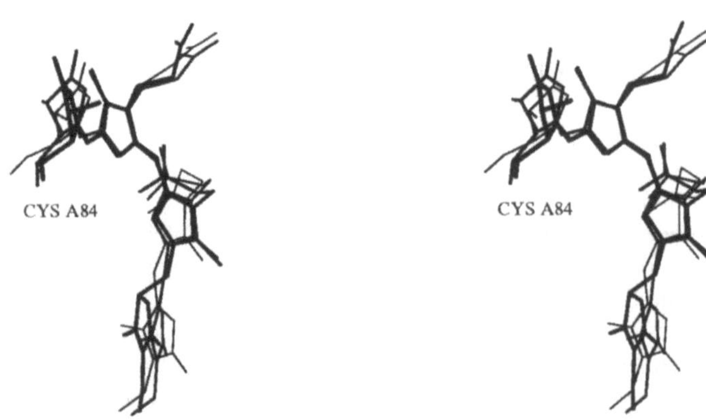


**Fig. 3.** Best fit superposition of the α84 chromophores of APC (very thin lines), PCFR (thin lines) and PEPC (thick lines) structures.

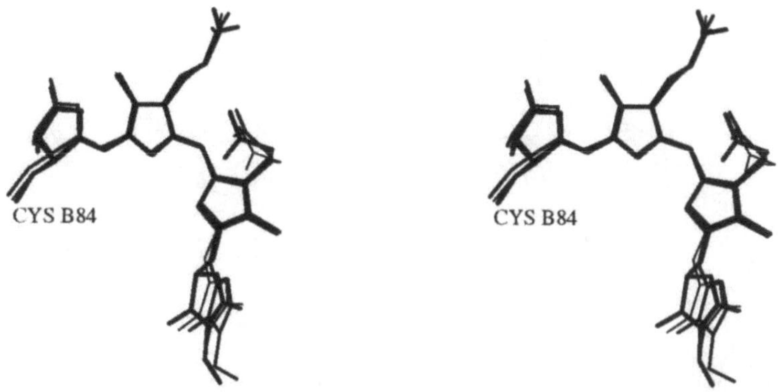


**Fig. 4.** Best fit superposition of the β84 chromophores of APC (very thin lines), PCFR (thin lines) and PEPC(thick lines) structures.

chromophore: amino acid residues from the 1α-subunit and 2β-subunit (Fig.7).

The regions enveloping the chromophore binding site in the α-subunit differ between APC, PCFR and PEPC (Fig. 2a). The α84 binding pocket is formed by the C-terminal part of the B helix, the BE loop, the N-terminal part of the E helix and the N-terminal part of the G helix. Specific protein-pigment contacts arise from the different amino acid sequence and changed polypeptide conformation. The deletion of two amino acid residues in the BE loop of APC results in novel interactions not present in PCFR or PEPC structures.

The second major difference arises from the changed conformation of the BE loop of the neighbouring 2β-subunit. Although the BE loop is highly conserved, the deletion at the beginning of this loop in APC has a strong impact on the protein-pigment interactions with the 1α84 chromophore. In PCFR only two residues (Arg2β57 and Thr2β77) and in PEPC one residue (Thr2β77) make contact with this chromophore. In APC the following amino acid residues participate in protein-pigment interactions: Tyr2β62, Thr2β67, Met2β75 and Thr2β77.

A superposition of the α84 chromophore (Fig. 3) reveals large differences between the three structures. These differences are also confirmed by comparison of the dihedral angles of the chromophores. The largest deviations are observed for rings A and D. In PEB, rings C and D are connected by a methylene bridge (a methine bridge in PCB) allowing free rotation around the bond. Considering only the PCB chromophore it is evident that the PCB rings in APC have smaller deviations from the ideal planar conformation than the rings of the chromophore in PCFR.

### 3.2 Geometry and environment of the β84 chromophore

The β84 chromophore, interacting only with the β-chain, points towards the inner channel of the trimer and probably interacts with a linker molecule. The superposition of the β-subunits (Fig. 2b) shows high similarity in the

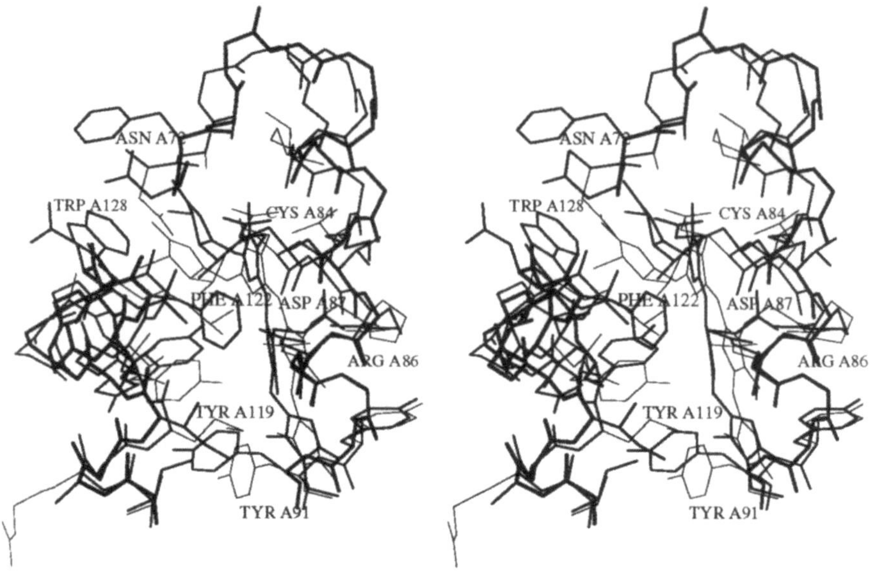


**Fig. 5.** Superposition of the C$^{\alpha}$ backbone of APC (thin lines) and PCFR (thick lines) structures. The α84 chromophores and their interaction with neighbouring residues in the monomer (note that the α84 chromophore is buried in the trimeric form).

conformation of the regions around the chromophore. From the residues which interact with the β84 chromophore, only two are not conserved in all three structures: β90 and β91. In PCFR and PEPC both residues are isoleucine and in APC both are tyrosine.

From the superposition of the β84 chromophores it is evident that the differences between the PCB chromophores are rather small (Fig. 4). The only two major differences are in ring D and in the conformation of the propionate side-chain of ring C which is exposed in the solvent.

## 4 Spectral differences between APC and PC

In PC, the absorption spectra of the monomers and trimers are almost identical (near 620 nm), so the oligomerisation do not influence the spectral properties of either chromophore. In APC, the monomer spectrum is very similar to that of PC monomer. In contrary, the spectrum of trimer is significantly different. The trimeric form of APC has a sharp maximum near 650 nm. The exciton interactions between the neighbouring chromophores [11] or a changed chromophore conformation due to subunit interaction [12] were proposed as the cause for the red shift in APC trimers.

Structure-based comparison of the APC and PCFR binding regions shows a significant conformational change in the α84 chromophore due to the different protein environment of the α and the neighbouring β-subunit (Fig. 5). The only conserved residues which surround and interact with the α84 chromophore are Argα86, Aspα87, Tyrα90 and Tyrα91. The amino acid substitutions Trpα128Ala and Tyrα129Ile reduce the hydrophobic interaction in APC compared to PCFR. The smaller deviation from planarity of the α84 chromophore in APC is probably contributing to the shift of the trimer absorption maximum to a longer wavelength.

As shown in Fig. 6. the difference between the chromophore environment in APC and PCFR structures are small in the β-subunit. The only not conserved amino acid residues are β90 and β91. In APC, one of the tyrosines Tyr (β90) has an almost parallel orientation to the D ring and may interact with the aromatic system of the β84 chromophore.

## 5 Conclusions

The phycobilins are conformational highly flexible molecules, whose spectral properties strongly depend on the environment imposed by the native protein. The spectral differences between phycobiliproteins generally depend on the type and number of pigments bound and different protein environment of the chromophores. Although the phycobiliproteins look alike in the overall structure small differences, occuring outside the conserved secondary structure elements, determine the spectral characteristics of the entire protein. Certainly, the implications of such differences could not be predicted. The phycobiliproteins are another example for the generation of the specifities within homologous proteins by insertions and deletions of amino acids in loop regions.

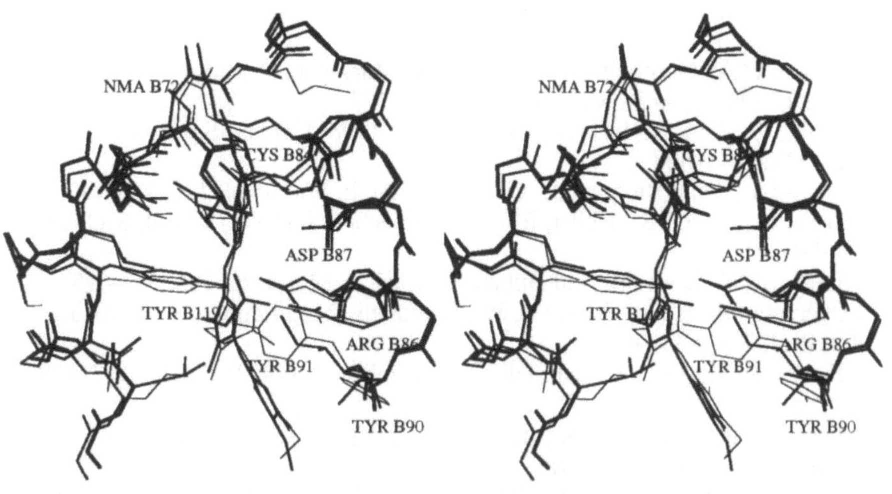


**Fig. 6.** Superposition of the $C^{\alpha}$ backbone of APC (thin lines) and PCFR (thick lines) structures. The β84 chromophores and their interaction with neighbouring residues in the monomer.

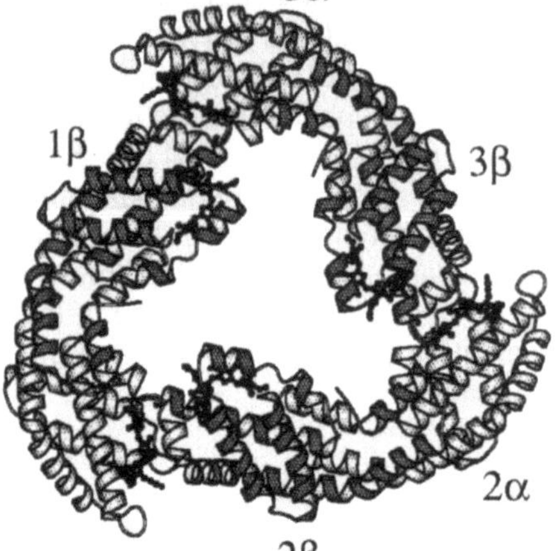


**Fig. 7.** Structure of the APC trimer projected down the crystallographic 3-fold axis. Ribbon presentation of the α-subunit (light grey), the β-subunit (dark grey) and chromophores (black coloured). The α84 chromophore is situated on the periphery of the trimer and the β84 chromophore is pointing towards the central channel.

## References

[1] Gantt, E. and Conti, S.F. (1966) J. Cell Biol. 29, 423-434.
[2] Sauer, K. (1975) in Bioenergetic of Photosynthesis (Govindjee, ed.), pp. 115, Academic Press, San Francisco.
[3] Mörschel, E., Koller, K.-P. and Wehrmeyer, W. (1980) Arch. Microbiol. 125, 43-51.
[4] Bryant, D.A., Guiglielmi, G., Tandeau de Marsac, N., Castets, A. and Cohen-Bazire, G. (1979) Arch. Microbiol. 123, 113-127.
[5] Sidler, W., Gysi, J., Isker, E. and Zuber, H. (1981) Hoppe-Seyler's Z. Physiol. Chem. 362, 611-628.
[6] Boussiba, S. and Richmond, A.E. (1979). Arch. Microbiol. 120, 155-159.
[7] Brejc, K., Ficner, R., Huber, R. and Steinbacher, S. (1995) J. Mol. Biol. 249, 424-440.
[8] Schirmer, T., Bode, W. and Huber, R. (1987) J. Mol. Biol. 196, 677-695.
[9] Duerring, M., Huber, R., Bode, W., Ruembeli, R. and Zuber, H. (1990) J. Mol. Biol. 211, 633-644.
[10] Ficner, R., Lobeck, K., Schimdt, G. and Huber, R. (1992) J. Mol. Biol. 228, 935-950.
[11] MacColl, R., Csatorday, K., Berns, D.S. and Traeger, E. (1981) Arch. Biochem. Biophys. 208, 42-48.
[12] Murakami, A., Mimuro, M., Ohki, K. and Fujita, Y. (1981) J. Biochem. 89, 79-86.

# New Insights into the X-Ray Structure of the Reaction Center from *Rhodopseudomonas viridis*

C. Roy D. Lancaster and Hartmut Michel
Max-Planck-Institut für Biophysik, Abteilung Molekulare Membranbiologie, Heinrich-Hoffmann-Straße 7, D-60528 Frankfurt am Main, Germany

**Abstract.** Light-induced electron transfer in the reaction center (RC) is coupled to the uptake of protons from the cytoplasm at the binding site of the secondary quinone ($Q_B$). With quinone reconstitution, a higher $Q_B$ occupancy has been obtained in the *Rhodopseudomonas (Rp.) viridis* RC crystals, which, after X-ray analysis, has resulted in a well-defined $Q_B$-site model. On the basis of this new model and the structure of a stigmatellin-RC complex, a new understanding of $Q_B$ binding, particularly concerning the role of Ser L223, has emerged. Furthermore, with high quality data collected on RC complexes with atrazine and two chiral atrazine derivatives, it has been possible to describe the exact nature of triazine binding and its effect on the structure of the protein. The new data are of sufficient quality to improve the original model also in other parts, e.g. regarding the asymmetry of binding of the primary quinone ($Q_A$), the structure of the carotenoid, and additional tightly-bound water molecules.

**Keywords.** Crystallography, refinement, $Q_B$, ubiquinone, stigmatellin, atrazine, triazine, $Q_A$, carotenoid, proton uptake

## 1 Introduction

The photosynthetic reaction centers (RCs) from purple bacteria are the best characterized membrane protein complexes (1-5). Their function, the storage of light energy in form of chemical energy, i.e. reduction potential, is associated with the uptake of protons from the cytoplasm into the membrane. This coupling of light-induced electron transfer to proton uptake takes place in the RC at the binding site of the secondary acceptor quinone, $Q_B$. In the original structure of the RC from the purple bacterium *Rhodopseudomonas (Rp.) viridis* (Brookhaven Protein Data Bank entry 1PRC, 6-9), the $Q_B$ site was poorly defined because it was only 30% occupied in the standard RC crystals. Recently, quinone reconstitution experiments have yielded crystals with full quinone occupancy of the $Q_B$ site. Subsequent X-ray analysis and refinement has led to a well-defined $Q_B$-site model, which shall be discussed below (cf. 2.1) and compared to the RC $Q_B$-site models deposited in the Protein Data Bank (cf. 2.2). Also, the structure of a $Q_B$-depleted RC will be presented (cf. 2.3). Furthermore, the complex of the RC with the antibiotic stigmatellin (10,11; cf. Fig. 1), a chromone-type electron transfer inhibitor, will be introduced (cf 2.4). Apart from inhibiting quinone reduction, both in photosystem II (12) and in photosynthetic RCs (13,14), stigmatellin is equally active at inhibiting quinol oxidation, for instance in the photosynthetic *bc*$_1$-complex (15), in the cytochrome *b*$_6$*f*-complex of chloroplasts (12), and in the mitochondrial

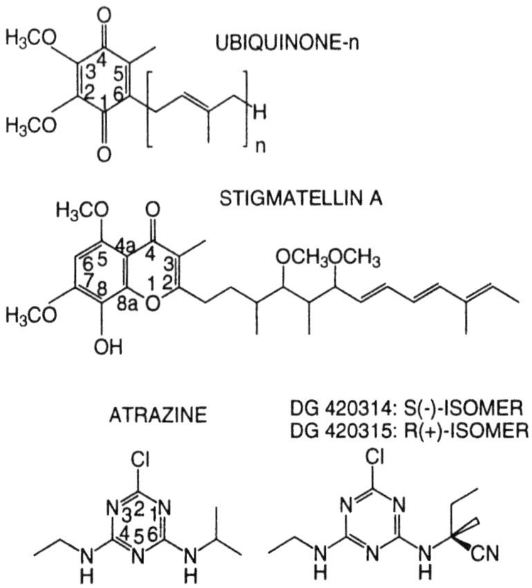


**Figure 1.** Chemical structures of $Q_B$ and $Q_B$-site inhibitors. The native $Q_B$ in the RC of *Rp. viridis* is ubiquinone-9. See text for details

*bc₁*-complex (16,17).

The $Q_B$ site is also a well-established site of herbicide action. Over 50% of commercially available herbicides function by inhibition of higher plants at the $Q_B$ site on the D1 polypeptide of the photosystem II (PSII) reaction center (18). Many herbicides are triazines (e.g. atrazine, terbutryn). Although the triazine binding site has been localized by X-ray crystallographic analysis of the RC-terbutryn complex at a resolution of 2.9 Å (19), description of the exact nature of triazine binding and its effect on the structure of the protein has had to await the refinement of high resolution structures. With high quality data collected recently on RC complexes with atrazine (2-chloro-4-ethylamino-6-isopropyl-amino-*s*-triazine) and two chiral atrazine derivatives (cf. Fig 1), such a detailed description is now possible (cf. 2.5).

The data collected to resolve the $Q_B$-site related issues introduced above are of sufficient quality to improve the original model in parts other than the $Q_B$ site. Possibly biologically relevant modifications include the asymmetry of binding of the primary quinone $Q_A$ (cf. 3), the structure of the carotenoid (cf. 4), and the observation of more than a dozen additional tightly-bound water molecules between the cytoplasmic surface and the $Q_B$ site (will be presented elsewhere).

## 2 The Secondary Acceptor Quinone ($Q_B$) Binding Site

Photosynthetic reaction centers were purified and crystallized essentially as described earlier (20), with $Q_B$ removal achieved by a one hour illumination (900-1000 nm) of the protein in the presence of 20 mM ascorbate between column runs. Reaction centers were reconstituted with 50 μM ubiquinone-2, 170 μM atrazine (Riedel-de-Haën, Seelze), 200 μM chiral atrazine derivative, and 400 μM stigmatellin (Fluka, Neu-Ulm), respectively. Diffraction data were collected using monochromatic synchrotron radiation at wavelengths of 0.92-1.0 Å at the EMBL X11 and MPG/GBF-BW6 beamlines at HASYLAB/DESY Hamburg. Data were processed with the MOSFLM (21) and CCP4 (22) program packages. Crystallographic refinement was performed using iterative cycles of simulated annealing, conventional positional refinement, isotropic overall

B-factor refinement, and restrained refinement of individual B-factors with the program X-PLOR (23). Manual inspection and refitting were done with O (24). Details of the above methods will be given elsewhere.

The crystallographic R-factors ($R = \Sigma||F_o|-|F_c||/\Sigma|F_o|$) for the structures are currently betreen 17.6% and 19.2% (cf. Table 1). Of particular interest is the DGM (DG-420314) complex, in that it is defined by over 100,000 unique relections to a high resolution limit of 2.3 Å. This is the best-defined reaction center structure to date and the R-factor currently stands at 18.4% for all data between 10.0 and 2.30 Å. The deviations from ideal stereochemistry of the models are minute.

**Table 1.** Data collection and refinement statistics for *Rp. viridis* RC crystals modified at the $Q_B$ site.

| | Q286[a] | $Q_B$-dpl. | STG | DGP | DGM | ATZ |
|---|---|---|---|---|---|---|
| **DATA** | | | | | | |
| High resol. limit [Å][b] | 2.45 | 2.40 | 2.40 | 2.65 | 2.30 | 2.35 |
| No. of unique refl. | 79,397 | 87,694 | 76,640 | 72,314 | 102,117 | 92,322 |
| Completeness [%] | 76.5 | 79.4 | 69.4 | 93.9 | 81.4 | 78.5 |
| $R_{sym}$[c] [%] | 9.6 | 8.7 | 8.2 | 7.4 | 7.3 | 7.9 |
| **REFINEMENT** | | | | | | |
| R-factor[d] [%] | 18.0 | 17.6 | 19.1 | 18.5 | 18.4 | 19.2 |
| 10.0-2.70 Å | 17.0 | 16.6 | 17.7 | 17.8 | 16.8 | 17.7 |
| r.m.s. dev. from ideal values[e] | | | | | | |
| bonds [Å] | 0.011 | 0.011 | 0.011 | 0.012 | 0.011 | 0.011 |
| angles [°] | 1.38 | 1.39 | 1.46 | 1.50 | 1.37 | 1.40 |

a) Q286 = ubiquinone-2 reconstituted RCs; $Q_B$-depl. = $Q_B$-depleted RCs;
STG = stigmatellin-RC complex; DGP = RC complex with DG-420315;
DGM = RC complex with DG-420314; ATZ = RC complex with atrazine;
b) Low resolution limit = 10.0 Å in all cases.
c) $R_{sym} = \Sigma_{i,hkl}(|I(i,hkl)-\langle I(hkl)\rangle|)/\Sigma_{i,hkl}(I(i,hkl))$ = R-factor on symmetry-related intensities
d) $R = \Sigma||F_o|-|F_c||/\Sigma|F_o|$ = The crystallographic R-factor, calculated for the full resolution range (upper row) and for the 10 to 2.7 Å range (lower row).
e) based on X-PLOR version 3.1 parameter files (parhcsdx.pro (25), param19.sol, param19.ion) and newly developed cofactor parameter files (26).

## 2.1 Quinone Reconstitution

The electron density at the $Q_B$-binding site in the Q2-reconstituted complex is displayed in Figure 2. All protein residues are well defined in terms of electron density, as is also the case for the ubiquinone-2 molecule itself. The overall shape of the density correlates well with the dimensions of the molecule and thus allows unequivocal placement of all atoms.

Compared to the previous model (1PRC; 9), the hydrogen bonds assigned between the proximal carbonyl $Q_B$ oxygen (O4) and the His L190

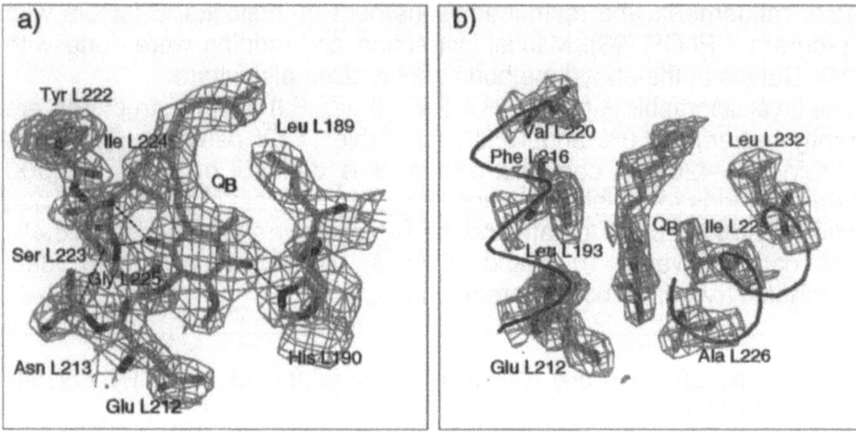


**Figure 2. a)** Electron density at the $Q_B$-binding site: The map was calculated with the coefficients 2lFol-lFcl and phases $\alpha_C$. Contouring is at a level of 1.3 $\sigma$ above the mean density of the map. Hydrogen bonds are indicated as dashed thin black lines. **b)** Side view of Figure 3a. The phenylalanine ring is oriented approximately parallel to the quinone ring system.

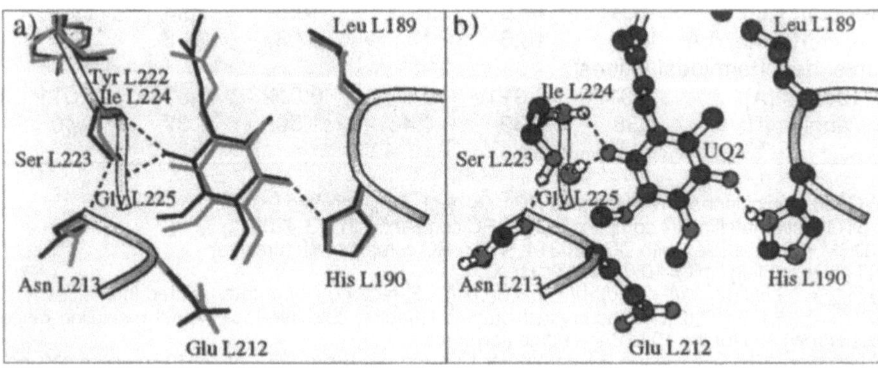


**Figure 3. a)** Comparison of the structure at the $Q_B$ site based on Q2-reconstitution (in black) to the original structure 1PRC (9) in green. **b)** Hydrogen bonding at the secondary quinone ($Q_B$) binding site. Hydrogen bonds are represented as dashed black lines.

---

Nδ and between the distal carbonyl oxygen (O1) and the peptide backbone nitrogen of Gly L225 are maintained (cf. Fig 3a). However the distal side of the molecule has moved by approximately 0.7 Å. Such a movement is consistent with calculations on favoured quinone carbonyl binding regions of the Rp. viridis RC (27). As a consequence of this movement, the hydrogen bond from the side chain Oγ of Ser L223 is more favourably donated to the side chain Oδ of Asn L213 (cf. Fig 3b) and not to the distal carbonyl oxygen of $Q_B$, as was suggested earlier (8). Other discrepancies to the previous model include different side chain orientations for Glu L212 and Asn L213 (cf. Fig 3a) and the reduction of the interplanar angle between the aromatic ring of Phe L216 and the quinone ring of $Q_B$ from 12.9° to 5.6°.

## 2.2 Comparison to Other $Q_B$ Site Models

The RC coordinate files which are available from the Brookhaven Protein Data Bank (PDB; 28,29) are listed in Table 2 as are two of the new structures based on quinone reconstitution and the DG-420314-complex, respectively. Structures are listed in terms of PDB entry code, where applicable, of maximum resolution, the number of unique reflections, and the resulting completeness of data. When comparing the structures (cf. (30) for an overall comparison), it is important to keep in mind the ratio between the number of unique reflections used for refinement, i.e. the number of independent observations, $n_{obs}$, and the number of parameters necessary to define the model, $n_{par}$. The latter includes three parameters (x,y,z) per non-zero occupancy atom plus one parameter for the isotropic atomic B-factor, where applicable. The *Rp. viridis* RC structures and the new *Rb. sphaeroides* RC structure 1PCR, based on the trigonal crystal form, all display $n_{obs}/n_{par}$ ratios of approximately 2 to 2.5, whereas the ratios are smaller than 1 for all *Rb. sphaeroides* RC structures based on orthorhombic crystal forms.

The profile of the $Q_B$ pocket is conserved in all reaction center structures. The $Q_B$ molecule in the 1PCR model is displaced by approximately 5 Å from the other $Q_B$ models and displays high isotropic temperature factors (32). It is mostly omitted from the following discussion, because it seems possible that this $Q_B$ site contains the (reduced) quinol and not the quinone. Since it combines the requirements of high quinone occupancy and a sufficiently high $n_{obs}/n_{par}$ ratio (cf. Table 2), the structure of the ubiquinone-2-reconstituted *Rp. viridis* RC complex (Q286) will be used as a guideline for the following discussion of the $Q_B$ site. Apart from new insights regarding the role of Ser L223, the hydrogen bonding pattern is similar to Q286 in the structures 1PRC, 4RCR, and 1PSS (cf. Fig 4a), although in the latter structure the hydrogen bond from Gly L225 N to the distal $Q_B$ carbonyl oxygen, O1, is lost. In the 2RCR structure, the quinone ring plane is oriented nearly perpendicular (65°-80°) to the other $Q_B$ rings.

In the Q286 structure, the ring planes of Phe L216 and $Q_B$ are approximately parallel, with an interplanar angle of only 5.6°. This angle is 12.9° (1PRC), 22.1° (1PSS), 68.4° (2RCR), and 78.7° (4RCR) in the other structures, indicating that it is smaller for the better-defined structures. The interplanar distance in the Q286 structure is approximately 3.5 Å. However, as the rings are displaced in the planes, the centroid-to-centroid distance of 5.3 Å is significantly larger. The importance of Phe L216 for quinone binding is illustrated by a reduced affinity for $Q_B$ in a Phe L216→Ser herbicide-resistant *Rp. viridis* mutant (T6; 38). The most striking discrepancy concerning the Phe L216 residues of the other structures is the 60°-70° plane deviation in the 4RCR structure (cf. Fig. 4b).

**Table 2.** Reaction Center Structures[a] (from (30); updated)

| | max. res.[b] (Å) | No. of unique reflections | compl.[c] (%) | No. of non-zero occupancy atoms | $n_{obs}/n_{par}$ [d] | R-factor (%) | mean coord. error (max.)[e] | Reference |
|---|---|---|---|---|---|---|---|---|
| *Rp. viridis* | | | | | | | | |
| 1PRC | 2.30 | 95,762 | 75.4 | 10,045 | 2.38 | 19.3[f] | 0.26 | (6-9) |
| Q286 | 2.45 | 79,397 | 76.5 | 10,102 | 1.96 | 18.0 | 0.3 | (26) |
| DGM | 2.30 | 102,117 | 81.4 | 10,307 | 2.47 | 18.4 | 0.25 | inhib. compl. (31) |
| *Rb. sphaeroides* | | | | | | | | |
| 1PCR | 2.65 | 56,141 | 90.4 | 7340 | 1.91 | 18.6 | 0.3 | (32) |
| 1PSS | 3.0 | 21,518 | 68.9 | 6789 | 0.79 | 22.3 | 0.5 | (33) |
| 1YST | 3.0 | 19,630 | 66.0 | 7136 | 0.69 | 23.4 | 0.4 | (34) |
| 2RCR | 3.1 | 13,493 | 50.8 | 6921 | 0.63 | 22.0 | 0.5 | (35) |
| 4RCR | 2.8 | 21,992 | 60.0 | 6764 | 0.81 | 22.7 | 0.4 | (36) |

a) Structures 1PRC, 1PCR, 1PSS, 1YST, 2RCR, and 4RCR are named according to their four-character entry code in the Brookhaven Protein Data Bank (PDB). Names Q286 and DGM are unofficial and refer to the structures based on ubiquinone-2 reconstituted RCs the complex with DG-420314, respectively
b) maximum resolution limit
c) Data completeness;These values were calculated for the number of reflections actually used in refinement and may differ from those given in the appropriate references.
d) $n_{obs}$ = the number of observed unique reflections (only the reflections used for refinement are considered); $n_{par}$ = the number of parameters necessary to define the model. This includes 3 parameters (x,y,z coordinates) per non-zero-occupancy atom plus one (isotropic atomic B-factor) where applicable (the 2RCR structure is reported without atomic B-factors), plus additional parameters (occupancy) for alternate conformations.
e) estimate of the upper limit of the mean corrdinate error from a Luzzati (37) plot.
f) In our system of R-factor calculation , which does not account for the solvent contribution to diffuse scattering at low resolution, the R-factor of the 1PRC structure (9) is not 19.3%, but 20.2% at 2.3 Å resolution, indicating that the R-factor of the DGM structure is nearly 2 percentage points better than that of the original 1PRC structure.

In summary, the improved model of the *Rp. viridis* $Q_B$ site (Q286) displays only small but important differences to the original (1PRC) model. In particular, this applies to the hydrogen bonding of the distal carbonyl oxygen of the quinone and the role of Ser L223. The discrepancies between the four *Rb. sphaeroides* $Q_B$ sites based on orthorhombic crystal forms are substantially greater than those between the two *Rp. viridis* $Q_B$-site models. Considering the conservation of the overall profile of the $Q_B$ site, it may well turn out that the $Q_B$ sites of the two species are more similar than can presently be discussed.

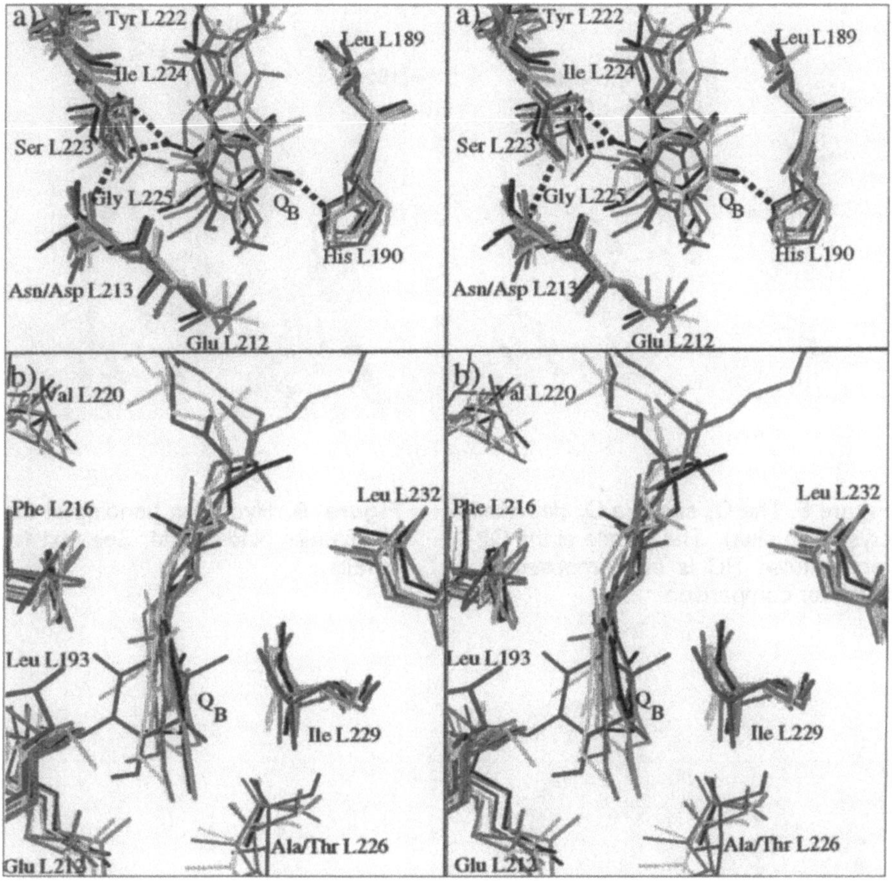


**Figure 4.** Orthogonal stereo views of the $Q_B$ site: Seven different structures were superimposed by a least-squares procedure using only Cα atoms as implemented in O (24). The colour coding is as follows: Q286 (black), 1PRC (green), 1PCR (pink), 2RCR (dark blue), 4RCR (brown), 1PSS (light blue) and 1YST (orange). The dashed black lines indicate hydrogen bonds in the Q286 structure.

## 2.3 Quinone Depletion

The $Q_B$-site model of the $Q_B$-depleted preparation is shown in Figure 5. Apparently, six water molecules and a detergent replace the quinone and form hydrogen bonds with the respective protein residues. Differences between the quinone-reconstituted and the quinone-depleted structure consist of a small (~0.3 Å) movement of the main chain of the loop (L220-L226) connecting the cytoplasmic DE helix (L209-L220) and the transmembrane E-helix (L226-L249).

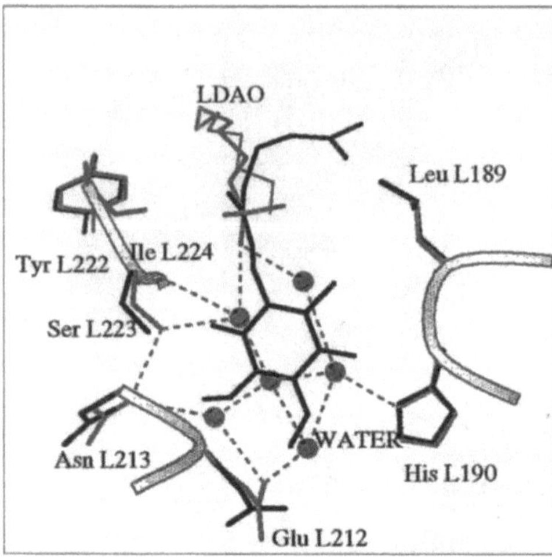


**Figure 5.** The $Q_B$ site in a $Q_B$-depleted crystal (in pink). The $Q_B$ site of the Q2-reconstituted RC is superimposed in black for comparison.

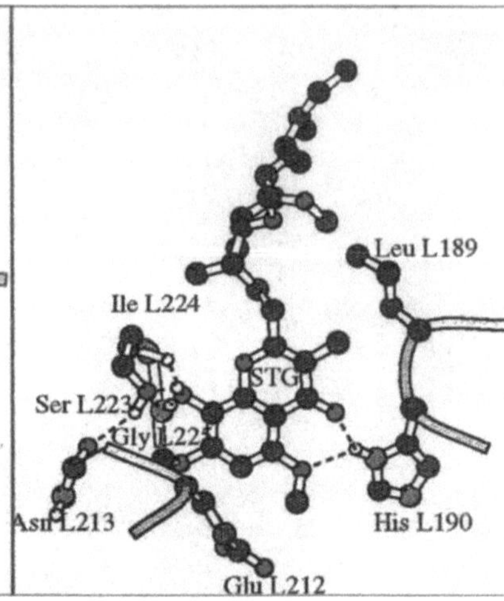


**Figure 6.** Hydrogen bonding at the stigmatellin binding site. See text for details.

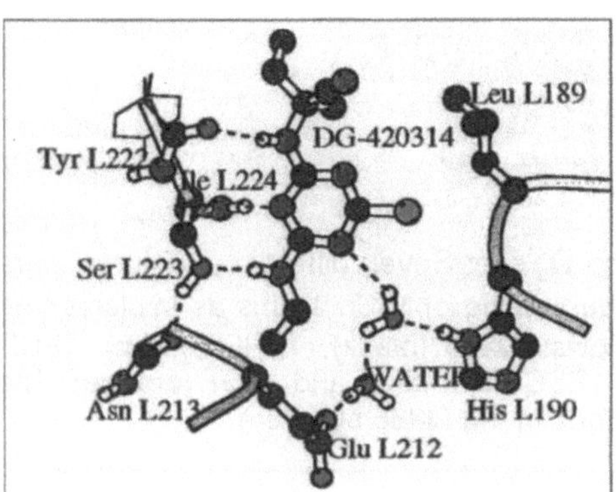


**Figure 7.** Hydrogen bonding at the triazine binding site. See text for details.

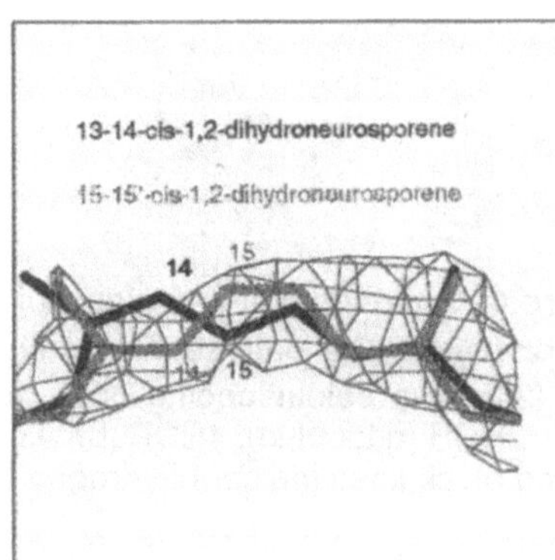


**Figure 9.** Simulated annealing (SA) omit map $(2|F_o|-|F_c|)$ of the carotenoid. Contouring is at a level of1.3 $\sigma$ above the mean density of the map. The 13,14-cis-carotenoid is shown in brown, the 15-15' cis carotenoid is depicted in pink.

The most marked side chain changes are found for Glu L212, which forms a hydrogen bond to one of the new waters, and Asn L213 (approx. 0.8 Å), causing a slight contraction of the pocket. The orientations of Glu L212 and Asn L213 in the $Q_B$-depleted structure are very similar to the original 1PRC model, indicating that the structure at the $Q_B$ site in the original model is dominated by the 70% $Q_B$-depleted RCs in the crystal.

## 2.4 Stigmatellin Binding

As is the case for the quinone-reconstituted RC, the $2|F_o|-|F_c|$ electron density at a contour level of 1.3 $\sigma$ (to be presented elsewhere) of the stigmatellin complex allows the unequivocal placement of all atoms relevant for the binding of the compound. In analogy to quinone binding, the chromone ring of stigmatellin lies in the same plane as does the quinone ring, with the phenyl ring of Phe L216 oriented roughly parallel to it.

Compared to quinone binding, the stigmatellin carbonyl oxygen O4 substitutes for the proximal carbonyl O4 in accepting a hydrogen bond from the His L190 δNH (cf. Fig. 6). The hydroxyl O8 replaces the distal carbonyl as an acceptor in a bifurcated hydrogen bond from the backbone amides of Ile L224N and Gly L225. Additionally, a hydrogen bond is donated by the hydroxyl O8 to the Oγ of Ser L223. Also, the imidazole Hδ of His L190 lies exactly in the plane formed by the stigmatellin carbonyl oxygen, its proximal methoxy oxygen, and the Nδ of His L190 and is located equidistantly from each of these two stigmatellin oxygens, thus establishing a three-centre hydrogen bond. These two features demonstrate how both Ser L223 and the proximal methoxy group could be involved in stabilizing the binding of the monoprotonated doubly reduced $Q_BH^-$.

## 2.5 Triazine Binding

Refinement of the RC complex with DG-420314 has yielded a highly reliable model of triazine inhibitor binding. At a contour level of 1.5 standard deviations above the mean $2|F_o|-|F_c|$ electron density (to be displayed elsewhere), unequivocal assignment of the substituents at the chiral centre of the inhibitor is possible, as is the case for the rest of the triazine molecule and all other atoms relevant for triazine binding. In addition, novel and, also considering the other two triazine complexes (cf. Table 1), most probably general principles of triazine class inhibitor binding have emerged. These include, firstly, two tightly-bound water molecules, one which is involved in accepting a hydrogen bond from His L190 Nδ and in donating a second hydrogen bond to the N3 of the triazine ring, the other completes the hydrogen-bonding web with hydrogen bonds to the first water molecule and to the Oε1 of Glu L212 (cf. Fig. 7). Secondly, a subtle change (~0.3 Å) in the conformation of the L220-L226 loop region connecting helices DE and E is observed, which brings the backbone carbonyl oxygen of Tyr L222 within hydrogen bonding distance of the triazine cyanobutylamino nitrogen. In the original terbutryn complex (19), only the two hydrogen bonds from the backbone amide of Ile L224 to the N5 of the triazine ring and from the aminoethyl NH to Ser L223 Oγ had been observed. This means that the triazine molecule is bound to the protein directly by three hydrogen bonds on the distal side, and only indirectly via water molecules (and four hydrogen bonds) on the proximal side. Finally, compared to the quinone-reconstituted complex of the RC, the aromatic ring of Phe L216 is reoriented by 18.2° to be aligned approximately parallel (with an interplanar angle of 2.4°) to the plane of the triazine ring which is tilted by 14.5° compared to the quinone ring plane (30).

## 2.6 Conclusions

In all of the structures listed in Table 1, the side chain Oγ of Ser L223 is involved in the **donation** of a hydrogen bond to the side chain Oδ of Asn L213. The existence of this hydrogen bond in the wild type structure is further corroborated by the change in side chain orientation observed for Asn L213 in the X-ray structure of the triazine-resistant mutant T1, where one of the mutations is Ser L223→Ala (39). Ser L223 may act as an **acceptor** of a hydrogen bond in those cases where the ligand is able to donate a hydrogen bond to this position. This is the case for stigmatellin and also for triazine inhibitors, but not for $Q_B$ in its oxidized, i.e. quinone form. With the absence of a hydrogen bond, the presence of the Ser L223 Oγ so close to the distal carbonyl oxygen is electrostatically unfavourable for quinone binding. This finding explains the observed 9-fold increase in quinone affinity for the aforementioned SerL223-less *Rp. viridis* mutant T1 (38). While not favourable for quinone binding, the close association of the two oxygen atoms is compatible with the importance of Ser L223 in stabilizing the reduced quinone and for efficient proton donation to $Q_B$ as demonstrated for the *Rp. viridis* RC (40) and for the *Rb. sphaeroides* RC (41). Since the first proton is most probably taken up by the distal carbonyl oxygen, such a stabilization of a monoprotonated doubly reduced $Q_BH^-$ intermediate could be similar to the hydrogen bonding pattern observed for the stigmatellin complex.

All of these six structures are important for the modelling of the $Q_B$ site in the D1 protein of the chloroplast photosystem II (42-47).

# 3 The Primary Acceptor Quinone ($Q_A$) Binding Site

Recent studies on the asymmetric binding of the primary acceptor quinone in the reaction center of *Rb. sphaeroides* as probed by EPR spectroscopy (48) and light-induced FTIR difference spectroscopy (49,50) indicate a stronger hydrogen bonding of the proximal carbonyl oxygen of $Q_A$ than that of the distal carbonyl oxygen. None of the structures deposited so far in the Brookhaven Data Bank (cf. Table 3) are compatible with these findings, since those that do indicate a stronger, i.e. shorter, proximal hydrogen bond assign this to Thr M220/222[1], which in better-defined structures is not hydrogen-bonded to $Q_A$, but to Trp M250/252. In our newer series of structures, the proximal hydrogen bond to His M217/219 is in all cases shorter than the distal bond to Ala M258/260 which is in agreement with the aforementioned spectroscopic data.

**Table 3.** Distances (DA) and angles (DHA) between prospective H-bonding donors (D) and acceptors (A) in the $Q_A$-site. Hydrogen positions were taken from X-PLOR output (Q286) or added with the program HBPLUS (52)

| Donor | His M217/219 δN | | Thr M220/222 γO | | Ala M258/260 N | |
|---|---|---|---|---|---|---|
| Acceptor | $Q_A$ O4 | | $Q_A$ O4 | | $Q_A$ O1 | |
| Structure | DA (Å) | DHA (°) | DA (Å) | DHA (°) | DA (Å) | DHA (°) |
| *Rp. viridis* | | | | | | |
| Q286 | 2.9 | 168 | | | 3.2 | 167 |
| 1PRC | 3.1 | 168 | | | 3.1 | 162 |
| *Rb.sphaeroides* | | | | | | |
| 1PCR | 3.2 | 157 | | | 2.8 | 160 |
| 1PSS | 3.3 | 164 | | | 2.7 | 130 |
| 1YST | 4.6 | | 3.1 | 141 | 2.8 | 136 |
| 2RCR | | | 2.4 | 166 | 2.5 | 139 |
| 4RCR | | | 2.8 | 170 | 2.7 | 142 |

# 4 The Carotenoid

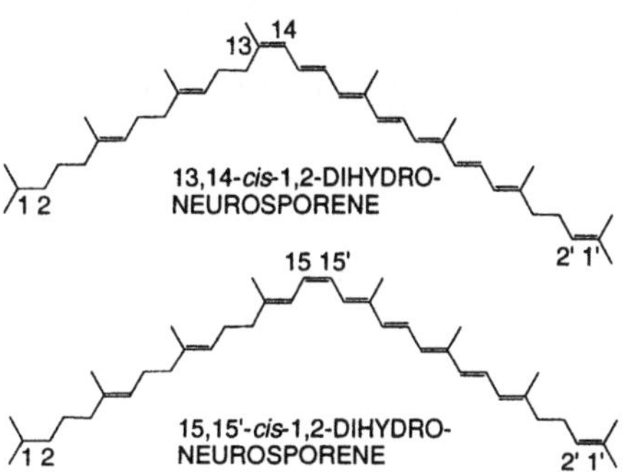


**Figure 8.** Dihydroneurosporene in cis-13-14 and cis-15-15' conformations

In the original (1PRC) structure (9), the model contains a 13,14-cis-1,2-dihydroneurosporene carotenoid (cf. Fig. 8). A 13-cis bond would contradict the results obtained by Raman (53) and NMR (54) spectroscopy on the *Rb. sphaeroides* RC, which indicate a cis bond at position 15.

In the DGM structure based on over 100,000 unique reflections (cf. Table 1), parallel refinements were performed on models with a 13-14-cis-carotenoid and a 15-15'-cis-carotenoid, respectively. Using the techniques of simulated annealing omit maps for the removal of model bias (55), individual B-factor refinement and analysis, and analysis of the deviations from ideal geometry, it was concluded that the 15-15'-cis- carotenoid fitted the SA omit map density better than the model with a 13,14-cis bond (cf.

[1] Because of insertions, the amino acid sequences of the M subunits of *Rp. viridis* and *Rb. sphaeroides* are out of register by one residue from M38 and by two residues from M104/105 onwards (51). The smaller numbers refer to *Rp. viridis*.

Fig. 9). Consequently, both crystallographic and spectroscopic data indicate a (symmetrical) 15-15'-cis-carotenoid.

## 5 Acknowledgements

Figures 2 and 9 were made using O (24). Figures 3 to 7 were generated with MolScript (56). Ubiquinone-2 was kindly provided by Dr. Thomas Link, Universität Frankfurt. DG-420314 and DG-420315 were generously provided by Dr. Schäfer, Degussa AG, Hanau. Molecular coordinates from the X-ray structure analysis of DG-420314 and DG-420315 have kindly been made available prior to publication by Prof. W. Frank, Universität Kaiserslautern.

## 6 References

1 Michel-Beyerle, M.E., ed. (1985) Antennas and Reaction Centers of Photosynthetic Bacteria. Structure, Interactions and Dynamics. Springer Verlag, Berlin.
2 Michel-Beyerle, M.E., ed. (1990) Reaction Centers of Photosynthetic Bacteria.. Springer Verlag, Berlin.
3 Breton, J. and Verméglio, A., eds. (1988) The Photosynthetic Bacterial Reaction Center: Structure and Dynamics. Plenum Press, New York.
4 Breton, J. and Verméglio, A., eds. (1992) The Photosynthetic Bacterial Reaction Center II, NATO-ASI Series A, Life Sciences 237, Plenum Press, New York.
5 Deisenhofer, J. and Norris, J.R., eds. (1993) The Photosynthetic Reaction Center. Academic Press, San Diego
6 Deisenhofer, J., Epp, O., Miki, K., Huber, R. and Michel, H. (1984) J Mol Biol 180, 385-398.
7 Deisenhofer, J., Epp, O., Miki, K., Huber, R. and Michel, H. (1985) Nature 318, 618-624.
8 Deisenhofer, J. and Michel, H. (1989) EMBO J 8, 2149-2169.
9 Deisenhofer, J., Epp, O., Sinning, I. and Michel, H. (1995) J. Mol. Biol. 246, 429-457.
10 Kunze, B., Kemmer, T., Höfle, G. and Reichenbach, H. (1984) J. Antibiotics 37, 454-461.
11 Höfle, G., Kunze, B. Zorzin, C. and Reichenbach, H. (1984) Liebigs Ann. Chem. 1984, 1883-1904.
12 Oettmeier, W., Godde, D., Kunze, B., Höfle, G. (1985). Biochim. Biophys. Acta 807, 216-219.
13 Oettmeier, W. and Preuße, S. (1987) Z. Naturforsch. 42c, 690-692.
14 Giangiacomo, K.M., Robertson, D.E., Gunner, M.R., Dutton, P.L. (1987) In: Biggins, J. (ed.) Progress in Photosynthesis Research Vol. II, Martinus Nijhoff Publishers, Dordrecht, pp. II.6.409-II.6.412.
15 Atta-Asafo-Adjei, E. and Daldal, F. (1991) Proc. Natl. Acad. Sci. USA 88, 492-496.
16 Thierbach, G., Kunze, B., Reichenbach, H., & Höfle, G. (1984) Biochim. Biophys. Acta 765, 227-235.
17 von Jagow , G. and Link, Th.A. (1986) Methods Enzymol. 126, 253-271.
18 Percival, M.P. and Baker, N.R. (1991) In: Baker, N.R. and Percival, M.P. (eds.) Herbicides, pp. 1-26. Elsevier Science Publishers, Amsterdam, The Netherlands.
19 Michel, H., Epp, O., & Deisenhofer, J. (1986) EMBO J. 5, 2445-2451.
20 Michel, H. (1982) J. Mol. Biol. 158, 567-572.
21 Leslie, A.G.W. (1993) MOSFLM (Version 5.10), MRC Laboratory of Molecular Biology, Cambridge, UK.
22 Bailey, S. (1994) Acta Cryst D50, 760-763.
23 Brünger, A.T. (1993) X-PLOR (Version 3.1), The Howard Hughes Medical Institute and Department of Biophysics and Biochemistry , Yale University, New Haven, CT, USA.
24 Jones, T.A,. Zou, J.Y., Cowan, S.W. and Kjeldgaard ,M. (1991) Acta Cryst A47, 110-119.
25 Engh, R.A.. and Huber, R. (1991) Acta Cryst A47, 392-400.
26 Lancaster, C.R.D. and Michel, H., manuscript in preparation.
27 O'Malley, P.J. and Braithwaite, C.J. (1994) Photosynth. Res. 39, 51-56.
28 Bernstein, F.C., Koetzle, T.F., Williams, G.J.B., Meyer, E.F., Brice, M.D., Rodgers, J.R., Kennard, O., Shimanouchi, T. and Tasumi, M. (1977) J. Mol. Biol. 112, 535-542.
29 Stampf, D.R., Felder, C.E. and Sussmann, J.L. (1995) Nature 374, 572-574.
30 Lancaster, C.R.D., Ermler, U. and Michel, H. (1995) In: Blankenship RE, Madigan MT, and Bauer CE (eds.), Anoxygenic Photosynthetic Bacteria. Kluwer Academic Publishers, The Netherlands, Chapter 23, in press
31 Lancaster, C.R.D. and Michel, H., manuscript in preparation.
32 Ermler, U., Fritzsch, G., Buchanan, S. and Michel, H. (1994) Structure 2, 925-936.

33 Chirino, A.J., Lous, E.J., Huber, M., Allen, J.P., Schenck, C.C., Paddock, M.L., Feher, G. and Rees, D.C. (1994) Biochemistry 33, 4584-4593.
34 Arnoux, B., Gaucher, J.-F., Ducruix, A., Reiss-Husson, F. (1995) Acta Cryst D51, 368-379.
35 Chang, C.H., El-Kabbani, O., Tiede, D., Norris, J. and Schiffer, M. (1991). Biochemistry 30, 5353-5360.
36 Yeates, T.O., Komiya, H., Chirino, A., Rees, D.C., Allen, J.P. and Feher, G. (1988). Proc Natl Acad Sci USA 85, 7993-7997.
37 Luzzati, V. (1952) Acta Cryst 5, 802-810.
38 Sinning, I., Michel, H., Mathis, P. and Rutherford, A.W. (1989). Biochemistry 28, 5544-5553.
39 Sinning, I., Koepke, J. and Michel, H. (1990) In: (2), pp. 199-208.
40 Leibl, W., Sinning, I., Ewald, G., Michel, H. and Breton, J. (1993). Biochemistry 32, 1958-1964.
41 Paddock, M.L., McPherson, P.H., Feher, G. and Okamura, M.Y. (1990) Proc Natl Acad Sci USA 87, 6803-6807.
42 Michel, H. and Deisenhofer, J. (1988) Biochemistry 27, 1-7.
43 Trebst, A., (1987) Z. Naturforsch. 42c, 742-750.
44 Egner, U., Hoyer, G.-A. and Saenger, W. (1992) Eur. J. Biochem. 206, 685-690.
45 Egner, U., Hoyer, G.-A. and Saenger, W. (1993) Biochim. Biophys. Acta 1142, 106-114.
46 Sobolev, V. and Edelman, M. (1995) Proteins 21, 214-225.
47 Mackay, S.P. and O'Malley, P.J. (1993). Z. Naturforschung 48c, 474-481.
48 van den Brink, J.S., Spoyalov, A.P., Gast, P., van Liemt, W.B.S., Raap, J., Lugtenberg, J., Hoff, A.J. (1994) FEBS Lett. 353, 273-276.
49 Brudler, R., de Groot, H.J.M., van Liemt, W.B.S., Steggerda, W.F., Esmeijer, R., Gast, P., Hoff A.J., Lugtenberg, J., and Gerwert, K. (1994) EMBO J. 13, 5523-5530.
50 Breton, J., Boullais, C., Burie, J.-R., Nabedryk, E., Mioskowski, C. (1994) Biochemistry 33, 14378-14386.
51 Michel, H., Weyer, K.A., Gruenberg, H., Dunger, I., Oesterhelt, D., Lottspeich, F. (1986) EMBO J. 5, 149-1158.
52 McDonald, I.K. and Thornton, J.M. (1994) J. Mol. Biol. 238, 777-793.
53 Arnoux, B., Ducruix, A., Reiss-Husson, F., Lutz, M., Norris, J., Schiffer, M. and Chang, C.H. (1989). FEBS Lett 258, 47-50.
54 De Groot, H.J.M., Gebhard, R., van der Hoef, I., Hoff, A.J., Lugtenburg, J., Violette, C.A. and Frank, H.A. (1992) Biochemistry 31, 12446-12450.
55 Hodel, A., Kim, S.-H., Brünger, A.T. (1992) Acta Cryst A48, 851-858.
56 Kraulis, P.J. (1991) J. Appl. Cryst. 24, 946-950.

# Effect of Orbital Asymmetry in $P^{\bullet+}$ on Electron Transfer in Reaction Centers of *Rb. sphaeroides*

J. Rautter[1], F. Lendzian[1], X. Lin[2], J. C. Williams[2], J. P. Allen[2] and W. Lubitz[1]

[1]Max-Volmer-Institut für Biophysikalische und Physikalische Chemie, Technische Universität Berlin, Straße des 17. Juni 135, D-10623 Berlin, Germany
[2]Department of Chemistry and Biochemistry, Center for the Study of Early Events in Photosynthesis, Arizona State University, Tempe, Arizona 85287-1604, USA

**Abstract.**The cation radical of the primary donor, $P^{\bullet+}$, has been investigated in a set of mutants of *Rb. sphaeroides* with different hydrogen bonds to the special pair, using EPR/ENDOR spectroscopy. The results show that these mutants exhibit large variations in the distribution of the unpaired electron between the dimer halves of P dependent on the hydrogen bond situation. The comparison with the electron transfer rate from cytochrom $c_2$ to $P^{\bullet+}$ reveals a correlation of this rate with the fraction of the unpaired electron on the L-half of the dimer. The implications of this finding for the electron pathway of this reaction are discussed.

**Keywords.**EPR/ENDOR spectroscopy, spin density distribution, primary donor, hydrogen bond mutants, electron transfer rates, cytochrom $c_2$,

# 1 Introduction

Electron transfer (ET) reactions in bacterial photosynthesis are controlled by both the spatial and electronic structure of the involved cofactors [1-3]. The spatial structure of the reaction center (RC) has been determined for the two photosynthetic bacteria *Rhodopseudomonas (Rps.) viridis* [4] and *Rhodobacter (Rb.) sphaeroides* [5-8] and revealed the arrangement of the protein and the chromophores.

The electronic structure, in particular the spin density distribution in the valence orbital of the pigment radicals that are formed in the ET-process, is accessible from the hyperfine coupling constants (hfc's) determined by Electron Paramagnetic Resonance (EPR) and Electron Nuclear DOuble Resonance (ENDOR) spectroscopy. Due to its critical function at the interface of exciton and electron transfer, the electronic structure of the primary donor P is of particular interest. Intensive EPR and ENDOR investigations of the cation radical of the primary donor $P^{\bullet+}$ in RC single crystals of *Rb. sphaeroides* R-26 revealed an unequal distribution of the unpaired electron between the two dimer halves with a ratio of approximately 2:1 in favor of the L-half of the special pair $P_L$ [9]. This asymmetric distribution of the $\pi$-spin density with a ratio of 2:1 was found to be conserved in $P^{\bullet+}$ of the BChl *a* containing purple bacteria *Rb. capsulatus*, *Rhodospirillum (Rs.) rubrum* and *Rs. centenum* [10], as well as for *Rps. viridis* in which a dimer of BChl *b* molecules constitutes the primary donor [11].

The asymmetric distribution of the unpaired electron within $P^{\bullet+}$ seems to be a general property of the special pair in the RCs of purple bacteria. Therefore the question arises whether this asymmetry influences any of the ET-rates. Since the EPR/ENDOR results probe the highest occupied molecular orbital (HOMO) of P, any electron transfer reaction that involves this orbital may be directly influenced by the asymmetry of $P^{\bullet+}$. In particular, we discuss in this manuscript the electron transfer reaction from the exogenous cytochrome $c_2$ ($cyt^{2+}P^{\bullet+} \rightarrow cyt^{3+}P$).

The search for a possible correlation requires a variation in the degree of delocalization of the unpaired electron in $P^{\bullet+}$ and therefore experimental tools to alter the spin density distribution of the special pair in purple bacteria. We have recently shown that for several purple bacteria one possibility lies in the use of different detergents in the RC preparation which leads to the stabilization of a second distinct state of the primary donor cation radical with a spin density distribution of approximately 5:1 in favor of $P_L$ [10, 12]. Another approach to change the electronic structure of the primary donor is provided by site directed mutagenesis. For example, in the so-called heterodimer mutants HL(M202) and HL(L173) of *Rb. sphaeroides* the unpaired electron is localized in each case on the Bchl-half of the primary donor [13]. In order to verify a possible correlation between ET-rates and the spin density distribution in $P^{\bullet+}$ one needs to investigate more than these two extreme cases.

Recently, a set of single, double and triple mutants of *Rb. sphaeroides* has been designed, in which hydrogen bonds to the carbonyl groups of the special pair are altered [14-16] (see Figure 1).

The changes in hydrogen bond interactions between the protein and the primary donor are accompanied by large changes in the $P/P^{\bullet+}$ redox midpoint potential, ranging from 410 to 765 mV in these mutants, compared with a value of 505 mV in the wild type [16]. Since the $P/P^{\bullet+}$ midpoint potential determines the driving force of the ET-reactions involving the special pair, studies were performed to investigate the relation between the altered driving force and the rate $k_{AP}$ for the reduction of $P^{\bullet+}$ from the primary quinone

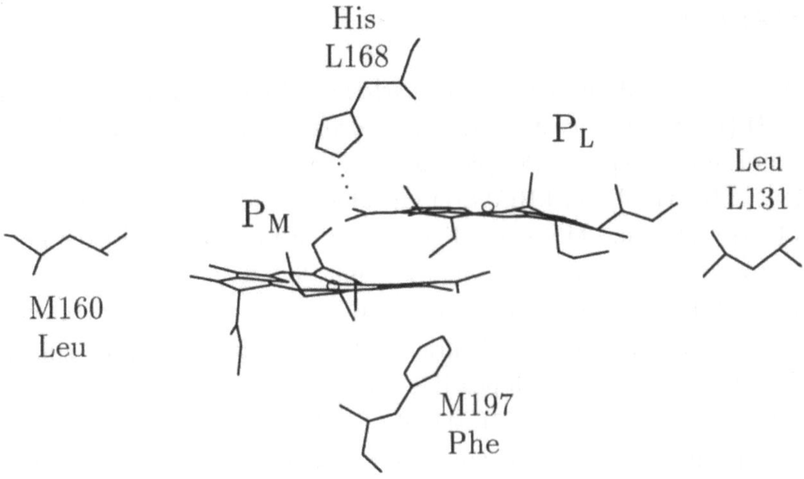


Figure 1: Structure of the two BChls of the primary donor of *Rb. sphaeroides* with the amino acid residues Leu-L131, Leu-M160, His-L168, Phe-M197 which are exchanged by site directed mutagenesis. The hydrogen bond between His L168 and the acetyl oxygen of $P_L$ present in the wild type is indicted by dotted lines. View along the approximate $C_2$-symmetry axis of the RC. Structure from Brookhaven Protein Data Bank, File 1 PCR, see [8].

[16] as well as the rate $k_{CP}$ from the soluble cytochrome $c_2$ [17].
ENDOR investigations of these mutants showed that the asymmetric hydrogen bonding between His L168 and the acetyl group of $P_L$ is a very important factor for the asymmetric spin density distribution encountered in the wild type. If this bond is broken as in the mutant HF(L168), the spin density distribution is changed to approximately 1:1. Depending on the number of hydrogen bonds, the unpaired electron can be gradually shifted from a predominant localization on $P_L$ to a localization on $P_M$ [18]. Therefore, this set of mutants presents the ideal system to study the relationship between the electronic structure of $P^{\bullet+}$ and the observed ET-rates in the RC.
In this contribution we demonstrate that the asymmetry of the distribution of the unpaired electron in purple bacteria has a pronounced influence on the rate of the reduction of $P^{\bullet+}$ by the soluble cytrochrome $c_2$.

## 2 Results

The investigated mutants include the exchange of residue His L168 to Phe, mutant HF(L168), that cleaves the only hydrogen bond to a carbonyl group present in the wild type of *Rb. sphaeroides.* The other three point mutations LH(L131), LH(M160) and FH(M197) introduce histidine residues in the positions L131, M160 and M197 (see Figure 1), that form new hydrogen bonds to the keto carbonyl groups of $P_L$ and $P_M$ and the acetyl oxygen of $P_M$. The formations of the intended hydrogen bonds were verified by Fourier transform resonance Raman (FTRR) and Fourier transform infrared (FTIR) spectroscopy [19, 20]. Furthermore, all possible combinations of double mutants and the triple mutant LH(L131)+LH(M160)+FH(M197), in which all four carbonyl groups of the special pair are hydrogen bonded [16], were studied.

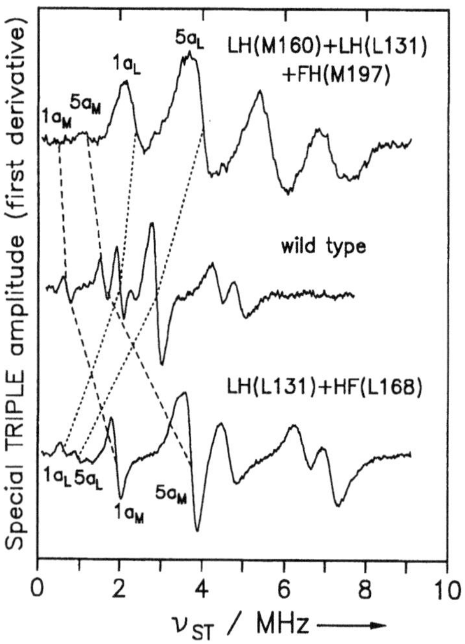


Figure 2: Comparison of the $^1H$ Special TRIPLE spectra in liquid solution of $P^{\bullet+}$ in RCs of two of the investigated mutants with the respective spectrum of the wild type. The assignment of the hfc's of the methyl protons at positions 5a and 1a, to the individual dimer halves, which is based on the different ratios of these couplings in the two dimer halves, is indicated for $P_L$ (dotted lines) and $P_M$ (dashed lines), for details see [18]. The hfc's A (in MHz) for the triple mutant are: $A(5a_L) = 8.05$; $A(1a_L) = 4.9$; $A(5a_M) = 2.4$; $A(1a_M) = 0.9$, $A(\beta_L) = 14.20$ , 11.30; for details see [21]. Experimental conditions: T = 288 K; microwave power, 10 mW ; rf power, 2 × 150 W; rf modulation depths, 150 kHz (frequency 12.5 kHz); $\nu_{1H} = 14.6$ MHz .

The large variations in the spin density distribution of the primary donor in the investigated mutants are due to the alteration of the energies of the HOMOs of the two halves of the dimer, $P_L$ and $P_M$, caused by the specific hydrogen bonds, which lower the energy of the involved molecular orbitals (for details see refs. [18, 22]). As an example for the different spin density distributions of $P^{\bullet+}$ in the mutant RCs, Figure 2 displays the Special TRIPLE spectra of two of the investigated mutants in comparison with the wild type. The Special TRIPLE spectrum of $P^{\bullet+}$ in the triple mutant which has four hydrogen bonds to the special pair, LH(M160)+LH(L131)+FH(M197), shows an almost complete localization (80 %) of the unpaired electron on the L-half of the dimer whereas the double mutant LH(L131)+HF(L168), where only one hydrogen bond to the 9-keto carbonyl of $P_L$ is present, exhibits an excess of the spin density (78 %) on $P_M$ (see Table 1). The variation of the distribution of the unpaired electron of $P^{\bullet+}$ in the eleven investigated mutants with respect to the wild type is shown in Figure 3.

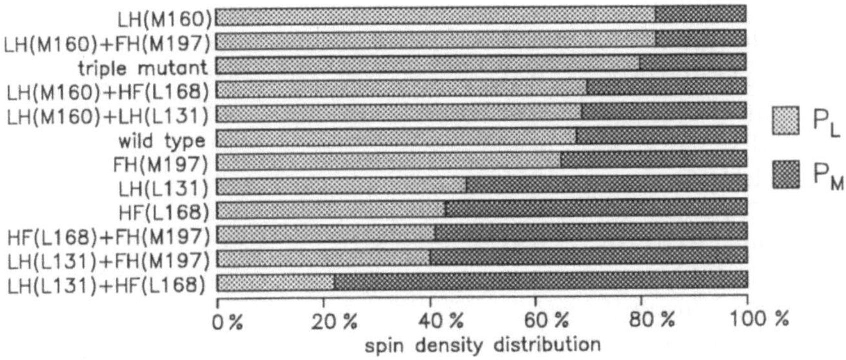


Figure 3: Histogram to illustrate the degree of variation in the asymmetric distribution of the unpaired electron between the two dimer halves $P_L$ and $P_M$ in the mutants of *Rb. sphaeroides* with altered hydrogen bond interactions. The triple mutant is LH(M160)+LH(L131)+FH(M197).The bars indicate the amount of spin density on the L-half (light gray) and on the M-half (dark gray) of the dimer (see Table 1).

Since the spin density distribution in $P^{\bullet+}$ also reflects the distribution of the hole, i.e. the positive charge in the supermolecule, the rates of the secondary ET-reactions, which lead to the reduction of $P^{\bullet+}$ may be affected by the different asymmetries encountered in these mutants.

In the presence of exogeneous cytochrome $c_2$ from *Rb. sphaeroides*, $P^{\bullet+}$ decays rapidly with a first-order exponential component, which is attributed to the ET within a RC-cyt complex (proximal configuration) with a characteristic decay time of $\sim 1\ \mu s$ for the wild type [17,23-25].

The rates $k_{CP}$ for the reduction of $P^{\bullet+}$ by cytochrome $c_2$, have been measured for all eleven histidine mutants using time-resolved optical spectroscopy [17] and the characteristic times for the major fast component, $\tau_{CP}$, are given in Table 1.

It was found that $\tau_{CP}$ varies by two orders of magnitude in the different mutants and that in general the values decrease with increasing midpoint

Table 1. Comparison of the spin density distribution of $P^{\bullet+}$ in the RCs of the hydrogen bond mutants and the wild type of *Rb. sphaeroides* with the $P/P^{\bullet+}$ redox midpoint potentials and the electron transfer rates for the reduction of $P^{\bullet+}$ by the soluble cytochrome $c_2$.

| mutant strain | $\frac{\rho_L}{\rho_L+\rho_M}$ [a] | Number of H-bonds | $E_m$ ($P/P^{\bullet+}$) (mV) [b] | $\tau_{CP}$ (ns)[c] |
|---|---|---|---|---|
| HF(L168) | 0.43 | 0 | 410 | 7730 |
| LH(M160)+HF(L168) | 0.70 | 1 | 485 | 1545 |
| LH(L131)+HF(L168) | 0.22 | 1 | 485 | 4820 |
| wild type | 0.68 | 1 | 505 | 960 |
| HF(L168)+FH(M197) | 0.41 | 1 | 545 | 1515 |
| LH(M160) | 0.83 | 2 | 565 | 345 |
| LH(L131) | 0.47 | 2 | 585 | 470 |
| FH(M197) | 0.65 | 2 | 630 | 185 |
| LH(M160)+LH(L131) | 0.69 | 3 | 635 | 210 |
| LH(M160)+FH(M197) | 0.83 | 3 | 700 | 130 |
| LH(L131)+FH(M197) | 0.40 | 3 | 710 | 165 |
| LH(M160)+LH(L131) +FH(M197) | 0.80 | 4 | 765 | 80 |

[a]Fraction of the spin density on the L-half of $P^{\bullet+}$ determined from the methyl proton hfc's, for details see [18].

[b]Data from [16]

[c]Times for the major fast component of the reduction of $P^{\bullet+}$ by the soluble cytochrome $c_2$ at room temperature, data from [17]

potential [17] (see Table 1). However, there are some exceptions from this behavior, the most pronounced are the double mutants LH(L131)+HF(L168) and HF(L168)+FH(M197). An inspection of the electronic structures (see Table 1) reveals that both these double mutants have an excess of the spin density on the M-half of the dimer.

The most striking effect is observed when one compares data of the two double mutants LH(L131)+HF(L168) and LH(M160)+HF(L168). Both mutants have the same redox midpoint potential of 485 mV, i.e the same driving force for the reaction, but the measured time constants for the ET differ by a factor of three. Interestingly, the same factor of three is also observed when comparing the fraction of unpaired $\pi$-spin density on $P_L$ for the two double mutants (see Table 1).

It was found that the binding constants and second-order rate constants for the formation of the RC-cyt complex for the mutants are similar to the respective values of the wild type, which indicates that the binding of the cyt

to the RC is not changed in the mutants [17].
These results indicate that the electronic distribution of $P^{\bullet+}$ also influences the reduction rates of this species.

# 3 Discussion

For a nonadiabatic ET-process as assumed here, the rate is given by

$$k = \frac{2\pi}{\hbar}|V_{\mathrm{DA}}|^2 \times FC \tag{1}$$

where $\hbar$ is Planck's constant, $V_{DA}$ is the matrix element of the electronic coupling between donor (D) and acceptor (A) molecules and FC is the nuclear Franck-Condon factor. In the semi-classical approach introduced by Marcus, the FC depends exponentially on the free energy difference, $\Delta G^0$, between the initial and final states and the reorganisation energy. $\lambda$, of the reaction [26].

$$k = \frac{2\pi}{\hbar}|V_{\mathrm{DA}}|^2(4\pi\lambda k_B T)^{-\frac{1}{2}} \times exp\left[-\frac{(\Delta G^0 + \lambda)^2}{4\lambda k_B T}\right] \tag{2}$$

where $\hbar$ is Planck's constant, $k_B$ is the Boltzman constant and T is the temperature.
Figure 4 shows the experimental ET-rates for the reduction of $P^{\bullet+}$ by cyt $c_2$ and includes a fit according to eq.1 (taken from Lin *et al.* [17]). The emphasis in Figure 4 is put on the spin density distribution. Mutants with an excess of the $\pi$-spin density on $P_L$ and $P_M$ are marked with different symbols, respectively. As is evident from the Figure, some of the mutants show pronounced deviations from the fit. The inspection of the spin density distribution of $P^{\bullet+}$ reveals, that all mutants which exhibit slower recombination kinetics as predicted by the fit of eq.1, have an excess of spin density on the M-half of the primary donor (full triangles in Figure 4). In contrast, the points representing the mutants with an asymmetric spin density in favor of $P_L$ (open circles in Figure 4) lie slightly above the theoretical curve. This indicates that the reduction rates correlate with the fraction of spin density on $P_L$.
Since the theoretical curve according to eq.1 does not take the electronic structure of $P^{\bullet+}$ into account, the fit in Figure 4 is effectively performed for a mean spin density distribution for all mutants, which can be taken as the average of the different distributions found in the mutants. This average spin density on $P_L$, $\bar{\rho}_L$, has a value of 59 % in favor of $P_L$. The correlation of the factor by which $\rho_L$ deviates from this average in each mutant, $\rho_L/\bar{\rho}_L$, with the factor of the deviation of the measured rates from the theoretical rate given by the curve of Figure 4, k(exp)/k(fit), is shown in Figure 5.

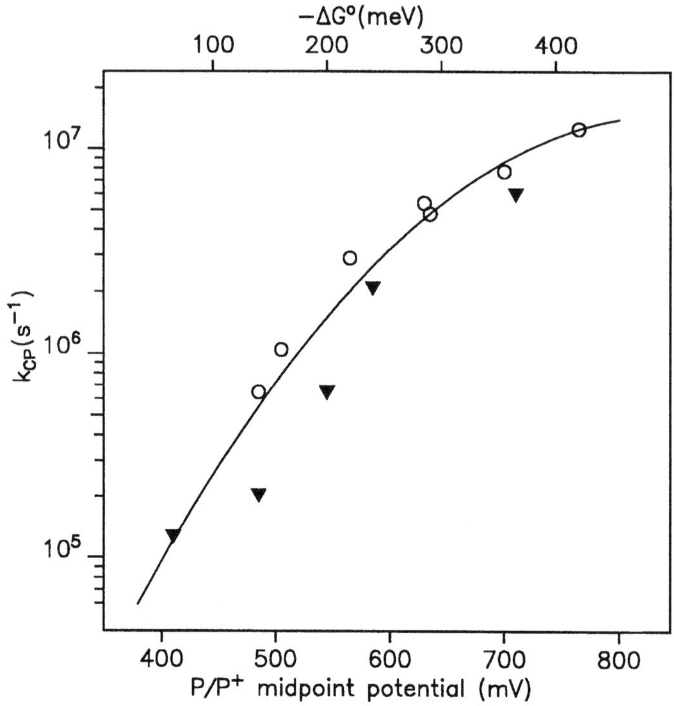


Figure 4: Relationship of the electron transfer rate $k_{CP}$ from cytochrome $c_2$ to $P^{\bullet+}$ and the $P/P^{\bullet+}$ midpoint potential in RCs of the wild type and the histidine mutants of *Rb. sphaeroides* from [17]. The curve shown represents the favored fit of [17] using eq.2 with a value of $\lambda$ = 500 meV and a maximum rate of 1.4 × $10^7$ $s^{-1}$. The mutants in which the spin density is predominately on $P_L$ or $P_M$ are indicated by open circles or full triangles, respectively.

Figure 5 clearly shows that the rate for the reduction of $P^{\bullet+}$ correlates with the fraction of the spin density, $\rho_L$, on the dimer half $P_L$. The straight line included in Figure 5 would be expected for a linear correlation between $\rho_L$ and the ET-rate

$$k(exp) \sim \frac{\rho_L}{\bar{\rho}_L} \times k(fit) \tag{3}$$

The finding that the rate for this reaction correlates with the fraction of the spin density on $P_L$, k(exp) ~ $\rho_L$, has interesting implications for the ET-pathway of this reaction.

According to eq.1 for non-adiabatic electron transfer the rate is expected to be proportional to $|V_{DA}|^2$, where $V_{DA}$ is the electronic coupling between donor and acceptor. A dependence of the ET-rate on the fraction of the spin

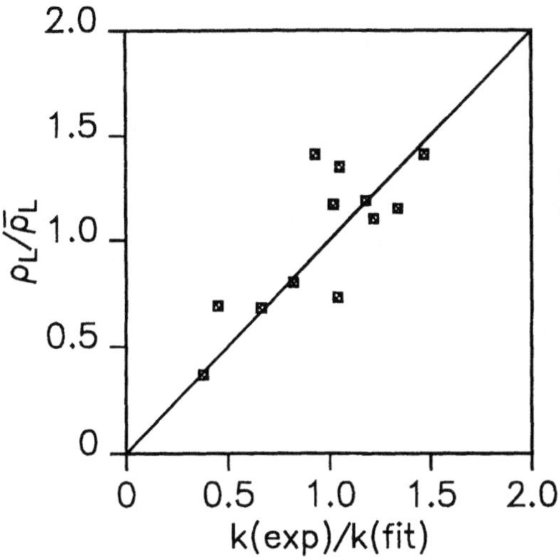


Figure 5: Correlation of the deviation of the experimental rates for the reduction of $P^{\bullet+}$ by cytochrome $c_2$ [17] from the fit according to eq.2 with the deviation of the fraction of the unpaired $\pi$-spin density on $P_L$, $\rho_L$, from the 'mean' spin density, $\bar{\rho}_L$, for the investigated mutants. The straight line would be obtained for a direct proportionality of the electron transfer rate and the fraction of the spin density on the L-half of P.

(and charge) [1] on one dimer half would be expected, if the coupling between the heme in the cyt and $P_L$ and $P_M$ is different. If the electron transfer proceeds through the HOMO of $P^{\bullet+}$, i.e. the $cyt^{2+}$ state lies between $P^{\bullet+}$ and the first excited single state of P, $P^*$, which is most likely, our results indicate a more favorable coupling between the heme and the L-half of the dimer [2]. In the most simple picture the coupling is described by an exponential dependence on the distance between donor and acceptor molecules [28]

$$|V_{DA}|^2 = |V_{DA}^0|^2 e^{-\beta R} \tag{4}$$

where $V_{DA}^0$ is the electronic coupling element at contact distance, which includes the orientation dependence and R is the edge-to-edge distance between donor and acceptor. For a variety of ET-reactions in proteins, Moser *et al.* found $\beta = 1.4$ Å$^{-1}$ for the exponential coefficient of the decay. This coefficient corresponds to a dramatic tenfold decrease in rate for every 1.7 Å increase in distance between the edges of the cofactors in the RC [28]. Since the rates $k_{CP}$ differ by a factor of three for mutants with the same driving force but a localization of the unpaired electron on different dimer halves (see Table 1), one can roughly estimate that in this model the distance of the heme to $P_M$ is about $\sim 0.8$ Å larger than to $P_L$. Therefore the observed correlation could be simply due to slightly different edge-to-edge distances between the heme and

[1] Note that the spin density distribution also reflects the distribution of the positive charge (hole) in the HOMO of $P^{\bullet+}$

[2] For the situation that the ET goes through the LUMO state of $P^{\bullet+}$, our results show a more favorable coupling between the heme and the M-half of P, because the charge distribution in the LUMO is reversed with respect to the HOMO [27]. However this is not very likely, even for the mutants with the highest midpoint potentials

$P_L$ and $P_M$. Alternatively, the difference in rates may reflect other factors, such as differences in relative orientation or the involvement of intermediate states, as we shall now discuss.

One possibility would be a more favorable orientation of $P_L$ with respect to the heme at a given distance. Plato *et al.* [1] evaluated electronic coupling integrals using the so-called intermolecular overlap approximation. In this model the wavefunctions are represented by a linear combination of atomic orbitals of the donor and acceptor molecules. Application of this model yields for the matrix element of the electronic coupling

$$V_{\mathrm{CP}} = K \sum_i \sum_j c_i^C c_j^P S_{ij} \tag{5}$$

where K is a constant, $c_i^C$ and $c_j^P$ are the coefficents of the HOMOs on atoms i of the heme and j of $P^{\bullet +}$ and $S_{ij}$ are the intermolecular overlap integrals of the ith and jth atomic orbitals. Since the ET-rate is proportional to $V^2$ it is also expected to be proportional to the squares $(c_j^P)^2$ of the respective orbital coefficients. These are directly related to the spin density at the corresponding nuclei. In this model our data would indicate that the overlap integrals $S_{ij}$ are larger for $P_L$ than for $P_M$.

Since the edge-to-edge distance obtained from the maximum rate has been shown to be in the range between 9-14 Å [17], which is in agreement with the distances obtained by modelling of the RC-cyt $c_2$ complex (10-12 Å) [29, 30], a direct overlap is not very likely. Therefore amino acid residue(s) could be involved in a virtual coupling (superexchange) mechanism. In the superexchange model the coupling of the redox components is mediated by a bridge state B, that is not involved in the electron transfer as a real intermediate and the coupling can be expressed by [31]

$$V_{\mathrm{CP}} \approx V_{\mathrm{CP}}^0 + \frac{V_{\mathrm{CB}} V_{\mathrm{BP}}}{\delta E_{\mathrm{CB}}} \tag{6}$$

where $V_{CB}$ and $V_{BP}$ are the electronic couplings between the bridge and the heme and the bridge and the primary donor respectively, and $\delta E_{CB}$ is the energy difference between the involved donor and bridge orbitals. Our results point out that the overlap of the bridging residue(s) has to be larger with $P_L$ than with $P_M$. Tyrosine L162, which lies approximately in the middle between the heme $c_{559}$ and the primary donor $P_{960}$ in RCs of *Rps. viridis* has been suggested to play such a role [32], but recent experiments on mutants of *Rps. viridis* in which tyrosine was exchanged against the aliphatic leucine exhibited no large disturbances in the reduction rates of $P^{\bullet +}$ [33].

Certainly, the picture is more complicated and probably involves more than one residue in the electron transfer forming one or more distinct pathways between the heme and the dimer. Elaborate models for tunneling pathways in proteins show that dependent on the number of involved pathways, the electron transfer can either vary exponentially with distance or exhibit a more complex behavior [34, 35]. Pathway models explicitly account for covalent bonds, hydrogen bridges and van der Waals contacts that link donor and acceptor molecules [36]. The critical role of the protein in providing different tunneling pathways for ET-reactions involving cytochromes has for example been shown for a cytochrome c/cytochrome c peroxidase complex, in which three distinct pathways contribute to the electronic coupling and the dominant one occurs along the protein backbone [37].

A recent calculation of the electron pathways for two model structures of the reaction center/cytochrom $c_2$ complex from *Rb. sphaeroides* showed that dependent on the docking of the cytochrome, different pathways inside the RC are possible [38]. Calculations for the structure of Tiede *et al.* [30], which is based on the electrostatic complementarity between charged residues on the surface of the cytochrome and the RC, predict a favorable pathway involving atoms of residues Leu M191 and Ser M190 of the M-subunit and of $P_M$ [38]. In contrast, the strongest electron pathway in a different structural model, which is based on maximum surface coupling, includes atoms of residues Gly L161 and Tyr L162 of the L-subunit and of $P_L$ [38]. Our results support the existence of a favorable pathway between the heme and the L-half of P.

An additional effect on the rate would be expected, if the reorganization energy $\lambda$ is different for a localization of the spin (and charge) on $P_L$ or $P_M$. This is, however, not very likely since no correlation between the rates and the number of exchanged polar residues in the vicinity of $P_L$ and/or $P_M$ is observed.

# 4 Conclusion

The comparison of the reduction kinetics of the primary donor by cytochrome $c_2$ with the electronic structure of $P^{\bullet+}$ in a set of eleven mutants of *Rb. sphaeroides* showed a correlation of the rate for this reaction with the fraction of the unpaired electron on $P_L$. This interesting finding implies that the reduction of $P^{\bullet+}$ occurs mainly via the L-half BChl, assuming that the reorganization energy stays constant. This preference could either simply be due to a different edge-to-edge distance of the heme to $P_L$ and $P_M$ or due to specific protein residues which provide a more efficient pathway for the electron transfer to $P_L$. The x-ray structure analysis of co-crystals from the RC-cytochrome $c_2$ complex of *Rb sphaeroides* is currently under investigation in the group of G. Feher [39]. The analysis of the protein medium between the heme and the two dimer halves of the special pair should provide further insight into the exact electron pathway to $P^{\bullet+}$. A correlation of the charge recombination rate from the primary quinone $Q_A$ to $P^{\bullet+}$ ($k_{AP}$) with the fraction of the unpaired electron on $P_L$ has been revealed at low temperatures. This is currently under investigation in our laboratories.

# 5 Acknowledgments

The authors want to thank G. Feher for helpful discussions of possible electron pathways to $P^{\bullet+}$ and J. Onuchic for communicating results prior to publication. This work was supported by DFG (Sfb 312, TP A4), NATO (CRG910468), Fonds der Chemischen Industrie to W.L., NIH (GM45902) to J.C.W., and NSF (MCB-9404925) to J.P.A.

# References

[1] Plato, M., Möbius, K., Michel-Beyerle, M. E., Bixon, M., and Jortner, J. (1988) J. Am. Chem. Soc. 110, 7279–7285.

[2] Warshel, A., Creighton, S., and Parson, W. W. (1988) J. Phys. Chem. 92, 2696–2701.

[3] Bixon, M., Jortner, J., and Michel-Beyerle, M.-E. (1991) Biochim. Biophys. Acta 1056, 301–315.

[4] Deisenhofer, J., Epp, O., Miki, K., Huber, R., and Michel, H. (1984) J. Mol. Biol. 180, 385–398.

[5] Allen, J. P., Feher, G., Yeates, T. O., Komiya, H., and Rees, D. C. (1987) Proc. Natl. Acad. Sci. USA 84, 5730–5734.

[6] El-Kabbani, O., Chang, C. H., Tiede, D., Norris, J., and Schiffer, M. (1991) Biochemistry 30, 5361–5369.

[7] Chirino, A. J., Lous, E. J., Huber, M., Allen, J. P., Schenck, C. C., Paddock, M. L., Feher, G., and Rees, D. C. (1994) Biochemistry 33, 4584–4593.

[8] Ermler, U., Fritzsch, G., Buchanan, S. K., and Michel, H. (1994) Structure 2, 925–936.

[9] Lendzian, F., Huber, M., Isaacson, R. A., Endeward, B., Bönigk, B., Möbius, K., Lubitz, W., and Feher, G. (1993) Biochim. Biophys. Acta 1183, 139–160.

[10] Rautter, J., Lendzian, F., Wang, S., Allen, J. P., and Lubitz, W. (1994) Biochemistry 33, 12077–12084.

[11] Lendzian, F., Lubitz, W., Scheer, H., Hoff, A. J., Plato, M., Tränkle, E., and Möbius, K. (1988) Chem. Phys. Lett. 148, 377–385.

[12] Wang, S., Lin, S., Lin, X., Woodbury, N. W., and Allen, J. P. (1994) Photosynth. Res. 42, 203–215.
[13] Huber, M., Lous, E. J., Isaacson, R. A., Feher, G., Gaul, D., and Schenck, C. C. (1990) in Reaction Centers of Photosynthetic Bacteria, (Michel-Beyerle, M. E., ed.), pp. 219–228, Springer, Berlin.
[14] Williams, J. C., Alden, R. G., Murchison, H. A., Peloquin, J. M., Woodbury, N. W., and Allen, J. P. (1992) Biochemistry 31, 11029–11037.
[15] Murchison, H. A., Alden, R. G., Allen, J. P., Peloquin, J. M., Taguchi, A. K. W., Woodbury, N. W., and Williams, J. C. (1993) Biochemistry 32, 3498–3505.
[16] Lin, X., Murchison, H. A., Nagarajan, V., Parson, W. W., Williams, J. C., and Allen, J. P. (1994) Proc. Natl. Acad. Sci. USA 91, 10265–10269.
[17] Lin, X., Williams, J. C., Allen, J. P., and Mathis, P. (1994) Biochemistry 33, 13517–13523.
[18] Rautter, J., Lendzian, F., Schulz, C., Fetsch, A., Kuhn, M., Lubitz, W., Lin, X., Williams, J. C., and Allen, J. P. (1995) Biochemistry 34, 8130–8143.
[19] Mattioli, T. A., Williams, J. C., Allen, J. P., and Robert, B. (1994) Biochemistry 33, 1636–1643.
[20] Nabedryk, E., Allen, J. P., Taguchi, A. K. W., Williams, J. C., Woodbury, N. W., and Breton, J. (1993) Biochemistry 32, 13879–13885.
[21] Rautter, J. (1995), Doctoral Thesis, Technische Unversität Berlin.
[22] Plato, M., Lendzian, F., Lubitz, W., and Möbius, K. (1992) in The Photosynthetic Bacterial Reaction Center II: Structure, Spectroscopy and Dynamics, (Breton, J. and Vermeglio, A., eds.), pp. 109–118, Plenum Press, New York.
[23] Tiede, D. M., Vashita, A.-C., and Gunner, M. R. (1993) Biochemistry 32, 4515–4531.
[24] Overfield, R. E. and Wright, C. A. (1986) Photosyn. Res. 9, 167–179.
[25] Moser, C. C. and Dutton, P. L. (1988) Biochemistry 27, 2450–2461.
[26] Marcus, R. A. and Sutin, N. (1985) Biochim. Biophys. Acta 811, 265–322.
[27] Plato, M., Möbius, K., Lubitz, W., Allen, J. P., and Feher, G. (1990) in Perspectives in Photosynthesis, (Jortner, J. and Pullman, B., eds.), pp. 423–434, Kluwer, Dordrecht, The Netherlands.
[28] Moser, C. C., Keske, J. M., Warncke, K., Farid, R. S., and Dutton, P. L. (1992) Nature 355, 796–802.
[29] Allen, J. P., Feher, G., Yeates, T. O., Komiya, H., and Rees, D. C. (1987) Proc. Natl. Acad. Sci. USA 84, 6162–6166.

[30] Tiede, D. M. and Chang, C.-H. (1988) Isr. J. Chem. 28, 183–191.

[31] E.Michel-Beyerle, M., Plato, M., Deisenhofer, J., Michel, H., Bixon, M., and Jortner, J. (1988) Biochim. Biophys. Acta 932, 52–70.

[32] Michel, H., Epp, O., and Deisenhofer, J. (1986) EMBO J. 5, 2445–2452.

[33] Mathis, P. (1994) Biochim. Biophys. Acta 1187, 177–180.

[34] Onuchic, J. N., de Andrade, P. C. P., and Beratan, D. N. (1991) J. Chem. Phys. 95, 1131–1138.

[35] Evanson, J. W. and Karplus, M. (1992) J. Chem. Phys. 96, 5272.

[36] Beratan, D. N., Betts, J. N., and Onuchic, J. N. (1992) J. Phys. Chem. 96, 2852–2855.

[37] Beratan, D. N., Onuchic, J. N., Winkler, J. R., and Gray, H. B. (1992) Science 258, 1740–1742.

[38] Aquino, A. J. A., Beroza, P., Beratan, D. N., and Onuchic, J. N. (1995) Chem. Phys. 197, 277–288.

[39] Adir, N., Okamura, M. Y., and Feher, G. (1994) Biophys. J. 66, A137.

# ENDOR on the Primary Electron Donor Cation Radical of Heterodimer-Double-Mutants of *Rb. sphaeroides*. Effect of Hydrogen Bonding on the Electronic Structure of Bacteriochlorophyll *a*-type Radicals

M. Huber[1*], M. Plato[2], S. T. Krueger-Koplin[3], C.C. Schenck[3]

[1] Institut für Organische Chemie, Freie Universität Berlin, Takustr. 3, D-14195 Berlin
[2] Instritut für Experimentalphysik, Freie Universität Berlin, D-14195 Berlin
[3] Department of Biochemistry, Colorado State University, Ft. Collins, USA

**Abstract.** The primary electron donor (D) of photosynthetic bacteria (*Rhodobacter sphaeroides*) is a dimer of bacteriochlorophyll *a* (BChl *a*) molecules. The molecular origin of the electronic asymmetry of the cation radical of D ($D^{+\bullet}$) is investigated by EPR/ENDOR on $D^{+\bullet}$ in a heterodimer hydrogen bond double mutant HL (M202)/LH (L131). HL (M202) is the heterodimer mutant, in which $D_L$ is a BChl and $D_M$ a bacteriopheophytin. In $D^{+\bullet}$ of this mutant the unpaired electron is localized on $D_L$. The LH (L131) mutation introduces a hydrogen bond to $D_L$, which shifts spin density from ring I into ring III. Molecular orbital calculations on BChl $a^{+\bullet}$ are performed in order to predict the effect of acetyl group rotation and hydrogen bonding on the spin density distribution. Interaction of the newly introduced His L131 with $D_L$ seems weaker than that of the acetyl hydrogen bond formed by His L168 in the native structure. Additional EPR signals observed in HL (M202)/LH (L131) suggest further paramagnetic species, which might be a result of the high oxidation potential of D in this mutant.

**Keywords.** Bacterial photosynthesis, primary electron donor, heterodimer mutant, hydrogen bonding, ENDOR spectroscopy, bacteriochlorophyll *a*

## 1 Introduction

The primary electron donor D in bacterial photosynthesis is a dimer of bacteriochlorophyll (BChl) molecules [1,2,3,4,5]. EPR/ENDOR methods have been used to investigate the electronic structure of the cation radical of D ($D^{+\bullet}$), revealing that the unpaired electron is delocalized over the two halves of the dimer [6,7,8]. The spin density distribution is asymmetric, however, with a ratio of 2:1 in favour of the BChl on the L side of the dimer ($D_L$) [9]. Given that D is made up of two BChl molecules the asymmetry must be due to specific inter-

* To whom correspondence should be addressed.

actions of the protein and/or pigment with the two BChl's of D. The origin of this asymmetry, which might have a functional role [10], is the focus of the present study.

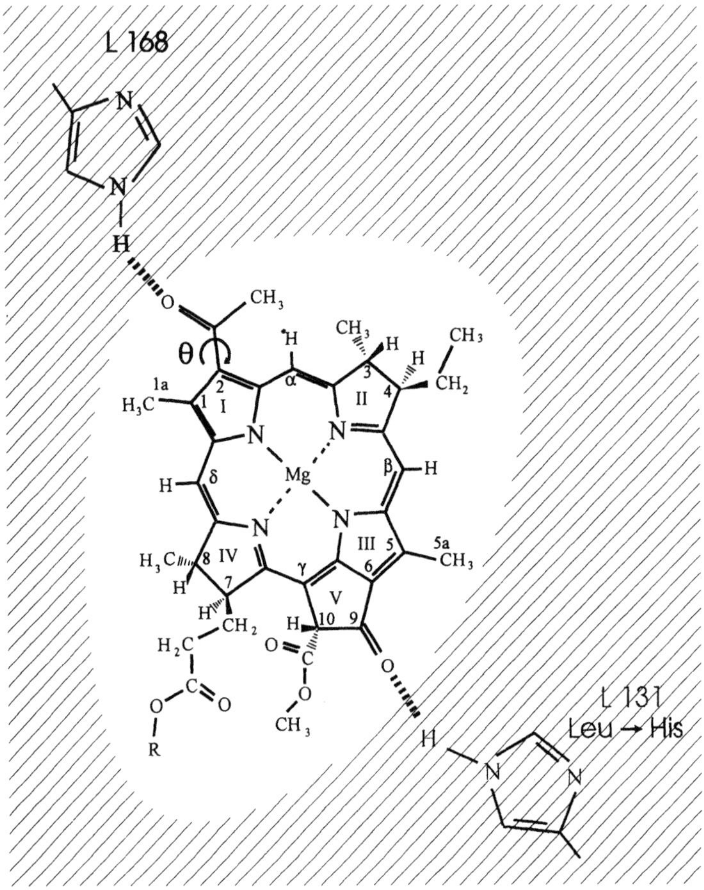


**Figure 1** Structure and numbering scheme of Bacteriochlorophyll *a*. Imidazole rings represent histidines relevant for hydrogen bonding. Labels of residues refer to the BChl on the L-side of the primary electron donor in *Rb. sphaeroides*. His L168: native structure, Leu→His L131 mutation: introduces hydrogen bond at ring V (see text).

Heterodimer mutants [11,12], whose electronic structures have been investigated previously [13,14] are of particular interest in this context. In these mutants one BChl of D is replaced by a bacteriopheophytin, the Mg-free analogue of BChl [12,15]. In $D^{+\bullet}$ of heterodimer mutants the unpaired electron is localized on the BChl, i.e. on $D_L$ in HL (M202) (a histidine to leucine mutation at position M202) or on $D_M$ in HL (L173) [13,14,16]. It was shown that the electronic structure of $D^{+\bullet}$ of the two heterodimers differs significantly, which shows that the BChl's at $D_L$ and $D_M$ are not equivalent with respect to their conformation and/or environment [13,16]. Hydrogen bonding to the BChl *a* system (see Figure 1) is the most likely candidate to cause the difference between $D_L$ and $D_M$. For example, the acetyl group of the BChl at $D_L$ was found to be hydrogen bonded to a histidine residue (His L168) - schematically shown in Figure 1, whereas no corresponding hydrogen bond is observed in $D_M$ [3,4,8].

We were thus interested in the effect of hydrogen bonding on the electronic structure of the BChl. There are two keto groups in BChl *a*, which can act as hydrogen bond acceptors: the acetyl C=O group at ring I, and the C=O group at ring V (see Figure 1). With respect to $D_L$ in the RC, the removal of the acetyl group hydrogen bond (formed by His L168) and the introduction of a ring V C=O hydrogen bond (through the Leu→His L131 mutation) would change the hydrogen bonding situation. (Residues were first suggested in [17]). These and other hydrogen bond mutants were investigated in the homodimer system, where the main effect was found to be a redistribution of spin density over the two macrocycles [18,19]. Thus, to investigate the effect of hydrogen bonding on the *individual* chromophores introducing these modifications in the appropriate ($D_L$) heterodimer system was of interest. The corresponding mutants are the HL (M202)/HL (L168) and the HL (M202)/LH (L131) double mutants. Both double mutants were made, in the present publication EPR/ENDOR results on $D^{+\bullet}$ of HL (M202)/LH (L131) will be reported. An independent line of investigation was pursued by the groups of J. Allen (Arizona State University, Tempe, USA), and W. Lubitz (Technical University Berlin, FRG), leading to interest in the same mutants. Their results are being published independently [19].

## 2 Materials and Methods

Site directed mutagenesis, growing of bacteria and isolation of RC's were performed as described in [12], with modifications to be published elsewhere (C.C. Schenck, et al., to be published). The EPR and ENDOR spectrometer used is described in [20]. $D^{+\bullet}$ was created by *in situ* illumination of the RC's using 100-200 W halogen lamps with a $\lambda \geq$ 750 nm cutoff, and a 5-cm $H_2O$ filter.

In contrast to wild type and HL (M202) RC's illumination at 18°C in the double mutant caused sample deterioration as evidenced by the disappearance of the liquid solution Special TRIPLE signals after illumination times of 1-2 hours. EPR spectra in frozen solution (130 K) show that HL (M202)/LH (L131) has additional signals, which are not observed in HL (M202). Some of these signals partially overlap with the signal attributed to $D^{+\bullet}$ (see 'Discussion'). The EPR signals in this field region can be separated on the basis of their different relaxation behaviour and lineshapes. The additional signals are broader and do not saturate at microwave- (mw-) powers below 150 mW (i.e. the EPR signal intensity is approximately proportional to the square root of the mw-power). Since the ENDOR effect critically depends on the ability to saturate the corresponding EPR signal [21], the additional signal cannot contribute to the

ENDOR and Special TRIPLE spectra obtained at 40 mW (130 K), and 10 mW (285 K). Further EPR signals were observed in the g=1.9-1.8 region, i.e. outside the field region where $D^{+\bullet}$ is to be expected. The origin of the additional signals, i.e. the chemical nature of the radicals or paramagnetic metal ions causing these signals is presently investigated.

To theoretically assess the effect of hydrogen bonding on BChl $a^{+\bullet}$ MO calculations of the RHF-INDO/SP type [22] were performed. To model BChl *a*, standard bond lengths and angles [23] were used. The effect of acetyl group rotation (angle Θ in Figure 1) and hydrogen bonding were investigated. Hydrogen bonds were modelled using partial positive charges located at a fixed distance (1.6 Å) from the oxygen atom in the line of the C=O bond.

## 3 Results

### 3.1 EPR Spectra

EPR spectra of $D^{+\bullet}$ in RC's of the HL (M202) mutant in frozen solution consist of an inhomogenously broadened Gaussian line at $g=2.0027\pm0.0001$ with a peak-to-peak linewidth ($\Delta B_{pp}$) of $(1.23\pm0.02)$mT. These values agree within experimental error to liquid solution EPR results reported for this mutant previously [16]. In the HL (M202)/LH (L131) double mutant additional EPR signals are observed. In the g=2 region several signals are superimposed, one of which is similar ($g=2.0026\pm0.0002$, $\Delta B_{pp}=(1.26\pm0.04)$mT) to the EPR signal of $D^{+\bullet}$ in HL (M202). The remaining features in this spectral region can be interpreted as a powder pattern of a radical species with a larger spread in g values. In contrast to the EPR signal of $D^{+\bullet}$, this signal does not saturate at mw-powers below 150 mW. In HL (M202)/LH (L131) further signals are observed in the g=1.9 and 1.8 region (see 'Materials and Methods'), which is outside the magnetic field region where $D^{+\bullet}$ is to be expected, and thus does not interfere with the ENDOR experiments.

### 3.2 ENDOR and Special TRIPLE Spectra

Individual hfc's can be resolved using ENDOR methods. The frequency of ENDOR lines ($\nu_{ENDOR}$) is given by

$$\nu_{ENDOR} = | \nu_H \pm A/2 | \qquad (1)$$

where $\nu_H$ is the Larmor frequency of the free proton. Thus, in ENDOR spectra two lines symmetrically displaced about $\nu_H$ are observed for each hfc A, provided that $\nu_H > A/2$. The hfc consists of an anisotropic (dipolar, $A_{dip}$) and an isotropic (scalar, $A_{iso}$) part. In liquid solution only the isotropic part is observed. In frozen solution the dipolar part of the hfc causes broadening of the ENDOR lines, and only protons having small anisotropies, in particular those belonging to rotating methyl groups yield well resolved powder patterns.

In Figure 2 ENDOR (frozen solution) and Special TRIPLE (liquid solution - see below) spectra of the HL (M202)/LH (L131) double mutant are shown. In the frozen solution (130 K) spectrum the lineshape of lines 4 and 4' is characteristic of a powder pattern, where 4 is the parallel ($A_{\parallel}$) and 4' the perpendicular component ($A_{\perp}$) of a hyperfine tensor (see [13] for illustration). Line 6' is the perpendicular component of a second hf tensor, whose parallel component overlaps line 4'. Thus, two methyl group hfc's are identified, their hfc's are given in Table 1.

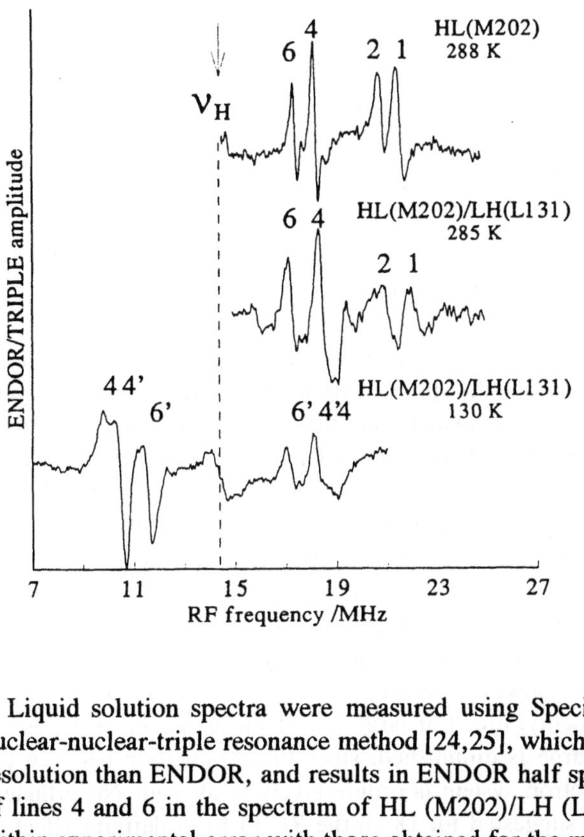


**Figure 2** ENDOR (130 K) and Special TRIPLE spectra of $D^{+\bullet}$ in the HL (M202) heterodimer (upper panel, modified from [13,16]), and the HL (M202)/LH (L131) double mutant (lower two panels). Experimental conditions: for HL (M202), see [16]; for HL (M202)/LH (L131), 283 K (Special TRIPLE): mw-power 10 mW, rf-power 150 $W_{tot}$, rf-frequency modulation: deviation $\pm$100 kHz at 15 kHz, total measuring time: 6 hr; for HL (M202)/LH (L131), 130 K (ENDOR): mw-power 40 mW, rf-power 150 W, rf-frequency modulation: deviation $\pm$100 kHz at 15 kHz, total measuring time: 1.5 hr.

Liquid solution spectra were measured using Special TRIPLE, an electron-nuclear-nuclear-triple resonance method [24,25], which has higher sensitivity and resolution than ENDOR, and results in ENDOR half spectra. The isotropic hfc's of lines 4 and 6 in the spectrum of HL (M202)/LH (L131) (see Figure 2) agree within experimental error with those obtained for the methyl groups in the frozen solution ENDOR spectra (cf. Table 1). Consequently, lines 4 and 6 are assigned to methyl groups. Lines 1 and 2 are assigned to the non-methyl *β* protons (e.g. 4 and 3 at ring II of BChl *a* - see Figure 1 for numbering of protons) in analogy to $D^{+\bullet}$ in HL (M202) [14,16]. The isotropic hfc's are given in Table 2. The spectrum of $D^{+\bullet}$ in HL (M202) [16] is shown for comparison in Figure 2. It reveals that $A_{iso}$ of lines 1, 2, and 4 in HL (M202)/LH (L131) are larger, whereas $A_{iso}$ of line 6 is smaller than in HL (M202).

**Table 1.** Methyl Proton Hfc's of $D^{+\bullet}$ in Heterodimer and Heterodimer Double Mutants in Frozen and Liquid Solutions. Isotropic ($A_{iso}$) and Anisotropic ($A_{\parallel}$, $A_{\perp}$) Hfc's in MHz. $A_{\parallel}$, $A_{\perp}$: Parallel and Perpendicular Components of Hf tensor; $A_{iso} = 1/3(2A_{\perp} + A_{\parallel})$.

| | Assign-ment | Liquid Solution | | Frozen Solution | | |
|---|---|---|---|---|---|---|
| | | Line | $A_{iso}$ [a] | $A_{\perp}$ [b] | $A_{\parallel}$ [b] | $1/3(2A_{\perp} + A_{\parallel})$ |
| HL (M202)/ LH (L131) | $1a_L$ | 6 | 5.7 | 5.5 | (6.9) [c] | 6.0 |
| | $5a_L$ | 4 | 8.1 | 7.7 | 9.2 | 8.2 |
| HL (M202) [d] | $1a_L$ | 6 | 6.0 | 5.6 | (7.2) | 6.1 |
| | $5a_L$ | 4 | 7.6 | 7.0 | 8.4 | 7.6 |

[a] errors: ±80 kHz (HL (M202)/LH (L131)), ±50 kHz (HL (M202)) [16]
[b] errors: ±100 kHz
[c] estimated from ($A_{\parallel}$-$A_{\perp}$) of methyl group $5a_L$
[d] from [14,16]

## 3.3 Molecular Orbital Calculations on BChl $a^{+\bullet}$

The RHF-INDO/SP method is used to calculate the hfc's of BChl $a^{+\bullet}$. Two major parameters which influence the spin density distribution in the macrocycle (see for example [26,27]) are investigated: the effect of rotating the acetyl group at ring I (angle Θ in Figure 1) and the effect of hydrogen bonding to oxygen atoms of the acetyl group and the ring V keto group (see also 'Materials and Methods'). Rotating the acetyl group out of the plane of the $\pi$ electron system causes a shift in spin density from ring I to ring III, evidenced by an increasing 5a methyl group hfc and a decreasing 1a methyl group hfc. This is in agreement with the results of MO calculations by Hanson [27]. Point charges near the oxygen atom of the acetyl group (corresponding to a hydrogen bond to the acetyl group) counteract this effect, however, leading to an increase in the 1a methyl hfc at the expense of the 5a methyl hfc. For large out of plane angles of the acetyl group the effect of the charge is diminished, since the conjugation between the C=O group and the $\pi$ electron system is progressively weakened. Nevertheless at angles of up to 45°, as suggested for $D_L$ in the RC [4,8], a definite effect of the point charge is predicted. For $\Theta = 45°$ and a charge of +0.2 the ratio of methyl group hfc's calculated agrees with the experimental one for HL (M202). Point charges at the ring V keto group result in a decrease of the 1a methyl group hfc, and an increase of the 5a methyl group, suggesting a shift of spin density from ring I into ring III, i.e. opposite to the effect of the point charge at the acetyl group.

**Table 2.** Isotropic Hfc's of $D^{+\bullet}$ in Heterodimer and Heterodimer Double Mutants, and of BChl $a^{+\bullet}$

| | HL(M202)/ LH (L131) | HL (M202)[a] | | BChl $a^{+\bullet}$ [b] | |
|---|---|---|---|---|---|
| Line[c] | $A_{iso}$[d] [MHz] | $A_{iso}$[e] [MHz] | *Assignment*[f] | $A_{iso}$ [MHz] | *Assignment*[f] |
| 7 | 3.0 | | | +2.34 | $\alpha$ |
| 6 | 5.7 | +6.0 | $1a_L$ | +4.93 | $1a$ |
| 5 | | 6.6 | $8_L$ | | |
| 4 | 8.1 | +7.6 | $5a_L$ | +9.62 | $5a$ |
| | | | | +11.76 | $8$ |
| 3 | 9.1 | 8.3 | $7_L$ | +13.11 | $7$ |
| 2 | 13.2 | +12.8 | $3_L$ | +13.47 | $3$ |
| 1 | 15.8 | +14.2 | $4_L$ | +16.35 | $4$ |

[a] Hfc's and assignment from [16].
[b] $A_{iso}$ and assignment from [22].
[c] Line numbering according to Figure 2.
[d] From liquid solution spectra, errors: ±80 kHz.
[e] From [16], errors: ±50 kHz.
[f] Numbering of protons, see Figure 1.

## 4 Discussion

The most striking difference between the HL (M202)/LH (L131) double mutant and HL (M202) is the larger number of EPR signals in the former. While the origin of these signals is interesting in itself (see below), it complicates the analysis, since, in order to investigate $D^{+\bullet}$ we have to identify which EPR signal is due to $D^{+\bullet}$ and establish whether the ENDOR signals observed are due to this species. One of the EPR signals in the double mutant has similar characteristics as the signal of $D^{+\bullet}$ in HL (M202), therefore, it is attributed to $D^{+\bullet}$ in the double mutant. ENDOR signals observed in HL (M202)/LH (L131) (see Figure 2 and Table 2) can only be due to the EPR signal identified with $D^{+\bullet}$, since the other EPR signals cannot be saturated at the mw-powers used in the ENDOR experiment (see 'Materials and Methods').

### 4.1 Electronic Properties of $D^{+\bullet}$ in HL (M202)/LH (L131)

The isotropic hfc's of $D^{+\bullet}$ in HL (M202)/LH (L131) RC's are compared to those of HL (M202) (see Figure 2 and Table 2) in order to determine the effect of the additional mutation on the electronic structure of $D^{+\bullet}$. Since methyl group hfc's are directly proportional to the spin density in the $\pi$ electron system [28] we will focus on those. To characterize the electronic structure of $D^{+\bullet}$ and BChl $a^{+\bullet}$ two parameters have been introduced before [14,16], the sum of methyl group hfc's ($\Sigma A_i(CH_3)$) as a measure for the localization of the unpaired electron in the heterodimer, and the ratio of the 5a and 1a methyl group hfc's ($A(CH_3)_{III}/A(CH_3)_I$), which characterizes the spin density distribution within the BChl macrocycle. As shown in Table 3, the sum of methyl group hfc's in HL (M202)/LH (L131) agrees within experimental error with $\Sigma A_i(CH_3)$ in HL (M202), indicating that the hydrogen bond does not significantly change the degree of localization. This is in contrast to the effect of hydrogen bonding in the homodimer, where the dominant effect of hydrogen bonding is a redistribution of spin density over the two macrocycles [18]. Consequently, the heterodimer double mutant allows to focus on the effect of the hydrogen bond on the individual BChl system, in our case on $D_L$.

**Table 3.** Methyl Group Hfc's of BChl $a^{+\bullet}$, and $D^{+\bullet}$ of Heterodimer and Heterodimer Double Mutants

| | $\Sigma A_i(CH_3)$ | $A(CH_3)_{III}/A(CH_3)_I$ |
|---|---|---|
| BChl $a^{+\bullet}$ [a] | 14.5(1) | 1.95(2) |
| HL (M202)[b] | 13.6(1) | 1.27(2) |
| HL (M202)/LH (L131) | 14.0(2) | 1.41(4) |
| HL (L173)[b] | 13.7(1) | 1.69(2) |

[a] using $A_{iso}$ from [22]
[b] from [14,16]

The ratio $A(CH_3)_{III}/A(CH_3)_I$ is larger in HL (M202)/LH (L131), than in HL (M202). In other words the additional hydrogen bond causes a shift in spin density from ring I into ring III in the BChl. The magnitude of the shift observed can be interpreted using the MO calculations referred to earlier (see 'Results'). $A(CH_3)_{III}/A(CH_3)_I$ of $D^{+\bullet}$ in the HL (M202) mutant could be reproduced in the calculation using an acetyl group out of plane rotation of 45° and a partial charge at the acetyl group of +0.2. An additional positive charge in the vicinity of the ring V keto group should model the hydrogen bond introduced by His L131 in the double mutant. A charge of 0.05 increases $A(CH_3)_{III}/A(CH_3)_I$ to match the experimental value. Thus, in the framework of our model the newly introduced hydrogen bond appears to be weaker than the acetyl group hydrogen bond. This suggests, either, a larger hydrogen bond distance, or a weaker hydrogen bond donor group at the ring V keto group. Since both hydrogen bonds are formed by histidines the former possibility seems more likely.

## 4.2 Additional EPR Signals Observed in the HL (M202)/LH (L131) Mutant

In principle, the additional EPR signals in the HL (M202)/LH (L131) mutant could be due to (i) paramagnetic impurities e.g. an artefact of the preparation, or they could (ii) be due to a photoreaction or chemical reaction associated with the radical species generated in the light induced charge separation in the RC's.

The fact that the additional signals were not observed in an analogously prepared HL (M202) sample seems to argue against the first suggestion, but given the high sensitivity of the EPR method, the present state of the experiments does not allow to fully exclude that possibility. The second possibility is intriguing, since the introduction of the hydrogen bond is expected to increase the oxidation potential of $D^{+\bullet}$ in this mutant relative to HL (M202) [29]. This could result in a tendency of $D^{+\bullet}$ to oxidatively react with its environment by forming radicals which would cause (some of) the observed EPR signals, but could also provide a pathway by which sample decay is initiated or aided.

# 5 Summary and Outlook

Heterodimer-hydrogen bond mutants have been prepared and, in the case of the HL (M202)/LH (L131) mutant, $D^{+\bullet}$ could be investigated by EPR/ENDOR methods. In this double mutant the spin density is found to be localized to approximately the same degree as in the HL (M202) mutant. The spin density within the chromophore is shifted, however, increasing the spin density in ring III while decreasing it in ring I. This shift is attributed to the effect of an additional hydrogen bond to the ring V keto group of $D_L$ in this mutant.

Results of MO calculations of the RHF-INDO/SP type are reported, which illustrate the effect of the rotation of the acetyl group, and of hydrogen bonding. Using these calculations the hfc's observed in the HL (M202) mutant can successfully be modelled assuming an out of plane orientation of the acetyl group by 45° and a hydrogen bond to its keto group. The shift in spin density distribution due to the additional Leu→His mutation at L131 in HL (M202)/LH (L131) can be modelled satisfactorily assuming a slightly weaker hydrogen bond.

For HL (M202)/HL (L168) removing the hydrogen bond to the acetyl group is expected to have a potentially complex effect on the spin density distribution: If only the hydrogen bond was removed, the predicted result would be an increase in $A(CH_3)_{III}/A(CH_3)_I$. It is conceivable, however, that the acetyl group rotates upon removal of the hydrogen bond, either due to its equilibrium position being at a different angle $\Theta$ when the hydrogen bond is not present, or due to steric interaction with the mutated residue. In this case, the change in $A(CH_3)_{III}/A(CH_3)_I$ would be diagnostic of the rotation of the acetyl group.

Thus, in principle, the ENDOR results could be used to determine the strength of hydrogen bonds, and the conformation of selected groups in the radical via their effect on the spin density distribution. The viability of the results depends on the reliability of the MO method used, and the way the hydrogen bonds are modelled. Investigations addressing these questions are presently under way.

Finally, initial results are reported here, which suggest the presence of additional paramagnetic components in the RC's of the HL (M202)/LH (L131) mutant. In particular we are currently trying to establish whether the tempting hypothesis of radical formation due to an increase in the oxidation potential of $D^{+\bullet}$ holds.

## 6 Acknowledgements

Particular thanks go to G. Feher (Department of Physics, University of California, San Diego, USA), for initiating this project. Spectrometers, material and moral support by K. Möbius (Institut für Molekülphysik, Freie Universität Berlin) is gratefully acknowledged. Funding was through grants of the Deutsche Forschungsgemeinschaft (SFB 337) and the NIH (GM 48254).

**References.**

1. J. P. Allen, G. Feher, T.O Yeates, D.C. Rees, D.S. Eisenberg, J. Deisenhofer, H. Michel, R. Huber (1986) Biophys. J. 49, 538a
2. T.O. Yeates, H. Komiya, A. Chirino, D.C. Rees, J.P. Allen, G. Feher (1988) Proc. Natl. Acad. Sci. U.S.A. 85, 7993 and references therein
3. C.H. Chang, D. Tiede, J. Tang, U. Smith, J. Norris and M. Schiffer (1986) FEBS Letters 205, 82
4. U. Ermler, G. Fritzsch, S. Buchanan, H. Michel (1994) Structure 2, 925-936
5. C.-H. Chang, O. El-Kabbani, D. Tiede, J. Norris, M. Schiffer (1991) Biochemistry 30, 5352
6. G. Feher, A.J. Hoff, R.A. Isaacson, L.C. Ackerson (1975) Ann. N.Y. Acad. Sci. (U.S.) 244, 239-259
7. J.R. Norris, H. Scheer, J.J. Katz (1975) Ann. N.Y. Acad. Sci USA 244, 260
8. M. Plato, K. Möbius, W. Lubitz, J. P. Allen, G. Feher (1990) in Perspectives in Photosynthesis (J. Jortner, P. Pullman, eds.), p. 423, Kluver Academic Publishers, Netherlands
9. F. Lendzian, M. Huber, R.A. Isaacson, B. Endeward, M. Plato, B. Bönigk, K. Möbius, W. Lubitz, G. Feher (1993) Biochim. Biophys. Acta 1183, 139-160
10. M. Plato, K. Möbius, M.E. Michel-Beyerle, M. Bixon, J. Jortner (1988) J. Am. Chem. Soc. 110, 7279-7285
11. E.J. Bylina, S.V. Kolaczkowski, J.R. Norris, and D.C. Youvan (1990) Biochemistry 29, 6203
12. L.M. McDowell, D. Gaul, C. Kirmaier, D. Holten, C.C. Schenck, (1991) Biochemistry 30, 8315
13. M. Huber, E.J. Lous, R.A. Isaacson, G. Feher, D. Gaul, C.C. Schenck (1990) in Reaction Centers of Photosynthetic Bacteria (M.-E. Michel-Beyerle ed.), pp. 219-228, Springer Series in Biophysics, Berlin.
14. G. Feher (1992) J. Chem. Soc. Perkin Trans 2, 11, 1861
15. A.J. Chirino, E.J. Lous, M. Huber, J.P. Allen, C.C. Schenck, M.L. Paddock, G. Feher, D.C. Rees (1994) Biochemistry 33, 4584

16. M. Huber, R.A. Isaacson, E.C. Abresch, D. Gaul, C.C. Schenck, G. Feher (1995) Biochim. Biophys. Acta, submitted
17. J.C.Williams, R.G. Alden, H.A. Murchison, J.M. Peloquin, N.W. Woodbury, J.P. Allen (1993) Biochemistry 31, 11029-11037
18. J. Rautter, Ch. Geßner, F. Lendzian, W. Lubitz, J.C.Williams, H.A. Murchison, S. Wang, N.W. Woodbury, J.P. Allen (1992) in The Photosynthetic Reaction Center (J. Breton, A. Vermeglio eds.), pp. 99-108, Plenum Press, New York
19. J. Rautter, F. Lendzian, C. Schulz, A. Fetsch, M. Kuhn, X. Lin, J.C.Williams, J.P. Allen, W. Lubitz (1995) Biochemistry, in press
20. K. Möbius, W. Lubitz, M. Plato (1989) in Advanced EPR (A. Hoff, ed.) pp. 441-499 Elsevier, Amsterdam
21. R. Allendoerfer, A.H. Maki (1970) J. Magn. Res. 3, 396
22. M. Plato, K. Möbius, W. Lubitz (1991) in The Chlorophylls (H. Scheer ed.) pp. 1015-1046, CRC Press, Boca Raton
23. J.A. Pople, D.L. Beveridge (1970) Approximate Molecular Orbital Theory, pp. 111-112 McGraw-Hill Inc., New York.
24. G. Feher (1958) Physica XXIV, 80
25. K.P. Dinse, R. Biehl, K. Möbius (1974) J. Chem. Phys. 61, 4335
26. M.S. Davis, A. Forman, L.K. Hanson, J.P. Thornber, J. Fajer (1979) J. Phys. Chem. 83, 3325
27. L.K. Hanson (1991) in The Chlorophylls (H. Scheer ed.) pp. 993-1014, CRC Press, Boca Raton
28. C. Heller, H.M. McConnell (1960) J. Chem. Phys. 32, 1535
29. X. Lin, H.A. Murchison, V. Nagarajan, W.W. Parson, J.P. Allen, J.C.Williams (1994) Proc. Natl. Acad. Sci. U.S.A. 91, 10265-10269

# Structure Information on the Bacterial Primary Donor $P^{+\bullet}$, Acceptor $Q_A^{-\bullet}$, and Radical Pair $P^{+\bullet}$-$Q_A^{-\bullet}$ as Obtained from High-Field EPR/ENDOR and MO Studies

K. Möbius and M. Plato

Dept. of Physics, Free University Berlin, Arnimallee 14, D-14195 Berlin, Germany

**Abstract.** Recent 3 mm (W-band, 95 GHz) high-field EPR and ENDOR studies are presented on the primary donor cation radicals $P_{865}^{+\bullet}$ (bacteriochlorophyll dimer) in single crystals of reaction centers (RC's) and on frozen solutions of acceptor anion radicals $Q^{-\bullet}$ (quinones) and of the charge-separated radical pair ($P_{865}^{+\bullet}$-$Q_A^{-\bullet}$) in photosynthetic bacteria *Rb. sphaeroides* R-26 and biomimetic model systems. Both the hyperfine tensors of various protons and the g-tensor of $P_{865}^{+\bullet}$ have been determined and compared with tensor values calculated by INDO type MO methods that are based on recent X-ray structure data. The results consistently reveal a breaking of the local $C_2$ symmetry of the electronic structure at the primary donor side of the reaction center. This might be relevant for the vectorial electron transfer along the protein complex. Among the quinone radical anions studied are frozen solutions of the electron acceptors of bacterial and plant reaction centers (ubiquinone and plastoquinone, respectively). The increased electron Zeeman interaction in high-field EPR leads to almost completely resolved g-tensor components, and single crystal-like hyperfine couplings could be measured even in disordered samples. The g-tensor values and component linewidths are sensitive probes for specific anisotropic interactions with the environment. Pulsed high-field EPR reveals anisotropic contributions to $T_2$ relaxation by librational motion of the primary acceptor in its protein binding site. In the case of the transient correlated coupled radical pair $P_{865}^{+\bullet}$-$Q_A^{-\bullet}$ (Fe replaced by Zn) the spin-polarized high-field EPR spectra allow an unambiguous determination of the relative orientation of the g-tensors of the donor and acceptor parts. Thereby high-precision structure information is obtained on the electron transfer pigments after light-induced charge separation.

**Keywords.** Bacterial photosynthesis, high-field EPR/ENDOR, structure determination, electron transfer pigments

## 1 Introduction

Over the last 25 years important progress in bacterial photosynthesis research has been accomplished regarding both sample characterization and spectroscopic information. Concerning the sample this progress is high-lighted by the isolation of the reaction center (RC) pigment-protein complex [1-3], the determination of the RC composition [4,5], the crystallization of an RC complex [6] and its subsequent X-ray crystal structure determination at atomic resolution [7,8], and by the manipulation of RC's by site-directed mutagenesis of cofactors and amino acids. Building on these accomplishments in sample preparation and characterization, an enormous increase of spectroscopic information, both on structure and dynamics, could be achieved from applying ps laser spectroscopy and multiresonance EPR (electron paramagnetic resonance) techniques. These experiments are now setting the stage for elucidating the relation between the electron transfer (ET) kinetics and the spatial and electronic structures of the cofactors embedded in their protein environment.

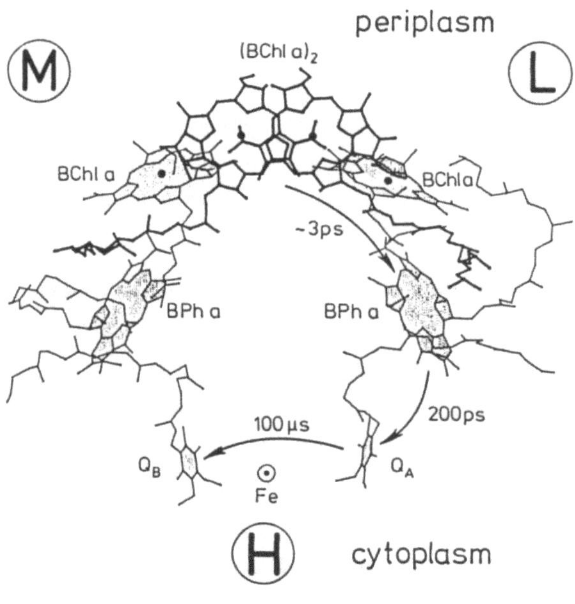


**Fig. 1.** Cofactor arrangement in the reaction center of *Rb. sphaeroides* R-26 as determined by X-ray diffraction [11]. The side chains of the BChl *a* monomers and of the quinones $Q_A$ and $Q_B$ have been partly truncated in the figure for better clarity. *L, M,* and *H* indicate the protein subunits. Arrows show ET steps and their time constants.

At present, the best characterized RC's are those of the purple bacteria *Rb. sphaeroides* and *Rps. viridis* with their primary donors $P_{865}$ (a bacteriochlorophyll a (BChl a) dimer) and $P_{960}$ ( a bacteriochlorophyll b (BChl b) dimer), respectively. Together with the transient electron acceptors, two monomeric bacteriochlorophylls, two bacteriopheophytins (BPh), two ubiquinones (*Rb. sphaeroides*) or a menaquinone and a ubiquinone (*Rps. viridis*), they are anchored in the L- M-, and H protein subunits of the transmembrane RC complex. Fig. 1 gives the X-ray structure of the cofactor arrangement in the RC of *Rb. sphaeroides* R-26 [9-13]. A striking aspect of this arrangement in both bacteria is the approximate local two-fold symmetry, the $C_2$ axis being aligned vertically to the membrane surfaces and running through the (BChl a)$_2$ primary donor near the periplasmic side of the membrane and the iron ($Fe^{2+}$) near the cytoplasmic side. In view of this $C_2$ symmetry relation between the cofactors within the L- and M protein subunits it is most surprising that the light-induced ET proceeds preferentially along the pigments of the L branch with an L/M ratio for the time constants of more than 10 [14]. Pertinent questions are still being asked: What are the factors that control this "unidirectionality" of primary ET and its high quantum yield? Is the symmetry of the electronic structure of the special pair donor an important factor or is it the dielectric asymmetry of the protein environment along the two potential ET pathways? Are there specific amino acid residues that provide a "fine tuning" of the energetics and wavefunctions for an optimized long-range ET in terms of Marcus theory?

In the following we report on recent cw and pulsed 95 GHz (W-band) high-field/high-frequency EPR/ENDOR experiments on the primary donor cation radical $P_{865}^{+\bullet}$, the ubiquinone primary acceptor radical anion $Q_A^{-\bullet}$ and other quinone acceptors, and the correlated coupled charge-separated radical pair $P_{865}^{+\bullet}$-$Q_A^{-\bullet}$ of *Rb. sphaeroides*. These experiments were done in an attempt to contribute to a better understanding of the structure-function relationship of photosynthetic ET by providing - via measurements of g- and hyperfine-tensor interactions - detailed information about electronic and spatial structures not accessible by X-ray crystallography or ultra-fast laser spectroscopy.

## 2 Experimental

### 2.1 High-Field/High-Frequency EPR and ENDOR

For many bioorganic systems, in particular when they are in frozen solution samples, conventional X-band (9.5 GHz) EPR runs into problems with spectral resolution because of different magnetic "sites" of rather similar g-values in the sample or because of small g-factor anisotropies. By analogy with modern NMR spectroscopy, the spectral resolution can be dramatically improved by applying high magnetic Zeeman fields and correspondingly high microwave frequencies:

$$\Delta B_0 = \frac{h\nu}{\beta}\left(\frac{1}{g_1} - \frac{1}{g_2}\right) \qquad (1)$$

In eq. (1) $\Delta B_0$ is the difference in resonance field positions for g-values $g_1$, $g_2$; $\nu$ is the microwave frequency; h and $\beta$ are the Planck constant and Bohr magneton, respectively. Fortunately, for many bioorganic systems the increase of $\Delta B_0$ with increasing Zeeman field directly translates into an increase of spectral resolution, at least up to 300 GHz ($B_0 \approx 10$ T for $g = 2$ systems), since no noticeable line broadening occurs with increasing $B_0$. Double resonance extension to high-field ENDOR has the additional advantage of allowing single crystal-like spectra to be taken even from disordered samples with very small g-factor anisotropies [15].

### 2.2 Spectrometer Performance

Only very few laboratories so far have constructed high-field EPR spectrometers. Details of the Berlin 3-mm (W-band) EPR/ENDOR spectrometer are given by Burghaus et al. [16] for the cw mode and by Prisner et al. [17] for the pulsed mode of operation. Here, only the key features will be summarized: The Zeeman field is provided by a superconducting magnet ($B_0^{max}$ = 6 T) with field sweep capability, the 95 GHz microwave radiation by a klystron (300 mW output power) or a Gunn diode oscillator (6 mW). As microwave resonance structure we use either a tunable Fabry-Perot resonator, which can incorporate a 2-axes goniometer, or a tunable cylindrical $TE_{011}$ cavity. ENDOR experiments can be performed by using an external rf coil energized by a 1 kW rf amplifier. The EPR/ENDOR signals are detected in a homodyne scheme by a helium-cooled InSb hot-electron bolometer and lock-in amplification (cw mode [16]), or in a heterodyne scheme using 4 GHz as the intermediate frequency (pulsed mode [17]). Variable temperature capability between $30 \leq T \leq 300$ K is achieved by a temperature-controlled helium or nitrogen gas stream in contact with the sample capillary. The sample can be photoexcited *in-situ* by quartz light fibers connected to cw lamp or pulsed (10 ns) laser sources. The ringing time of the detection channel after the 1 ns microwave pulses is of the order of 5 ns and, for a 90° pulse, a pulse length of 40 ns is achieved with a microwave power of only 10 mW at the cavity. The sensitivity of the W-band EPR spectrometer is about $4 \cdot 10^8$ spins/mT, both in the cw and pulsed mode of operation [16,17].

## 3 Results and Discussion

### 3.1 W-Band EPR on the $P_{865}^{+\bullet}$ Primary Donor in RC Single Crystals

In our group, over the last years, we have put some effort into the determination of the electronic structure of $P^{+\bullet}$ of *Rps. viridis* and *Rb. sphaeroides*, in particular addressing the question, whether the spin density distribution is symmetric or asymmetric over the dimer halves. We started with X-band EPR, ENDOR and TRIPLE resonance [18] on $P_{960}^{+\bullet}$ of *Rps. viridis* in RC solutions under physiological conditions. By analyzing the measured hyperfine couplings of protons and $^{14}N$ by large-scale MO calculations based on the known X-ray structure [8] we found that the electron spin density distribution is asymmetric over the two dimer halves, favoring the L-half by approx. 2 : 1 [19,20].

In a second series of experiments, partly in collaboration with the group of G. Feher in San Diego who had independently performed similar investigations, we studied $P_{865}^{+\bullet}$ of *Rb. sphaeroides* R-26, both in liquid RC solution and in RC single crystals near room temperature. The X-band EPR/ENDOR/ TRIPLE experimental data of the hyperfine tensors and their MO interpretation again show that the symmetry of the electron spin density is broken and that also in this organism the L-half is favored by approx. 2 : 1 [20-22].

These results show that the hyperfine tensor – being a local probe for the electronic wave function – is very sensitive with regard to symmetry properties of the electronic structure, even for large biomolecules. The measurement of the hyperfine couplings and their assignment to molecular positions is quite a formidable task for systems like $P^{+\bullet}$ and, therefore, the question arose whether precise measurements of the electronic g-tensor – being a more global probe for the electronic wave function – would also reflect the symmetry properties of the electronic structure of large biomolecules. Conventional X-band EPR cannot do the job because it cannot resolve the exceedingly small g-factor anisotropy of $P^{+\bullet}$ and, thus, we had to resort to high-field (W-band) EPR with its 10 times higher Zeeman magnetoselection. By using single crystal RC's not only the principal values of the g-tensor of $P_{865}^{+\bullet}$ could be measured with high precision, but also the orientation of the g-tensor with respect to the molecular axes system.

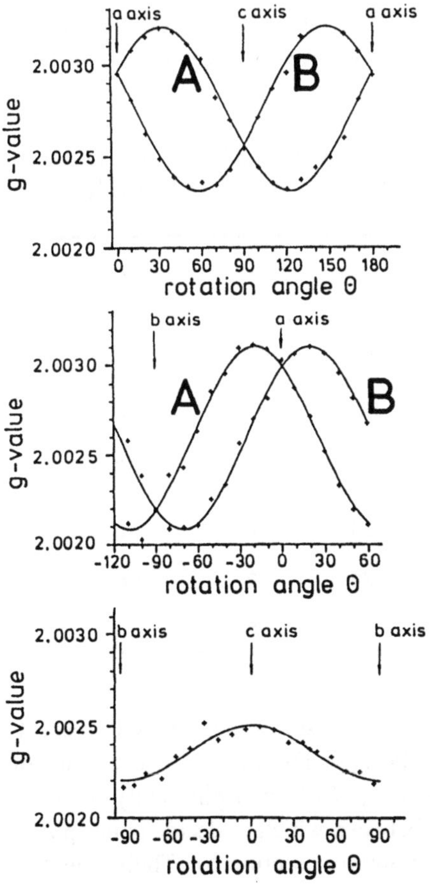


**Fig. 2.** W-band EPR rotation patterns of the g-factor of $P_{865}^{+\bullet}$ in the three symmetry planes of an RC single crystal of *Rb. sphaeroides* R-26 [24]. Θ is the angle between the respective crystallographic axis and the $B_0$ direction. The two solid lines in each pattern are the least squares fits with eq. 2 (see text). They belong to the pairwise magnetically inequivalent RC sites, A, B. The crossings of the curves correspond to orientations, where $B_0$ lies parallel to one of the crystallographic axes; (a) *ac* plane, (b) *ab* plane, (c) *bc* plane. For details, see [24].

The space group of the orthorhombic *Rb. sphaeroides* RC's studied in our laboratory is $P2_12_12_1$ and a complication is encountered in that there are four RC's ("sites") per unit cell, pairwise related by a twofold symmetry axis [9]. Fortunately, for $B_0$ lying in the symmetry planes ab, ac, bc there are only two magnetically inequivalent sites, A and B, each of them consisting of two magnetically equivalent RC's; and for $B_0 \parallel a$, $B_0 \parallel b$, $B_0 \parallel c$ all four RC's are magnetically equivalent [23].

Fig. 2 shows the rotation pattern $g(\Theta)$ of the g-values in the three crystallographic symmetry planes. The site splitting (A, B) is clearly resolved in the ac and ab planes, but not in the bc plane [24]. Since $g = 2 + \delta g$ with $\delta g$ in the order of $10^{-3}$, the angular distribution function $g(\Theta)$ can be linearized to the form

$$g(\Theta) = g_{ii} \cos^2 \Theta + g_{jj} \sin^2 \Theta + 2g_{ij} \sin \Theta \cos \Theta \tag{2}$$

where the labels i an j denote pairs of crystal axes a, b, c. The solid lines in Fig. 2 are the least-squares fits of the experimental g-values (error $\Delta g = \pm 5 \cdot 10^{-5}$) with eq. 2, from which the g-tensor elements $g_{ii}$, $g_{jj}$ and $g_{ij}$ can be directly determined in the crystal axes system. Diagonalization of this yields the principal g-values ($g_{\alpha\alpha} = 2.00329$, $g_{\beta\beta} = 2.00239$, $g_{\gamma\gamma} = 2.00203$) in the g-tensor axes system $\{\alpha, \beta, \gamma\}$ together with the principal axes directions of the g-tensor.

**Table 1:** Angles (deg, error $\pm 4°$) between the principal axes $\{\alpha,\beta,\gamma\}$ of the g-tensors $\mathbf{g_1}$, $\mathbf{g_2}$ and the dimer axes system, $\{x,y,z\}$ [24].

| tensor | angle between axes | x | y | z |
|---|---|---|---|---|
| $\mathbf{g_1}$ | $\alpha$ | 45 | 69 | 128 |
| ($g_{ab} < 0$, | $\beta$ | 121 | 32 | 96 |
| $g_{ac} < 0$) | $\gamma$ | 61 | 67 | 38 |
| $\mathbf{g_2}$ | $\alpha$ | 20 | 90 | 70 |
| ($g_{ab} > 0$, | $\beta$ | 86 | 12 | 101 |
| $g_{ac} > 0$) | $\gamma$ | 109 | 78 | 23 |

There remains a fourfold ambiguity in the g-tensor orientation with respect to the crystal (or molecular) axes system, because the signs of the off-diagonal elements cannot be determined by EPR [24]. Only two physically reasonable solutions prevail, the tensors $\mathbf{g_1}$ and $\mathbf{g_2}$, between which to decide is not possible by cw EPR, not even at 95 GHz (see, however, section 5). Their orientation with respect to the molecular dimer axes system {x, y, z} is given in Table 1 and depicted in Fig. 3 [24]. The dimer system is constructed in the following way: The z axis is the average dimer normal (defining the average monomer planes as having the least sum of squares of distances to the four nitrogen atoms on the respective dimer half); the x axis is the projection of the Mg-Mg direction onto the plane normal to z; the y axis then describes the approximate local $C_2$ axis of the dimer.

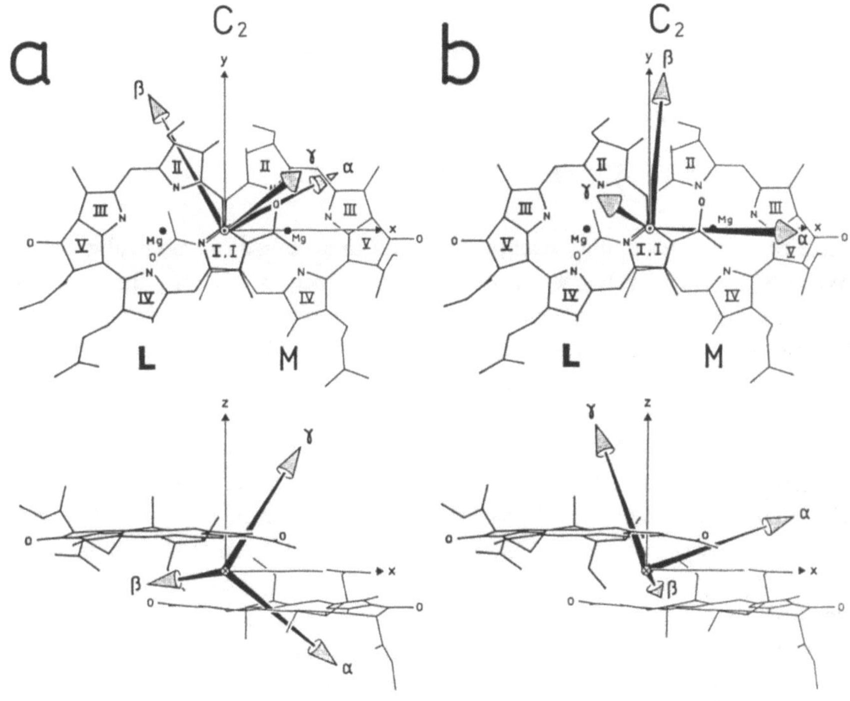


**Fig. 3.** Tensor principal axes system {α, β, γ} and average dimer system {x, y, z} of $P_{865}^{+\bullet}$. Left (a) for the $\mathbf{g}_1$ tensor, right (b) for the $\mathbf{g}_2$ tensor. Top: View from the average dimer normal (z direction) onto the monomer planes. Bottom: Side view into the direction of the local $C_2$ symmetry axis (y axis). For details, see [24].

The main result of this exercise is that the principal axes system {α, β, γ} of the g-tensors is rotated with respect to the molecular dimer system {x, y, z} by angles well outside the error margin of the W-band EPR experiment. Coinciding axes systems would reflect local $C_2$ symmetry also for the electronic structure. This means that also the g-tensor is sensitive to the breaking of $C_2$ symmetry – in agreement with the conclusion from ENDOR/ TRIPLE measurements of the hyperfine structure mentioned above.

### 3.2 MO Calculations of the g-Tensor of $P_{865}^{+\bullet}$

In the following we summarize the results of our first attempt of a quantum mechanical interpretation of high-precision g-tensor measurements on monomeric and dimeric BChl *a* cation radicals [25]. This MO study is based on the available spatial structure of the primary donor $P_{865}$ [26] and the LCAO-MO scheme of Stone [27] for calculating the g-tensor components of free radicals in doublet states.

For a non-degenerate doublet state system, the g-tensor components can be written [27]

$$g_{qr} = g_e\delta_{qr} + \Delta g_{qr}^{o} + \Delta g_{qr}^{u}, \qquad q, r = x, y, z \tag{3}$$

where $g_e = 2.002319$ is the free electron g-factor and

$$\Delta g_{qr}^{o,u} = -2 \sum_{n}^{o,u} \sum_{p,k} \frac{\xi_p \langle \psi_o | \hat{l}_{pq} | \psi_n \rangle \langle \psi_n | \hat{l}_{kr} | \psi_o \rangle}{E_n - E_o} \tag{4}$$

with n running over doubly occupied (o) and unoccupied (u) molecular orbitals (MO's) $\Psi_n$, respectively; $\Psi_o$ is the unpaired electron MO, $\xi_p$ the spin orbit coupling constant for atom p and $\hat{l}_{pq}$ the orbital momentum operator component q on atom p. The energy denominator in eq. 4 is strictly defined as the difference of corresponding electronic state energies. In the present study, we take the simplified approach of using a semiempirical doublet ground state (single configuration) RHF-INDO MO [28]. The calculations were carried out by a modified closed-shell procedure (half-electron concept) introduced by Dewar [29]. The parametrization is the original INDO/1 parametrization of Pople and Beveridge [28]. The energy differences $E_n$-$E_o$ were replaced by differences in orbital energies $\epsilon_n$-$\epsilon_o$ as in the work of Stone [27]. This requires the introduction of empirical scaling factors to eliminate systematic errors arising from the energy denominator [30]. Such scaling factors will also tend to compensate for multi-center contributions to the electronic g-tensor which are excluded by the zero-differential overlap assumed by Stone [27]. We have found it advantageous to introduce two such scaling factors, $\lambda_o$ and $\lambda_u$, in the form

$$g_{qr} = g_e\delta_{qr} + \lambda_o \Delta g_{qr}^{o} + \lambda_u \Delta g_{qr}^{u} \tag{5}$$

to give a good correlation with experimental data from a variety of free radicals [25]. The values $\lambda_o = 3.0$ and $\lambda_u = 1.0$ have been used throughout this work. Matrix elements in eq. 4 were calculated using expressions based on MO coefficients [31]. The following values for spin-orbit coupling constants were used: $\xi^{2p} = 28\ cm^{-1}$ (C), $76\ cm^{-1}$ (N), $151\ cm^{-1}$ (O) [32].

The monomeric BChl *a* molecule has been modelled according to standard rules and limited geometry optimization by energy minimization procedures [20]. Its $\pi$ system is assumed to be planar (excluding peripheral groups coupled to the $\pi$ system by hyperconjugation), except for the acetyl group which was allowed to rotate around the $C_2$-$C_{2a}$ bond by an angle $\theta_{ac}$. This should model environmental effects (e.g. hydrogen bonding in the protein environment) by which the acetyl group might stabilize at different positions $\theta_{ac}$ due to its low rotational energy barrier. The rotational angle $\theta_{ac}$ is defined as the dihedral angle between atoms 1, 2, 2a, $O_6$ (for the standard numbering scheme, see [25]. For *monomeric* BChl $a^{+\bullet}$ the results of the MO calculations can be summarized as follows [25]:

The closest agreement with the principal g-values measured by W-band EPR on frozen organic solutions of BChl $a^{+\bullet}$ (2.0033(1); 2.0026(1); 2.0022(1) [33]) are obtained for $\theta_{ac} = 0°$ and $\theta_{ac} = 180°$. In the intermediate $\theta_{ac}$ ranges, however, the calculated principal g-tensor axes show pronounced tilt angles with respect to the molecular axes system. In other words, small environmental perturbations changing $\theta_{ac}$ can produce small off-diagonal contributions to the g-tensor thereby creating quite large tilt angles. Furthermore, it turned out that the g-tensor is mainly determined by the spin density distribution on the many "light" carbon atoms of the BChl $a^{+\bullet}$ $\pi$ system although they have the smallest spin-orbit coupling parameter. The spin densities on the heavier, but comparatively rare nitrogen and oxygen atoms are too small ($<0.05$) to play a decisive role. This is in contrast to the quinones, where up to 60 % of the total spin density is concentrated on the O atoms and, consequently, the g-tensor is predominantly determined by the local electronic environment of these heavy atoms [33,34].

The g-tensor calculations on the BChl *a dimer* $P_{865}^{+\bullet}$ were performed using the X-ray structure of *Rb. sphaeroides R-26* [26]. Some structural refinements on orientations of acetyl groups, methyl group positions and ring distortions ("ring puckering") of the saturated rings have been added [22]. The influence of polar amino acid residues in the near vicinity of $P_{865}$ was also considered in a first order approach using atomic point charges as calculated by semiempirical MO methods.

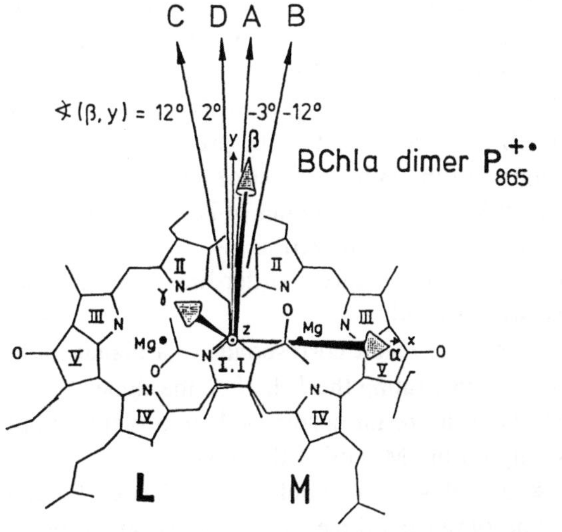


**Fig. 4.** Bacteriochlorophyll *a* dimer $P_{865}^{+\bullet}$ (top view) with experimental g-tensor axes system $\{\alpha, \beta, \gamma\}$ (bold arrows, from ref. [24]) with respect to the dimer coordinate system $\{x, y, z\}$. Long arrows show calculated $\beta$ axes for four different spin density distributions (cases A, B, C, D) (see text). Case A corresponds to the energy optimized dimer with $\rho_L : \rho_M = 2$, as is in accordance with measured hyperfine tensors [22].

Table 2. Calculated and experimental[d] g-tensors for the dimer radical cation $P_{865}^{+\bullet}$

| Case | | Angle [a] between axes | | | $(g\text{-}2)^b$ |
|---|---|---|---|---|---|
| | | x | y | z | $\cdot 10^4$ |
| | $\alpha$ | 3 | 93 | 80 | 30 |
| A | $\beta$ | 87 | 3 | 95 | 25 |
| | $\gamma$ | 100 | 85 | 10 | 24 |
| | $\alpha$ | 12 | 102 | 90 | 31 |
| B | $\beta$ | 78 | 12 | 90 | 25 |
| | $\gamma$ | 90 | 90 | 0 | 23 |
| | $\alpha$ | 12 | 78 | 90 | 31 |
| C | $\beta$ | 102 | 12 | 90 | 25 |
| | $\gamma$ | 90 | 90 | 0 | 23 |
| | $\alpha$ | 2 | 88 | 90 | 30 |
| D | $\beta$ | 92 | 2 | 90 | 25 |
| | $\gamma$ | 90 | 90 | 0 | 24 |
| | $\alpha$ | 20(2)[c] | 90(88)[c] | 70 | 33 |
| Exp. | $\beta$ | 86(85)[c] | 12(5)[c] | 101 | 24 |
| [24] | $\gamma$ | 109 | 78 | 23 | 21 |

A: Energy minimized dimer including amino acid residues, here $\theta_{ac}^{L} = 117°$, $\theta_{ac}^{M} \simeq -7°$; B: spin density $\rho_L \gg \rho_M$, $\theta_{ac} = 0°$; C: $\rho_L \ll \rho_M$, $\theta_{ac} = 0°$; D: $\rho_L = \rho_M$, $\theta_{ac} = 0°$. [a] between principal and molecular axes (in degrees); [b] principal values; [c] projected angles on xy plane in brackets, for details, see text; [d] As experimentally determined g-tensor only $g_2$ is considered (see section 3.1).

The result of the g-tensor calculations on the energy-minimized dimer including amino acid residues is presented in Table 2 (case A) and Fig. 4 together with the experimental result [24]. The experimental tensor $g_2$ has been selected from a set of four symmetry-related g-tensors, $g_1 \ldots g_4$, corresponding to the four sites of the RC crystal per unit cell (see section 3.1). The preferred assignment of $g_2$ to the particular site belonging to the published set of atomic coordinates is based on our recent time-resolved W-band EPR studies of Zn-substituted RC's of *Rb. sphaeroides R-26* [35] described below in section 5.

In Table 2 and Fig. 4 we have also included three border-line cases (B,C and D) in the calculations for which we assumed:

Case B: $\rho_L \gg \rho_M$, $\theta_{ac} = 0°$; here the result corresponds to a planar monomeric (model) BChl $a^{+\bullet}$ constituting the L-half of the dimer.

Case C: $\rho_L \ll \rho_M$, $\theta_{ac} = 0°$; here the result corresponds to a planar monomeric BChl $a^{+\bullet}$ residing on the M-half of the dimer.

Case D: $\rho_L \cong \rho_M$, $\theta_{ac}^{L} \cong \theta_{ac}^{M} \cong 0°$; this is the result for the "bare" (without protein environment) non-optimized dimer of the X-ray structure.

The calculated *principal g-values* (see Table 2) do not show significant departures from the values for the monomeric BChl $a^{+\bullet}$ in all four cases, A to D. Moreover, considering the fairly large errors of the experimental values ($\Delta g = \pm 1 \cdot 10^{-4}$) for BChl $a^{+\bullet}$ in frozen solution, no significant difference between the principal values of the monomer and the dimer can be safely stated. At present, any distinction between the two structures can thus only be made using additional information on *principal axes orientations*.

A graphical presentation of the experimental g-tensor principal axes system is given in Fig. 4 which shows a top view of the dimer $P_{865}$ along the z direction. As stated already in section 3.1, none of the principal axes coincides with the $C_2$ axis of the dimer (y axis) within the experimental error of $\pm 4°$ [24], and therefore a clear breakage of $C_2$ symmetry is apparent.

The computational result (see Table 2 and Fig 4 where the directions of the calculated $\beta$ axes for the cases A to D are given) is rather convincing with respect to the $C_2$ symmetry breakage: For case A the deviation of the calculated $\beta$ axis from the $C_2$ dimer axis is only 3°. That the calculated tensor corresponds very well with the experimental tensor becomes obvious if one looks at the projections of its $\alpha$ and $\beta$ axes onto the xy plane (see Fig. 4 where this is shown for the $\beta$ axes): The symmetrical case D, where $\rho_L \cong \rho_M$, $\theta^L_{ac} \cong \theta^M_{ac} = 0°$, is significantly further removed from the experimental result than the asymmetrical case A, where $\rho_L : \rho_M = 2 : 1$ is assumed according to the result from the hyperfine structure analysis [20-22], and for the monomer-type cases B ($\rho_L \gg \rho_M$) and C ($\rho_L \ll \rho_M$) the tensor axes $\alpha$, $\beta$ deviate even stronger from the experimental ones.

A particularly interesting aspect is the experimentally observed tilt angle ($\gamma$, z) of 23°. From the theoretical point of view this angle should be close to zero for a completely planar $\pi$ system [27]. As we have found for the monomer BChl $a^{+\bullet}$, such a tilt of the $\gamma$ axis might be expected also for $P^{+\bullet}_{865}$ from an out-of-plane movement of an acetyl group [25]. In fact, this manifests itself in the calculated g-tensor of case A for the energy minimized X-ray dimer (see Table 2), where the two acetyl groups on $P_L$ and $P_M$ are asymmetrically rotated out of the plane of the core system by $\theta^L_{ac} = 117°$ and $\theta^M_{ac} = -7°$. This asymmetrical behavior of the acetyl groups results from the presence of a histidine (L 168) in the L protein subunit which is able to form an N-H...$O_{ac}$ hydrogen bond with the oxygen atom of the L acetyl group. In other words: In spite of some quantitative deficiencies of our present approximate approach in calculating g-tensors, it nevertheless appears safe to conclude that the observed tilt angle of the $\gamma$ axis in $P^{+\bullet}_{865}$ is caused by an asymmetrical rotational position of the acetyl groups on the two halves of the dimer.

## 4. W-Band EPR/ENDOR on Quinone Acceptors in Frozen Solution

Quinones play an important role in many biological systems where they are involved in electron transfer and proton translocation processes. In the photosynthetic bacteria *Rb. sphaeroides* and *Rps. viridis*, for example, the primary and secondary quinones, $Q_A$ and $Q_B$, act as one- and two-electron gates, respectively. In *Rb. sphaeroides* $Q_A$ and $Q_B$ are the same ubiquinones-10, and their different functions in the primary ET processes are obviously induced by differences in their protein environment.

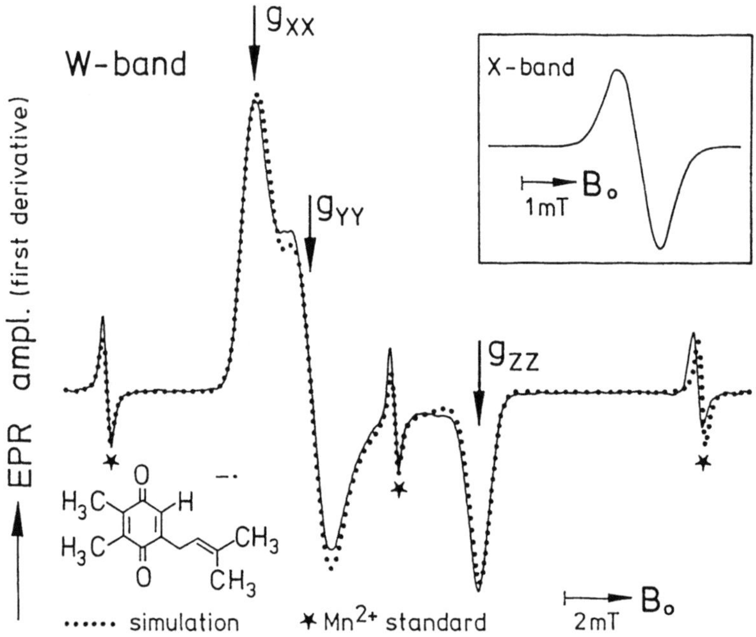


**Fig. 5.** EPR spectra (X- and W-band) of plastoquinone anion PQ-$1^{-\bullet}$ [36]. Solid line: W-band cw EPR spectrum of frozen solution of PQ-$1^{-\bullet}$ in perdeuterated isopropanol (T = 130 K). The three lines marked with * correspond to $Mn^{2+}$ of the field calibration standard. Dotted line: Computer simulation. Field positions where ENDOR experiments were performed are marked by arrows, they correspond to the principal values $g_{xx}$, $g_{yy}$, $g_{zz}$, respectively. Inset: X-band cw EPR spectrum. Temperature and other conditions comparable with W-band experiment. For details, see [36].

The aim of our high-field EPR/ENDOR studies on quinones was to perform precise measurements on anisotropic g- and hyperfine tensor components in frozen semiquinone solutions in order to learn about anisotropic interactions with the protein environment, such as hydrogen bonding to specific amino acid residues. From more than a dozen quinone anion radicals, both natural and model systems, we have recorded powder EPR spectra that, due to the high Zeeman magnetoselection capability of W-band EPR, show a high degree of orientational selectivity [33]. In all cases studied, the measured g-tensor components are consistently assigned to the molecular axes as follows: $g_{xx} > g_{yy} > g_{zz}$, where x is along the C-O bond direction and z is perpendicular to the quinone plane. When varying the solvent with and without perdeuteration, characteristic changes of linewidth components (predominantly along the y-direction) and g-tensor components (predominantly along the x-direction) could be discerned which we attribute to hydrogen-bond formation at the lone-pair orbitals on the oxygens [33]: Dipolar hyperfine interactions with the solvent protons will result in line broadening along the oxygen lone-pair direction, i.e. they broaden the $g_{yy}$ part of the EPR spectrum, while changes in the lone-pair excitation energy $\Delta E_{n\pi^*}$ and/ or spin density $\rho_O^\pi$ at the oxygen due to H-bonding will predominantly shift the $g_{xx}$ component of the g-tensor [33].

As an example, Fig. 5 shows X-band (inset) and W-band EPR spectra of the radical anion of plastoquinone-1 (PQ-1) in frozen perdeuterated isopropanol solution [36]. In contrast to X-band the W-band spectrum exhibits well-separated canonical orientation peaks belonging to $g_{xx}$, $g_{yy}$, and $g_{zz}$. From the computer simulation of the $PQ\text{-}1^{-\bullet}$ spectrum and the simultaneously recorded $Mn^{2+}$ field calibration signals we obtained (error $\Delta g = \pm 3 \cdot 10^{-5}$) : $g_{xx} = 2.00610$; $g_{yy} = 2.00512$; $g_{zz} = 2.00226$ [36].

When the W-band ENDOR spectra are recorded at the $g_{xx}$, $g_{yy}$, and $g_{zz}$ field positions (see Fig. 6), various degrees of angle selectivity are achieved which allow single crystal-like information about the principal values of the hyperfine tensors to be extracted by spectra simulation procedures The hyperfine data for $PQ\text{-}1^{-\bullet}$ are collected in ref. [36].

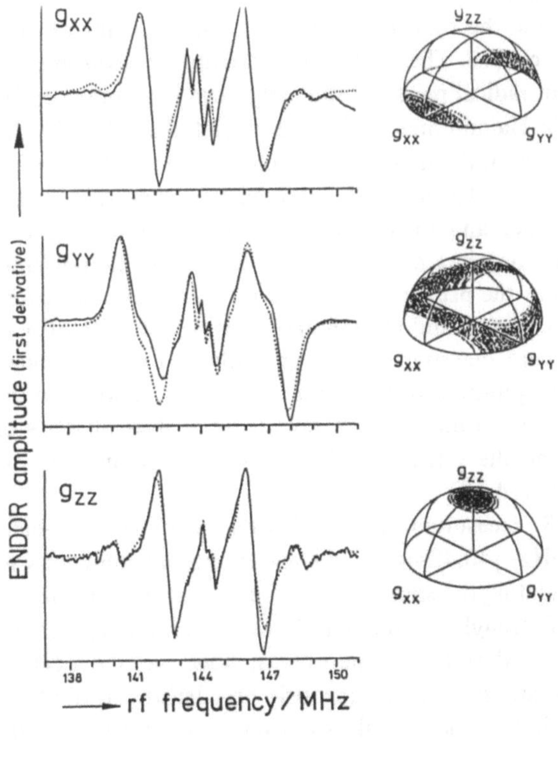


**Fig. 6.** W-band cw ENDOR spectra of $PQ\text{-}1^{-\bullet}$ at the principal g-tensor values for angle selections [36]. Left: Solid lines: Spectra of frozen solution of $PQ\text{-}1^{-\bullet}$ in perdeuterated isopropanol (T = 130 K) taken at the three field positions marked in Fig. 5. Dotted lines: Simulated ENDOR spectra. Right: Schematic representation of the calculated angle selections as determined at the EPR field positions of $g_{xx}$, $g_{yy}$, $g_{zz}$, respectively. For details, see [36].

So far we have described our cw W-band EPR/ENDOR experiments; in the following we will briefly summarize our most recent results obtained by pulsed W-band EPR. In the case of the $Q_A^{-\bullet}$ acceptor of *Rb. sphaeroides* we were able to discern anisotropic $T_2$ relaxation processes due to librational motion of the ubiquinone-10, hydrogen-bonded to specific amino acids of the $Q_A$ binding site [37]. This motion occurs on a time scale around 1 $\mu$s, i.e. fast in comparison with the protein relaxation processes in the 100 $\mu$s time range, and the question is whether it plays a functional role in the formation of the quinone anions. In another series of experiments, pulsed W-band EPR on the coupled spin-correlated $P_{865}^{+\bullet}$-$Q_A^{-\bullet}$ radical pair of *Rb. sphaeroides* enabled us to solve the ambiguity problem of the two g-tensor orientations, $\mathbf{g_1}$ and $\mathbf{g_2}$, of $P_{865}^{+\bullet}$, which was left open in section 3.1.

## 5. Light-Induced Spin-Correlated Radical Pair $P_{865}^{+\bullet}$-$Q_A^{-\bullet}$ of Zn-Substituted RC's of *Rb. sphaeroides* R-26

In these experiments [35] the high-field W-band EPR spectra were recorded using a field-swept two-pulse echo technique. The RC solution was frozen and studied around 150 K. In order to avoid fast spin relaxation of the $Q_A^{-\bullet}$, the non-heme $Fe^{2+}$ ion had been replaced by $Zn^{2+}$. Furthermore, also fully deuterated samples were used to reduce the inhomogeneous linewidth due to unresolved hyperfine interactions. The charge separated radical pairs $P_{865}^{+\bullet}$-$Q_A^{-\bullet}$ were generated by 10 ns laser flashes, and the first microwave pulse had a delay of only 250 ns with respect to the laser pulse. The EPR spectra are highly electron spin-polarized because the transient radical pairs are suddenly born in a spin-correlated non-eigenstate of the spin Hamiltonian with pure singlet character [35,38-40]. The spin-polarized spectrum of the radical pair contains important structural information of the molecule-fixed g-tensors of the two radical partners, $P_{865}^{+\bullet}$ and $Q_A^{-\bullet}$, with respect to each other and to the dipolar axis connecting the two radicals (see Fig. 7). Several other parameters critically determine the polarization pattern [35,38-40], such as the principal values of the g- and dipolar coupling tensors, the exchange coupling J and the inhomogeneous linewidth of both radicals. These parameters were determined independently in order to obtain meaningful simulations of the spin-polarized spectra. For instance, we used the cofactor arrangement as obtained from recent X-ray structure data of RC's of *Rb. sphaeroides* R-26 [41] and the g-tensors of $P_{865}^{+\bullet}$ and ubiquinone-$10^{-\bullet}$ as determined by high-field EPR [24,33].

Earlier time-resolved EPR measurements at X-band (9.5 GHz), K-band (24 GHz) and Q-band (35 GHz) [40,42-44] could not extract unambiguous g-tensor orientations when simulating the spin-polarized spectra of the radical pair $P_{865}^{+\bullet}$-$Q_A^{-\bullet}$, mainly because of strongly overlapping lines, even when deuterated samples were used. In the pulsed W-band experiments, however, the Zeeman field is strong enough to separate the inhomogeneously broadened contributions from $P_{865}^{+\bullet}$ and $Q_A^{-\bullet}$ so much that the overall spectrum is dominated by the anisotropies and differences of the two g-tensors. Thus the interpretation of the polarized spectrum is simplified and allows an unambiguous analysis of the tensor orientations.

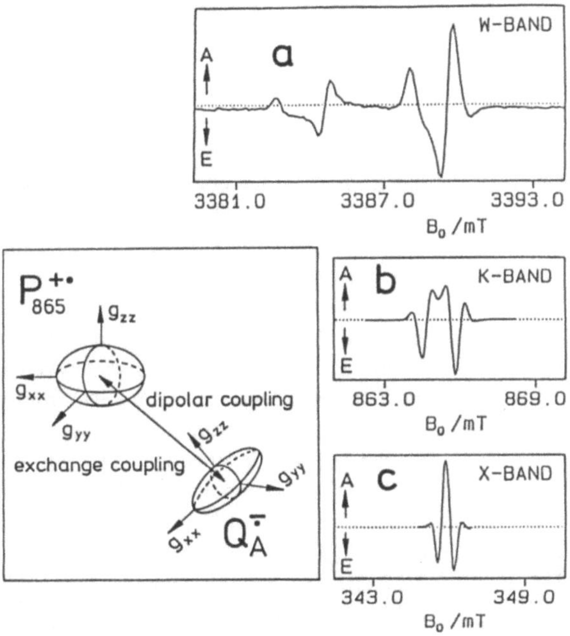


**Fig. 7.** Molecule-fixed g-tensors and spin-polarized EPR spectra of the correlated coupled radical pair $P_{865}^{+\bullet}$-$Q_A^{-\bullet}$ in deuterated Zn-substituted RC of *Rb. sphaeroides* R-26 in frozen solution at three microwave frequencies: (a) 95 GHz (W-band [35]), (b) 24 GHz (K-band [40]), (c) 9.5 GHz (X-band [40]). **A** stands for absorption, **E** for emission of the non-derivative EPR lines.

To illustrate the drastically improved resolution of the transient EPR spectra of the $P_{865}^{+\bullet}$-$Q_A^{-\bullet}$ correlated coupled radical pair at high-field, Fig. 7 compares the spin-polarized spectra at X-, K-, and W-band microwave frequencies. In Fig. 8 the simulations of the transient W-band EPR spectrum for the two possible orientations of the g-tensor of the primary donor $P_{865}^{+\bullet}$ ($\mathbf{g}_1$, $\mathbf{g}_2$ tensors, see section 3.1) with respect to the dipolar axis of the radical pair are shown. (For the simulation parameters used, see caption of Fig. 8.) As can be seen, the choices $\mathbf{g}_1$ or $\mathbf{g}_2$ become now clearly distinguishable: The simulation for the orientation of g-tensor $\mathbf{g}_2$ agrees much better with the experimental spectrum at the high-field side ($P_{865}^{+\bullet}$ part) than the simulation based on the orientation of $\mathbf{g}_1$, i.e. $\mathbf{g}_2$ is the correct tensor.

We believe that by this kind of pulsed high-field EPR experiments on spin-correlated coupled radical pairs $P^{+\bullet}$-$Q^{-\bullet}$, possible ET-induced changes in the relative orientation of P and Q can be detected, provided X-ray structural data of the neutral ground state become available with sufficient accuracy. Such information is highly desirable for a detailed understanding of the ET characteristics on the molecular level, but hitherto the X-ray data are not yet accurate enough for photosynthetic RC's. The situation is more favorable for biomimetic porphyrin-quinone model systems for which very recently the X-ray structure could be determined with much higher accuracy than for RC's [45].

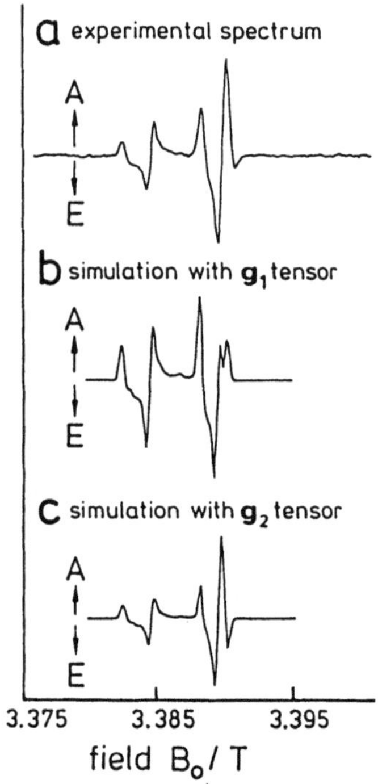


**Fig. 8.** Spin-polarized pulsed W-band EPR spectrum (a) of the correlated coupled radical pair $P_{865}^{+\bullet}$-$Q_A^{-\bullet}$ in deuterated Zn-substituted RC of *Rb. sphaeroides* R-26 in frozen solution together with the simulations of (a) for the two possible orientations ($\mathbf{g}_1$ and $\mathbf{g}_2$) of the g-tensor of $P_{865}^{+\bullet}$ with respect to the dipolar axis of the radical pair. The calculations are based on X-ray structural data of the RC from ref. [41], the exchange couplings J is set to 0, the dipolar coupling is taken from ref. [40] (D = -0.124 mT), the principal g-values of $P_{865}^{+\bullet}$ are taken from ref. [24] (2.0034, 2.0025, 2.0021) and of $Q_A^{-\bullet}$ from ref. [33] (2.0066, 2.0054, 2.0022), the inhomogeneous linewidth is 0.33 mT for $P_{865}^{+\bullet}$ and 0.3 mT for $Q_A^{-\bullet}$.

## 6. Concluding Remarks

In concluding we want to state that high-field/high-frequency EPR- and ENDOR spectroscopies, both in their cw and pulsed modes of operation, have proven to be very powerful tools in photosynthesis research. Their strength is to provide detailed information about electronic and geometrical structures and mobility of the primary reactants that is complementary to structure information obtainable by X-ray crystallography and kinetic information obtainable from fast laser spectroscopy. This statement holds both for the donor side and the acceptor side of the photoinduced electron transfer chain. The MO calculations of the g-tensor of the primary donor cation $P_{865}^{+\bullet}$ are in qualitative agreement with the W-band EPR results. It is anticipated that the remaining quantitative discrepancies can be removed by improving the required excited state energies in the frame of configuration interaction (CI) methods. Work along these lines is in progress for the bacteriochlorophyll donor and quinone acceptor ions.

## Acknowledgements

The high-field EPR/ENDOR work described in this contribution has been done in collaboration with O. Burghaus, A. van der Est, R. Klette, M. Plato, T.F. Prisner, M. Rohrer, D. Stehlik, J.T. Törring from the Free University Berlin and R. Bittl, B. Bönigk, W. Lubitz and F. MacMillan from the Technical University Berlin. We thank G. Feher (San Diego), J.P. Allen (Tempe) and D.C. Rees (Pasadena) and U. Ermler, G. Fritzsch, S. Buchanan and H. Michel (Frankfurt/Main) for providing X-ray structural data of the RC of *Rb. sphaeroides* R-26. Helpful discussions with B. Endeward and M. Huber (FU Berlin) about properties of RC single crystals are gratefully acknowledged. Expert help in solving problems of mechanical construction and radio-frequency matching of our W-band EPR and ENDOR components was provided by Y. Grishin (Novosibirsk) and C. Claus (FU Berlin). The cooperation with H. Kurreck (FU Berlin), H. Levanon (Jerusalem) and J. Fajer (Brookhaven) on P-Q model systems is acknowledged with gratitude. This work was supported by the Deutsche Forschungsgemeinschaft (SFB 337).

### References

1 Reed D.W., Clayton, R.K.: Biochem.Biophys.Res.Commun. **30**, 471 (1968).
2 Feher G.: Photochem.Photobiol. **14**, 373 (1971).
3 Clayton R.K., Wang R.T.: Methods Enzymol. **23**, 696 (1971).
4 Feher G., Okamura M.Y., Raymond J.A., Steiner L.A.: Biophys.Soc.Abstr. **11**, 38 (1971).
5 Feher G., Okamura M.Y. in: *The Photosynthetic Bacteria* (Clayton R.K., Sistrom W.R., eds.), chapter 19, Plenum, New York 1978.
6 Michel H.: J.Mol.Biol. **158**, 567 (1982).
7 Deisenhofer J., Epp O., Miki K., Huber R., Michel H.: J.Mol.Biol. **180**, 385 (1984).
8 Michel H., Epp O., Deisenhofer J.: EMBO J. **5**, 2445 (1986).
9 Allen J.P., Feher G., Yeates T.O., Rees D.C., Eisenberg D.S., Deisenhofer J., Michel H., Huber R.: Biophys.J. **49**, 583a (1986).
10 Chang C.H., Tiede D., Tang J., Smith U., Norris J.R., Schiffer M.: FEBS Lett. **205**, 82 (1986).
11 Allen J.P., Feher G., Yeates T.O., Komiya H., Rees D.C.: Proc.Natl.Acad.Sci. USA **84**, 5730 (1987).
12 Allen J.P., Feher G., Yeates T.O., Komiya H., Rees D.C.: Proc.Natl.Acad.Sci. USA **84**, 6162 (1987).
13 Allen J.P., Feher G., Yeates T.O., Komiya H., Rees D.C.: Proc.Natl.Acad.Sci. USA **85**, 8487 (1988).
14 Michel-Beyerle M.E., Plato M., Deisenhofer J., Michel H., Bixon M., Jortner J.: Biochim.Biophys. Acta **932**, 52 (1988).
15 Möbius K.: Biol.Magn.Reson. **13**, 253 (1993).
16 Burghaus O., Rohrer M., Götzinger T., Plato M., Möbius K.: Meas.Sci.Technol. **3**, 765 (1992).
17 Prisner T.F., Rohrer M., Möbius K.: Appl.Magn.Reson. **7**, 167 (1994).
18 Möbius K., Biehl R., in: *Multiple Electron Resonance Spectroscopy* (Dorio M.M.,

Freed J.H., eds.), chapter 14, Plenum, New York 1979.
19 Plato M., Lubitz W., Lendzian F., Möbius K.: Israel J.Chem. **28**, 109 (1988).
20 Plato M., Möbius K., Lubitz W. in: *Chlorophylls* (Scheer H., ed.), chapter 4.10, CRC Press, Boca Raton 1991.
21 Lendzian F., Bönigk B., Plato M., Möbius K., Lubitz W. in: *The Photosynthetic Bacterial Reaction Center II* (Breton J., Vermeglio A., eds.) p. 89, Plenum, New York 1992.
22 Lendzian F., Huber M., Isaacson R.I., Endeward B., Plato M., Bönigk B., Möbius K., Lubitz W., Feher G.: Biochim.Biophys. Acta **1183**, 139 (1993).
23 Budil D.E., Taremi S.S., Gast P., Norris J.R., Frank H.A.: Israel J.Chem. **28**, 59 (1988).
24 Klette R., Törring J.T., Plato M., Möbiuis K., Bönigk B., Lubitz W.: J.Phys.Chem. **97**, 2015 (1993).
25 Plato M., Möbius K.: Chem.Phys., in print (1995).
26 Allen J.P., Feher G., Rees D.C.: Brookhaven data bank entry P4RCR.
27 Stone A.J.: Proc.Roy.Soc. **A271**, 424 (1963); Mol.Phys. **6**, 509 (1963).
28 Pople J.A., Beveridge D.L.: *Approximate Molecular Orbital Theory*, McGraw-Hill, New York, 1970.
29 Dewar M.J.S., Hashmall J.A., Vernier C.G.: J.Am.Chem.Soc. **90**, 1953 (1968).
30 Chuvylkin N.D., Zhidomirov G.M.: Mol.Phys. **25**, 1233 (1973).
31 Minegishi A.: J.Phys.Chem. **81**, 1688 (1977).
32 Carrington A., McLachlan A.D.; *Introduction to Magnetic Resonance*, Chapman and Hall, London 1979.
33 Burghaus O., Plato M., Rohrer M., Möbius K., MacMillan F., Lubitz W.: J.Phys. Chem. **97**, 7639 (1993).
34 Prabhananda B.S., Felix C.C., Hyde J.S., Walvekar A.: J.Chem.Phys. **83**, 6121 (1985).
35 Prisner T.F., van der Est A., Bittl R., Lubitz W., Stehlik D., Möbius K.: Chem. Phys. **194**, 361 (1995).
36 Rohrer M., Plato M., MacMillan F., Grishin Y., Lubitz W., Möbius K.: J.Magn. Reson., in print (1995).
37 Prisner T.F., Rohrer M., Törring J.T., Möbius K.: to be published.
38 Hore P. in: *Advanced EPR* (Hoff A.J., ed.), chapter 12, Elsevier, Amsterdam 1989.
39 Salikhov K.M., Bock C., Stehlik D.: Appl.Magn.Reson. **1**, 195 (1990).
40 Van der Est A., Bittl R., Abresch E.C., Lubitz W., Stehlik D.: Chem.Phys.Lett. **212**, 561 (1993).
41 Ermler U., Fritsch G., Buchanan S., Michel H. in: *Research in Photosynthesis*, Vol. 1 (Murata N., ed.), p. 341, Kluwer, Dordrecht 1992.
42 Norris J.R., Morris A.L., Thurnauer M.C.: J.Chem.Phys. **92**, 4239 (1990).
43 Snyder S.W., Thurnauer M.C. in: *The Photosynthetic Reaction Center*, Vol. II (Deisenhofer J., Norris J.R., eds.), chapter 11, Academic Press, San Diego 1993.
44 Thurnauer M.C., Gast P.: Photobiochem.Photobiophys. **9**, 29 (1985).
45 Barkigia K.M., Melamed D., Sweet R.M., Kurreck H., Möbius K., Fajer J.: to be published.

# Molecular Triads that Mimic the Spin-Polarized Triplet State In Photosynthetic Reaction Centers

Scott R. Greenfield[1], Walter A. Svec[1], Michael R. Wasielewski[1,2]
Kobi Hasharoni[3], and Haim Levanon[3]

[1]Chemistry Division, Argonne National Laboratory, Argonne, Illinois 60439
[2]Department of Chemistry, Northwestern University, Evanston, Illinois 60208
[3]Department of Physical Chemistry and The Farkas Center for Light-Induced Processes, The Hebrew University of Jerusalem, Jerusalem 91904, Israel

**Abstract.** The electron transfer and spin dynamics of a photosynthetic model system, that departs from the paradigm of using chlorophylls, porphyrins, and quinones for this purpose are described. This model consists of a linear structure with a 4-aminonaphthalene-1,8-dicarboximide (B) chromophore positioned at a fixed distance between a *p*-methoxyaniline donor (A) and a 1,8:4,5-naphthalene tetracarboxydiimide acceptor (C). Photoexcitation of B results in two-step electron transfer to yield the radical ion pair, $A^+$-B-$C^-$. Charge recombination within $A^+$ -B-$C^-$ reproduces, for the first time, the unique, spin-polarized triplet state that has been observed only in photosynthetic reaction center proteins.

**Keywords.** Radical pair, ultrafast spectroscopy, EPR, spin polarization, triplet

## 1 Introduction

One of the main goals of supramolecular chemistry is to mimic the primary charge separation that occurs in bacterial and green plant photosynthesis [1,2]. In photosynthetic proteins the primary charge separation chemistry is a sequence of efficient electron transfer reactions that utilize chlorophylls, pheophytins, and quinones, whose mutual distances and orientations are restricted by the surrounding protein[3,4]. Biomimetic modeling studies of photosynthetic charge separation involve the synthesis of supermolecules consisting of linear arrays, A-B-C, of several electron donor and acceptor molecules A, B, and C separated by appropriate spacer molecules. These models serve as vehicles to increase our understanding of photosynthetic primary events by attempting to reproduce the relevant electronic states, the electronic couplings between these states, the free energies of reaction, and thus, the rate constants and efficiencies for electron transfer in photosynthetic reaction center proteins. The relevant states are the lowest excited singlet and triplet states $^{1,3}$(A-B-C), as well as the singlet and triplet radical pair (RP) states, $^{1,3}$($A^+$-$B^-$-C) and $^{1,3}$($A^+$-B-$C^-$). Time-resolved optical spectroscopy can be used establish the identities and rate constants for formation of the singlet state intermediates on a sub-nanosecond time scale. In addition, time-resolved electron paramagnetic resonance (TREPR) can be used as a powerful tool to detect the presence of the longer-lived triplet and radical pair states, and to monitor their spin dynamics to gain insight into the ET dynamics [5,6].

Most model systems reported in the literature are based on chromophores that are related to the ones found in the photosynthetic reaction center, such as: chlorophylls, porphyrins, and quinones [1,2,7]. Such assemblies are designed to reproduce several key properties of the reaction center protein: 1) multi-step ET to increase the lifetime of the RP product, 2) high quantum yield charge separation initiated from a photoexcited singlet state, 3) fast charge separation and slow charge recombination rates, 4) temperature independent ET rates, and 5) non-Boltzmann spin population of the RP states. Thus far, reports of molecules which satisfy successfully *all* of these aspects, and which have been detected by TREPR are scarce [8]. Most reaction center models fulfill only a subset of these criteria. In addition, there is one key property of the reaction center primary photochemistry that has not been successfully reproduced by any model system until now. This property is the unique ability of the spin-polarized RP intermediate within the photosynthetic reaction center to yield upon charge recombination a triplet state that retains a memory of the precursor RP spin state. We now report results on a photosynthetic model system, **1**, that closely mimics the spin dynamics of triplet state formation found only in photosynthetic reaction centers. It is the *first and only* system to date that exhibits the electron spin polarization properties of the triplet state found in reaction centers for which electron transfer to the quinone acceptor has been blocked by pre-treatment with a reducing agent [9].

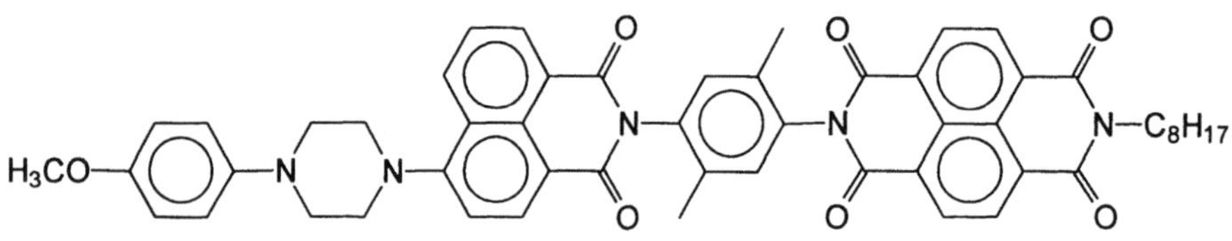


**1**

## 2 Results and Discussion

Structure **1** was prepared in three steps: 4-nitro-1,8-naphthalenedicarboxylic anhydride (Aldrich) was refluxed in ethanol with an equivalent of 2,5-dimethyl-*p*-phenylenediamine to give an 85% yield of the monoimide. Reaction of the monoimide with an equivalent of *N*-(*p*-methoxyphenyl)piperazine in *N*-methylpyrrolidinone at 120° gives an 56% yield of the A-B unit with a terminal amino group on B. Reaction of the A-B amine with an equivalent each of 1,4,5,8-naphthalenetetracarboxylic dianhydride and n-octylamine in *N,N*-dimethylformamide at 120° gives a 38% yield of **1** [10]. Transient optical absorption spectroscopy was carried out on **1** in toluene to determine the nature of the intermediates and the rate constants for electron transfer between the electronic states given in the energy level diagram in Figure 1.

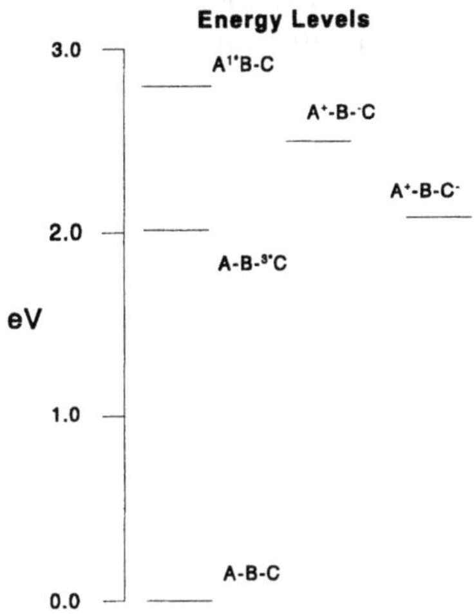


Figure 1.

Pulsed laser excitation of **1** in toluene with 416 nm light selectively excites the charge transfer state of chromophore B [11]. The lowest excited singlet state of B accepts an electron from A in $\tau = 8$ ps. The formation of $A^{+}$-$B^{-}$-C can be monitored near 500 nm, Figure 2. At this wavelength the cation radical of *p*-methoxyaniline absorbs [12]. The transient optical spectrum of the $A^{+}$-$B^{-}$-C radical pair is shown in Figure 3. A subsequent dark ET reduces the naphthalenediimide acceptor. The reduction of the naphthalenediimide is monitored at 480 nm. The final radical pair $^{1,3}(A^{+}$-B-$C^{-})$ forms with $\tau = 430$ ps, Figure 4. The transient optical spectrum of this radical pair is shown in Figure 5. The $A^{+}$-B-$C^{-}$ radical pair has a 300 ns lifetime, Figure 6. Thus, the lifetime of the second RP is sufficiently long to be readily observable by TREPR [6].

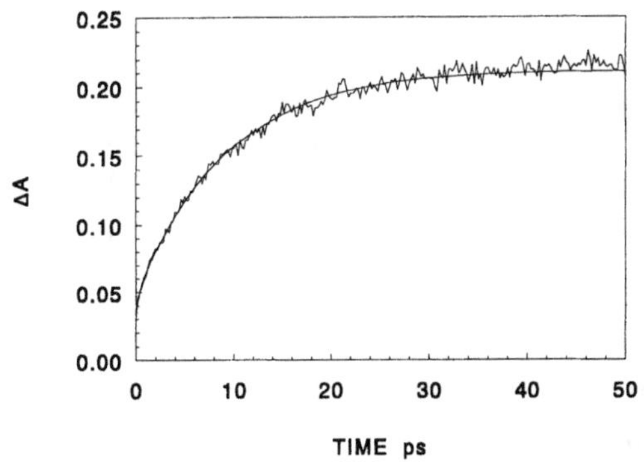


Figure 2. Transient absorption kinetics for **1** at 500 nm following a 130 fs, 416 nm laser flash.

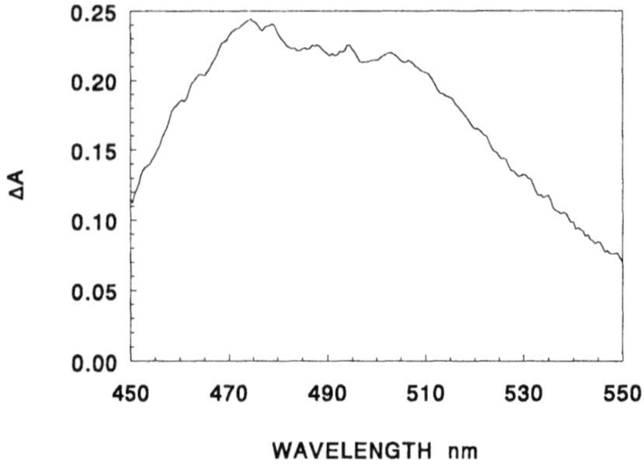


Figure 3. Transient absorption spectrum for **1** at 40 ps following a 130 fs, 416 nm laser flash.

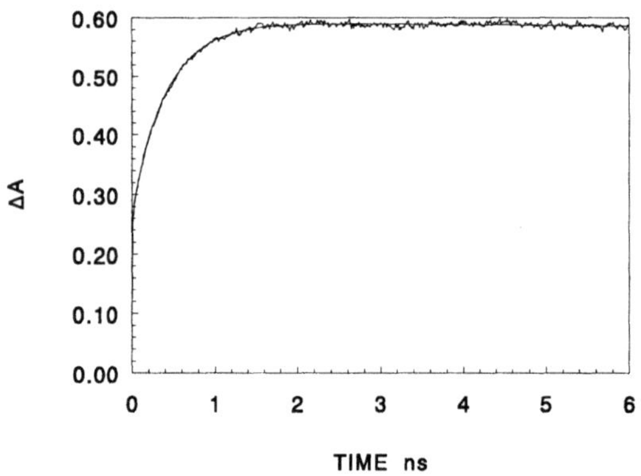


Figure 4. Transient absorption kinetics for **1** at 480 nm following a 130 fs, 416 nm laser flash.

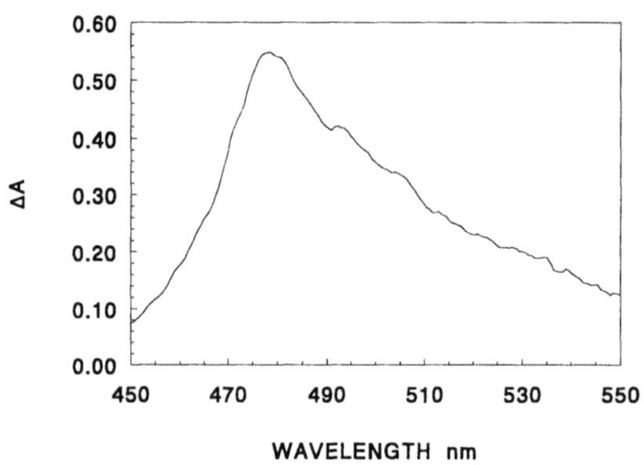


Figure 5. Transient absorption spectrum for **1** at 4 ns following a 130 fs, 416 nm laser flash.

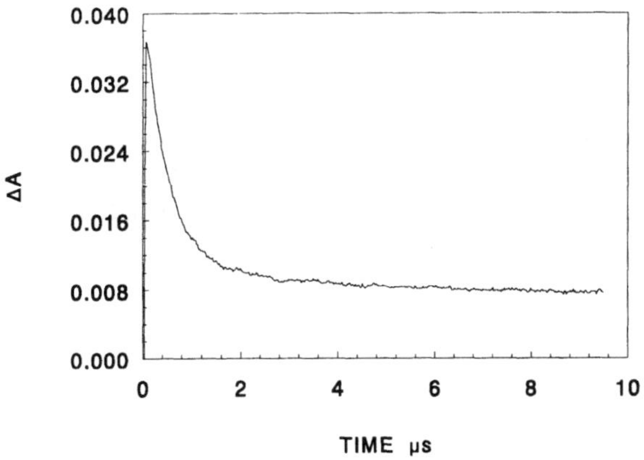


Figure 6. Transient absorption kinetics for **1** at 480 nm following a 4 ns, 416 nm laser flash.

Liquid crystals can be used to study the dynamics of electron transfer and triplet states [13]. It has been shown that the LC environment permits the study of these phenomena over a broad temperature range. We have studied the photochemistry of **1** in the nematic LC Merck E-7. In Figure 7, we show the TREPR spectra of A-B-$^{3*}$C in a LC. The upper trace was taken with the external magnetic field, **B**, parallel to the LC director **L**, i.e., (**B** || **L**) and the lower trace is when **B** ⊥ **L** [14]. As shown previously, the relative phases of the EPR transitions in a triplet spectrum are indicative of both the population mechanism and of the sign of the zero-field splitting parameter, *D*. Unlike ordinary triplet states, which are formed via spin orbit intersystem crossing, SO-ISC, the triplet state A-B-$^{3*}$C develops from a RP precursor (Fig. 1) via the radical pair intersystem crossing, RP-ISC, mechanism. These two mechanisms can be differentiated via the polarization pattern of the six transitions at the canonical orientations of the triplet spectrum. In SO-ISC, the three zero-field levels are populated with different rates due to the selective nature of the triplet <- singlet ISC route, and thus, a selective population of triplet sub-levels is created. This selectivity is carried over to the high-field levels and is distributed between them through the combination of the zero-field wave functions [15]. RP-ISC on the other hand, is also a selective population mechanism, but it acts directly on the high-field triplet sublevels, via S-$T_0$ (or S-$T_{+1}$) mixing [16,17]. Inspection of the triplet spectra in the parallel and perpendicular orientations, show that the *a,e,e,a,a,e* pattern cannot arise from SO-ISC. Such a phase pattern was found in pre-reduced bacterial reaction centers [9] and in pre-reduced photosystem I reaction centers [18], and was assigned to RP-ISC [16,17].

## 3 Conclusion

The observation of this unique triplet state in model system **1** by TREPR demonstrates for the first time, that all of the electronic states found in the primary photochemistry of photosynthetic reaction centers can be reproduced successfully in synthetic models. This makes it possible to design new models that can be used with confidence to probe the mechanism of the primary events of photosynthesis.

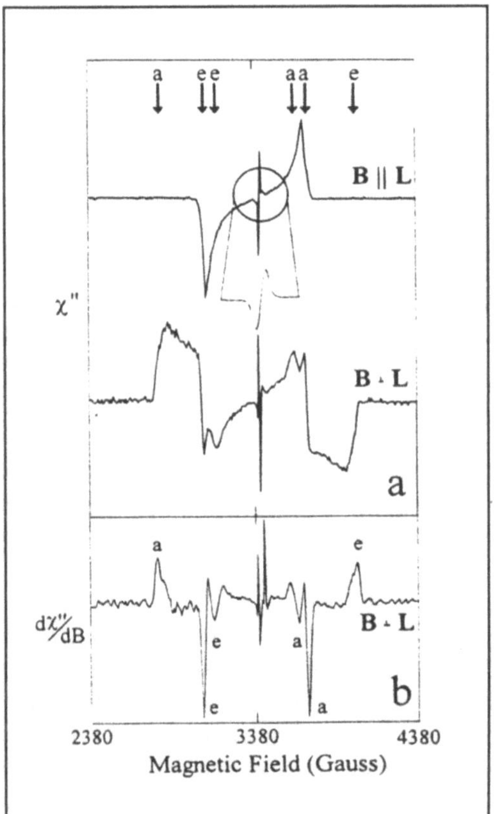


Figure 7. (a) Direct detection TREPR spectra of A-B-$^{3*}$C in a LC at two orientations of the LC director, **L**, taken 700 ns after the 416 nm laser flash. (b) Numerical integration of the **B** ⊥ **L** spectrum.

## 4 Acknowledgment

This work was partially supported by a U. S. - Israel BSF grant (HL), by the Deutsche Forschungsgemeinschaft (SFB 337) (HL), by a Volkswagen grant (HL), and by the Division of Chemical Sciences, Office of Basic Energy Sciences, U. S. Department of Energy under contract W-31-109-Eng-38 (MRW). SRG is supported by a Distinguished Postdoctoral Fellowship from the U. S. Department of Energy. A special grant of the Erna and Victor Hasselblad Foundation (HL) is highly acknowledged. The Farkas Research Center is supported by the Minerva Gesellschaft fur die Forschung, GmbH, Muchen, FRG. This work is in partial fulfillment of the requirements for a Ph. D. degree (KH) at the Hebrew University of Jerusalem.

## 5 References


1 Wasielewski, M. R. (1992) Chem. Rev. 92, 435.
2 Gust, D., Moore, T. A., Moore, A. L. (1993) Accounts Chemical Res. 26, 198.
3 Bixon, M. et. al., (1992) Isr. J. Chem. 32, 369.
4 Norris, J. R. and Deisenhofer, J. Eds. (1993) The Photosynthetic Reaction Center, Academic Press, San Diego.
5 Lendzian, F. and von Maltzan, B. (1991) Chem. Phys. Lett. 180, 191.
6 Hasharoni, K., Levanon, H., von Gersdorff, J., Kurreck, H., Mobius, K. (1993) J. Chem. Phys. 98, 2916.
7 (a) Johnson, D. G., Niemczyk, M. P., Minsek, D. W., Wiederrecht, G. P., Svec, W. A., Gaines, III, G. L., and Wasielewski, M. R. (1993) J. Am. Chem. Soc. 115, 5692-5701. (b) Gust, D., Moore, T. A., Moore, A. L., Lee, S. J., Bittersmann, E., Luttrull, D. K., Rehms, A. A., DeGraziano, J. M., Ma, X. C., Gao, F., Belford, R. E., Trier, T. T. (1990) Science, 248, 199.
8 Wasielewski, M. R., Gaines, G. L. Wiederrecht, G. P., Svec, W. A., Niemczyk, M. P. (1993) J. Am. Chem. Soc. 115, 10442.
9 Dutton, P. L., Leigh, J. S., Seibert, M. (1972) Biochim. Biophys. Res. Commun. 46, 406.
10 All synthetic intermediates and **1** were completely characterized by 300 MHz proton NMR and laser desorption mass spectroscopy.
11 The apparatus and techniques for measuring the transient absorption spectra and kinetics of **1** following a 420 nm, 130 fs laser flash are described in Frank, H., Cua, A., Young, A., Gosztola, D., Wasielewski, M. R. (1994) Photosynthesis Res. 41, 389.
12 Hester, R. E., and Williams, K. P. J. (1982) J.C.S. Perkin II, 559.
13 Regev, A., Levanon, H., Murai, T., Sessler, J. L. (1990) J. Chem. Phys. 92, 4718.
14 Levanon, H. (1987) Rev. Chem. Intermed. 8, 287.
15 Carrington, A. and McLauchlan, A. D. (1967) Introduction to Magnetic Resonance, Harper & Row, New York.
16 Levanon, H. and Norris, J. R. (1978) Chem. Rev. 78, 185.
17 Salikhov, K. M., Molin, Y. N., Sagdeev, R. Z., and Buchachenko, A. L. (1984) Spin Polarization and Magnetic Effects in Radical Reactions, Elsevier, Amsterdam.
18. Regev, A., Nechushtai, R., Levanon, H., Thornber, J. P. (1989) J. Phys. Chem. 93, 2421.

# Quantum Calculations for the Special Pair Dimer and for Hetero Dimers: Analysis of the Internal Charge Transfer States

P.O. J. Scherer and Sighart F. Fischer
Physik Department, Technische Universität München D-85747 Garching

**Abstract.**
Quantum calculations of the INDO-S-MRCI-type including single, double and triple excitations are performed for the low lying excited singlet states of the native bacteriochlorophyll dimer and the heterodimers in which one of the two bacteriochlorophylls is replaced by a bacteriopheophytin. Furthermore the transient spectra originating from the intense special pair dimer state P* are evaluated. These spectra help to identify the internal charge transfer states. Their splittings and relative transition intensities are good measures of the degree of symmetry breakage of the system. As a symmetry breaking coordinate relevant also to the native dimer we discuss a rotation of the acetyl group of the M half $P_M$ by 180 °. The results are analyzed with respect to the Stark spectra. They point to a relatively small change of the permanent dipole due to excitation of $P^*_-$ but to a rather high polarizability. It is induced via a double excited state which interacts indirectly via the internal charge transfer states with this dimer band $P^*_-$.

**Key words**: INDO calculations, Charge transfer states, special pair, hetero dimers

# 1 Introduction

The initial charge separation within the reaction center after excitation of the special pair is unique with regard to its efficiency, that is the high yield and small energy loss during the first step. Even though experimental observations /1,2/ have helped to clarify the involvement of the bacteriochlorophyll monomer on the active branch $B_L$ as an intermediate electron acceptor, the detailed mechanism in terms of the couplings is not fully understood. In particular it is not clear if the internal charge transfer states of the dimer play an essential part in the charge separation process.

To elucidate the situation we present in this contribution quantum calculations of the INDO/S type which include those single, double and triple excited configurations , which contribute significantly to the singlet excitations of the dimers below 3eV./3/ Apart from the two lowest dimer bands $P^*_-$ and $P^*_+$ which originate from excitonic coupling of the Qy transitions of the two dimer halves we analyze the energy splitting of the internal charge transfer states and their coupling to $P^*_-$ and $P^*_+$,We compare results for the native $BChl_2$ dimer and the hetero dimers BChl-BPh and BPh-BChl./4/.Our calculations suggest that the internal charge transfer states of the hetero dimers do not fall below the state $P^*_-$ unless an additional relaxation takes place. This relaxation must be more effective for the M-hetero dimer than for the L-hetero dimer. We also find that the rotation of the acetyl group of the dimer half $P_M$ is similarly effective in terms of symmetry breakage as the hetero formation. A unique feature of the dimer is found in the appearance of a double excited singlet state $P^{**}$, which results from the combination of two triplets located at the two dimer halves. This state comes out close in energy to the dimer state $P^*_-$ . It couples strongly to the charge transfer states, and causes a high polarizability of $P^*_-$ and might effect indirectly the electronic coupling for the charge separation.

# 2 Methods

In previous calculations /3/ we used Zerner's parametrization of the INDO hamiltonian which is optimized to reproduce spectra of aromatic hydrocarbons in the single CI approximation. . As now double and higher order excitations are incorporated the semiempirial parameters have to be modified. To reproduce the absorption spectra of benzene and pyridine we had to use the Ohno-Klopman /5/ parametrization of the two center Coulomb integrals and to readjust the parameters $\gamma_i$ as well as the scaling factors $\beta_i$ of the resonance integrals. The optimized parameter values are listed in Table 1. We like to mention however, that for the Chlorophyll molecules the differences compared to Zerner's parameterization are not really essential

**table 1:** *INDO parameters used for calculations including multiple excited configurations*

| element | H | C | N | O | Mg |
|---|---|---|---|---|---|
| $\beta$ | -9.0 | -19.0 | -26.4 | -31.2 | -15.4 |
| $\gamma$ | 12.4 | 9.75 | 9.75 | 13.7 | 5.62 |

$f_\pi = 0.585, f_\sigma = 1.267, Weiss - factor = 1.0,$
two center integrals by Ohno-Klopman scheme

**table 2:** *electron density of the highest occupied (HOMO) and lowest unoccupied (LUMO) orbitals on $P_L$ in percent and energy splitting of the internal charge transfer states in eV*

| $P_L$ | $P_M$ | $P_M$ acetyl rotated | HOMO on $P_L$ | LUMO on $P_L$ | $E(P_L^+P_M^-) - E(P_M^+P_L^-)$ |
|---|---|---|---|---|---|
| BP | BC | o | 5 | 76 | -0.61 |
| BP | BC | x | 12 | 62 | -0.33 |
| BC | BC | o | 46 | 62 | +0.05 |
| BC | BC | x | 85 | 50 | +0.38 |
| BC | BP | o | 87 | 61 | +0.50 |
| BC | BP | x | 93 | 41 | +0.70 |

**fig.1:** *INDO results for the dimer of rps.viridis. The bars show the excitonic and CT states without resonance interaction between the two dimer halves. The arrows and numbers show the coupling between excitonic and CT states due to the resonance interaction.*

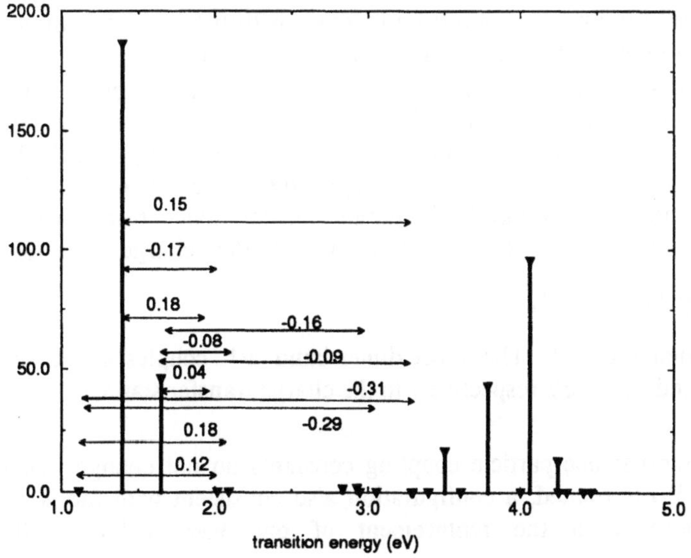

with regard to our major conclusions. We are also aware of some deficiencies still present in the program. Transition dipoles tend to be overestimated and as a result excitonic couplings as well. Also the calculated lowest triplet states of the Bchl and Bph molecules come out too low by about 0.25eV, so that the 'double triplet' P** should be corrected also towards higher energies by about 0.5 eV.

Our calculations are based on the X-ray structure of rps. viridis and include besides the two dimer chromophores $P_L$ and $P_M$ also the most important protein residues which are in bonding interaction to the dimer. For the heterodimers the central magnesium atom of one dimer chromophore was replaced by two hydrogens and the position of the hydrogens and the four nitrogens were optimized by an MNDO calculation. The histidine attached to the magnesium was removed.

# 3 Results from the quantum calculations

In order to analyze the internal couplings the full INDO hamiltonian of the dimer is partitioned by extracting all resonance interaction between pairs of atoms which are not located on the same chromophore

$$H = H_0 + \sum_{\substack{i \in PM \\ j \in PL}} \beta_{i,j} \left( a_i^+ a_j + a_j^+ a_i \right)$$

The Hartree-Fock orbitals of $H_o$ provide a basis set of one electron functions which are localized on one of the dimer halves. From these pure excitonic states as well as pure charge transfer states are constructed. Finally the matrix elements of the intermolecular resonace interaction are evaluated which mediate the coupling between excitonic and charge separated states.

The resulting excitonic spectrum together with the coupling parameters is shown in figure 1 for the X-ray structure of Rps viridis.

The lowest state, which carries almost no intensity is the double triplet P**, followed by the two excitonic components $P^*_-$ and $P^*_+$ of the dimer Qy transition and the two internal charge transfer states $P_M^+P_L^-$ and $P_L^+P_M^-$. First we notice that the dimer band $P^*_-$ couples quite strongly with 0.18 eV and -0.17 eV to the two localized charge transfer states respectively. These couplings admix the odd combination of the charge transfer states

$$CT_- = \frac{1}{\sqrt{2}} \left( P_M^+ P_L^- - P_L^+ P_M^- \right)$$

to the dimer state $P^*_-$ .The upper dimer band $P^*_+$ couples much weaker by 0.04 eV and - 0.08 eV respectively to the charge transfer states.

The internal one-particle coupling constants do not change much if the acetyl of $P_M$ is rotated, a configuration also consistent with the X-ray data. Furthermore, even the replacement of one bacteriochlorophyll by a bacteriopheophytin does not change these coupling by more than 10 % as long as the overall geometry, in particular the distance between the two pyrrole rings stays the same. So we conclude, the changes in transition intensities and permanent dipoles for the states of the modified dimers are due to changes of their energetics, mainly the splitting $\Delta = E(P_M^+P_L^-) - E(P_L^+P_M^-)$ of the two lowest charge transfer states.

In the following discussion we will focus on the parameter Δ and its influence on the dipole change relative to the ground state for the prominent two dimer bands $P^*_-$ and $P^*_+$.

Fig. 2 shows the calculated absorption spectrum in the range between 1eV and 3eV. The numbers refer to the dipole change in units of Debye. and the angle between the transition dipole of $P^*_-$ and the permanent dipole of the state of interest. The lowest excited state is the „double triplet" $P^{**}$. It carries almost no oscillator strength but a sizeable permanent dipole of 6 Debye. The strong transition relates to $P^*_-$ . Its dipole is only 1 Debye and the angle of 89° reflects the approximate $C_2$-symmetry in which only a component of the permanent dipole parallel to the $C_2$ axis is allowed. The experimental Stark spectra point to a much stronger dipole between 6 and 8 Debye and a smaller angle which deviates substantially from 90°. Also in conflict with the experiments is the dipole change of 3 Debye for $P^*_+$. It is known from the experiments /6/ that this state shows a much weaker Stark intensity than $P^*_-$ and couples only weakly to vibrations. The larger change in dipole for $P^*_+$ as compared to that of $P^*_-$ is at first sight surprising, since we have seen from Fig. 1 that the coupling to the charge transfer states is much weaker. This effect must be overcompensated by the smaller energy difference. The splitting of the charge transfer states of Δ=0.05 eV is rather small, so that they are not strongly localized. This is seen in the transient spectrum of $P^*_-$, where the lower charge transfer state, which is dominated by the even combination $CT_+$ of the charge transfer transitions, carries almost all of the intensity. The results are puzzling also with regard to the direction of the asymmetry. The HOMO of the dimer is more localized on the M side (see table 2) but the experimental spin density distribution of the dimer cation, which should correlate with the density of this orbital is by the ratio 2:1 localized on the L - side /7/.

A more consistent picture develops if we consider a rotation by 180° of the acetyl group as an alternative structure. The x-ray data cannot clearly discriminate against it and quantum calculations would favor this confirmation.

Fig. 3 shows the corresponding absorption and transient spectra. Now the charge transfer splitting is increased to Δ=0.38 eV. The HOMO is localized

**fig.2:** *INDO results for the dimer of rps.viridis. The bars show optical transitions from the groundstate P (lower panel) and from the excited state P* (upper panel). The numbers give magnitudes and angles of the dipole moment change*

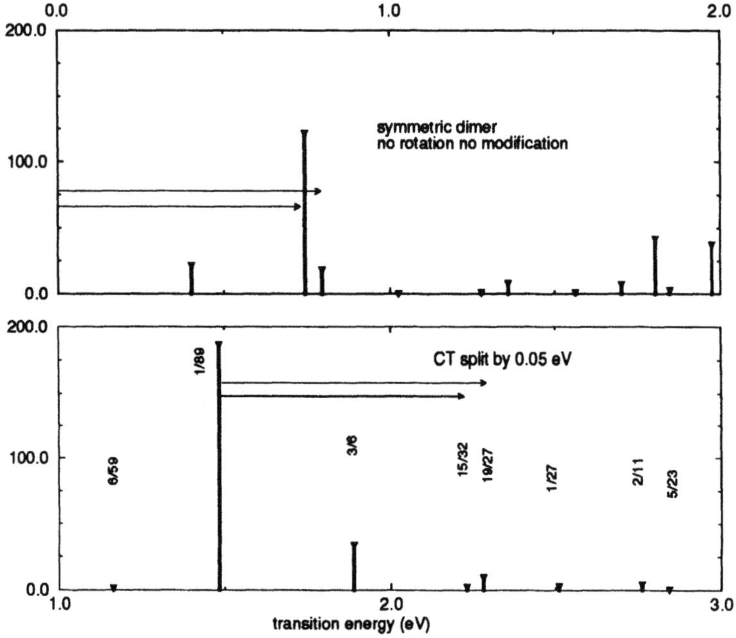


**fig.3:** *INDO results for the dimer of rps.viridis with the acetyl of $P_M$ rotated by 180 degrees*

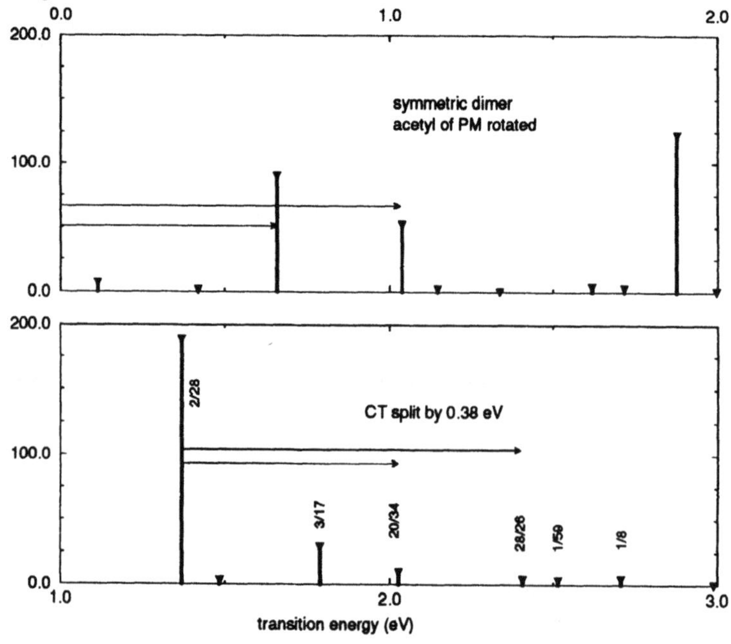

more on the L-side and the dipole change is increased. Furthermore, the angle is reduced to 28° a value even too small compared to the experimental of about 40° . Still the upper dimer band $P^*_+$ has a stronger dipole change of 3 Debye as compared to 2 Debye for $P^*_-$. Later we will see that an alternative interpretation of the Stark spectrum based on a high polarizability can remove this discrepancy. It is also interesting to notice, that the double triplet P** is now located above $P^*_-$ .

Next we would like to turn to the L-hetero dimer with the acetyl not rotated Fig. 4. This spectrum shows an entirely new feature. There are two states with intensity in the $P^*_-$ region. Actually, the state P** almost accidentally has the same energy as $P^*_-$ but it is now interacting relatively strong with $P^*_+$ . The intrinsic asymmetry makes this coupling possible. The asymmetry is also evident in the strong splitting of the charge transfer states by Δ=0.61 eV. The two mixed states $P^*_-$ and P** show also both a relatively strong dipole change but the upper dimer band $P^*_+$ does not. We would have expected such a situation for the native $Bchl_2$ dimer. In a model study we found due to vibronic couplings a scenario for the homo dimer which resembles indeed some of the features predicted here for the L-hetero dimer. The spectrum changes strongly as we consider the acetyl rotation for the L hetero dimer (Fig. 5). Now P** falls below $P^*_-$ and the latter again has very little dipole change (1.6 Debye) and the charge transfer splitting is reduced to 0.33 eV. The transient spectrum distributes the intensity over both charge transfer states. The M-hetero dimers (Fig. 6) and its acetyl rotated version (Fig. 7) show again strong charge transfer splittings, whereby the rotated conformation has the largest value of 0.7 eV. In this case the lower charge transfer state which is of the form $BC_L^+$ $BP_M^-$ falls below the upper dimer band $P^*_+$ but not below $P^*_-$. So we see that the experimental observation of a very broad band for $P^*_-$ with a long red shifted part which shows a strong Stark effect and which has been interpreted as the $BC_L^+BP_M^-$ state must involve some configurational relaxation. This effect must be more pronounced for the M hetero dimer as compared to the L hetero dimer, since the latter deviates much less in the Stark and in the absorption spectrum from those of the homo dimer.

In the following we will see that the Stark spectrum can be better understood if polarization effects induced via the double triplet P** are included.

**fig.4:** *INDO results for the $BP_LBC_M$ heterodimer. bottom:groundstate absorption, top:absorption of P**

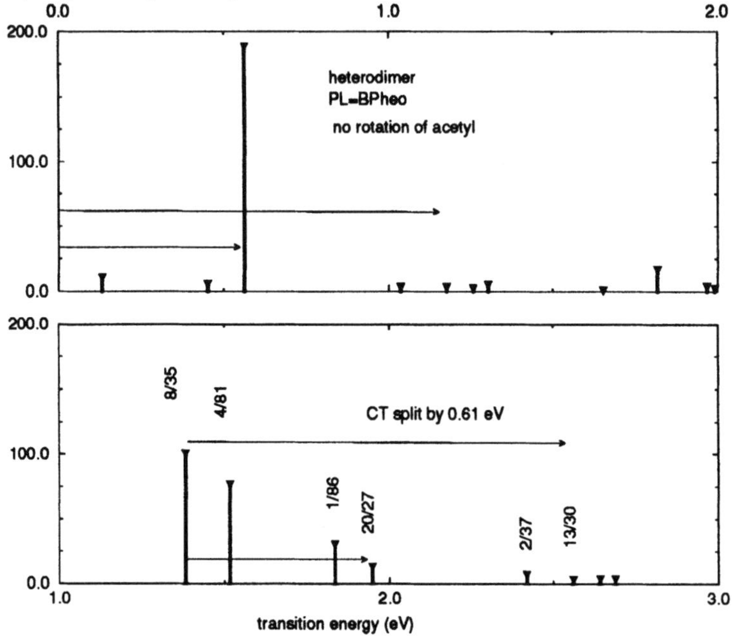


**fig.5:** *INDO results for the $BP_LBC_M$ heterodimer with the acetyl of $BC_M$ rotated by 180 degrees*

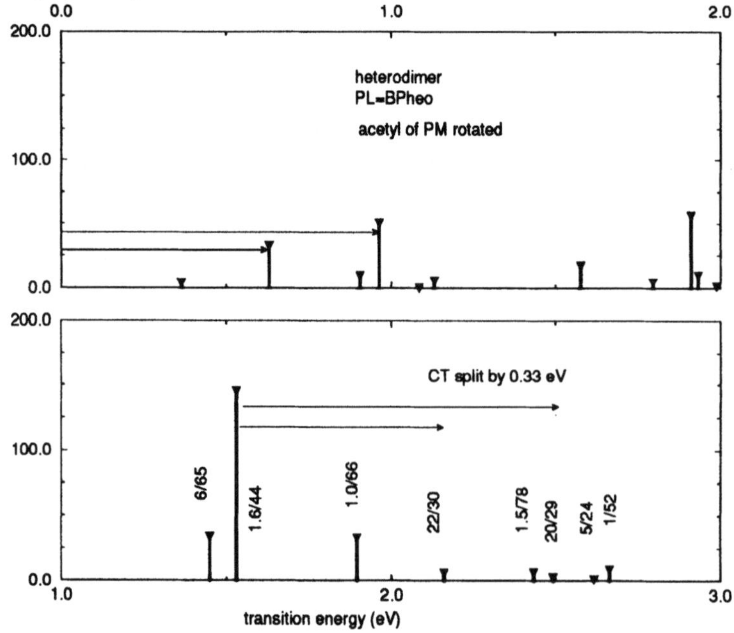

**fig.6:** *INDO results for the $BC_LBP_M$ heterodimer bottom:groundstate absorption, top:absorption of $P^*$*

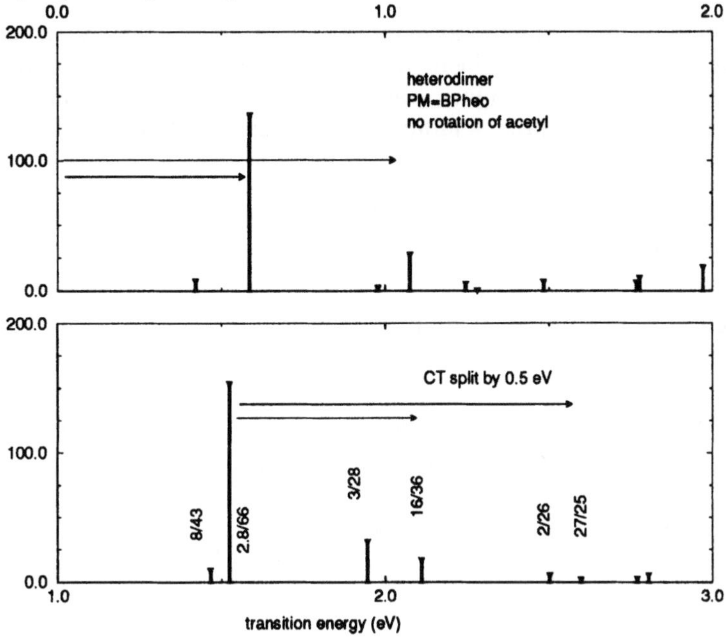


**fig.7:** *INDO results for the $BC_LBP_M$ heterodimer with the acetyl of $BP_M$ rotated by 180 degrees*

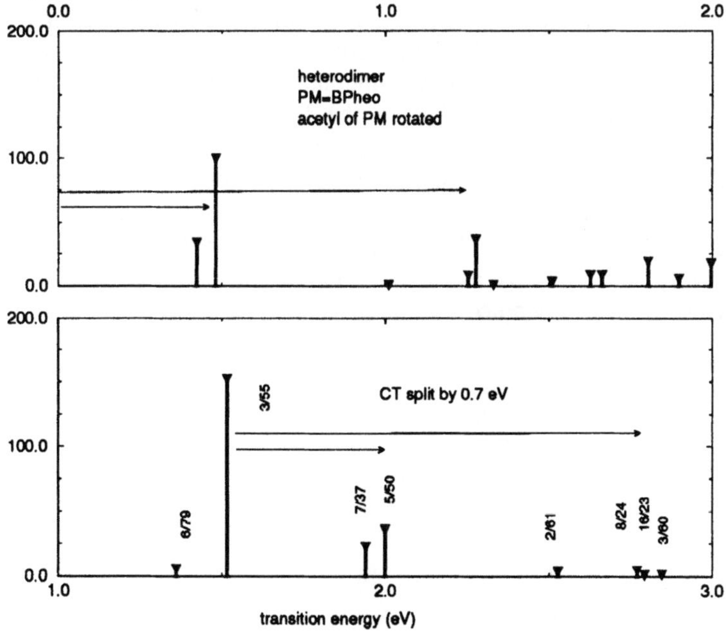

# 4 A simplified model for the dimer excitations

The results of the quantum calculations can be interpreted in terms of a simplified exciton model which treats only the lowest dimer excitations explicitly and takes all higher states into account by renormalizing the effective interaction matrix. A minimum model has to include

- the excitonic states $P_L*$ and $P_M*$
- the charge separated states $P_L^+P_M^-$ and $P_M^+P_L^-$
- the double triplet state P**

The effective interaction matrix is given by

$$H_0 = \begin{bmatrix} E(P_M^*) & V & -U_0 & U_1 & \circ \\ V & E(P_L^*) & U_1 & -U_0 & \circ \\ -U_0 & U_1 & E(P_L^+P_M^-) & \circ & U_2 \\ U_1 & -U_0 & \circ & E(P_M^+P_L^-) & U_2 \\ \circ & \circ & U_2 & U_2 & E(P**) \end{bmatrix}$$

where V is the excitonic coupling of the two optical excitations and $U_0$,$U_1$ are approximately the resonance integrals between the HOMOs and LUMOs of the two halves. The negative sign of $U_0$ results from the evaluation of one electron matrix elements involving a hole transfer. $U_2$ involves a resonace integral between the HOMO of one half and the LUMO of the other. The HOMO and LUMO of the single chromophores are very similar in the region of the I-rings where the overlap of the two dimer halves has its major contribution. Therefore one has approximately $U_o \approx U_1$. From INDO calculations for the dimer we deduce values of $U_0$=0.12 eV and $U_1$=0.15 eV. In the case of perfect $C_2$ symmetry we have vanishing energy differences $d=E(P_M*)-E(P_L*)$ and $\Delta=E(P_L^+P_M^-)-E(P_M^+P_L^-)$. Then the excitations of the dimer can be characterized as even or odd and the energy matrix becomes in the basis of symmetry adapted excitations

$$\tilde{H}_0 = \begin{bmatrix} E(*)-V & \circ & U_- & \circ & \circ \\ \circ & E(*)+V & \circ & U_+ & \circ \\ U_- & \circ & E(CT) & \circ & \circ \\ \circ & U_+ & \circ & E(CT) & \sqrt{2}U_2 \\ \circ & \circ & \circ & \sqrt{2}U_2 & E(P^{**}) \end{bmatrix}$$

$$U_+ = U_1 - U_0 \quad ; \quad U_- = -(U_1 + U_0)$$

We note that the coupling to the CT state is enhanced for the lower dimer band P*+ as $U_o$ and $U_1$ add up constructively to give a coupling of $|U_-| = (U_o+U_1) \approx 0.27$ eV. For the upper dimer band P*+ the coupling to the CT state almost vanishes as $|U+| = (U1-U0) \approx 0.03$ eV

Formation of a heterodimer or rotation of the M-acetyl mainly effects the energy differences $d=E(P_M^*)-E(P_L^*)$ and $\Delta=E(P_L^+P_M^-)-E(P_M^+P_L^-)$. An applied electric field effects the diagonal energies. As the dipole moments of the charge separated states are the largest, these states are more strongly affected than the remaining ones. The combined effect of this symmetry breaking perturbations is summarized by the matrix

$$\tilde{H}_1 = \begin{bmatrix} \vec{p}^{\,*}_{//}\vec{F} & \frac{d}{2}+\vec{p}^{\,*}_{\perp}\vec{F} & \circ & \circ & \circ \\ \frac{d}{2}+\vec{p}^{\,*}_{\perp}\vec{F} & \vec{p}^{\,*}_{//}\vec{F} & \circ & \circ & \circ \\ \circ & \circ & \vec{p}^{\,CT}_{//}\vec{F} & \frac{\Delta}{2}+\vec{p}^{\,CT}_{\perp}\vec{F} & \circ \\ \circ & \circ & \frac{\Delta}{2}+\vec{p}^{\,CT}_{\perp}\vec{F} & \vec{p}^{\,CT}_{//}\vec{F} & \circ \\ \circ & \circ & \circ & \circ & \vec{p}^{\,**}_{//}\vec{F} \end{bmatrix}$$

where $\vec{F}$ is the electric field and $\vec{p}^{\,i}_{//}, \vec{p}^{\,i}_{\perp}$ is the component of the dipole change projected on an axis parallel or perpendicular to the $C_2$ axis. If the CT states are far above P*, the apparent permanent dipole and polarizability can be obtained from perturbation theory expanding the eigenvalues of the energy matrix in powers of the small quantities $U_\pm/(E_{CT} - EP^*_\pm)$ , $d/(E_{CT} - EP^*_\pm)$ and $\Delta/(E_{CT} - EP^*_\pm)$ as

$$\vec{p}(P_\pm^*) = \vec{p}_{//}^* \pm \frac{d}{2V}\vec{p}_\perp^* + \frac{U_\pm^2}{(E_{CT}-E_{P*\pm})^2}(\vec{p}_{//}^{CT} - \vec{p}_{//}^*) - \frac{U_\pm^2 \ \Delta}{(E_{CT}-E_{P*\pm})^3}\vec{p}_\perp^{CT} + \cdots$$

$$\alpha(P_\pm^*) = \pm\frac{p_\perp^* p_\perp^*}{2V} - \frac{1}{(E_{CT}-E_{P*\pm})}\frac{U_\pm^2}{(E_{CT}-E_{P*\pm})^2}\left(1+\frac{3}{2}\frac{\Delta^2}{(E_{CT}-E_{P*\pm})^2}\right)$$

$$\times \ \left(p_\perp^{CT} p_\perp^{CT} + (p_{//}^{CT} - p_{//}^*)(p_{//}^{CT} - p_{//}^*)\right) + \cdots$$

We note already here that all terms proportional to $U_+^2$ are very small so that $P^*_+$ achieves neither a strong dipole nor a high polarizability via the CT states.

The permanent dipole of $P^*_-$ can only be parallel to the $C_2$ axis. It consists of one part due to the intrinsic contributions of the dimer halves and another from the coupling to the CT state of the same symmetry. However, the permanent dipoles of the isolated molecules are small (about 2Debye. ) and the dipole of the CT state is nearly pependicular to the symmetry axis making an angle with the $P^*_-$ band of about 35°. Thus the permanent dipole for the symmetric dimer (i.e. $d=\Delta=0$) is small. However, the contributions induced by the CT states become important , if the CT states are split. Then the dimer bands get a perpendicular dipole component which in the limit of large $\Delta$ approaches a maximum value of

$$\vec{p}_{P*\pm} \approx \frac{U_\pm^{\ 2}}{2(E_{CT}-E_{P*\pm})^2}\vec{p}_\perp^{CT}.$$

The polarizability is less sensitive to symmetry violations. Its main contribution is the term containing $p^{CT}\perp$ which leads to

$$\alpha = -\frac{U_\pm^2}{(E_{CT}-E_{P*\pm})^3} p_\perp^{CT} p_\perp^{CT}$$

The low order perturbation theory may become questionable if the state P** is in the energy region between the excitonic and the CT states. Due to the large coupling of P** with the symmetric state $CT_+$ the symmetric combination of P** and $CT_+$ may come close to the P* bands and the energy denominators involved in a simple perturbation expansion diverge. In this case we use basis states which diagonalize the largest couplings of $\widetilde{H}_0$, i.e. $U_-$ and $U_2$

$$\begin{bmatrix} c_- \\ \circ \\ s_- \\ \circ \\ \circ \end{bmatrix}, \begin{bmatrix} -s_- \\ \circ \\ c_- \\ \circ \\ \circ \end{bmatrix}, \begin{bmatrix} \circ \\ 1 \\ \circ \\ \circ \\ \circ \end{bmatrix}, \begin{bmatrix} \circ \\ \circ \\ \circ \\ c_+ \\ s_+ \end{bmatrix}, \begin{bmatrix} \circ \\ \circ \\ \circ \\ -s_+ \\ c_+ \end{bmatrix}$$

with

$$c_- = \frac{\sqrt{\sqrt{4U_-^2 + (E_{P*-} - E_{CT-})^2} + (E_{P*-} - E_{CT-})}}{\sqrt{2}(E_{P*-} - E_{CT-})^{1/4}} \qquad s_- = sign(U_-)\sqrt{1-c_-^2}$$

$$c_+ = \frac{\sqrt{\sqrt{8U_2^2 + (E_{P**} - E_{CT+})^2} + (E_{P**} - E_{CT+})}}{\sqrt{2}(E_{P**} - E_{CT+})^{1/4}} \qquad s_+ = sign(U_2)\sqrt{1-c_+^2}$$

Symmetry breaking interactions couple states of different symmetry. Therefore the lower dimer band $P^*_-$ may interact strongly with the lower combination of $P^{**}$ and $CT_+$ by a coupling of

$$V_{eff} = -s_+ s_- \left( \frac{\Delta}{2} + \vec{p}_\perp^{CT} \vec{F} \right)$$

We point to the fact that a similar coupling does not exist for the upper dimer band $P^*_+$ as the state $P^{**}$ has no counterpart of odd symmetry. The interaction between $P^*_+$ and the remaining two even states of $U_+ / \sqrt{2} \approx 0.02\, eV$ is quite small and there is no term proportional to $p_\perp^{CT}$ F. This is consistent with the experimental observation of a much smaller Stark intensity of $P^*_+$ as compared to $P^*_-$

Consider now the case that $P^{**}$ with its admixture of CT+ is within the bandwith of the lower dimer band. An applied electric field induces a strong interaction of the two states inducing apparent dipoles of

$$\pm s_+ s_- \vec{p}_\perp^{CT} + \vec{p}_{//}^{CT} \left( \frac{1}{4} + \frac{s_-^2}{2} \right) + \vec{p}_{//}^{*} \left( \frac{1}{4} + \frac{c_-^2}{2} \right).$$

As the two states shift in opposite directions the main effect to the dimer band is a field dependent broadening. A more detailed analysis shows that the resulting Stark spectrum looks essentially like a second derivative despite the linear filed dependence of the transition energies.

# 5 Analysis of the Stark intensities

Let us now compare the magnitude of the Stark effect for several cases of interest.

We consider a Gaussian broadened transition with a change of permanent dipole p and a change of polarizability α. If the shift of the transition energy is small compared to the width Γ of the transition, the electrochromicity is given by

$$\left\langle f(x+\vec{p}\vec{F}+\vec{F}\alpha\,\vec{F}\right\rangle_{\vec{F}} - f(x) = \left\langle \vec{p}\vec{F}+\vec{F}\alpha\,\vec{F}\right\rangle f'(x) + \left\langle \frac{1}{2}\left(\vec{p}\vec{F}\right)^2\right\rangle f''(x)$$

For an isotropic sample Terms containing odd powers of the electric field F are zero on the average. For the remaining terms we have to consider the polarization of the optical electric field. For parallel and perpendicular polarization with respect to the static field the averages are given by

$$\left\langle \left(\vec{p}\vec{F}\right)^2\right\rangle_{\perp} = \frac{3-2\cos^2(\beta)}{15}p^2F^2 \qquad \left\langle \left(\vec{p}\vec{F}\right)^2\right\rangle_{//} = \frac{1+2\cos^2(\beta)}{15}p^2F^2$$

$$\left\langle \vec{F}\alpha\vec{F}\right\rangle_{\perp} = \frac{\alpha_1\left(3-2\cos^2(\gamma_1)\right)+\alpha_2\left(3-2\cos^2(\gamma_2)\right)+\alpha_3\left(3-2\cos^2(\gamma_3)\right)}{15}F^2$$

$$\left\langle \vec{F}\alpha\vec{F}\right\rangle_{//} = \frac{\alpha_1\left(1+2\cos^2(\gamma_1)\right)+\alpha_2\left(1+2\cos^2(\gamma_2)\right)+\alpha_3\left(1+2\cos^2(\gamma_3)\right)}{15}F^2$$

where β is the angle between p and the optical transition and $\gamma_i$ are .the angle between the axes of the polarization tensor α and the optical transition. We note that if the polarizibility is large essentially only along an axis parallel to the permanent dipole than the dependence on the angle to the optical transition is the same for the polarizability contribution and the dipole contribution.

For a Gaussian profile $exp(-x^2/\Gamma^2)$ the maximum of the first derivative is ≈ $1/\Gamma$ and that of the second derivative is $2/\Gamma^2$ relative to an absorption maximum of unity.

To obtain a quantitative measure of the relative absorption changes we drop the information on the angles with respect to the optical transition and average the angles β and $\gamma_i$ This yields

$$\left\langle \left(\vec{F}\vec{p}\right)^2\right\rangle = \frac{p^2}{3}F^2 \qquad \left\langle \vec{F}\alpha\,\vec{F}\right\rangle = \frac{trace(\alpha)}{3}F^2$$

Hence the maximum of the first derivative contribution is $\approx \frac{trace(\alpha)}{3\Gamma}F^2$ and that of the second derivative is $\approx \frac{p^2}{3\Gamma^2}F^2$. Let us now compare some numbers for the special pair dimer using a width of $\Gamma$=250cm$^{-1}$ and a field of F=$10^6$ V/cm.

- For a permanent dipole of 7Debyes which is about the apparent experimental value for rps.viridis, the energy shift is pF=120 cm$^{-1}$ giving a second derivative of $7\cdot 10^{-2}$
- For a symmetric dimer we consider the polarizability of the P*– band induced by the coupling to the CT-states. For a coupling to the asymmetric CT-state of $U_-$=1500 cm$^{-1}$, an energy difference of 2500 cm$^{-1}$ and a permanent dipole of the CT states of $p_{CT} \approx$30 Debye. we have an energy shift of $\alpha F^2$ = 36 cm$^{-1}$ giving a first derivative contribution of $5\cdot 10^{-2}$
- For an asymmetric dimer with strongly split CT states we consider only the coupling to the lower localized CT-state. We find

$$p=\frac{U^2}{2(E_{P*}-E_{CT})^2}p_{CT} \quad , \quad \alpha=\frac{U^2}{2(E_{P*}-E_{CT})^3}p_{CT}^2 .$$

- The permanent dipole has a value of 5.4 Debyes. Thus the magnitude of first and second derivative are $3\cdot 10^{-2}$ and $4\cdot 10^{-2}$. The two contributions add up constructively on the long wavelength side of the band.
- If the Triplet-Triplet-State is within the dimer band and is mixed strongly with the symmetric CT-state then the P* state shifts linearly with $\alpha=\frac{U}{\sqrt{2}(E_{P*}-E_{CT})}p_{CT}$ giving an energy shift of $\alpha F^2$=210 cm$^{-1}$ However the P** state also contributes as it gains intensity and is shifted oppositely. Together the shape of the Stark spectrum looks more like a second derivative and reaches a maximum of $\frac{1}{3}\frac{\alpha F^2}{\Gamma^2}=24\cdot 10^{-2}$ which is considerably larger than in the other cases.

# 6 Conclusions

We have found that the quantum calculations based on the structure data of *Rps. viridis* reflect in first approximation the $C_2$ symmetry for the electronic states. In particular the lower dimer state $P^*_-$ gets a weak almost symmetry preserving permanent dipole, which cannot explain the observed Stark intensities with regard to its strength or its direction./7/ This situation is not improved if the field caused by the protein surrounding is incorporated. Even more, the partial localization of the spin density of the cation $P^+$ is predicted to be on the M side in contradiction to the observed localization by 2:1 on the

L-side. The proper change could be achieved if the acetyl from $P_M$ was rotated by 180°. In the case of the hetero dimers, a large splitting of the charge transfer states is found but not strong enough to place the lowest charge transfer state below P*. So we concluded that some relaxation must take place in the native dimer as well as in the hetero dimers after excitation of P*–. A possible clue for an instability of the structure in the P*– state can be found in the appearance of the double triplet state P**. It is in first order symmetric and interacts strongly with the internal charge transfer states, which can induce via an asymetric mode an interaction with the state P*–. With this mechanism one can understand that the Stark intensity of P*– is largely due to a polarization, that the strong coupling to vibrations is due to the instability to an asymetric mode, that the upper dimer band P*+ shows a weak Stark intensity and, that the relaxed conformation can have an unusual admixture of charge transfer character and the state P**. This in turn is a completely new perspective for the charge separation mechanism, since the state P** can induce the charge separation via the Coulombic rather than the one particle interaction. Recent experiments on a triple mutant /8/ provide evidence for a mechanism of the charge separation which is controlled by an internal relaxation of the dimer in the state P*–. Our results may form a basis for such a mechanism.

**Acknowledgement**
This work has been supported by the DFG (SFB 143).

**References**

/1/ T. Arlt, S. Schmidt, W. Kaiser, C. Lauterwasser, M. Meyer, H. Scheer and W. Zinth, Proc. Natl. Acad. Sci. USA 90 (1993) 11757.

/2/ S. Schmidt, T. Arlt, P. Hamm, H. Huber, T. Nägele, J, Wachtveitl, M. Meyer, H. Scheer and W. Zinth, Chem. Phys. Letters 223 (1994) 116.

/3/ P.O.J. Scherer, C. Scharnagl, Sighart F. Fischer,Chemical Physics 197 (1995) 333-341.

/4/ D. Holten and C. Kirmaier, Photosynth. Res. 13 (1987) 225.

/5/ K. Ohno, Theoret. Chim. Acta 2 (1964) 219.

/6/ G.J. Small, Chemical Physics 197 (1995) 239-257.

/7/ F. Lendzian, M. Huber, R.A. Isaacson, B. Endeward, M. Plato, B. Bönigk, K. Möbius, W. Lubitz and G. Feher, Biochim. Boiphys. Acta 1183 (1993) 139.

/8/ D.J. Lockhart, C. Kirmaier, D. Holten and S.G. Boxer, J. Phys. Chem. 94 (1991) 6987.

/9/ Neal W. Woodbury, Su Lin, Xiaomei Lin, Jeffrey M. Peloquin, Aileen K.W. Taguchi, JoAnn C. Williams, James P. Allen, Chemical Physics 197 (1995) 405-421.

# Macroscopic and Microscopic Estimates of the Energetics of Charge Separation in Bacterial Reaction Centers

R. G. Alden[1], W. W. Parson[1], Z. T. Chu[2] and A. Warshel[2]

[1] Department of Biochemistry, University of Washington, Seattle, WA 98195-7350, USA

[2] Department of Chemistry, University of Southern California, Los Angeles, CA 90007, USA

**Abstract.** Two approaches for calculating the free energies of transient radical-pair states in bacterial reaction centers are discussed. Although macroscopic models that assign a homogenous dielectric constant to the protein and solvent are major oversimplifications, they help to clarify the importance of considering the self-energies of the charged species, and to put limits on the energetics of the charge-separation processes. The microscopic Protein-Dipoles-Langevin-Dipoles (PDLD) approach provides a much more realistic treatment of dielectric effects, but requires lengthy calculations that depend on numerous interrelated factors. Calculations by both approaches indicate that, in *Rhodopseudomonas viridis* reaction centers, the state $P^+B^-$ generated by movement of an electron from the primary electron donor (P) to a neighboring bacteriochlorophyll (B) lies close to the excited state P* in energy, where it possibly could serve as an intermediate in electron transfer to the bacteriopheophytin (H). This conclusion agrees with previous free-energy-perturbation calculations and indicates that any model (macroscopic or microscopic) that includes all the relevant contributions and reproduces the energy of the relaxed $P^+H^-$ should find the relaxed $P^+B^-$ state near P*. In addition, the macroscopic model shows that electron transfer from P* to H is likely to be exothermic even in the absence of a strong field from the atomic charges of the protein.

Key words: Electron transfer, Energetics, Electrostatics, Computer calculations

## 1 Introduction

The bacterial photosynthetic reaction center efficiently traps the energy of light through a series of charge-separation reactions. In the first of these reactions, a bacteriochlorophyll dimer (P) in an excited singlet state (P*) transfers an electron to a bacteriopheophytin (H). The mechanism of this reaction has remained a topic of controversy. Some experimental studies suggest that the reaction proceeds in two distinct steps through an intermediate state, $P^+B^-$, in which an electron resides on a bacteriochlorophyll (B) situated between P and H. Other studies suggest that $P^+B^-$ is not formed as a distinct kinetic intermediate, but rather that this state mediates electron transfer between P and H through quantum mechanical coupling to the donor and acceptor states. The latter mechanism is termed "superexchange."

Because the charge-separation reaction proceeds rapidly at temperatures as low as 4 K, it evidently does not require formation of an intermediate that lies significantly above P* in energy. Thus, if $P^+B^-$ is located more than about 2kT above P*, the electron-transfer reaction probably must rely on superexchange. On the other hand, if $P^+B^-$ lies close to or below P*, the sequential mechanism would dominate. Because experimental measurements of the energy of $P^+B^-$ are problematic, it is of interest to try to calculate the relative energies of P*, $P^+B^-$ and $P^+H^-$ on the basis of the crystal structure of the reaction center.

In previous theoretical studies, Creighton et al. [1] and Parson et al. [2] concluded that $P^+B^-$ lies within about 3 kcal/mol of P*. Dielectric effects were treated microscopically by assigning an atomic polarizability to each atom of the protein and to grid points representing the surrounding membrane or solvent. By contrast, calculations by Marchi et al. [3] placed $P^+B^-$ approximately 20 kcal/mol above P*. The latter authors used a simpler treatment of dielectric effects, in which all charge-charge interactions were screened by a homogeneous effective dielectric constant. In subsequent work [4,5] we showed that much of the disagreement between the results of the two studies probably resulted from underestimates by Marchi et al. [3] of the role of the solvent and counterions in screening electrostatic interactions with ionized amino acid side chains, and of the contribution of induced dipoles of the protein to the self-energies of $P^+$, $B^-$ and $H^-$.

Although fully microscopic approaches provide the most realistic treatments of dielectric effects in proteins, the calculations are lengthy and the results depend on a large number of interdependent terms. It is therefore useful to consider macroscopic models that, though less rigorous, lend themselves to simpler, more intuitive pictures. Here we explore such a model and relate it to the picture provided by microscopic calculations.

## 2 Basic Macroscopic Treatments of Charge Separation in a Uniform Medium

Proteins are intrinsically heterogeneous systems that should, in general, have highly non-uniform dielectric behavior. Any treatment that ascribes a homogeneous dielectric constant to the reaction center clearly represents a severe oversimplification. Such macroscopic treatments can, however, provide instructive models, as long as they rest on a reliable theoretical approach such as the generalized Born model rather than the reaction field model [6-9].

The energetics of any electron-transfer process can be written

$$\Delta G^o = \Delta E^{gas} + \Delta G_{sol} + \Delta V_{QQ} \quad (1)$$

where $\Delta E^{gas}$ is the energy change of the electron-transfer reaction when the donor and acceptor are separated to infinite distance in a vacuum, $\Delta V_{QQ}$ is the change in vacuum charge-charge interactions between the donor and acceptor in their actual positions (e.g., in a protein), and $\Delta G_{sol}$ is the change in the solvation of the donor and acceptor by the protein and/or other surroundings. If the protein and solvent are modeled as a medium with a homogeneous dielectric constant $\varepsilon$, then (see pp. 314-315 and 327-329 in [6])

$$\Delta G_{sol}^{\varepsilon;R} \approx \Delta G_{sol}^{\varepsilon;\infty} - \Delta V_{QQ}(1 - 1/\varepsilon) + \Delta V_{Q\mu}/\varepsilon \qquad (2)$$

Here $\Delta G_{sol}^{\varepsilon;R}$ represents the solvation energy when the donor and acceptor are in a solution with dielectric constant $\varepsilon$ but are still at their actual coordinates (R), $\Delta G_{sol}^{\varepsilon;\infty}$ is the solvation energy at infinite separation, and $\Delta V_{Q\mu}$ is the change in vacuum interactions with any other charges in the system. $\Delta G_{sol}^{\varepsilon;\infty}$ can be related to the solvation energy for a reaction in water or a polar organic solvent ($\Delta G_{sol}^{w;\infty}$):

$$\Delta G_{sol}^{\varepsilon;\infty} \approx \Delta G_{sol}^{w;\infty}(1 - 1/\varepsilon)/(1 - 1/80) \approx \Delta G_{sol}^{w;\infty}(1 - 1/\varepsilon) \qquad (3)$$

Equation 3 follows from the Born expression [see, *e.g.*, 6-8]:

$$\Delta G_{sol}^{\varepsilon;\infty} \approx -\frac{C}{2}\left\{\frac{Q_{D+}^2}{a_{D+}} - \frac{Q_D^2}{a_D} + \frac{Q_{A-}^2}{a_{A-}} - \frac{Q_A^2}{a_A}\right\}(1 - 1/\varepsilon) \qquad (4)$$

where $Q_{D+}$ and $a_{D+}$ are the charge and radius of the oxidized electron donor, $Q_D$ and $a_D$ are those of the reduced donor, $Q_{A-}$, $Q_A$, $a_{A-}$ and $a_A$ are the corresponding parameters for the reduced and oxidized acceptor, and $C$ = 332 (kcal/mol)(Å/unit charge).

Combining eqns. 1 - 3 gives

$$\Delta G^o \approx \Delta E^{gas} + \Delta G_{sol}^{w;\infty}(1 - 1/\varepsilon) + (\Delta V_{QQ} + \Delta V_{Q\mu})/\varepsilon$$

$$= (\Delta E^{gas} + \Delta G_{sol}^{w;\infty}) + (\Delta V_{QQ} + \Delta V_{Q\mu} - \Delta G_{sol}^{w;\infty})/\varepsilon \qquad (5)$$

The "self-energy" or "reaction-field" term $\Delta G_{sol}^{w;\infty}(1 - 1/\varepsilon)$ in this expression can be quite large, particularly in charge-separation reactions. This term was omitted in the treatment used by Marchi et al. [3].

$\Delta E^{gas}$ can be obtained by quantum calculations such as those described by Thompson and Zerner [10] and Scherer and Fischer [11]. An alternative approach that may be more accurate for molecules as large as bacteriochlorophyll (BChl) and bacteriopheophytin (BPh) makes use of the measured midpoint redox potentials of the acceptor and donor redox couples ($E_m^A$ and $E_m^D$). The $E_m$ values are related to $\Delta E^{gas}$ by the expression

$$\Delta E^{gas} = -\mathcal{F}[E_m^A - E_m^D] - \Delta G_{sol}^{ref} \tag{6}$$

where $\mathcal{F}$ is the Faraday constant and $\Delta G_{sol}^{ref}$ is the sum of the changes in solvation energies for separate oxidation of the electron donor and reduction of the acceptor under the particular conditions of the redox measurements [1,2,5]. Equations 5 and 6 can be combined and simplified if $\Delta G_{sol}^{ref} = \Delta G_{sol}^{w;\infty}$, which often is the case.

To use eqn. 6, we need values for $\Delta G_{sol}^{ref}$. We have calculated these solvation energies by the protein-dipoles-Langevin-dipoles (PDLD) method [5-9], in which the electron carrier is surrounded by a grid of points that represent solvent dipoles. Each non-hydrogen atom of the protein and each of the grid points acquires an induced dipole that is proportional to the electric field at that point and to the atomic polarizability of the protein or the surroundings. The field includes contributions from the atomic charges of the electron carrier, the induced dipoles at all the other points and, in some of the calculations, the charges of electrolytes in the solvent. (In the calculations on proteins described below, ions were included in the solvent to model an ionic strength of 0.1 M; when the amino acid side chains were ionized additional anions or cations were added as needed to give overall electrical neutrality in the initital state of the system. The mean electrolyte charges at the solvent grid points were evaluated by an iterative procedure [5] for each oxidation state of the electron carrier. Electrolytes were not included in the calculations for free BChl or BPh in solution.) A self-consistent set of induced dipoles is obtained by evaluating the dipoles and fields iteratively until the energy converges. Finally, the PDLD treatment uses a Born-type expression for the effect of the bulk solvent outside the spherical region that is treated microscopically. For calculations on free BChl, we included an acetonitrile axial ligand as part of the solvent.

Table I shows the results of calculations of $\Delta E^{gas}$ for the process $B^-H \rightarrow BH^-$. For electron transfer from the BChl-*b* anion radical to BPh-*b* in polar organic solvents, $\mathcal{F}[E_m^A - E_m^D] \approx 7.8$ kcal/mol [12] and $\Delta G_{sol}^{ref}$ is calculated to be -12.1 kcal/mol [5]. Combining $-\mathcal{F}[E_m^A - E_m^D]$ and $-\Delta G_{sol}^{ref}$ gives $\Delta E^{gas} \approx$ 4.3 kcal/mol. Calculations that included mobile ions in the solvent gave a similar value of 3.4 kcal/mol [5].

The procedure for evaluating $\Delta G_{sol}^{ref}$ must be modified slightly for the charge-separation reaction $PH \rightarrow P^+H^-$ because P cannot be removed from the protein for a redox titration in solution. P and H can, however, be titrated in *Rp. viridis* reaction centers (see [2] for references and discussion). $\Delta G_{sol}^{ref}$ can be calculated for titrations in the reaction center using the PDLD approach essentially as outlined above for BChl and BPh in solution, except that in addition to solvent and electrolytes, the calculations must consider the atomic charges and induced dipoles of the protein. A realistic treatment also should consider that the protein normally is embedded in a phospholipid membrane or surrounded by a belt of detergents. To model such an environment, we divided the solvent grid embedding the protein into three regions. The volume elements in a central slab of 25 Å, surrounding the hydrophobic region of the protein, were given a low atomic polarizability consistent with phospholipid side chains, while the regions on either side had higher polarizability typical of water. Electrolyte charges were distributed among the grid points in the two outer regions but were excluded from the hydrophobic region.

**Table I.** Estimates of $\Delta E^{gas}$ for PH → $P^{+}H^{-}$ and $B^{-}H$ → $BH^{-}$ in *Rp. viridis* reaction centers [5].[a]

| Reaction | ionizable residues | $\Delta G^{ref}_{sol}$ | $\mathcal{F}[E^{A}_{m} - E^{D}_{m}]$ | $\Delta E^{gas}$ |
|---|---|---|---|---|
| PH → $P^{+}H^{-}$ | neutral | -51.8[b] | -25.8 | 77.6 |
| | ionized | -63.5[b] | -25.8 | 89.3 |
| $B^{-}H$ → $BH^{-}$ | - | -12.1[c] | 7.8 | 4.3 |

[a]Energies are in kcal/mol. Calculations with 10 randomly centered solvent grids were carried out for each half-reaction; the standard errors of the means of the individual solvation energies were less than ±0.5 kcal/mol. [b]The calculations for P → $P^{+}$ and H → $H^{-}$ included all amino acid residues, crystallographic waters and pigments with atoms within 32 Å of either the center of P or the center of H in the *Rp. viridis* reaction center [13]; the solvent grid radius was 48 Å. The axial ligands of the two BChls of P were treated as part of the protein. The membrane was modeled by a 25-Å thick region with a low atomic polarizability; mobile water and ions were restricted to the regions on either side of this space. The ion charges were allowed to redistribute on the water grid in response to oxidation or reduction of the electron carrier. [c]The calculations for $B^{-}$ → B and H → $H^{-}$ are for BChl-*b* and BPh-*b* in a polar solvent without electrolytes. The solvent grid radius was 40 Å. The axial ligand of BChl (acetonitrile) was treated as part of the solvent.

Although most of the ionizable amino acid side chains in the reaction center are likely to be charged at physiological pH, some probably are not, particularly in the more hydrophobic regions of the protein. Since we presently cannot be certain exactly which groups are ionized, calculations were done both on models in which all the ionizable residues were in their neutral states, and also with all these residues other than Glu L104 ionized. Barring structural changes in P, the actual $\Delta E^{gas}$ for PH → $P^{+}H^{-}$ must be independent of the ionization state of the protein, whereas the calculated $\Delta G^{ref}_{sol}$ will depend on our assumptions concerning this state. (The measured $E_m$ values that enter into eqn. 6 also are likely to depend on the ionization state, but this point has not been explored experimentally in *Rp. viridis*. Redox titrations of H are necessarily performed at a high pH.) Our estimate of $\Delta E^{gas}$ thus differs in the two treatments of the ionizable residues. Treating all the ionizable residues as neutral gives $\Delta E^{gas}$ = 77.6 kcal/mol; treating them all as ionized gives 89.3 (see Table I). It is likely that the PDLD calculations somewhat overestimate the effects of the ionized groups, and that the smaller of these values is the more accurate. Fortunately, as we will show below, the uncertainty in the true value of $\Delta E^{gas}$ cancels out in a fully microscopic treatment if the ionizable groups are handled consistently in all parts of the analysis. However, to obtain a single number for use in the macroscopic model, we will simply assume that $\Delta E^{gas}$ lies somewhere between the two estimates described above, and will use the average value of 83 kcal/mol.

The estimate of 83 kcal/mol for the gas-phase energy of forming $P^+H^-$ can be compared with the results of INDO/S quantum calculations reported by Thompson and Zerner [10]. Thompson and Zerner obtained a vacuum energy of 56.4 kcal/mol for the reaction PH → $P^+H^-$ in a model that included all four BChls, both BPhs and the imidazole ligands of the BChls. The quantum calculation includes the vacuum electrostatic interactions between P and H when the electron carriers are in their positions in the reaction center, which we treat separately as $\Delta V_{QQ}$. With the *Rp. viridis* crystal structure [13] and the set of atomic charges that we use [2], this term amounts to -19.0 kcal/mol. In addition, because we treated the imidazole ligands and the photochemically inactive BChl and BPh molecules as part of the protein, their interactions with P and H do not contribute to our $\Delta E^{gas}$. Interactions with the four histidine imidazoles and the inactive BChl and BPh contribute about -8.9 kcal/mol to the electrostatic energy of transferring an electron from P to H, mainly as a result of favorable interactions of $P^+$ with its axial ligands. The imidazoles and the inactive pigments should have a similar effect in the quantum calculations, because their resonance interactions with P and H are relatively small. The INDO/S estimate of $\Delta E^{gas}$ for PH → $P^+H^-$ thus is approximately 56.4 + 19.0 + 8.9 = 84.3 kcal/mol, which is very close to the estimate of 83 kcal/mol given above. The quantum calculations [10] put P* too high by about 4 kcal/mol, suggesting that the true vacuum energy of $P^+H^-$ could be somewhat lower than the calculated value. Relative to $P^+H^-$, our gas-phase energy for $P^+B^-$ is lower than that obtained by Thompson and Zerner [10] but closer to that obtained by Scherer and Fischer [11].

Table II gives the additional terms that contribute to $\Delta G^o$ for the reactions PH → $P^+H^-$ and $P^+B^-H$ → $P^+BH^-$ in the macroscopic model. $\Delta V_{QQ}$ for electron transfer from $B^-$ to H is found to be -1.8 kcal/mol [2,5]. If we include the atomic charges of $P^+$, the additional BChl and BPh molecules and the imidazole axial ligands of all four BChls, but neglect the charges of all the other protein atoms, $\Delta V_{Q\mu}$ for $P^+B^-H$ → $P^+BH^-$ is 4.4 kcal/mol. $\Delta G_{sol}^{w;\infty}$ is the same as $\Delta G_{sol}^{ref}$ (-12.1 kcal) because the redox titrations were done with the individual components in polar solutions. The macroscopic model for this reaction (eqn. 5) thus gives $\Delta G^o \approx$ -7.8 + 14.7/ε kcal/mol. Electron transfer from $B^-$ to H is predicted to be endothermic when ε = 1 because $\Delta E^{gas}$ is positive and the unscreened Coulombic interactions with $P^+$ favor the closer

$B^-$ radical. It becomes exothermic when $\varepsilon$ exceeds 1.9 because $\Delta G_{sol}^{\varepsilon;R}$ then stabilizes the larger dipole of $P^+H^-$.

**Table II.** $\Delta G^o$ for PH → $P^+H^-$ and $P^+B^-H$ → $P^+BH^-$ in *Rp. viridis* reaction centers as calculated with a macroscopic model[a]

| Reaction | $\Delta E^{gas}$ | $\Delta G_{sol}^{w;\infty}$ | $\Delta V_{QQ}$ | $\Delta V_{Q\mu}$ | $\Delta G^o$ |
|---|---|---|---|---|---|
| PH → $P^+H^-$ | 83[b] | -76.7 | -19.0 | -8.9[c] | 6.3 + 48.8/ε |
| $P^+B^-H$ → $P^+BH^-$ | 4.3 | -12.1 | -1.8 | 4.4[c] | -7.8 + 14.7/ε |

[a]All energies are in kcal/mol. $\Delta G_{sol}^{w;\infty}$ was obtained by PDLD calculations on BChl-*b*, BPh-*b* and a BChl-*b* dimer in polar solvents without electrolytes; 10 randomly centered solvent grids were used for each half-reaction. Axial ligands of the BChls (imidazoles for the BChl dimer and acetonitrile for monomeric BChl) were included as part of the solvent. [b]Average of the two estimates given in Table I. [c]Interactions with the other BChls and BPhs and the imidazole axial ligands of the four BChls in the reaction center. The other atomic charges of the protein are ignored in this model.

The term $\Delta G_{sol}^{w;\infty}(1 - 1/\varepsilon)$ in eqn. 5 is much larger for the charge-separation process PH → $P^+H^-$ than for the charge-shift reaction $B^-H$ → $BH^-$ because now we have the sum of the solvation energies for forming two charged species rather than the difference. The calculated $\Delta G_{sol}^{w;\infty}$ for oxidation of a BChl dimer and reduction of BPh in polar solvents is -76.7 kcal/mol (Table II). In addition, $\Delta V_{QQ}$ is larger (-19 kcal/mol) because this term, rather than $\Delta V_{Q\mu}$, now represents the unscreened charge-charge interactions between $P^+$ and $H^-$. With $\Delta E^{gas}$ = 83 kcal/mol and $\Delta V_{Q\mu}$ = -8.9 (*i.e.*, neglecting all the atomic charges of the protein except for those of the imidazole axial ligands of the BChls), eqn. 5 gives $\Delta G^o \approx 6.3 + 48.8/\varepsilon$ kcal/mol (Table II). Experimental studies of *Rb. sphaeroides* reaction centers [14-17] place $P^+H^-$ about 6 kcal/mol below P* or 23 kcal/mol above the ground state. The macroscopic model reproduces this result when $\varepsilon \approx 2.9$.

Figure 1a shows the relative free energies of $P^+B^-$ and $P^+H^-$ in the macroscopic model with several values of $\varepsilon$. Values of $\varepsilon$ that position $P^+H^-$ at approximately the right level relative to P* put $P^+B^-$ between these two states.

The stability of $P^+B^-$ and $P^+H^-$ in the macroscopic model owes partly to the favorable electrostatic interactions of $P^+$ with the reduced electron acceptor. Figure 1b shows the energies calculated for a macroscopic model in which the electron carriers are moved infinitely far apart so that $\Delta V_{QQ}$ for each electron-transfer step is zero and $\Delta V_{Q\mu}$ includes only the interactions of P or B with its own axial ligands. The macroscopic expression for P + H → $P^+$ + $H^-$ becomes $\Delta G^o = 6.3 + 64.8/\varepsilon$, and that for $B^-$ + H → B + $H^-$ becomes $\Delta G^o$ = -7.8 + 2.6/ε kcal/mol. Because P and B are located very close together in the reaction center, the effect of separating the electron carriers is greater for $P^+B^-$ than for $P^+H^-$; electron transfer from $B^-$ to H now is driven by the unfavorable interaction of $B^-$ with the axial imidazole and is exothermic even for $\varepsilon = 1$. However, at large $\varepsilon$, the calculated free energy difference between $P^+B^-H$ and $P^+BH^-$ becomes independent of the distance between the electron carriers and is the same for a complex with the geometry of the reaction center as for the electron carriers in solution (-7.8 kcal/mol). The insensitivity of ion-pair energies to the distance between the ions in solvents of high $\varepsilon$ is in accord with experimental information on the $pK_a$s of dicarboxylic acids and amino acids (see [6] and references therein). The reaction-field treatment used by Thompson and Zerner [10] does not capture this general feature of solvation effects (see Appendix A of [9]).

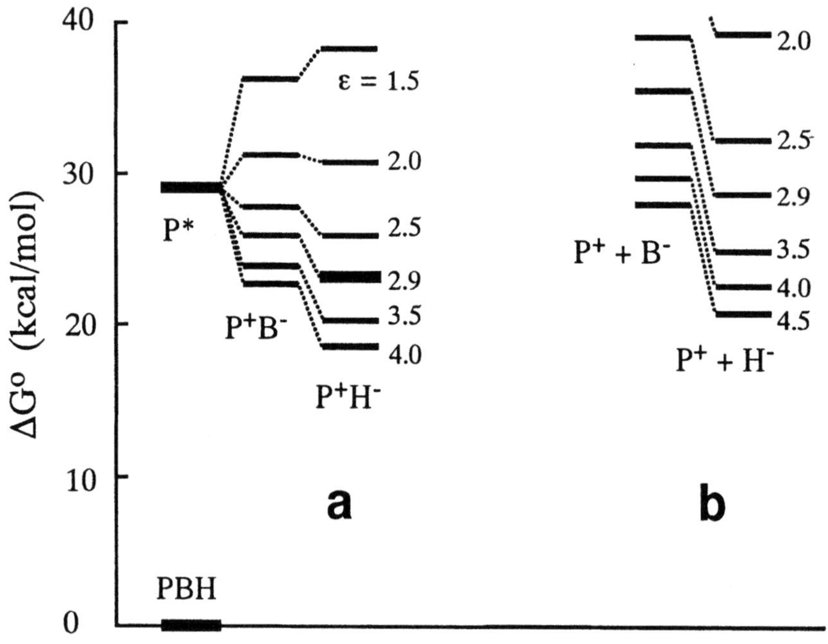


**Figure 1.** Free energies of P*, $P^+B^-$ and $P^+H^-$ relative to the ground state (PBH) in *Rp. viridis* reaction centers (*a*), and with P, B and H separated to infinite distances (*b*), as calculated with a macroscopic model using several values of the dielectric constant ($\varepsilon$). $\Delta E^{gas}$ was taken to be 83 kcal/mol for PH $\rightarrow$ $P^+H^-$ and 4.3 kcal/mol for $P^+B^-H$ $\rightarrow$ $P^+BH^-$. Changing $\Delta E^{gas}$ for PH $\rightarrow$ $P^+H^-$ would shift the calculated free energies of $P^+B^-$ and $P^+H^-$ up or down in parallel. Experimentally measured free energies in the reaction center are indicated with heavy bars.

Values of $\Delta G^0$ obtained with the macroscopic model obviously cannot be expected to be very accurate for the reaction center. Interactions with the charges of protein atoms and electrolytes in the solvent will raise or lower the relative energy of $P^+B^-H$, depending on the locations and signs of the charges. Although these charges could be included in the macroscopic model by incorporating the additional interaction energies in $\Delta V_{Q\mu}$, more refined methods such as the microscopic treatment described in the following section are needed to explore their effects in any detail because charges in the solvent or in polar regions of the protein will be screened more effectively than charges in the reaction center's central, hydrophobic region. However, the macroscopic model shows clearly that, with a homogeneous dielectric and in the absence of strong interactions with fixed charges, $P^+B^-H$ will lie relatively close to P* and $P^+BH^-$ in energy. In addition, if the estimate of $\Delta E^{gas}$ for PH → $P^+H^-$ obtained by either eqn. 6 or the INDO/S quantum calculations is approximately correct, a strong electrostatic field from components other than the pigments and their axial ligands is not essential in order to position $P^+H^-$ below P*, even in a medium with a relatively low dielectric constant (see Fig. 1a). This is not to say that such a field is necessarily absent or unimportant in the reaction center, but just that electron transfer from P* to H would be exothermic by about 6 kcal/mol if the pigments were embedded in a homogenous medium with a dielectric constant on the order of 3 and no fixed charges. However, if $\varepsilon$ is restricted to the high-frequency or "optical" dielectric constant of approximately 2.0, $P^+H^-$ would lie slightly above P*, and an additional field from the protein or solvent atoms would be needed to stabilize the radical-pair. Several investigators have suggested that a static field from the protein could play an important role in the charge-separation reaction [18; M. Gunner and B. Honig, personal communication].

## 3 Fully Microscopic Treatments

The PDLD approach [5-9] provides a more realistic, but also more computationally intensive, way to calculate the solvation energy of a radical-pair state. In the PDLD treatment, the change in solvation energy for the reaction of interest ($\Delta G_{sol}^{pro}$) is written as a sum of all the terms mentioned above in connection with the calculations of $\Delta G_{sol}^{ref}$ and $\Delta G_{sol}^{w;\infty}$:

$$\Delta G_{sol}^{pro} = \Delta V_{Q\mu} + \Delta V_{ind} + \Delta V_{H_2O} + \Delta V_{memb} + \Delta V_{ions} + \Delta V_{bulk} \quad (8)$$

As in eqn. 2, $\Delta V_{Q\mu}$ includes the vacuum electrostatic interactions with the atomic charges of the protein, crystallographic water molecules and the pigments that are not involved directly in the electron-transfer step. $\Delta V_{ind}$ represents interactions of the donor and acceptor with induced dipoles in the protein and pigments not involved in the reaction; $\Delta V_{H_2O}$, interactions with mobile solvent molecules; $\Delta V_{memb}$, interactions with the sidechains of membrane phospholipids or detergents; $\Delta V_{ions}$, interactions with ions in the solvent; and $\Delta V_{bulk}$, interactions with the bulk solvent outside the region that is treated microscopically. $\Delta V_{ions}$, $\Delta V_{ind}$, $\Delta V_{H_2O}$ and $\Delta V_{memb}$ are evaluated iteratively as outlined above. The standard free energy change for an electron-transfer reaction is obtained by eqn. 1 with $\Delta G_{sol}^{pro}$ replacing $\Delta G_{sol}$.

As discussed above, PDLD calculations on the individual redox reactions of P and H lead to different estimates of $\Delta E^{gas}$ for the reaction PH → $P^+H^-$, depending on the treatment of the ionizable side chains. For the macroscopic treatment, we averaged two of these values to obtain a single estimate of the actual $\Delta E^{gas}$. In the PDLD treatment, it is preferable to use the individual values for $\Delta E^{gas}$ in combination with the values of $\Delta G_{sol}^{pro}$ obtained by the same treatment of the ionizable groups. This is because variations in the assumptions concerning the ionization state or other aspects of the model tend to have similar (though not identical) effects on $\Delta G_{sol}^{ref}$ and $\Delta G_{sol}^{pro}$, so that the final calculated values of $\Delta G^0$ for the charge-separation reaction become relatively insensitive to these details. As shown in Table III, the two limiting treatments of the ionizable groups both put $P^+H^-$ 22-23 kcal/mol above PH, or 6-7 kcal/mol below P*H, in good accord with experiment.

**Table III.** $\Delta G^o$ for PH → $P^+H^-$ and $P^+B^-H$ → $P^+BH^-$ in *Rp. viridis* reaction centers as calculated by the microscopic PDLD approach [5].[a]

| Reaction | ionizable residues | $\Delta E^{gas}$ | $\Delta G_{sol}^{pro}$ | $\Delta V_{QQ}$ | $\Delta G^o$ |
|---|---|---|---|---|---|
| PH → $P^+H^-$ | neutral | 77.6 | -36.1 | -19.0 | 22.5 |
| | ionized | 89.3 | -48.0 | -19.0 | 22.3 |
| $P^+B^-H$ → $P^+BH^-$ | neutral | 4.3 | -2.1 | -1.8 | 0.4 |
| | ionized | 4.3 | -6.4 | -1.8 | -3.9 |

[a]Energies are in kcal/mol. The model for calculations for PH → $P^+H^-$ was as in Table I. The model for $P^+B^-H$ → $P^+BH^-$ was the same except that it included all amino acids, crystallographic waters and pigments with atoms within 32 Å of a point midway between B and H. In the calculations presented in this table, the ion charges were not allowed to move in response to electron transfer. Each value of $\Delta G_{sol}^{pro}$ represents an average of 10 calculations with randomly centered solvent grids; standard errors of the means were typically ±0.5 kcal/mol. The entries for $\Delta E^{gas}$ and $\Delta V_{QQ}$ are from Tables I and II.

Table III also shows PDLD results for the reaction $P^+B^-H$ → $P^+BH^-$. These calculations put $P^+B^-H$ about 3 kcal/mol above $P^+BH^-$ with an estimated uncertainty of ±3 kcal/mol [5]. The calculated free energy difference between $P^+B^-H$ and $P^+BH^-$ is more sensitive to the treatment of the protein's ionizable groups than is the calculated free energy of $P^+H^-$, because now the reference calculations are done with BChl and BPh in solution rather than in the reaction center. The largest errors in the calculations result from uncertainties in the ionization states of two arginine residues and an aspartic acid residue that are within 16 Å of B or H; in the PDLD model, most of the more distant ionizable goups are strongly shielded by counterions and induced dipoles in the solvent and protein [5].

## 4 Discussion

In spite of its obvious limitations, the simple macroscopic model is useful for examining how the energy of the system varies qualitatively as a function of the dielectric constant. In particular, the macroscopic picture shows that a strong electric field from ionized amino acid side chains is not essential in order to make electron transfer from $P^*$ to $P^+H^-$ exothermic by about the amount that is observed experimentally, even if the electron carriers are surrounded by a material with a relatively low dielectric constant. With a low value of $\varepsilon$, the charge-separation reaction would have a small reorganization energy, which would be consistent with the observation that the kinetics are insensitive to temperature.

It is important to emphasize that the $\varepsilon$ on the order of 2.9 that is needed in our macroscopic model in order to position $P^+H^-$ at its observed energy is not an estimate of the actual screening factor for charge-charge interactions in the reaction center. The macroscopic model does not address the magnitude of this screening, because it does not include either the fixed charges of the protein or the medium with high $\varepsilon$ around the protein. In the PDLD treatment, ionizing the amino acid side chains stabilizes $P^+H^-$ relative to $P^*$ and $P^+B^-$ (compare the values of $\Delta G_{sol}^{pro}$ for the ionized and neutral systems in Table III), but the charge-charge interactions of $B^-$ and $H^-$ with individual ionized side chains are shielded by factors ranging from about 4 to more than 40 [5]. The screening of ionized groups, particularly for those near the surface of the protein, results largely from the counterions and solvent surrounding the protein. Indeed, most of these groups probably would not be ionized in the absence of solvent. As we have argued elsewhere [4,5], a large screening of the protein charges is consistent with experimental observations on the interactions of $H^-$ with the anionic radical of the nearby quinone $Q_A$ [19]. However, the PDLD treatment models a system in which the solvent (though not the electrolytes) is able to relax extensively in response to charge separation. The shielding of charge-charge interactions could be weaker on the very short time scale of the initial electron-transfer reaction, particularly at low temperatures [5].

Both the microscopic PDLD model and the simple macroscopic model favor the view that the relaxed $P^+B^-H$ state is located within a few kcal/mol of $P^*$. Our previous free-energy-perturbation calculations [4,5] also are in accord with this conclusion. It thus appears that any treatment of dielectric effects that includes all of the major terms and gives the correct free energy for the relaxed $P^+H^-$ radical pair will put $P^+B^-$ in this region.

## References


1. Creighton, S., Hwang, J.-K., Warshel, A., Parson, W.W. and Norris, J. (1988) Biochem. 27, 774-781.
2. Parson, W.W., Chu, Z.T. and Warshel, A. (1990) Biochim. Biophys. Acta 1017, 251-272.
3. Marchi, M., Gehlen, J.N., Chandler, D. and Newton, M.J. (1993) J. Am. Chem. Soc. 115, 4178-4190.
4. Warshel, A., Chu, Z.T. and Parson, W.W. (1994) J. Photochem. Photobiol. A. Chem. 82, 123-128.
5. Alden, R.G., Parson, W.W., Chu, Z.T. and Warshel, A. J. Am. Chem. Soc. in press.
6. Warshel, A. and Russell, S.T. (1984) Quart. Rev. Biophys. 17, 283-422.
7. Russell, S.T. and Warshel, A. (1985) J. Mol. Biol. 185, 389-404.
8. Lee, F.S., Chu, Z.-T. and Warshel, A. (1993) J. Comp. Chem. 14, 161-185.
9. Luzhkov, V. and Warshel, A. (1992) J. Comp. Chem. 13, 199-213.
10. Thompson, M. and Zerner, M.C. (1991) J. Am. Chem. Soc. 113, 8210-8215.
11. Scherer, P.O.J. and Fischer, S.F. (1989) J. Phys. Chem. 93, 1633-1637.
12. Fajer, J., Davis, M.S., Brune, D.C., Spaulding, L.D., Borg, D.C. and Forman, A. (1976) Brookhaven Symp. Biol. 28, 74-103.
13. Deisenhofer, J., Epp, O., Sinning, I. and Michel, H. (1995) J. Mol. Biol. 246, 429-457.
14. Woodbury, N.W.T. and Parson, W.W. (1984) Biochim. Biophys. Acta 767, 345-361.
15. Goldstein, R.A, Takiff, L. and Boxer, S.G. (1988) Biochim. Biophys. Acta 934, 253-263.
16. Ogrodnik, A. (1990) Biochim. Biophys. Acta 1020, 65-71.
17. Peloquin, J.M., Williams, J.C., Lin, X., Alden, R.G., Taguchi, A.K.W., Allen, J.P. and Woodbury, N.W. (1994) Biochem. 33, 8089-8100.
18. Middendorf, T.R., Mazzola, L.T., Lao, K., Steffen, M.A. & Boxer, S.G. (1993) Biochim. Biophys. Acta 1143, 223-234.
19. Woodbury, N.W., Parson, W.W., Gunner, M.R., Prince, R.C., Dutton, P.L. (1986) Biochim. Biophys. Acta 851, 6-22.

# Electron Transfer in Proteins: Beyond the Single Pathway Approach.

J.J. Regan and J.N. Onuchic

Department of Physics, University of California at San Diego,
9500 Gilman Dr. La Jolla, CA 92093-0319

**Abstract:** A considerable amount of effort has gone into understanding how the electron tunneling coupling in a protein is controlled by the details of the protein's structure. We propose a new approach to decomposing this control into physically meaningful ET *tubes*. We use use the approach to analyze three recent ET experiments in Ru-modified azurin. The experiments are used in a new way to determine the proper effective electron tunneling energy to use in the model, as well as to show how tube interference can control ET rates. The critical role played by H bonds in electronic coupling is confirmed. ET reactions can best be understood in terms of interfering tubes. As tubes can be blocked or created by mutation, the theory suggests how experimental control of rates can be acheived.

**Key words:** azurin, bridge assisted tunneling, electron transfer, pathways, tubes

# 1 Introduction

Multiple site experiments[1] have led to a new theoretical approach for ET beyond the single-pathway picture[2, 3, 4]. In some cases, for a given donor $D$ and acceptor $A$, the transfer can be thought of as "pathway-like", wherein the protein bridge can be physically reduced to a *tube* or tightly grouped family of pathways, without changing the overall coupling. In other cases, the transfer is characterized by multiple tubes, which can interfere with one another. Reducing the protein to only the relevant parts (tubes) that mediate the tunneling matrix element is a useful tool for understanding ET in a biological medium.

# 2 The Electron Transfer Model

The ET model used here arises from the standard non-adiabatic expression

$$k_{ET} = \frac{2\pi}{\hbar} \, |T_{DA}|^2 \, F.C., \tag{1}$$

where $F.C.$ is the Franck-Condon density of nuclear states, and $T_{DA}$ is the tunneling matrix element

$$T_{DA}(E_{tun}) = \sum_{d,a} \beta_{Dd}\, G_{da}(E_{tun})\, \beta_{aA}, \tag{2}$$

$$G = \frac{1}{E_{tun} - H}. \tag{3}$$

In the couplings discussed here, differences in $T_{DA}$ are expected to be much larger than differences in the $F.C.$ term, because the rates reported here are $k_{max}$ with a $F.C.$ factor of order unity. Also, from experiment to experiment the local environment of $D$ and $A$ stays the same while the intervening medium changes. One therefore expects the relative rates to be determined by the tunneling matrix element $T_{DA}$ alone.

All electronic properties of the protein are contained in $H$, a single-electron tight-binding[5] Hamiltonian, representing the protein. A "state" in this system is an electron residing in a particular tight-binding site, and the only sites used are the $\sigma$-bonding orbitals and lonepair orbitals in the protein matrix, and an orbital centered on both the Cu and the Ru. These orbitals are simply localization sites, and are not treated in any further detail; $H$ is just a large extended-Hückel-like matrix, with a dimension equal to the number of orbitals recognized in the protein. An off-diagonal element in $H$ is the coupling between the two states, and is directly related to the probability that an electron will move between the two sites involved. Two of the sites in the protein are special, in that they are associated with the $D$ and $A$ states, while the remaining states in the protein are collectively referred to as the *bridge* ($H$ is partitioned into a $DA$ subsystem and a bridge subsystem). The sum in Eqn. 2 is over the bridge entrance and exit states, indiced by $d$ and $a$ respectively. These are the states with *direct* coupling to $D$ and/or $A$. When the energies of $D$ and $A$ are close to each other, relative to their distance in energy from the closest bridge state, and when coupling to the bridge is small relative to this distance, then[6] the $DA$ subsystem can be viewed as an effective two-state Hamiltonian with a coupling determined by virtual occupation of the bridge. The so-called tunneling energy $E_{tun}$ is an energy parameter indicative of the energies of the $D$ and $A$ states. All bridge states have energy zero on the energy scale used here. The direct coupling between two orbitals which share the same atom (a covalent link) is taken as a constant $\gamma$ (a pathway-like Hamiltonian).

In this model a "hydrogen bond" (H bond) is an interaction between a $\sigma$ bonding orbital (between a heavy atom and a hydrogen) and a lone-pair orbital on another heavy atom. If one arranges $H$ so that the diagonal is ordered like the amino acid sequence, then H bonds (and through-space jumps) are far-off diagonal elements of $H$. Previous experience has shown that H bonds are vital for mediating ET in proteins, and for strong H bonds, such as those involved in protein secondary structure, our current results indicate that their contribution to ET is comparable to that provided by covalent links. H bond couplings are treated as distance independent covalent links (providing a direct coupling of $\gamma$). Recent experimental results[7, 8] support this hypothesis. The Hamiltonian used here has only covalent links and H bonds; no through-space jumps are important.

The upshot of this is that the theory models the protein bridge with precisely one parameter; the ratio $\gamma/E_{tun}$. All of the covalent bonds and H bonds in the bridge are treated as equivalant (i.e., they are all $\gamma$), meaning that the Hamiltonian matrix $H$ used with $E_{tun}$ to compute $G = 1/(E_{tun} - H)$ is just $\gamma$ times a sparse matrix of 1's. The ratio $\gamma/E_{tun}$ appears in expansions of $G$ matrix elements. *This is a highly simplified picture of the protein.* Despite its simplicitly, the $H$ used here exhibits the primary features required for this problem. There is a rough exponential decay of coupling with distance (down a tube), the coupling sign changes[9] with each step from orbital to orbital (because $\gamma/E_{tun} < 0$), quantum interference effects (interfering tubes) arise that have a direct connection to the secondary and tertiary structure of the protein, and most importantly, the computed ratios of $T_{DA}$ are within an order of magnitude of experimental rate ratios (see Sec. 3). More complicated Hamiltonians could and should be used, but they will only be understood in terms of the basic features already present in this Hamiltonian.

## 2.1 $G_{da}$ as a Sum of Pathways

A *pathway* is a specific sequence of *bridge* orbitals, starting at a site $d$ (which is directly coupled to the donor $D$), and ending at a site $a$ (directly coupled to $A$) — e.g., the -N-$C_\alpha$-C-N-$C_\alpha$-C- bond sequence of a protein backbone could be a segment of a pathway. The determination of the *relevance* of a particular pathway is discussed below in Sec. 2.4; here paths are simply defined and discussed in general terms. Consider a bridge with only two states, as in Fig. 1, where the donor is coupled to only one of the states and the acceptor is coupled to the other. Because of this, the sum in Eqn. 2 has only one term; $T_{DA}(E_{tun}) = \beta_{Dd}\, G_{da}(E_{tun})\, \beta_{aA}$. Recognizing the bridge as a two-state subsystem, for which the desired Green's function matrix element has a simple form, we have

$$\begin{aligned} G_{da}(E) &= \frac{\gamma_{da}}{(E-E_1)(E-E_2)-\gamma_{da}^2} \\ &= \frac{\gamma_{da}}{(E-E_1)(E-E_2)} \\ &\quad \times \left(1+\frac{\gamma_{da}^2}{(E-E_1)(E-E_2)}+\frac{\gamma_{da}^4}{((E-E_1)(E-E_2))^2}+\cdots\right) \quad (4) \end{aligned}$$

The $G_{da}$ matrix element can be written as a sum of terms, each of which corresponds to a specific sequence of states (a *pathway*). The terms in this sum are depicted in Fig. 1. In this simple case of a two-state bridge, each term multiplies in another factor of $\gamma^2$, representing a trip from $a$ back to $d$ and back to $a$ again. More complicated bridges are harder to expand this way, but the same idea applies. This "sum of pathways" view shows the potential role of interference in this model. There will be interference effects buried in the calculation of an individual $G_{da}$ (as in this example)

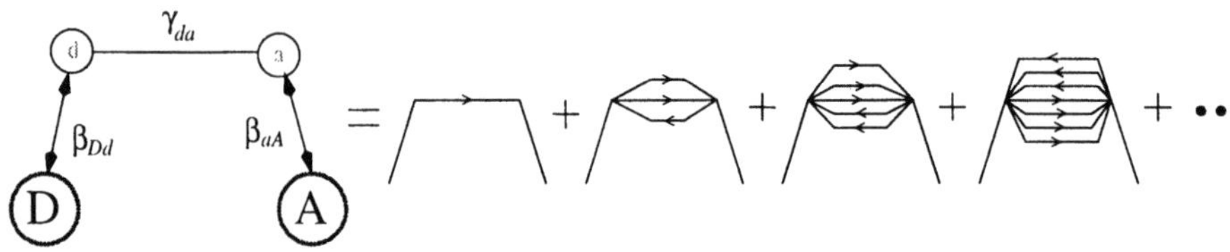


Fig. 1. A simple two-state bridge, with its effects in $G_{da}$ expanded as a sum of pathways.

and there will be interference effects in the $T_{DA}$ sum, where different $G_{da}$ matrix elements between different bridge exit and entrance points are added. Although these two categories are useful, it is more useful to define two categories of interference effects which are closer to each other than these two extremes.

## 2.2 The Pathway Tube

The first category of interference is called trivial interference, and this is the interference that arises from nearest-neighbor, next-nearest-neighbor and backscatter effects in propagation down a simple structure, like an ideal linear alkane chain or a protein backbone, centered on a core pathway. The coupling provided by a protein backbone is a much stronger function of backbone length than it is of the types of residues encountered along the backbone. The amide hydrogens, the lone-pairs on the oxygens, even the residues themselves, all provide small alternative pathways that interfere in a way that can easily be renormalized[10] into a much simpler set of states with the connectivity of a string of pearls. An example of this is shown in Fig. 2, to be discussed in Sec. 2.4.

A *pathway tube* is the set of states one finds by first identifying a core pathway (between some $d$ and $a$) which never visits the same state twice (Sec. 2.4), then adding to this set all nearest neighbors of the core states, then again adding the neighbors of these extra states. This catches all hanging orbitals off the core pathway, and this subset of the bridge is called a pathway *tube*. To find multiple tubes, a generalization of this method is used that differentiates between different tubes. Such tubes are shown in Fig. 6. A tube is a centrally useful concept from an experimental point of view, in that can sometimes be blocked, or created, via molecular replacement.

## 2.3 The Sum of Tubes

The second category of interference is that of interfering tubes. Once tubes have been defined, one can ask how they interfere. Whereas the latter kind of interference (trivial) is buried deep in the calculation of $G_{da}$, the effect of this kind of interference can be thought of as sitting at the coarse level of the $T_{DA}$ sum, or just below it. Sometimes there will be only one tube per *d*–*a* pair, in which case each $G_{da}$ itself represents a tube, and the tube interference is explicitly the $T_{DA}$ sum. Sometimes, however, there will be multiple tubes per $G_{da}$ matrix element.

## 2.4 The *Pathway* Approximation

The *pathways* approach[11] provides a way to find the *virtual* route through the protein matrix which contributes the most to the electronic coupling, and easily estimate the coupling provided by this route by turning the Green's function calculation into a simple scalar product along the path rather than an inversion of $H$. If a single tube dominates, then this product of decay factors is a good estimate of the coupling, as described below.

### 2.4.1 Renormalization

Any bridge can be "renormalized" using matrix partitioning[12]. The idea is to simply divide up the bridge into the states one wants to keep (one will always want to keep the states at the bridge exit and entrance points for example), and the states one wants to eliminate. These latter states are removed from the bridge, but their effect on the calculation of any matrix elements between the remaining states is perfectly accounted for by "renormalizing" the energies of these remaining states and the coupling between them. In general, one might want to eliminate states to clarify structural elements in the protein (e.g. reduce a residue to a single state), or to simplify later calculations (see Sec. 2.4.2).

For example, consider the alkane molecule of Fig. 2a. The $\sigma$-bonding orbitals between the C-C and C-H pairs have been identified with energies $\alpha_c$ and $\alpha_h$ respectively. Three classes of coupling have also been identified: $\gamma_c$, between C-C and C-C bonds; $\gamma_h$, between C-H and C-H bonds; and $\gamma_x$, between C-C and C-H bonds.

Consider the four orbital subsystem outlined in Fig. 2a, wherein the two C-H bonding orbitals physically located between the two C-C bonding orbitals are coupled to each other and to the C-C orbitals, but are otherwise isolated. From the point of view of the rest of the system, only the C-C bonding orbitals are visible, and one can define new energies and couplings for these orbitals which will include the effects of the C-H bonds implicitly. This permits one to reduce the system as shown in Fig. 2b-c. The original

Hamiltonian

$$H = \begin{bmatrix} \alpha_c & \gamma_x & \gamma_x & \gamma_c & & & & & \cdots \\ \gamma_x & \alpha_h & \gamma_h & \gamma_x & & & & & \\ \gamma_x & \gamma_h & \alpha_h & \gamma_x & & & & & \\ \gamma_c & \gamma_x & \gamma_x & \alpha_c & \gamma_x & \gamma_x & \gamma_c & & \\ & & & \gamma_x & \alpha_h & \gamma_h & \gamma_x & & \\ & & & \gamma_x & \gamma_h & \alpha_h & \gamma_x & & \\ & & & \gamma_c & \gamma_x & \gamma_x & \alpha_c & \gamma_x & \\ \vdots & & & & & & \gamma_x & & \ddots \end{bmatrix} \tag{5}$$

becomes the effective Hamiltonian

$$H^{eff} = \begin{bmatrix} \alpha_0 & \gamma & & & & & \\ \gamma & \alpha & \gamma & & & & \\ & \gamma & \alpha & \gamma & & & \\ & & \gamma & \alpha & \gamma & & \\ & & & \gamma & \ddots & \gamma & \\ & & & & \gamma & \alpha & \gamma \\ & & & & & \gamma & \alpha_0 \end{bmatrix} \tag{6}$$

whose matrix elements are functions of $E$:

$$\gamma = \gamma^{eff}(E) = \gamma_c + 2\gamma_x^2\, g(E), \tag{7}$$

$$\alpha = \alpha^{eff}(E) = \alpha_c + 4\gamma_c^2\, g(E), \tag{8}$$

$$\alpha_0 = \alpha_0^{eff}(E) = \alpha_c + 2\gamma_c^2\, g(E), \tag{9}$$

$$g(E) \equiv \frac{1}{E - \alpha_h - \gamma_h}.$$

In this simple example, only the endpoint energies are different, and the coupling to the endpoint is the same as the couplings in the middle ($\gamma_0 = \gamma$). The new "Hamiltonian" is a function of $E$; it is only meaningful in the context of finding $G = 1/(E - H)$ matrix elements. The $G$ matrix elements between the states which were retained (in this case the C-C bonding orbitals in an alkane bridge) in the reduced $H$ will match those computed using the full $H$.

The C-C to C-C coupling in the reduced alkane (Eqn. 7) already appears expanded in terms which correspond to the paths in Fig. 3. The first path is just the original $\gamma_c$, the second two terms (paths) correspond to side trips through the bonds to the H's, contributing $\gamma_x\alpha_h\gamma_x$ with a sign opposed to that of the first term, and the last two paths are three steps trips ($\gamma_x\gamma_h\gamma_x$) which contribute the same sign as the first term. A more careful expansion in $\gamma/E$ would reveal backscatter paths, but that level of detail is not needed for the point about to be made. If $\alpha_h$ is zero, $\gamma = \gamma_c + 2\gamma_x^2/(E - \gamma_h)$, a number which has a smaller magnitude than $\gamma_c$ because $\gamma_c$ is negative and $E = E_{tun}$ is positive. This trend will be seen in general; adding side trips reduces

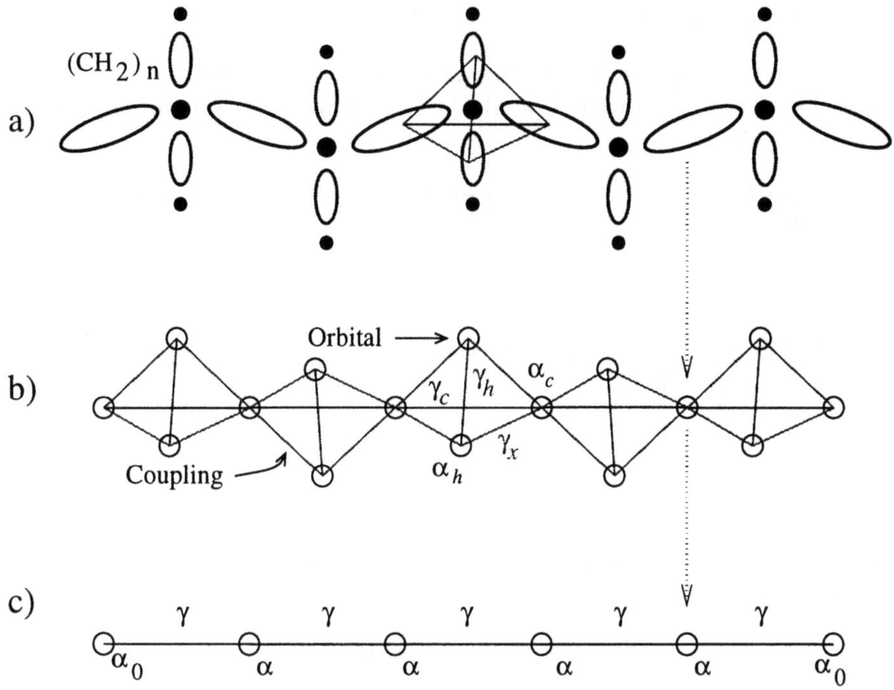


Fig. 2. This figure shows how alkane would be converted to a Hamiltonian used in this model (a→b), and then how this periodic Hamiltonian could be further reduced (b→c) to result in the "string of pearls" $H$ below it. The new energies and couplings in this chain have the effects of the side groups renormalized into them.

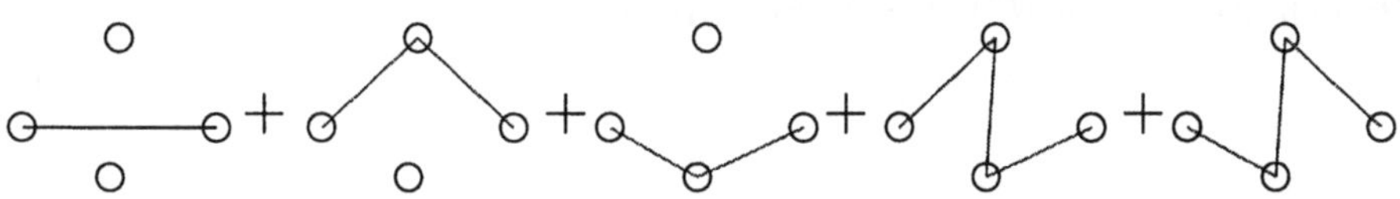

Fig. 3. A path breakdown of the effective C-C coupling (Eqn. 7) in the string of pearls renormalized from alkane.

overall coupling because of this "trivial renormalization" due to destructive interference.

Obviously, one cannot address coupling between the C-H bonds in the reduced $H$, because they no longer exist. This is why, in practice, one must not "normalize away" the states which are directly coupled to $D$ and $A$. Further, one must not "normalize away" so much of the protein that there is nothing left in the Hamiltonian that makes structural sense; this restraint is exercised in the next section, where a very specific partition and renormalization is done.

### 2.4.2 Decay per Step

The previous example suggested how a complicated, periodic bridge could be reduced to a simple, "string of pearls"*. The simplicity of this bridge can be further exploited.

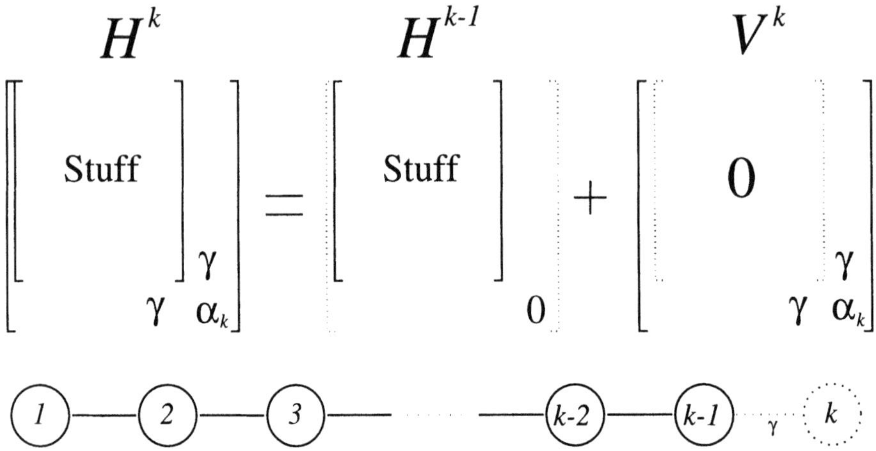


Fig. 4. A "growing" Hamiltonian.

Consider a "growing" Hamiltonian as in Fig. 4 and the following equation, where the superscript $k$ indicates the stage of growth:

$$H^k = H^{k-1} + V^k \tag{10}$$

By definition, $(E - H^k)G^k = 1$ (Eqn. 3), so

$$G^{k-1}(E - H^{k-1} - V^k)G^k = G^{k-1} \tag{11}$$

thus

$$G^k = G^{k-1} + G^{k-1}V^kG^k \tag{12}$$

This is a recurrence relation for $G(E)$, wherein $G^k$ appears on both sides of the equation. Now suppose one is interested only in the matrix elements of $G$ between the endpoints 1 and $k$. By taking matrix elements of the above one finds:

$$G^k_{1,k} = f_k G^{k-1}_{1,k-1} \tag{13}$$

where

$$f_k \equiv \frac{\gamma}{E} \frac{1}{1 - \frac{\gamma}{E} f_{k-1}} \qquad (f_0 \equiv 0) \tag{14}$$

(similar to the result obtained by da Gama[13]). Expanding $f$ in terms of $\gamma/E$ one finds

$$\frac{G^k_{1,k}}{G^{k-1}_{1,k-1}} = f_k \approx \begin{cases} f_2 + O[(\frac{\gamma}{E})^5] & k > 2 \\ f_3 + O[(\frac{\gamma}{E})^7] & k > 3 \end{cases} \tag{15}$$

*A *non-periodic* bridge with some isolated structural elements can also be reduced to a string of pearls, but the site energies and couplings down the chain won't all be the same.

In other words, once one goes beyond 2 or 3 states, the effect of adding another state to this chain-like bridge just multiplies the $G$ matrix element between the endpoints by a nearly constant factor of $f_2$ if $\gamma/E$ is small. This number is less than 1, so each additional step means a decay in the coupling. Further, for a particular choice of $E_{tun}$, this number should generally be negative, because $\gamma/E_{tun}$ is negative, so each step means a sign change — with obvious implications for pathway interference which a single path model cannot take into account.

The *pathways* scheme of Beratan and Onuchic[14] is the following. The protein is converted to a network of sites connected by "$\varepsilon$ couplings". This network is basically like the tight-binding Hamiltonians discussed earlier only the meaning and units of the direct couplings between the tight binding states are different. They define "pathway coupling" as

$$T_{DA} \propto \prod_{N-1} \varepsilon \tag{16}$$

where the $\varepsilon$ is exactly the $f_2$ or $f_3$ defined above, and $N$ is the number of states in the path with which the coupling is associated. The full pathway model allows for defining different $\varepsilon$'s for through-space jumps and H bonds, but this is covered elsewhere[14]. The network can be searched for a path which maximizes this pathway coupling, and in a network where all $\varepsilon$'s are equal, the maximum coupling path will also have the fewest number of steps because $\varepsilon < 1$. If $\varepsilon$ is chosen correctly, the path coupling will exhibit the decay which corresponds to that one would find down a simple covalently linked bridge *with the effects of hanging orbitals and simple back-scatter built in.* Likewise, the path coupling should correspond (up to a constant) to the $G_{da}$ one would compute over the tube whose core is the pathway.

### 2.4.3 Pathways vs. Tubes

The essential idea from the *pathway* approximation is that the Hamiltonian is converted to a searchable network, and one can compute estimates of $T_{DA}$ which are 1) simple products corresponding to the number of bonds involved and 2) expected to be correct (up to a prefactor), *if the coupling provided by the entire protein bridge is well approximated by only the states included on the pathway tube.* The concept of "finding a best path" is dependent on this concept of defining the coupling as a product of constant factors.

The first step in the *tube* approach is to use the pathways idea to isolate routes through the protein which are physically distinct (see example below), and then use these paths as *tube cores.* The analysis then proceeds with an emphasis on partitioning the protein into tubes; the tubes contain trivial interference effects, but can expose crucial interference between tubes. This is something which one cannot do in the single pathway approximation.

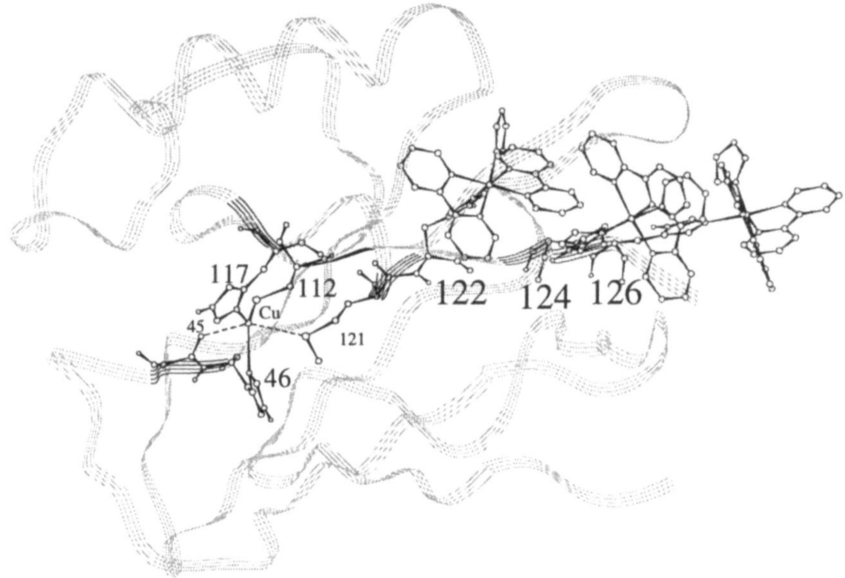


Fig. 5. A model of the azurin molecule showing the copper, its ligands, and $Ru(bpy)_2(im)$(histidine) groups at positions 122, 124 and 126. Each site represents a different experiment.

# 3 An Example in Azurin

The tube approach has been applied recently[15] to some recently measured couplings[1] in blue copper protein azurin. Some highlights of this analysis are provided in this section.

Azurin plays an important role as an electron carrier[16, 17, 18, 19] in biological systems. Based on extensive studies[20, 21, 22], it is likely that bimolecular ET reactions between azurin and other redox molecules take place via histidine 117, as it is directly coupled to the copper and close to the surface of the protein. Much less work has been done on long range ET through the protein, although Farver, Pecht and other investigators[23, 24, 25, 26] have made an impressive start by studying ET between the copper atom and a distant disulfide bridge.

We have applied the tube approach to analyze three distant electronic couplings in ruthenium modified derivatives of azurin. The atomic coordinates used in the calculations of Ru-Cu coupling (Fig. 5) are based on the crystal structure analyses of Nar[27]. The placement of the Ru electron acceptor in these experiments was such that the intervening $D$–$A$ medium was two $\beta$-strands. A best fit to an exponential decay with $D$–$A$ distance for these results yields a decay constant of $1.09Å^{-1}$[1]; this is the softest decay that can be expected for $\sigma$-bond tunneling through a protein because a $\beta$-strand covers the longest distance with the smallest number of bonds. By analyzing ET that goes via more than a single strand, we have been able to show the importance of H bonds in the coupling across a $\beta$-sheet. In addition, we have begun to investigate how the couplings between copper and its different ligands may influence the directionality of long range ET.

The pathway tubes for the case of 126 are shown in Fig. 6. These paths were found by first finding the best path, then finding all paths with a path coupling which offered at least 1% of the coupling of the best path. This search would typically generate *tens of thousands* of paths, each unique in its sequence of states - but most paths would simply be trivial variations of their neighbors. This unmanageable set of paths would then be scanned to extract "tube cores". To do so, a special set of states $S$ is initialized to zero. Then the paths are considered in order from highest path coupling to lowest. During consideration, a path is scanned for its *smallest continuous segment* of states which are not elements of $S$. If this segment is at least $X$% of the entire path (where $X$ is typically 10), then the path is retained, else it is deleted from the set of known paths. If it is retained, then all states on the path and within $N$ steps of the path are added to the set $S$ if they aren't in it already ($N$ is typically 1). This process continues until all known paths have been considered, and the set of paths has been dramatically reduced, usually to order 10. If one finds more than around 10 paths for a protein the size of azurin, it is probable that these paths are not "unique" enough, and that either $X$ or $N$ should be increased.

The tubes for 122 and 124 are subsets of those seen for 126 in Fig. 6. If one calculates $T_{DA}$ over a bridge that consists of just the states in pathway tubes, and if this coupling is the same as the number one finds over the entire protein, then[28] the protein can be reduced (from an ET point of view) to just the tubes, eliminating irrelevant superstructure to expose the important structural features.

The shortest distance, both through-space and through-bond, from the acceptor sites at 122, 124, and 126 back to the copper is directly through the $\beta$-strand to the MET at 121. This MET's sulfur is close 3.2Å[27] to the copper, but is not considered a Cu ligand. According to detailed analysis of the electron structure of the copper in this protein[29], the strongest coupling from the copper to any of its atomic neighbors is to the sulfur in the CYS at position 112. The coupling to the sulfur of 121 is relatively weak.

Tubes lead from the copper down the 112-111-110 strand, and cross over to the 121 strand via the standard H bonds (N–H$\cdots$O) of the anti-parallel $\beta$-sheet which these two strands comprise. Although the bridge tubes which go directly down the 121 strands are shorter, the strong coupling to 112 makes the 112 tubes (which must cross the H bonds) the dominant source of Cu–Ru electronic coupling. The multiple tubes coming down the 112 side all have the same length, and therefore all interfere with the same sign. These multiple

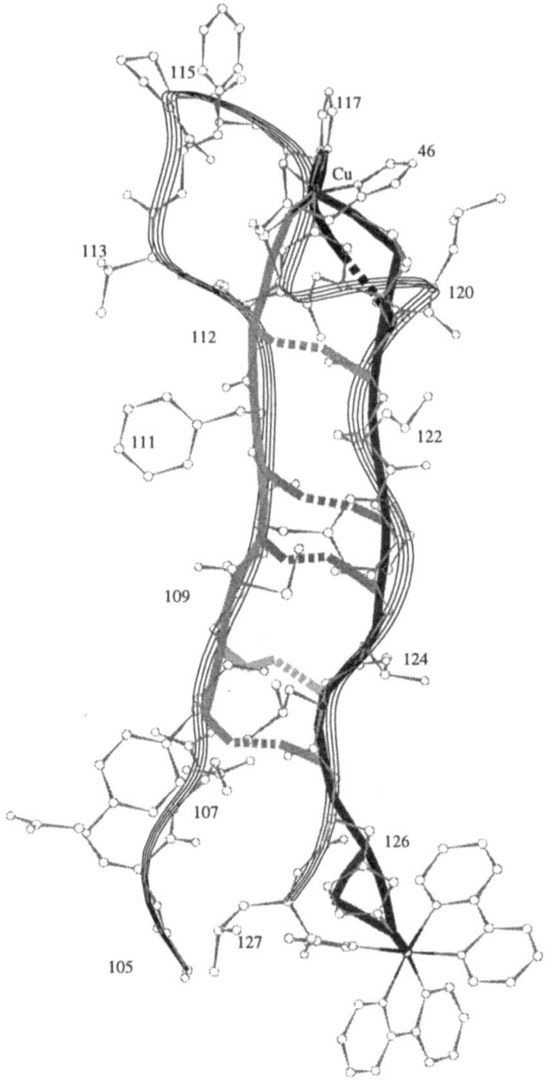


Fig. 6. Sets of pathway tubes for the ET coupling from Cu to $Ru(bpy)_2(im)$ in HIS126 modified azurin. The shaded lines (dashed lines are H-bonds) indicate the cores of the tubes which together are responsible for effectively all of the electronic coupling of the protein matrix. 122 and 124 are like 126, but with a subset of the tubes shown here. The copper ligands are 46, 117 and 112, and the coupling is dominated by the 112 $\beta$-strand.

tubes add together to increase the coupling to all sites on the 121 strand. If there were only one tube providing the coupling, one would expect that the rate ratio of 122/124 would be the same as the ratio 124/126, because both ratios represent roughly the same increase in donor—acceptor separation. However, since 124 has more tubes leading into it that 122, and 126 in turn has more than 124, the multiple tube prediction is that both 122/124 and 124/126 rate ratios will be smaller than what one would see with one tube, and further the rate ratio 124/126 will be smaller than the rate ratio 122/124. This analysis is discussed in more detail elsewhere[15].

# 4 The Tubular Breakdown

A straightforward goal of theories describing ET in proteins is to predict the value of the tunneling matrix element $T_{DA}$. This paper goes a step further, to define what it is about the structure of the protein which determines this value.

Our approach converts a protein, as represented by its atomic coordinates, to a very simple Hamiltonian which has just enough information to retain all essential features of the electron tunneling problem. The coupling derived from this model can be broken down into contributions from individual *tubes*, each of which is a a family of similar pathways through specific sequences of covalent or H bonds. The tubes encapsulate trivial interference effects and can expose crucial inter-tube interferences. The set of all tubes that are important to the coupling can be identified — and the rest of the protein can be dismissed. A purely quantum mechanical effect like tube interference is directly related to the secondary and tertiary structure of the protein, and H bonds play a central role in this effect, as they are the primary factor distinguishing a protein from what would otherwise be an effective (and uninteresting) one-dimensional chain. As proposed in earlier work, the treatment of H bonds on equal footing with covalent bonds leads to theoretical predictions which are consistent with rate measurements, indicating the critical role H bonds can play in ET.

This paper discusses the particular case of $\beta$-strand tubes. However, tube analysis is generally applicable to other motifs in different proteins. Different patterns of tube interferences will exist for different motifs, but the framework for analysis will remain the same. The step from a tube analysis of a protein to experimental design is obvious, as one anticipates that there will be situations where a tube can be created or blocked by an appropriate mutation.

**Acknowledgements:** We thank David Beratan and Spiros Skourtis for helpful comments about this work. This research supported by the NSF (grant MCB-93-16186) and the NIH (GM48043). JJR was also supported by the Berkeley Program in Mathematics in Biology (NSF grant DMS-94-06348).

# References

[1] Langen, R., Chang, I.J., Germanas, J.R., Richards, J.H., Winkler, J.R. and Gray, H.B. (1995) *Science.* In press.

[2] Beratan, D.N., Onuchic, J.N. and Hopfield, J.J. (1987) *J. Chem. Phys.* 86, 4488–4498.

[3] Beratan, D.N., Betts, J.N. and Onuchic, J.N. (1991) *Science* 252, 1285.

[4] Onuchic, J.N., Beratan, D.N., Winkler, J.R. and Gray, H.B. (1992) *Annu. Rev. Biophys. Biomol. Struct.* 21, 349–377.

[5] Ashcroft, N.W. and Mermin, N.D. *Solid State Physics.* W.B.Saunders, New York, (1976).

[6] Skourtis, S.S. and Onuchic, J.N. (1993) *Chem. Phys. Lett.* 209, 171–177.

[7] Turró, C., Chang, C.K., Leroi, G.E., Cukier, R.I. and Nocera, D.G. (1992) *J. Am. Chem. Soc.* 114, 4013.

[8] de Rege, P.J.F., Williams, S.A. and Therien, M.J. (1995). Submitted to *Science.*

[9] Beratan, D.N. (1986) *J. Am. Chem. Soc.* 108, 4321–4326.

[10] Onuchic, J.N. and Beratan, D. (1990) *J. Chem. Phys.* 92, 722–733.

[11] Beratan, D.N. and Onuchic, J.N. (1989) *Photosyn. Res.* 22, 173.

[12] Löwdin, P.O. (1962) *J. Math. Phys.* 3, 969.

[13] Gama, A.A.S.da (1990) *J. Theor. Biol.* 142, 251–260.

[14] Onuchic, J.N., Andrade, P.C.P. and Beratan, D.N. (1991) *J. Chem. Phys.* 95, 1131.

[15] Regan, J.J., Bilio, A.J. Di, Langen, R., Skov, L.K., Winkler, J.R., Gray, H.B. and Onuchic, J.N. (1995) *Chemistry and Biology.* Submitted for publication.

[16] Adman, E.T. and Jensen, L.H. (1981) *Isr. J. Chem.* 21, 8–120.

[17] Gray, H.B. (1986) *Chem. Soc. Rev.* 15, 17–30.

[18] Sykes, A.G. (1991) *Advances Inorg. Chem.* 36, 377–408.

[19] Nar, H., Messerschmidt, A., Huber, R., Van de Kamp, M. and Canters, G.W. (1991) *J. Mol. Biol.* 218, 427–447.

[20] Van de Kamp, M., Floris, R., Hall, F.C. and Canters, G.W. (1990) *J. Am. Chem. Soc.* 112, 907–908.

[21] Van de Kamp, M., Silvestrini, M.C., Brunori, M., Beeumen, J.van , Hall, F.C. and Canters, G.W. (1990) *Eur. J. Biochem.* 194, 109–118.

[22] Mikkelsen, K.V., Skov, L.K., Nar, H. and Farver, O. (1993) *Proc. Natl. Acad. Sci. USA* 90, 5443–5445.

[23] Farver, O. and Pecht, I. (1992) *J. Am. Chem. Soc.* 114, 5764–5767.

[24] Farver, O., Skov, L.K., Nar, H., Van de Kamp, M., Canters, G.W. and Pecht, I. (1992) *Eur. J. Biochem.* 210, 399–403.

[25] Farver, O., Skov, L.K., Pascher, T., Karlsson, B.G., Nordling, M., Lundberg, L.G., Vänngård, T. and Pecht, I. (1993) *Biochemistry* 32, 7317–7322.

[26] Broo, A. and Larsson, S. (1991) *J. Phys. Chem.* 95, 4925–4928.

[27] From the coordinates of chain "A", one of four azurin molecules in the unit cell of Brookhaven Protein Data Bank entry "4AZU"[19, 30].

[28] Regan, J.J., Risser, S.M., Beratan, D.N. and Onuchic, J.N. (1993) *J. Phys. Chem.* 97, 13083–13088.

[29] Solomon, E.I. and Lowery, M.D. (1993) *Science* 259, 1575–1581.

[30] Nar, H., Messerschmidt, A., Huber, R., Van de Kamp, M. and Canters, G.W. (1991) *J. Mol. Biol.* 221, 765–772.

# The role of the protein in the stabilization of the charge separated states in photosynthetic reaction centers

M.R. Gunner

Department of Physics, City College of New York, 138th St. and Convent Ave. New York, NY 10031 U.S.A.

**Abstract.** Electron transfer from the excited singlet state of a bacteriochlorophyll dimer (P) to a bacteriopheophytin ($H_L$) 18 Å away occurs within 2 picoseconds even at cryogenic temperatures in photosynthetic reaction centers. Estimates of the reaction field energy that would stabilize the ion pair $P^+H_L^-$, shows that a medium with a dielectric constant of 2 would be insufficient to place the energy of the ion pair below that of the excited singlet state of P. The contribution of additional dielectric screening appears to be small. Thus, the protein must play an active role in stabilizing the charge separated state. A static field due to pre-existing charges and dipoles is calculated to exist within reaction centers which is negative near P and positive near H. This field stabilizes $P^+H_L^-$ so that is lies below P*.

**Keywords.** electron transfer, electrostatics, reaction centers

## 1 Introduction

In the first steps in photosynthesis an absorbed photon provides the energy needed to transfer an electron across the reaction center protein (RC) embedded in a membrane. This process stores the photon's energy as a transmembrane electrical gradient. The formation of the first stable intermediate involves electron transfer from the excited singlet state of a dimer of bacteriochlorophyll (P) to a bacteriopheophytin ($H_L$), creating $P^+H_L^-$. The positive and negative charge in this ion pair are separated by 15 Å. The free energy difference between P* and $P^+H_L^-$ is approximately 250 meV

(1, 2). The reaction occurs in 2 ps and is slightly faster at cryogenic temperatures than at room temperature (3, 4). The protein plays several obvious roles in the process of charge separation. It provides a scaffolding holding the cofactors in the proper orientation. In addition, the protein and surrounding membrane keeps the cofactors isolated from water. The hydrophobicity of the amino acids at the exterior of the intra-membrane portion of the protein keeps a detergent sheath around the protein even when it is isolated from the membrane (5). The rigid orientation of the cofactors and their separation from the highly polarizable water keep the reorganization energy for electron transfer small. This is needed to maintain the fast electron transfer rate given the small driving force for the reaction (6). However, in this picture the protein plays only a passive role, which could in principle be duplicated by a bridged donor-acceptor complex immersed in a medium with a low dielectric constant. However, this paper will consider the difficulty of creating an ion pair rapidly, in a medium with a small dielectric constant. It will become clear that the protein plays a much more active role in ensuring the observed high quantum yield for the long distance electron transfer in reaction centers.

In any ion pair such as $P^+H_L^-$, the attraction between the unlike charges will always disfavor the charge separated states relative to neutral states such as P* or the ground state. Several studies in model systems have demonstrated the difficulty of forming stable, charge separated states in low dielectric, non-polar solvents (7-10). The particular difficulty of charge separation in frozen media has been shown to be due to the loss of dielectric relaxation as the solvent is frozen which destabilizes the charge separated state (7, 11). For example, in a porphryin-triptycene accepter system with an 11 Å donor-acceptor separation, the charge separated state is 800 meV higher in energy in frozen 2-methyltetra–hydrofuran ($\varepsilon = 2.6$) than in liquid butyronitrile ($\varepsilon = 20$) (11). In these model systems the difference in the electrochemical midpoint between donor and accepter must be greatly increased to ensure that the charge separated state is lower in energy than the initial excited state of the electron donor.

Thus, charge separation to create an ion pair requires free energy terms that can overcome the Coulomb attraction between the charges. The question is how RCs ensures that $P^+H_L^-$ is lower in energy than P*. Possible mechanism to make charge separation favorable are: (1) make the difference in intrinsic electron affinity of the electron donor and acceptor large; (2) make the reorganization of the medium sufficient to solvate and screen the ion pair; or (3) carry out charge separation in an appropriately oriented, preexisting, static field. The first two terms refer to the electrochemistry of the cofactors themselves and to the dielectric properties of a homogeneous medium. If the magnitude of these terms are sufficient, electron transfer could occur within a model system that duplicated the complex of the RC cofactors in the absence of the protein. In addition, the charges and dipoles in the protein can produce a static field. It is this last function of the protein that cannot be mimicked by a homogeneous medium. The problem is therefore to determine the relative contribution of the three factors to the free energy difference between $P^*$ and $P^+H_L^-$ in RCs.

## 2 Methods

Calculations were carried out on the Rps. viridis RC structure 1PRC from the Brookhaven Data Bank (12, 13). All crystallographic water molecules were treated as part of the protein. The influence of the membrane was estimated by including a cube 72x72x33 Å of low dielectric material surrounding the protein. Calculations reported for the isolated cofactors include all six cofactors and the 4 histidines that are axial ligands to the bacteriochlorophylls. The phytyl tails of the bacteriochlorophylls and bacteriopheophytins are truncated at the ether link between the atoms O2A and C1.

The program DelPhi solves Poisson's equation for the potential in and around a protein given a distribution of charges and of regions of water with a high dielectric constant and protein with a low dielectric constant (14, 15). The RC structure is combined with atomic radii to define the shape of the region that is removed from water. The charge distribution is defined by CHARMM charges placed on each atom center. The potential was calculated at each cofactor (see references 16 and 17 for a more complete account of the method). All ionizable residues except for Glu L 104 were assumed to be ionized. The charges on the neutral and ionized cofactors were taken from reference 18. DelPhi also provides the reaction field or solvation energy of the isolated cofactors, which is the free energy of transfering the isolated cofactors from vacuum to a medium with a higher dielectric constant (19).

## 3 Results and discussion

The free energy of charge separation in the protein can be estimated as the sum of three terms. These are (1) the free energy difference between reactant and product states in vacuum ($\Delta G_{vacuum}$), (2) the difference in the free energy for transfering the reactant and product states from vacuum into a homogeneous medium with the dielectric response of the protein ($\Delta G_{reaction\ field}$) and (3) the difference in the interaction of reactant and product states with charges on monopoles and dipoles in the protein ($\Delta G_{static\ field}$).

The free energy of charge separation in vacuum is appropriately determined by a quantum mechanical analysis. The calculations of Thompson and Zerner put $P^+H_L^-$ 1.13 eV higher in energy than P* (20). The estimates of Warshel and colleagues find comparable values (21, 22). Thus, charge separation would not occur in vacuum since the difference in electron affinity of the bacteriochlorophyll dimer and the bacteriopheophytin is not large enough to overcome the Coulomb attraction within the ion pair $P^+H_L^-$.

The reaction field energy of the medium will stabilize the ion pair. An estimate of $\Delta G_{reaction\ field}$ can be obtained from a calculation of the difference in the total energy of the system when the dielectric constant inside and outside of the cofactors is equal and when the complex is immersed in a medium with the appropriate external dielectric constant (19). Here we are interested in the free energy of transfering the cofactors from vacuum into a medium with the effective dielectric response of the protein.

Figure 1 shows the reaction field energy calculated with the 6 cofactors and the 4 histidines that are axial ligands to the BChls isolated from the protein. The region defined by these atoms has a dielectric constant of 1. An external dielectric constant of 2 provides an estimate of the instantaneous response of electrons in the surroundings to the charge transfer. The electronic response is expected to be independent of temperature. The reaction field energy with $\varepsilon$=2 stabilizes the ion pair $P^+H_L^-$ relative to P* by -0.78 eV. This is a significant value, but still places $P^+H_L^-$ -0.35 V higher in energy than P*. Clearly as shown in Figure 1, a relatively small increase in the effective dielectric response of the protein would be sufficient to make charge separation favorable. An exterior dielectric constant of 3 provides the additional -0.35 eV needed to make P* and $P^+H_L^-$ isoenergetic. If the dielectric constant is raised from 2 to 5 the added stabilization is -0.6 eV so that the ion pair would be -0.25 eV below P*, comparable to the measured value for this reaction (1, 2).

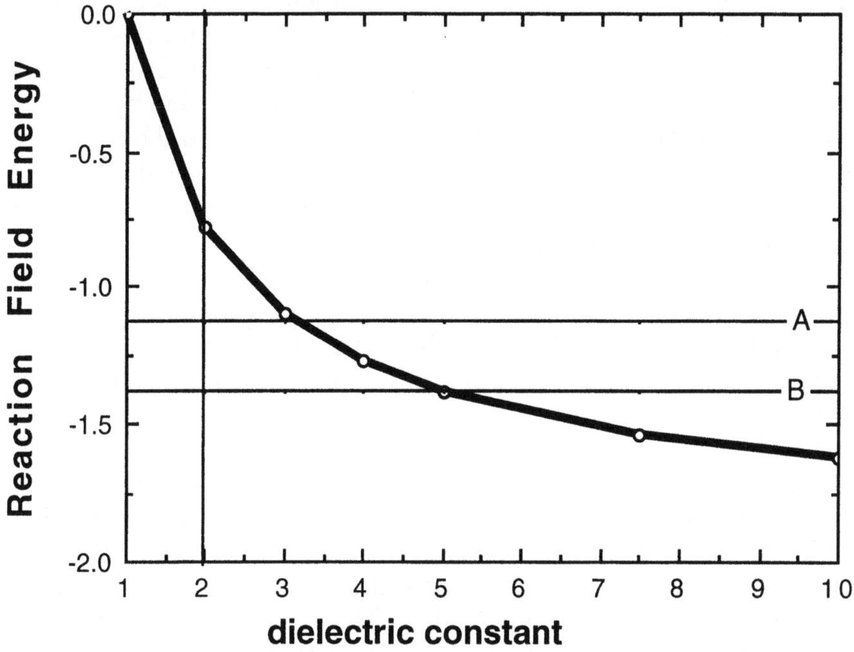


Figure 1: The difference in reaction field energy between P* and $P^+H_L^-$ as a function of the dielectric constant. The reaction field energy of P* is estimated from the ground state of P, which has the charge of a BChl on each half of the dimer. $P^+$ adds 1/2 the charge of $BChl^+$ to the charges on P*. Charges are taken from reference 18. The line A at -1.13 V shows the $\Delta G_{reaction\ field}$ needed to make P* and $P^+H_L^-$ isoenergetic. A reaction field energy of -1.38 eV (line B) would place $P^+H_L^-$ -0.25 eV below P*.

There is however evidence that charge separation is not significantly influenced by processes that can be modeled by a dielectric constant much higher than 2. The rearrangement of the medium that stabilizes the ion pair is also responsible for the outer sphere reorganization energy of the reaction ($\lambda_{out}$). The standard equation for $\lambda_{out}$ for two spherical ions with charge q, radii $r_D$ and $r_A$ separated by $R_{DA}$:

$$\lambda_{out} = q^2 \ (1/2r_D + 1/2r_A - 1/r_{DA}) \ (1/\varepsilon_{optical} - 1/\varepsilon_{medium})$$

is identical to the expression for calculating the additional reaction field energy when the dielectric constant is raised from 2 ($\varepsilon_{optical}$) to $\varepsilon_{medium}$. The total reorganization energy is generally larger than

$\lambda_{out}$ since an electron transfer reaction is expected to be accompanied by rearrangements of acceptor and donor themselves which are not included in this term (6). The reorganization energy for this electron transfer reaction is estimated to be 0.25 V or less (23, 24). Thus, the maximum reaction field energy beyond that provided by electronic polarization should be -0.25 eV. This would be sufficient to place $P^+H_L^-$ slightly above P*. However, this represents an upper limit for $\lambda_{out}$ and smaller values would result in a higher energy for the ion pair.

However, the protein can stabilize the charge separated state without increasing the reorganization energy by generating a preexisting static field. Figure 2 shows the potential at a slice through the protein due to the charges in the protein prior to electron transfer. P is found in a region of negative potential while both bacteriopheophytins are found in regions of positive potential. This arrangement clearly favors electron transfer. The difference in potential stabilizes the ion pair $P^+H_L^-$ by -0.78 eV. Combining this value of $\Delta G_{static\ field}$ with $\Delta G_{vacuum}$ of 1.13 and a $\Delta G_{reaction\ field}$ of -0.78 provides a $\Delta G$ for charge separation at -0.43 eV, -0.18 more negative than the experimental value. This error is certainly within the uncertainty of the analysis. Thus, the structure of the protein provides evidence for a mechanism that can provide efficient, long range electron transfer even in a medium with a very small dielectric response.

Figure 2 may also provide a simple reason for the observed direction of electron transfer that exclusively forms $P^+H_L^-$ and never $P^+H_M^-$ (see reference 17 for a more complete description). Both bacteriopheophytins are found in regions of positive potential, although the region at $H_L$ is 0.14 V more positive than $H_M$. Thus, $P^+H_L^-$ and $P^+H_M^-$ are lower in energy than P*, but reduction of $H_L$ is favored . However, there is an 0.35 V difference in potential at the bacteriochlorophyll monomers. $B_L$ is in a region of weak positive potential while $B_M$ is in a region which is 0.08 V more negative than P. Thus, a pathway of electron transfer via $B_L$ is much more favorable than one via $B_M$. The discrimination will hold if B is a real intermediate or facilitates electron transfer by super-exchange mechanism.

## 4 Conclusion

Reaction centers are designed to obtain a high quantum yield while storing a significant fraction of the free energy of the absorbed photon in the final reactants. If RCs have an effective, low dielectric

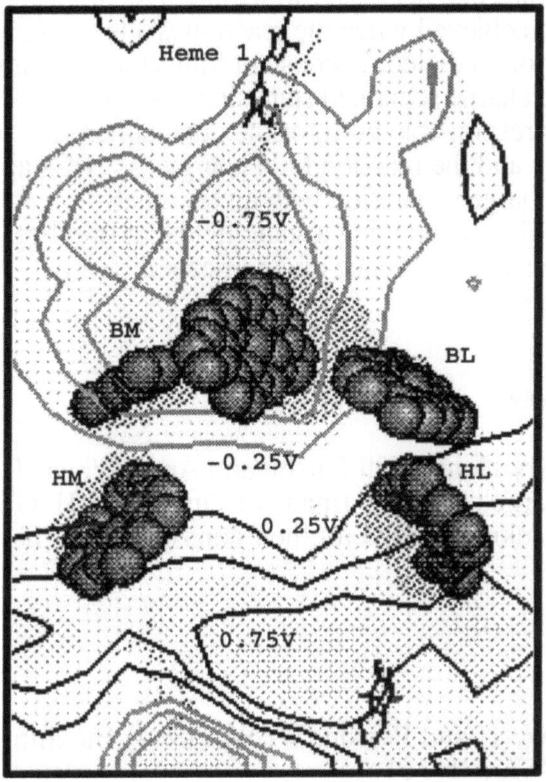


**Figure 2: Electrostatic potential at a slice through Rb. viridis RCs. The slice plane is semi–transparant. PL lies below the plane. Contour lines are drawn at +/–0.25, +/–0.50, and +/–0.75 V. Darker contours are for positive potentials. All ionizable residues except Glu L 104 are assumed to be ionized. The internal dielectric constant is 2 and external dielectric contant is 80. (Figure produced with GRASP A. Nicholls and B. Honig)**

constant the fast forward rate necessary to maintain the high quantum yield can be achieved when the reaction ΔG is small. However, any reduction in dielectric constant limits the mediums ability to stabilize the charge separated state since the solvent dielectric response is responsible both for the outer sphere reorganization energy and the reaction field energy (6). RCs appear designed to achieve charge separation with a very small reorganization energy by maintaining a static field that stabilizes the charge separated states in the absence of significant dielectric response of the protein.

## 5 Acknowledgements

Discussions with Barry Honig, Bill Parson and Arieh Warshel, the questions of Kurt Warncke and the financial support N.I.H. GM-48726 are gratefully acknowledged for their contribution to this work.

## 6 References

1. Woodbury, N. W. T. and Parson, W. W. (1984) Biochim. Biophys. Acta 767, 345-361
2. Horber, J. K. H., Gobel, W., Ogrodnik, A., Michel-Beyerle, M. E. and Cogdell, R. J. (1986) FEBS Lett. 198, 273-278
3. Woodbury, N. W., Becker, M., Middendorf, D. and Parson, W. W. (1985) Biochemistry 24, 7516-7521
4. Fleming, G. R., Martin, J. L. and Breton, J. (1988) Nature 333, 190-192
5. Roth, M., Lewit-Bentley, A., Michel, H., Deisenhofer, J., Huber, R. and Oesterhelt, D. (1989) Nature 340, 659-662
6. Marcus, R. A. and Sutin, N. (1985) Biochim. Biophys. Acta 811, 265-322
7. Overfield, R. E., Scherz, A., Kaufmann, K. J. and Wasielewski, M. R. (1983) J. Amer. Chem. Soc. 105, 5747-5752
8. Harrison, R. J., Pearce, B., Beddard, G. S., Cowan, J. A. and Sanders, J. K. M. (1987) Chem. Phys. 116, 439-448
9. Antolovich, M., Keyte, P. J., Oliver, A. M., Paddon-Row, M. N., Kroon, J., Verhoeven, J. W., Jonker, S. A. and Warman, J. M. (1991) J. Phys. Chem. 95, 1933-1941
10. Warman, J. M., Smit, K. J., deHaas, M. P., Jonker, S. A., Paddon-Row, M. N., Oliver, A. M., Kroon, J., Overering, H. and Verhoeven, J. W. (1991) J. Phys. Chem. 95, 1979-1987

11. Gaines, G. L., O'Neil, M. P., Svec, W. A., Niemczyk, M. P. and Wasielewski, M. R. (1990) J. Am. Chem. Soc. 113, 719-721
12. Bernstein, F. C., Koetzle, T. F., Williams, G. J. B., Meyer, E. F., Brice, M. D., Rodgers, J. R., Kennard, O., Shimanouchi, T. F. and Tasumi, M. (1977) J. Mol. Biol. 112, 535-542
13. Deisenhofer, J. and Michel, H. (1989) The EMBO Journal 8, 2149-2170
14. Sharp, K. A. and Honig, B. (1990) Annu. Rev. Biophys. Biophys. Chem. 19, 301-332
15. Honig, B. and Nicholls, A. (1995) Science 268, 1144-1149
16. Gunner, M. R. and Honig, B. (1991) Proc. Natl. Acad. Sci. USA 88, 9151-9155
17. Gunner, M. R., Nicholls, A. and Honig, B. (1995) J. Phys. Chem submitted.
18. Parson, W. W., Chu, Z.-T. and Warshel, A. (1990) Biochim. Biophys. Acta 1017, 251-272
19. Gilson, M. K. and Honig, B. (1988) Proteins 4, 7-18
20. Thompson, M. A. and Zerner, M. C. (1991) J. Am. Chem. Soc. 113, 8210-8215
21. Scherer, P. O. J. and Fischer, S. F. (1989) J. Phys. Chem. 93, 1633-1637
22. Alden, R. G., Parson, W. W., Chu, Z. T. and Warshel, A. (1995) J. Am. Chem. Soc. submitted,
23. Warshel, A. (1980) Proc. Natl. Acad. Sci. USA 77, 3105-3109
24. Woodbury, N. W., Peloquin, J. M., Alden, R. G., Lin, X., Lin, S., Taguchi, A. K. W., Williams, J. C. and Allen, J. P. (1994) Biochemistry 33, 8101-8112
25. Bockris, J. O. and Reddy, A. K. N., *Modern Eletrochemistry* (Plenum, New York, 1973), vol. 1.

# Shifts of the Special Pair Redox Potential of Mutants of *Rhodobacter sphaeroides* Calculated with DelPhi and CHARMM Energy Functions

**G.M. Ullmann[1], I. Muegge[1,2], and E.W. Knapp[1]**

[1] Freie Universität Berlin, Fachbereich Chemie, Institut für Kristallographie, Takustraße 6, D-14195 Berlin, Germany
[2] University of Southern California, Dept. of Chemistry, Los Angeles, CA 90089-1062, USA

**Abstract.** The shifts of the special pair redox potential of the photosynthetic reaction center of Rhodobacter sphaeroides are evaluated for point mutations in the neighborhood of the special pair and compared with experimental data. The shifts are computed from electrostatic energies using DelPhi and CHARMM. With the CHARMM energy function contributions of Van der Waals interactions are also considered. The influence of water molecules on the computed values of the shift is investigated. Agreement of the calculated experimental values of the shift can be obtained with DelPhi and CHARMM. With CHARMM the agreement is generally better if electrostatic energies are considered only. Including crystal waters and filling the protein cavities with water molecules is also improving the results obtained with the CHARMM energy function.

**Keywords.** Redox potential, mutants, special pair, electrostatic energies, Van der Waals energies, reaction center, *Rhodobacter sphaeroides*, DelPhi, CHARMM

## 1. Introduction

The photosynthetic reaction center (RC) catalyzes the primary events of photosynthesis. Since several years the crystal structures of RC's from two different organisms *Rhodopseudomonas (Rps.) viridis*[1,2] and *Rhodobacter (Rb.) sphaeroides*[3,4,5] are available. This knowledge has initiated many theoretical studies with the aim to understand the function and the specific role played by the cofactors and their environment[6-10]. Among the functionally relevant problems which have been investigated in the past are the following:

i) elucidation of the optical spectra of the chromophores in the RC which is essential for a detailed understanding of the excitation transfer from the antenna pigments to the special pair (D),
ii) simulation of the various electron transfer processes starting from the special pair, where the charge separation takes place,

iii) and of the proton up-take reactions at the secondary quinone $Q_B$.

The role of the cofactor environment on the functional processes can best be studied by replacing amino acids in the neighborhood of the cofactors with the help of site directed mutagenesis[11-15]. This allows to change the hydrogen bonding pattern and the hydrophobicity in the neighborhood of the special pair of the RC from *Rb. sphaeroides* in a definite way[16-22]. For several of these mutants crystals structures are also available with a resolution between 3-4Å[23]. Those structures are not very different from that of the wild-type RC.

These changes in the protein environment of the special pair have considerable consequences for the location of the special pair band in optical spectra[24], the spin density of the special pair radical $D^+$ measured by ENDOR spectroscopy[25] and lead to shifts of the special pair redox potential[21,22]. This redox potential determines the driving force of electron transfer processes involving the special pair in RC's.

In the present work the shifts of the redox potential of the special pair of RC's are calculated by solving Poisson's equation with the program DelPhi[26-28]. Alternatively the energy function from the molecular dynamics program CHARMM[29] is used to evaluate the shifts of the redox potential from electrostatic and Van der Waals energy terms. The role of explicit water molecules in protein cavities and at the surface of the RC is also investigated.

## 2. Methods

The DelPhi and CHARMM energy functions are used to calculate the shifts of the special pair redox potential of mutants from the RC. The computations are based on the new crystal structure of *Rb. sphaeroides*[5]. Hydrogen atoms are added with CHARMM, the cofactors bacteriochlorophyll a (BChl) and bacteriopheophytin a (BPh) are modeled in an all atom representation. Hydrogen atoms of the amino acid residues are added only in polar bonds. All other hydrogen atoms are represented by corresponding extended atom types. All atoms of residues and cofactors which are completely outside of a sphere of radius 18.5Å centered at the geometrical midpoint of residues L168 and M197 are spatially fixed. To minimize deviations from the crystal structure during energy minimizations the non-hydrogen atoms of the special pair are also fixed. For the same reason protein cavities inside the sphere of 18.5Å radius are filled with water molecules before energy minimization.

To probe the conformations of the ring V acetyl groups of the special pair BChl's ($D_L$, $D_M$) the acetyl group atoms are kept mobile with respect to their torsion angle $\gamma$. To increase the mobility of the acetyl groups the force constant k of the torsion potential

$$V_{torsion}(\gamma) = k(1 - \cos 2\gamma) \qquad (1)$$

is reduced from $k_S$=2kcal/mol, its value in the CHARMM energy function, to $k_W$=0.5 kcal/mol. Using the larger force constant $k_S$ the calculated redox potentials do not agree well with the experimental data[30]. The torsion potential eq. (1), possesses minima for the orientations $\gamma=0,\pi$ corresponding to conformations where the acetyl group atoms are in the porphyrin plane.

The atomic partial charges of the amino acids are taken from the CHARMM energy function. The charges at the special pair are taken from computations of Plato et al[31,32]. Accordingly for the oxidized special pair the fraction 0,28(0,72) of the unit positive charge is localized at $D_M$($D_L$) respectively. The charges at the other cofactors BChl, BPh and the carotenoid are adapted from computations of Scherer and Fischer[33].

The whole protein-water system of the RC including 165 added water molecules consists of 10125 atoms. It is energy minimized for the wild-type structure, each mutant structure and each special pair charge state (reduced: $D^0$, oxidized $D^+$) using the constraints mentioned above. Further details on the procedure of energy minimization are given in a forthcoming publication[30].

The shift of the midpoint potential $\Delta E_M$ of a mutant (mut) of the RC with respect to the wild-type (WT) RC is calculated from the double energy difference

$$\Delta\Delta G = \Delta G_{mut} - \Delta G_{WT} \qquad (2)$$

where the single energy differences are given by

$$\Delta G = G(0) - G(+). \qquad (3)$$

The individual energy terms in eq. (3), G($\alpha$), refer to the reduced ($\alpha$=0) and oxidized ($\alpha$=+) special pair state. Finally the $\Delta E_M$ values are obtained from the Nernst equation

$$\Delta E_M = -\frac{\Delta\Delta G}{nF}, \qquad (4)$$

where n=1 is the change of the charge in the considered redox reaction and F = 96485 C/mol is Faraday's constant.

The individual energy terms G($\alpha$) can be evaluated by computing the electrostatic energy of the special pair atomic partial charges $q_i(\alpha)$ in the local electrostatic potential $\Phi_i(\alpha)$ generated by the special pair environment according to

$$G(\alpha) = \sum_i \Phi_i(\alpha) q_i(\alpha), \qquad (5)$$

where $\alpha$ denotes the redox state of the RC ($\alpha$=0,+). The electrostatic potential $\Phi_i(\alpha)$ is calculated with DelPhi by solving Poisson's equation. Within the RC the dielectric constant was set to unity, $\varepsilon_p$=1. In a layer of 30Å thickness, where the membrane is localized the dielectric constant was also set to unity $\varepsilon_m$=1. All other parts and protein cavities not occupied by water molecules carry a dielectric constant of $\varepsilon_W$=80. Other combinations of dielectric constants have been tested elsewhere[30]. The calculated $\Delta E_M$ values do not vary considerably by changing the dielectric constant $\varepsilon_m$ from 1 to 4. By increasing the value of $\varepsilon_p$ from 1 to 4 the absolute values of the calculated $\Delta E_M$ decrease approximately by a factor of 4.

The energy terms for G($\alpha$) are also evaluated by using the CHARMM energy function. In this case all residues and cofactors which are completely outside of the sphere of 18.5Å radius are not considered. The dielectric constant is set to unity everywhere. Then the electrostatic energy of the interaction of the special pair charges with the environment can be calculated by simply adding the Coulomb energies of the corresponding atom pairs. To estimate the influence from Van der Waals interactions the contributions of the corresponding Lennard-Jones interaction terms are also considered.

## 3. Results and Discussion

The shifts of the midpoint potential are calculated for 14 different mutants as listed in Table 1. In the wild-type structure only one acetyl oxygen atom at $D_L$ forms a hydrogen bond with H(L168). The two keto oxygen atoms and the acetyl oxygen atom at $D_M$ are not involved in hydrogen bonding[5]. In the single point mutant HF(L168) and consequently also the double mutants FH(M197) + HF(L168), HF(L168) + LH(M160) and HF(L168) + LH(L131) the hydrogen bond of the acetyl oxygen atom at $D_L$ is removed. Similarly the mutants involving LH(M160) (No. 6,9,11,13,14) have an additional hydrogen bond to the keto oxygen atom at $D_M$. The mutants involving LH (L131) (No. 7,10,12,13,14) have an additional hydrogen bond to the keto oxygen atom at $D_L$. The mutants FH(M197) and FY(M197) (No. 3,4,8,9,10,14) are assumed to have an additional hydrogen bond to the acetyl oxygen atom at $D_M$. The mutations YF(L162) and YF(M210) do not change the hydrogen bonding pattern of the special pair but render the special pair environment more hydrophobic[34, 16, 18].

The locations of the mutated residues are depicted in Fig. 1. It shows the special pair of the energy minimized RC structures together with the mutated residues. The corresponding wild-type residues which are replaced in the mutation are also shown. The small deviations between equivalent atoms of wild-type and muta -

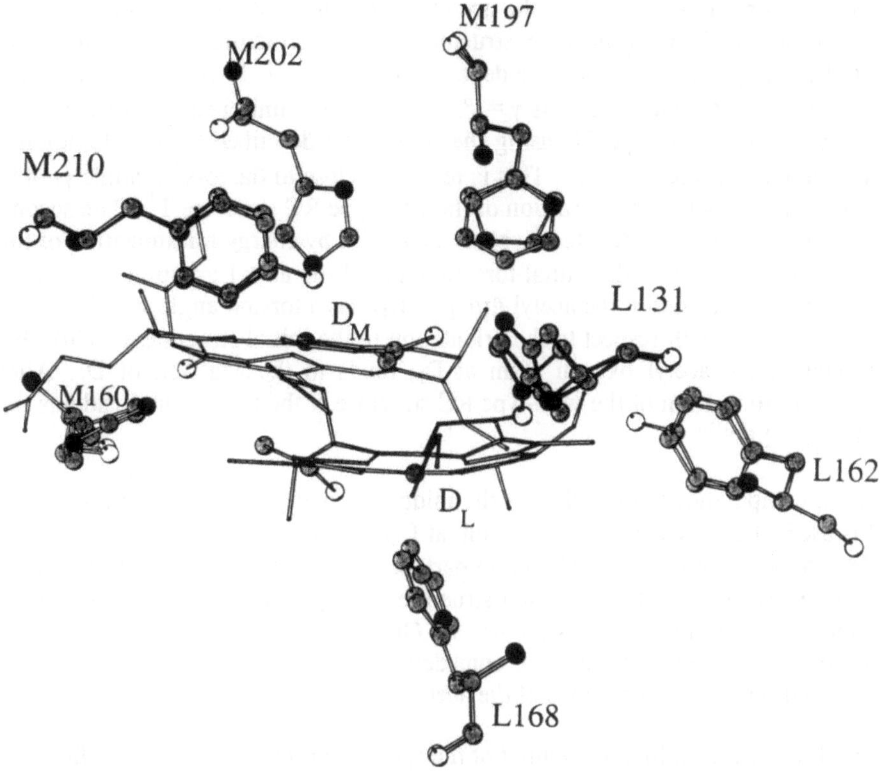


**Fig. 1:** Special pair of the RC *Rb. sphaeroides* with mutated residues. Energy minimized structures of the wild-type and mutated residues at the special pair are depicted together with a common special pair structure referring to the original crystal structure[5]. The oxygen atoms are represented by open the nitrogen atoms by black, the carbon atoms by light gray circles. With the exception of the keto oxygen atoms and the acetyl group atoms the atom skeleton of the porphyrin ring system is represented by sticks only. The BChl monomer $D_M$ from the special pair is on top of $D_L$. The cytochrome $c_2$ docking site is situated on the right side where residue L162 is located. The active branch (A) follows in the upper part of the figure.

ted residues demonstrate that the energy minimized structures of wild-type and mutant RC are very similar even at the mutated residues.

To probe the dependence of the calculated $\Delta E_M$ values on the orientations of the acetyl groups at the special pair two different orientations of the acetyl group at $D_M$ are considered as initial structures of the energy minimization. In the first case with torsion angle $\gamma_T$=47° the acetyl oxygen atom at $D_M$ points away from the $Mg^{2+}$ion of $D_L$ as in the x-ray structure[5]. All RC structures derived from this initial acetyl group orientation are denoted by the letter T. In the crystal structure the value of this torsion angle is $\gamma_c$=7°. After energy minimization of the crystal structure of the wild-type RC using the constraints described before, the torsion angle adopts a value of $\gamma_c$=33°. This is relatively close to the torsion angle $\gamma_T$=41° obtained after energy minimization of the wild-type RC structure T[30]. The second group of RC structures denoted by N are generated by energy minimization of the crystal structure where the initial torsion angle of the acetyl group at $D_M$ is $\gamma_N$=-133°. In this orientation the acetyl group at $D_M$ has a torsion angle which is 180° degrees rotated with respect to the orientation in the initial structure T. With this orientation the acetyl oxygen atom at $D_M$ binds to the $Mg^{2+}$ion of $D_L$. After energy minimization of the wild-type RC structure N the torsion angle adopts the value $\gamma_N$=-145° [30].

For all computations of the shift of the midpoint potential the wild-type structure N is used where the acetyl oxygen atom at $D_M$ points towards the $Mg^{2+}$ ion of $D_L$. Though there is no hydrogen bonding partner for the acetyl group oxygen atom at $D_M$ in the wild-type RC, the crystal structure has a preference for an acetyl group orientation corresponding to structure T[5]. Other orientations of the acetyl groups in the wild-type RC structure are considered in forthcoming publications[30]. For the mutations both orientations of the acetyl group at $D_M$ (T, N) are considered.

The shifts of the midpoint potential of the special pair calculated with DelPhi and CHARMM for 14 different mutants are compared with the experimental data in Table 1. For the first two mutants in Table 1 no agreement with the experimental data was possible. For mutant No 3 [FY(M197)] only the CHARMM energy function yields a $\Delta E_M$ value which correlates with the experimental value. In all other cases at least a semi-quantitative agreement with experiments was obtained if the acetyl group orientation at $D_M$ was N (T) in the absence (presence) of a hydrogen bond involving the acetyl oxygen atom at $D_M$. A detailed discussion of the results for the different mutants will follow[30].

The calculated $\Delta E_M$ values obtained from DelPhi by solving Poisson's equation are very close to the values obtained from the electrostatic part of the energy function of CHARMM (Table 1). This is surprising if one considers the simplicity of the treatment of the electrostatic interactions in the present application of the CHARMM energy function where the dielectric constant is unity everywhere. However, the special pair is located in the center of the sphere of 18.5 Å radius where all atoms of the RC and the water molecules are considered. Hence the dielectric environment used in the computations with DelPhi is quite distant from the special pair atoms. Therefore it can have only a moderate influence on the calculated shift of the midpoint potential. For a cofactor which is closer to the surface of its protein the computation of $\Delta E_M$ with a simple electrostatic energy function may not work.

**Table 1**. Shift of the $E_M$ in units of mV calculated with electrostatic energies using DelPhi and CHARMM

| # | Mutant | DelPhi [1] | | CHARMM [2] | | |
|---|---|---|---|---|---|---|
| | | *N* | *T* | *N* | *T* | *exp.* |
| 1 | YF(L162) | **-17** | 23 | **-24** | 9 | 28 |
| 2 | YF(M210) | **100** | 110 | **70** | 83 | 30 |
| 3 | FY(M197) | 99 | **96** | 82 | **52** | 31 |
| 4 | FH(M197) | -18 | **130** | -6 | **134** | 125 |
| 5 | HF(L168) | **-129** | -39 | **-81** | -59 | -95 |
| 6 | LH(M160) | **74** | 55 | **56** | 60 | 60 |
| 7 | LH(L131) | **85** | 94 | **94** | 77 | 80 |
| 8 | FH(M197)+HF(L168) | | **34** | 11 | **57** | 40 |
| 9 | FH(M197)+LH(M160) | | **174** | 103 | **152** | 195 |
| 10 | FH(M197)+LH(L131) | | **205** | 170 | **199** | 205 |
| 11 | HF(L168)+LH(M160) | **-46** | -4 | **-13** | 0 | -20 |
| 12 | HF(L168)+LH(L131) | **-11** | 40 | **19** | 45 | -20 |
| 13 | LH(L131)+LH(M160) | **159** | 144 | **146** | 144 | 130 |
| 14 | FH(M197)+LH(L131)+LH(M160) | | **339** | 209 | **275** | 260 |

[1] Only crystal water molecules are considered.
[2] Crystal and overlay water molecules are considered.
Experimental shifts are for the mutants No. 1 from Ref. 34, No. 2 from Ref. 16, 18; No. 3 from Ref. 19; No. 4,8-14 from Ref. 21,22; No. 5 from Ref. 35,21,22; No. 6,7 from Ref. 17,21,22. The data in bold digits are obtained with the relevant mutant structures. Mutant structure T is assumed to be relevant if the acetyl group oxygen atom at $D_M$ forms a hydrogen bond, mutant structure N is used, if no hydrogen bonding partner for the Acetyl oxygen atom is available. The wild-type structure N is used throughout. For more details see text.

Since the CHARMM energy function can be evaluated much faster than the electrostatic energy calculated by solving Poisson's equation several dependencies can be studied in more detail using CHARMM. One is the influence of water molecules on the calculated shift of the midpoint potential. The other is the influence of short range Van der Waals interactions which are not considered with DelPhi.

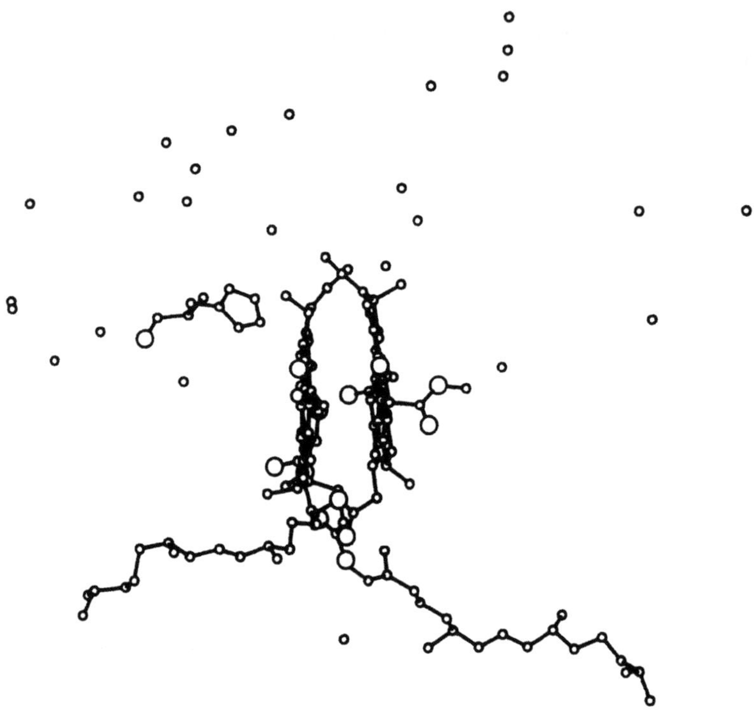

**Fig. 2** Special pair from energy minimized wild-type RC structure with crystal water molecules. The BChl monomers of the special pair $D_L$ ($D_M$) are on the left (right) side. The histidine L168 which is the hydrogen donor for the bonding with the acetyl oxygen of $D_L$ is also depicted. The majority of crystal water molecules are located at the cytochrome $c_2$ docking site which is in the upper half of the figure.

Before the dependence of the calculated $\Delta E_M$ values on the water content is investigated the distribution of water molecules in the RC is considered. Fig. 2 depicts the crystal water molecules found in the sphere with 18.5Å radius centered at the special pair. The majority of water molecules is located close to the docking site of the cytochrome $c_2$. In the wild-type RC structure the water molecule closest to the special pair has a non-hydrogen atom pair distance of 3.8Å only. Fig. 3 depicts the distribution of water molecules which where placed in the considered sphere of the RC in addition to the crystal water molecules using an overlay technique[36]. With the overlay procedure one water molecule could by placed as close as 3.4Å to the special pair. This overlay water molecule is located at the methyl group of the porphyrin ring system of $D_L$ which is close to the corresponding acetyl group. Its position does not change significantly between wild-type structures N and T as well as the charged and uncharged

**Fig. 3** Special pair with overlay water molecules, details as in Fig. 2.

special pair state. Also most of the overlay waters are near to the docking site of the cytochrome $c_2$. The large number of water molecules in this regime underlines the hydrophilic character of this part of the RC surface.

In Table 2 the influence of water molecules on the calculated shift of the midpoint potential is investigated by using the electrostatic part of the CHARMM energy function. In most cases the difference of the $\Delta E_M$ values between a computation with no water or only the crystal water is small. For all mutants not involving FH(M197) the influence from overlay water molecules is also small. The consideration of overlay water is however crucial for all mutants involving the point mutation FH(M197). In these mutants an overlay water molecule which is close to the acetyl group at $D_L$ is removed for both charge states of the special pair. In all other mutants and the wild-type structure this overlay water molecule remains approximately in the same location. The removal of this overlay water molecule in the mutants involving the point mutation FH(M197) improves the agreement with experimental $\Delta E_M$ values considerably.

**Table 2.** Shift of the $E_M$ in units of mV calculated with different water content using the electrostatic energy from CHARMM

| # | Mutant, structure* | no water | crystal water | crystal + overlay water | exp. |
|---|---|---|---|---|---|
| 1 | YF(L162), N | 4 | -4 | -24 | 28 |
| 2 | YF(M210), N | 69 | 85 | 70 | 30 |
| 3 | FY(M197), T | -27 | -36 | 52 | 31 |
| 4 | FH(M197), T | 69 | 58 | 134 | 125 |
| 5 | HF(L168), N | -79 | -72 | -81 | -95 |
| 6 | LH(M160), N | 60 | 70 | 56 | 60 |
| 7 | LH(L131), N | 90 | 97 | 94 | 80 |
| 8 | FH(M197)+HF(L168), T | -9 | -18 | 57 | 40 |
| 9 | FH(M197)+LH(M160), T | 90 | 77 | 152 | 195 |
| 10 | FH(M197)+LH(L131), T | 136 | 123 | 199 | 205 |
| 11 | HF(L168)+LH(M160), N | -12 | -2 | -13 | -20 |
| 12 | HF(L168)+LH(L131), N | 17 | 29 | 19 | -20 |
| 13 | LH(L131)+LH(M160), N | 141 | 154 | 146 | 130 |
| 14 | FH(M197)+LH(L131) +LH(M160), T | 204 | 149 | 275 | 260 |

* N,T denotes the RC structure used for corresponding mutants. The wild-type structure N is used throughout. The bold digits in Table 1 correspond to the third column where crystal and overlay water molecules are considered.

Finally the influence of Van der Waals interactions on the calculated $\Delta E_M$ values are studied (Table 3). For several mutants (No. 3,6,9,13,14) Van der Waals interactions lead to considerable deviations from $\Delta E_M$ values which were calculated with electrostatic energy terms only. In all of these cases the calculated $\Delta E_M$ values have agreed with the experimental values in the absence of Van der Waals interactions. In some cases the calculated shift is over- in other cases it is underestimated. No general trend can be observed. Presumably the strong distance dependence of the Van der Waals interactions at short atom pair distances renders the energy function very sensitive to small structural differences which result from the individual energy minimizations. The distance dependence of the electrostatic energy terms is much weaker such that small variations in the RC structure can not have a significant influence on the value of the electrostatic

**Table 3**. Shift of the $E_M$ in units of mV from electrostatic and Van der Waals energies using CHARMM

| # | Mutant | electro-static | electrostatic + Van der Waals | exp. |
|---|---|---|---|---|
| 1 | YF(L162), N | -24 | -2 | 28 |
| 2 | YF(M210), N | 70 | 116 | 30 |
| 3 | FY(M197), T | 52 | 78 | 31 |
| 4 | FH(M197), T | 134 | 140 | 125 |
| 5 | HF(L168), N | -81 | -104 | -95 |
| 6 | LH(M160), N | 56 | 94 | 60 |
| 7 | LH(L131), N | 94 | 72 | 80 |
| 8 | FH(M197)+HF(L168), T | 57 | 37 | 40 |
| 9 | FH(M197)+LH(M160), T | 152 | 99 | 195 |
| 10 | FH(M197)+LH(L131), T | 199 | 196 | 205 |
| 11 | HF(L168)+LH(M160), N | -13 | -16 | -20 |
| 12 | HF(L168)+LH(L131), N | 19 | 31 | -20 |
| 13 | LH(L131)+LH(M160), N | 146 | 188 | 130 |
| 14 | FH(M197)+LH(L131)+LH(M160), T | 275 | 219 | 260 |

For details see Table 1

energy. These problems may not occur if the RC structure the charged and uncharged special pair state are not individually minimized.

## 4. Conclusions

The shift of the redox potential of a cofactor in a protein molecule can be calculated also with a simple electrostatic energy function if the considered cofactor is completely surrounded by atoms from the protein and from water molecules. Water molecules located in protein cavities can be crucial to obtain agreement with the experimental shifts of the midpoint potential. Van der Waals interactions are very sensitive to small variations of short atom pair distances. Including these interactions in computations of the shift of the midpoint potential makes it difficult to obtain agreement with experimental data.

**Acknowledgment** The authors thank Prof. Wolfgang Lubitz for valuable discussions. We are grateful to Dr. Martin Plato, Dr. Philipp Scherer, and Bernd Melchers for calculating the atomic partial charges of the cofactors. The crystal structure of *Rb. sphaeroides* was provided by Dr. Ulrich Ermler and Dr. Günther Fritzsch prior to publication. The CHARMM source code has been provided by Prof. Martin Karplus and Molecular Simulations Inc.. This work is supported by the Deutsche Forschungsgemeinschaft SFB 312, Teilprojekt D7 and the Fonds of the German Chemical Industry. G. M. U. is grateful for a fellowship from the Studienstiftung des Deutschen Volkes.

## References

1 Deisenhofer, J., Epp, O., Miki, K., Huber, R., Michel, H. (1985) Nature **318**, 618-624.
2 Michel, H., Epp, O., Deisenhofer, J. (1986) EMBO J. **5**, 2445-2452.
3 Allen, J. P., Feher, G., Yeates, T. O., Komiya, H., & Rees, D. C. (1987) Proc. Natl. Acad. Sci. USA **84**, 6162-6166.
4 Chang, C. H., Tiede, D., Tang, J., Smith, U., Norris, J., & Schiffer, M. (1986) FEBS Lett. **205**, 82-86.
5 Ermler, U., Fritzsch, G., Buchanan, S. K., & Michel, H. (1994) Structure **2**, 925-936.
6 Michel-Beyerle, M. E., Ed. (1985), 'Antennas and Reaction Centers of Photosynthetic Bacteria', Springer Series in Chemical Physics, **42**, Springer, Berlin.
7 Breton, J., Vermeglio, A., Eds. (1988), 'The Photosynthetic Reaction Center - Structure and Dynamics', NATO ASI series A: Life Sciences, Plenum Press, New York.
8 Jortner, J., Pullmann, B., Eds. (1990), 'Perspectives in Photosynthesis', Kluwer, Dordrecht, The Netherlands.
9 Michel-Beyerle, M. E., Ed. (1990), 'Reaction Centers of Photosynthetic Bacteria', Springer Series in Biophysics Vol. **6**, Springer, Berlin.
10 Murata, N., Ed. (1992), 'Research in Photosynthesis' Kluwer, Dordrecht, The Netherlands.
11 Youvan, D. C., Ismail, S. & Bylina, E. J. (1985), Gene **38,** 19-30.
12 Farchaus, J. W. & Oesterhelt D. (1989) EMBO J. **8**, 47-54.
13 Paddock, M.L., Rongey, S.H., Feher, G. & Okamura, M.Y. (1989) Proc. Natl. Acad. Sci. U.S.A. **86**, 6602-6606.
14 Takahashi, E., Maroti P. & Wraight, C.A. (1989) in 'Current Research in Photosynthesis (Beltscheffsky, M. Ed.) pp 169-172, Kluwer, Dordrecht, The Netherlands.
15 Hammes, S. L., Mazzola, L., Boxer, S. G., Gaul, D. F. & Schenck, C. C. (1990) Proc. Natl. Acad. Sci. U.S.A. **87**, 5682-5686.
16 Gray, K. A., Farchaus, J. W., Wachtveitl, J., Breton, J., & Oesterhelt, D. (1990), EMBO J. **9**, 2091-2070.

17 Williams, J. C., Alden R. G., Murchison, H. A., Peloquin, J. M., Woodbury, N. W., & Allen, J. P. (1992) Biochemistry **31**, 11029-11037.
18 Nagarajan, V., Parson, W. W., Davis, D., & Schenck, C. C. (1993) Biochemistry **32**, 12324-12336.
19 Wachtveitl, J., Farchaus, J. W., Das, R., Lutz, M., Robert, B., & Mattioli, T. A. (1993) Biochemistry **32**, 12875-12886.
20 Murchison, H. A., Alden, R. G. Allen, J. P., Pelequin, J. M., Taguchi, A. K. W., Woodbury, N. W., & Williams, J. C. (1993) Biochemistry **32**, 3498-3505.
21 Lin, X., Murchison, H. A., Nagarijan, V., Parson, W. W., Allen, J. P., & Williams, J. C. (1994) Proc. Natl. Acad. Sci. U.S.A. **91**, 10265-10269.
22 Lin, X., Williams, J. C., Allen, J. P., & Mathis, P. (1994) Biochemistry **33**, 13517-13523.
23 Chirino, A. J., , Lous, E. J., Huber, M., Allen, J. P., Schenck, C. C., Paddock, M. L. Feher, G., & Rees, D. C. (1994) Biochemistry **33**, 4584-4593.
24 Mattioli, T. A., Lin, X., Allen, J. P., and Williams, J. C. (1995) Biochemistry **34**, 6142-6152.
25 Rautter, J., Lendzian, F., Schulz, C., Fetsch, A., Kuhn, M., Lin, X., Williams, J. C., Allen, J. P., and Lubitz, W. (1995) Biochemistry, in press.
26 Gilson, M. K., Rashin, A., Fine R., & Honig B. (1985) J. Mol. Biol. 1983, 503-516.
27 Gilson, M. K., Sharp, K. A., & Honig, B. (1987), J. Comp. Chem. **9**, 327-335.
28 Gilson, M. K. & Honig, B. (1988) Proteins **4**, 7-18.
29 Brooks, B. R., Bruccoleri, R. E., Olafson, B. D., States, D. J., Swaminathan, S., & Karplus, M. (1983) J. Comp. Chem. **4**, 187-217.
30 Muegge, I., Apostolakis, J., Ermler, U., Fritzsch, G., Lubitz, W. & Knapp, E. W., to be published.
31 Plato, M., Tränkle, E., Lubitz, W., Lendzian, F., & Möbius, K. (1986) Chem.Phys. **107**, 185-196.
32 Plato, M., Möbius, K., & Lubitz, W. (1991) in *Chlorophylls* (Scheer, H., Ed.) pp 1015-1046, Boca Raton, Florida.
33 Scherer, P. O. J., & Fischer, S. F. (1989) Chem. Phys. **131**, 115-127.
34 Wachtveitl, J., Farchaus, J. W., Mathis, P., Oesterhelt, D. (1993a) Biochemistry **32**, 10894-10904.
35 Murchison, H. A., Alden, J. P., Pelequin, J. M., Taguchi, A.. K. W., Woodbury, N. W., & Williams, J. C. (1993) Biochemistry **32**, 3498-3505.
36 Knapp, E. W., & Nilsson, L. (1990) in *Reaction Centers of Photosynthetic Bacteria* (Michel-Beyerle, M.E., Ed.) pp. 437-450, Springer, New York.

# PART 2
# Dynamics of Primary Events and the Role of Protein

# The First Femtoseconds of Primary Photosynthesis - The Processes of The Initial Electron Transfer Reaction

W. Zinth[1], T. Arlt[1], S. Schmidt[1], H. Penzkofer[1], J. Wachtveitl[1], H. Huber[1], T. Nägele[1], P. Hamm[1], M. Bibikova[2], D. Oesterhelt[2], M. Meyer[3], H. Scheer[3]

[1] Institut für Medizinische Optik der Universität München
Barbarastr.16, 80797 München, Germany
[2] Max-Planck-Institut für Biochemie
Am Klopferspitz, 82152 Martinsried, Germany
[3] Botanisches Institut der Universität München
Menzinger Str. 67, 80638 München, Germany

**Abstract.** Infrared and visible spectroscopy is used to investigate the primary processes in bacterial photosynthesis. Experiments are presented showing convincingly that the first electron transfer step in primary photosynthesis leads to the radical pair state $P^+B_A^-$ where the electron has reached the accessory bacteriochlorophyll $B_A$ of the active branch. Ultrafast spectroscopy in the mid infrared indicates that the electron transfer to the bacteriochlorophyll is preceded by a very fast (200 fs) reaction in the excited electronic state of the special pair.

**Keywords.** Femtosecond spectroscopy, infrared spectroscopy, reaction centers

## 1. Introduction

The structure analyses of photosynthetic reaction centers (RC) have given important information on the elements of primary electron transfer (ET) [1,2]. The primary donor in bacterial photosynthesis is the special pair (P) formed by two excitonically coupled bacteriochlorophyll (BChl) molecules. After excitation of the primary donor by direct light absorption or via excitation transfer from the antenna system an electron is transpürted through a chain of chromophores on the "A" chromophore branch to the accessory BChl $B_A$ [3]. The bacteriopheophytin (BPhe) $H_A$ is the secondary electron acceptor before the electron reaches the quinone $Q_A$ on the 200 ps time scale. The edge to edge distances of the chromophores involved are: P to $B_A$ = 5.5 A, P to $H_A$ = 9.5 A and $B_A$ to $H_A$ = 4.8 A [4]. The second chromophore branch "B" which is to a high degree symmetry related to branch "A" is apparently not used in primary ET.

At present the major interest on photosynthetic ET is focused on the following three topics:

(i) what are the molecular details of primary ET

(ii) Why does primary ET proceed only via branch "A"?

(iii) Why do photosynthetic bacteria use an excitonically bound special pair as the primary electron donor?

We will address in the following predominantly the first question. In addition we will present new experimental data taken in the mid IR spectral region which could bring new ideas related to questions (ii) and (iii).

## 1.1 General Remarks on Electron Transfer in Bacterial Reaction Centers

Most ET reactions have been treated in the picture of standard non-adiabatic electron transfer theory where the relatively slow ET reactions occur from vibrationally relaxed levels [5,6]. Since typical vibrational relaxation times are in the order of $10^{-12}$ s or below, non-adiabatic theory is well justified for slow reaction steps with time constants longer than a few picoseconds.

In bacterial photosynthesis the 200 ps transient, assigned to the ET from $H_A^-$ to $Q_A$, is certainly within non-adiabatic ET theory. The decay of the electronically excited state $P^*$ and the related ET away from $P^*$ with 3 ps at room temperature has also been explained by non-adiabatic ET theory [5,6]. However, in this case the applicability of non-adiabatic theory depends strongly on the relaxation times of the relevant vibrational modes. While relaxation of high frequency modes in large dye molecules has been found to be at room temperature in the subpicosecond regime, detailed experimental information on the relaxation of low frequency modes (and with that the information on the applicability of non-adiabatic ET theory) are lacking.

When vibrational relaxation is much slower than ET, the specific population of vibrational modes becomes important. Direct quantum mechanical coupling of $P^*$ and the product state $P^+A^-$ (A is used for the primary electron acceptor) leads to coherent oscillations of the population of $P^*$ and $P^+A^-$ with a frequency determined by the electronic coupling element **v** [7,8]. In contrast to that, calculations of the free energy changes of the electron acceptor state in reaction centers based on molecular dynamics simulations reveal a strong modulation of energy levels at room temperature [7]. Under these conditions the coherent processes are interrupted before the oscillation can built up; energy is transferred between the (closely spaced) vibronic levels coupled to electron transfer and coherent ET theory breaks down.

Coherent electron transfer reactions are not the only processes leading to oscillations in absorption changes: oscillation are also observed in many dye molecules if an ultrashort femtosecond laser pulse is used for excitation [9]. In these cases a superposition of low lying vibrational modes in the excited state is generated. The subsequent evolution of this wavepacket is reflected as a modulation of the absorbance changes. Indeed related oscillations are observed in RCs at low temperature and - with only a short period of visibility - also at room temperature [10,11]. Based on the wavelength dependence these oscillations were assigned to a wavepacket motion in the excited electronic state of P [10].

## 1.2 The Primary Electron Transfer in Reaction Centers at Room Temperature

Today important experimental information is available on the primary photosynthetic reactions. In the first femtosecond experiments the decay of $P^*$ was found to proceed with $\simeq$ 3 ps. With the same time constant the formation of the radical pair state $P^+H_A^-$ was observed [12,13]. Later experiments have shown that faster, but weak absorption changes occur in spectral regions where the accessory BChl or its anion absorb [14-16]. These changes grow in with 0.9 ps (*Rhodobacter* (*Rb.*) *sphaeroides*) or 0.65 ps (*Rhodopseudomonas* (*Rps.*) *viridis*) and they disappear with 3 ps. Kinetic models to describe these observations in non-adiabatic electron transfer theory are as follows:

The two step electron transfer model:

(1) $$P^* \xrightarrow{\tau_1} P^+H_A^- \xrightarrow{\tau_Q} P^+Q_A^-$$

The three step electron transfer model:

(2) $$P^* \xrightarrow{\tau_1} P^+B_A^- \xrightarrow{\tau_2} P^+H_A^- \xrightarrow{\tau_Q} P^+Q_A^-$$

In these models the assignment of time constants $\tau_1 \approx 3$ps to the decay of $P^*$ and $\tau_Q \approx 200$ps to the transfer of the electron to $Q_A$ was straightforward. However the role of the accessory BChl $B_A$ is different in the two models: From the first femtosecond experiments it was concluded that the electron is directly transferred from $P^*$ to the BPhe $H_A$. This long distance electron transfer is only possible if the accessory BChl is assumed to act as a high lying virtual electron carrier. ($P^+B_A^-$ should be more than 1.000 $cm^{-1}$higher than $P^*$) [5,6]. In this case very large values for electronic coupling are required in order to allow a 3 ps electron transfer reaction. It has been pointed out that superexchange forward electron transfer is difficult to be reconciled with the recombination rates in the reaction centers [17].

In the three step electron transfer model $P^+B_A^-$ is a real electron carrying state with a short life time $\tau_2$ = 0.9 ps (0.65 ps in *Rps. viridis*). Its free energy should be close to or below that of $P^*$. The slow population of $P^+B_A^-$ with $\tau_1$ and the faster depopulation $\tau_2$ leads to only weak absorbance changes $\Delta A$. Qualitatively the ratio $\tau_2/(\tau_1-\tau_2)$ describes the peak amplitude $\Delta A$. The amplitudes of the 0.9 ps kinetic component fit well into the three step model. They allow to reproduce the theoretically expected $P^+B_A^-$ spectrum if used in the three step kinetic model.

However, this three step model was not generally accepted in the literature. Objections were based on the weak amplitude related to $P^+B_A^-$, the previous inability to detect long-lived populations of $P^+B_A^-$ in mutated RCs and on the difficulty to explain the low free energy of $P^+B_A^-$theoretically.

In this paper we present experimental data where relatively large amplitudes caused by the radical pair state $P^+B_A^-$ are observed by selecting special wavelengths where absorption changes due to $P^+B_A^-$ prevail. In addition we show data taken on mutated reaction centers with faster initial electron transfer away from $P^*$. We present experimental results allowing to estimate the free energy of $P^+B_A^-$ and show that $P^+B_A^-$ is 450 $cm^{-1}$ below $P^*$ (data for reaction centers of *Rb. sphaeroides*). Finally we present results from experiments performed in the mid-infrared region showing an unexpected 200 fs transient which could allow to obtain a better understanding of the dimeric nature of $P^*$ and of the asymmetry of electron transfer.

## 2. Methods

The time resolved experiments are performed using the excite and probe technique. These room temperature experiments use subpicosecond laser-amplifier systems described recently [18]. Weak (<15% excitation) and short ($t_P \approx 150$fs) light pulses at $\lambda_{exc}=870$ nm (*Rb. sphaeroides*) and $\lambda_{exc} = 960$ nm (*Rps. viridis*) are used to excite the special pair. Properly delayed probing pulses at a variety of visible and

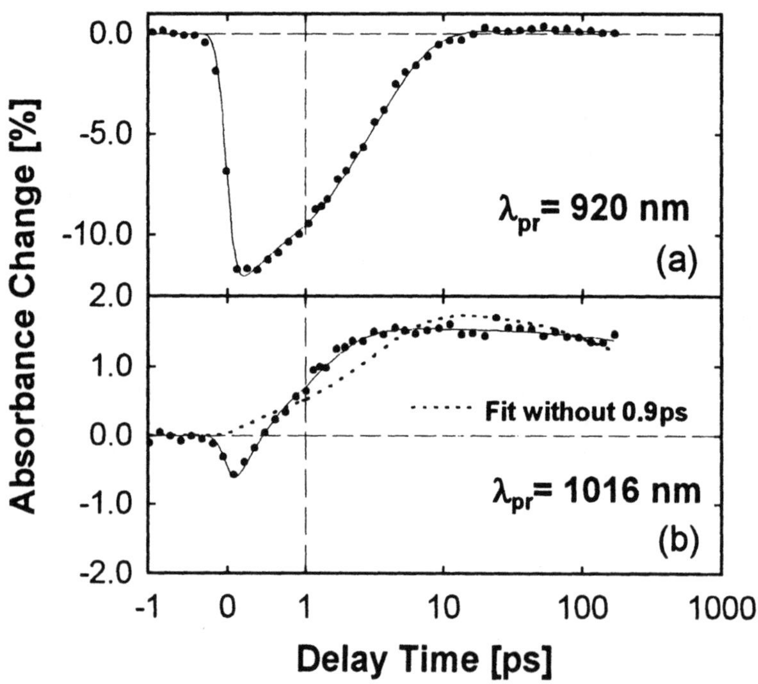


**Fig. 1.** Femtosecond absorption changes of wild type RC of *Rb. sphaeroides*. The decay of the excited electronic level $P^*$ and the appearance of intermediate $P^+B_A^-$ are well resolved at 920 nm and 1016 nm in (a) and (b) respectively. Note that the data are plotted on a linear scale for $t_D<1$ ps and on a logarithmic scale at $t_D>1$ps. Data are presented as symbols, model curves as lines.

near infrared probing wavelengths $\lambda_{pr}$ are obtained by continuum generation. They interrogate the absorbance changes as a function of time delay $t_D$. The sample concentration was adjusted to yield a transmission of T≈ 5 – 10% at the excitation wavelength.

A laser system which enables femtosecond mid-IR experiments in an extended frequency region (2000-1000 $cm^{-1}$) was described in detail in [19-21]. It is based on a standard Ti:sapphire regenerative amplifier system running at a repetition rate of 1 kHz. The RCs were excited by 150 fs pulses at 870 nm. The absorbance change of the sample was recorded by tunable probing pulses generated by difference frequency mixing between the output of the Ti:sapphire and that of a synchronized travelling wave dye laser in a $AgGaS_2$-crystal. The IR-pulses had a pulse duration of ≈300 fs and a spectral bandwidth of 60 $cm^{-1}$. For spectrally resolved detection, the probing pulses were dispersed in a grating spectrometer after passing the sample. A 10 element IR-detector array was used to cover the complete bandwidth of the probing pulses simultaneously.

For the room temperature experiments presented in this paper, we apply data evaluation based on the nonadiabatic ET theory. The reaction is described by a rate equation system [15]. As a consequence we discuss a reaction model with exponential kinetics where the intermediates i and j are connected by microscopic rates $\gamma_{ij}$.

# 3. Results and Discussion

## 3.1 Observation of $P^+B_A^-$ in Spectral Regions with Reduced Interference from Other Intermediates

In this chapter we present experimental results obtained from RC of *Rb. sphaeroides* R26.1. The preparation of the RC is described in [18]. Fig. 1 shows results taken at the probing wavelength 920 nm. Here the first state in the photosynthetic reaction sequence - the excited electronic state $P^*$ of the special pair P - is observed via its stimulated emission. The early signal points can well be described by an exponential model function with a decay time of 3.5 ps. At later times some deviations are observed from this simple picture. They point to a certain heterogeneity of the primary ET reaction. The use of a biexponential model function [22] with time constants of 2.3 s and 7 ps allows a perfect fit of the data points.

Of special interest are experiments where probing light pulses in the spectral region of the BChl anion absorption at 1020 nm are used (Fig. 1b). In this region no absorption bands from groundstate intermediates of the reaction center occur. Only weak gain from $P^*$ remains here in *Rb. sphaeroides* and the cation $P^+$ has increased absorption. Therefore the 0.9 ps kinetic component, related to the intermediate $P^+B_A^-$, should be well visible here. Indeed, the experiments show a strong contribution of the 0.9 ps kinetic component. After a weak absorption decrease due to stimulated emission, a rapid absorption rise occurs which is

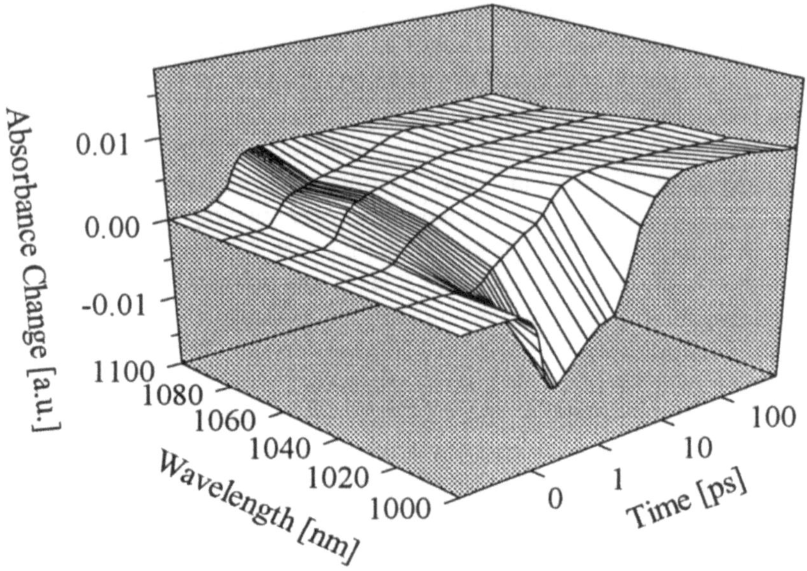


**Fig. 2.** Time resolved absorption differences $\Delta A(t_D)$ of wild type RCs in the spectral region between 1000 nm and 1100 nm.

mainly due to the 0.9 ps component. The dependence of the signal at longer wavelengths is shown in the 3-d-plot of Fig. 2. It displays the signal curves between 1000 nm and 1100 nm. From the data it is evident that the 0.9 ps kinetic component which is important at 1000 nm - 1040 nm vanishes at higher wavelengths (1100 nm). Analysing the data within the three step ET model (with ET from $P^*$ to $P^+B_A^-$ in 3.5 ps, and from $P^+B_A^-$ to $P^+H_A^-$ in 0.9 ps and transfer to $P^+Q_A^-$ in 200 ps) one obtains the expected spectrum for intermediate $P^+B_A^-$ with an absorption peak at 1020 nm[18]. Additional dichroic data allow to calculate the orientation of the transition dipole moment of $B_A^-$ . A value of 26° is obtained in perfect agreement with structure analyses [18]. In the frame of the two step model (superexchange) with a direct ET from $P^*$ to $P^+H_A^-$ there is no explanation for the strong 0.9 ps kinetic component.

### 3.2 Increased Population of $P^+B_A^-$ in Mutated Reaction Centers with Increased $P^*$ Decay Rates

In several publications [23,24] specific mutants from *Rb. sphaeroides* were presented where the introduction or removal of hydrogen bridges between amino acid residues and the special pair changed the redox potential $P/P^+$ and the free energy of $P^+$.

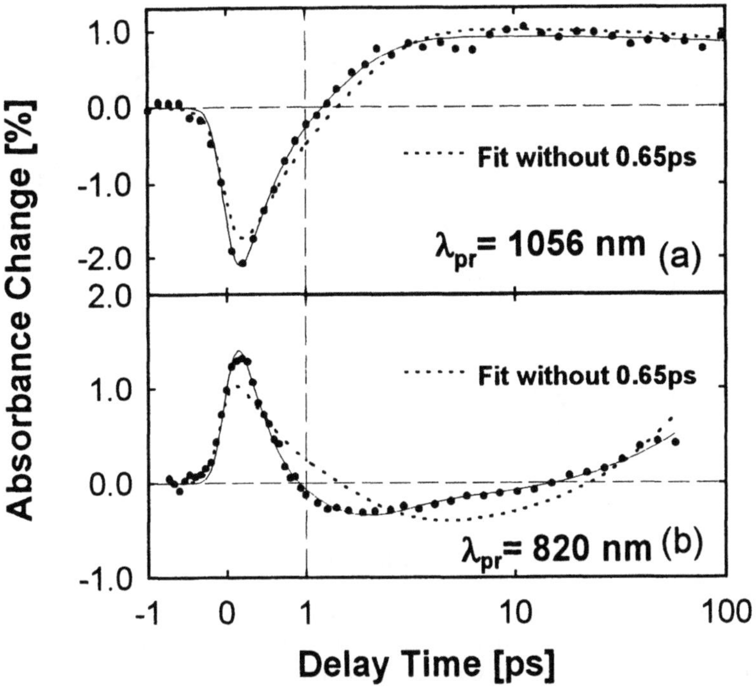


**Fig. 3.** Time resolved absorption differences $\Delta A(t_D)$ for mutated RC of *Rps. viridis* L168HF. Probing wavelengths are set to positions where $B_A^-$ (1056nm) and $B_A$ (820nm) strongly absorb.

These mutations are expected to be of special interest if introduced in RC of *Rps. viridis*, where - in contrast to *Rb. sphaeroides* (see below) - no direct information does exist for the free energy of the intermediate $P^+B_A^-$. Fig. 3 presents the results from experiments on the mutant L 168HF of *Rps. viridis*. Here the replacement of histidine L168 by a phenylalanine leads to the removal of one hydrogen bond. As a consequence, the energy of the cation radical $P^+$ is decreased relative to wild type RC by $\approx 700\ cm^{-1}$. The experiments testing the stimulated emission of $P^*$, e. g. at the probing wavelength 1000 nm clearly demonstrate an accelerated decay with a time constant of $\tau_1 = 1.1$ps (instead of 3.5 ps found with a monoexponential fit in wild type RC). This strong speed-up of $\tau_1$ should strongly increase the visibility of the $\tau_2 = 0.65$ ps process. A first indication is found around 1060 nm, where the BChl **b** anion is expected to absorb. Here wild type RC of *Rps. viridis* display a weak but visible 0.65 ps kinetic component. The mutated RC have a much stronger contribution of the 0.65 ps kinetic component (see Fig. 3a). Really striking is the 0.65 ps component in the $Q_y$ absorption band of BChl $B_A$. At 820 nm the 0.65 ps kinetic is by far the dominating contribution to the absorption changes. It causes even an absorption decrease at time $\approx$1 ps which can be related to a bleaching of the $B_A$ absorption band upon reduction.

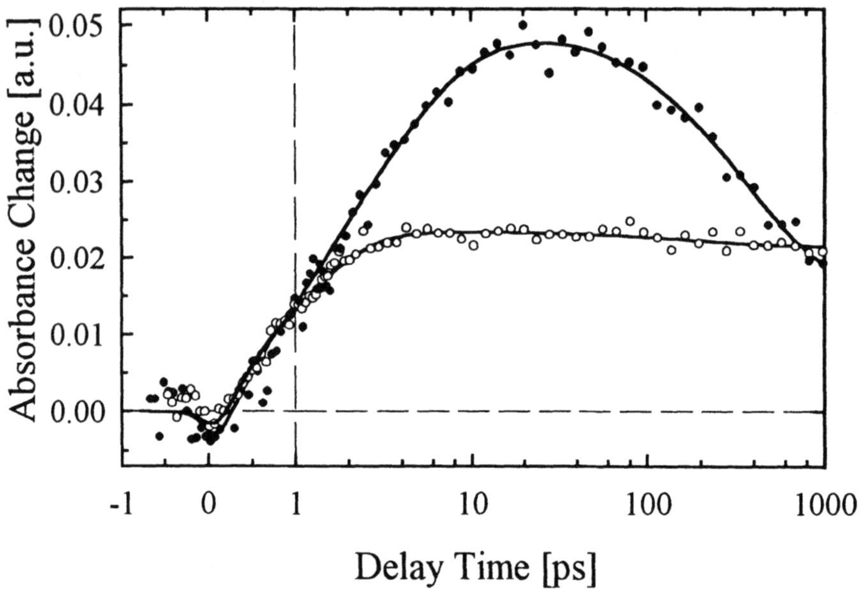


**Fig. 4.** Transient absorption changes at a probing wavelength of 1020 nm for modified RCs (filled symbols) and R26.1 RCs (hollow symbols) of *Rb. sphaeroides*. Experimental data are shown as points, the fits for the data as solid lines. The model function for R26.Pheo RCs contains time constants of 0.9 ps, 3.5 ps, 15 ps, 380 ps and infinity, the fit for native RCs time constants of 0.9 ps, 3.5 ps and infinity.

The results on the mutated RC convincingly demonstrate that the three step ET model is realised in photosynthetic RC. A further confirmation of this conclusion comes from the determination of the free energy of the radical pair state $P^+B_A^-$ given below.

### 3.3 The Determination of the Free Energy of Intermediate $P^+B_A^-$

Time resolved experiments in the BChl anion band of R26 RCs of *Rb. sphaeroides* showed the role of $B_A$ as a real electron carrier (see 3.1. and [18]). The energetics of the electron transfer intermediates can be investigated in detail in modified RCs containing pheophytins (Phe-a from plants) instead of bacteriopheophytins (BPhe-a). The *in vitro* difference of the redox potential of BPhe-a and Phe-a of 130-160 mV suggests a comparable rise of the free energy of $P^+H_A^-$ in the modified RCs [25]. The energy level of this radical pair state should then lie in the range of the free energy of P* or somewhere below. Assuming a stepwise electron transfer via $B_A$ (according to scheme (2)), one would expect a long-lived population of the radical pair state $P^+B_A^-$ in thermal equilibrium with $P^+H_A^-$. Indeed the signal trace at 1.020 nm (see Fig. 4.) shows a long lived increased absorption which can only be assigned to the bacteriochlorophyll anion $B_A^-$. The absorption subsequently decreases with a time constant of about 380 ps. This absorption

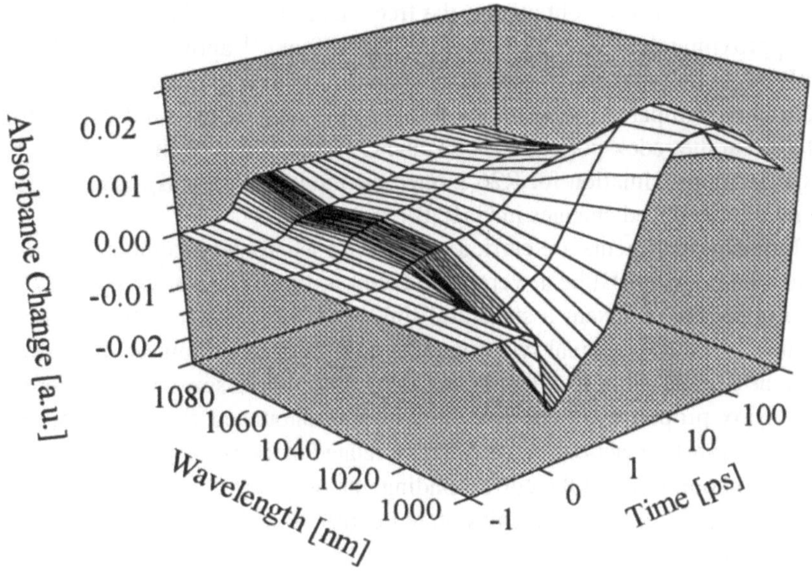


**Fig. 5.** Time resolved absorption differences $\Delta A(t_D)$ of R26.Pheo RCs in the spectral region between 1000 nm and 1100 nm. For the fit curves shown, five time constants, 0.9 ps, 3.5 ps, 15 ps, 380 ps and infinity were used. The time scale for $t_D < 1$ ps is linear; a logarithmic scale is applied at later delay times.

increase is connected with a long lived population of the BChl anion state $P^+B_A^-$ of approximately 30 %.

The characteristics of the modified RCs (R26.Pheo) in the spectral range between 1000 nm and 1100 nm are summarised in Fig. 5. The $B_A^-$ contribution has its maximum at 1020 nm and, as expected, the absorption increase vanishes until 1100 nm. Here only $P^+$ causes the remaining absorption change. Data obtained in other spectral regions allow to conclude [26] that the formation of the radical pair state $P^+Q_A^-$ is reduced by 30% compared to that of the native type RC and that a long-lived population of P* of about 3 % exists.

Combining these results one can estimate within the framework of a three step electron transfer model (Fig. 6.) the microscopic rates and the free energy levels of the participating intermediates. In the analysis for R26.Pheo, the recombination rate $\gamma_{20}$ from $P^+B_A^-$ to the ground state P and the free energies relative to $P^*$, $G(P^+B_A^-)$ and $G(P^+H_A^-)$ were varied as free parameters to reproduce the measured time constants and absorbance changes of the intermediates. The free energies $G_i$ of the different states are used to correlate forward and backward rates by the principle of detailed balance :

$$(3) \qquad \gamma_{ji} = \gamma_{ij} \ \exp(-(G_i - G_j)/(k_B T))$$

Consistency is only obtained if the free energy level of the intermediate $P^+B_A^-$ is approximately 450 $cm^{-1}$ below P* and 180 $cm^{-1}$ above $P^+H_A^-$. The rise of $G(P^+H_A^-)$ by 1400 $cm^{-1}$ - this corresponds to a level of 630 $cm^{-1}$ below $P^*$ - in R26.Pheo compared to the native RC is in the range one would expect from the change in the redox potential. The value $G(P^+B_A^-)$ of 450 $cm^{-1}$ can also be used as a first approximation for R26.1 since the modification at $H_A$ is not expected to lead to pronounced changes in $G(P^+B_A^-)$.

Further experiments are carried out on RCs where the chromophores 3-acetyl-Phe a and 3-vinyl-BPhe a were introduced at the BPhe a-sites $H_{A/B}$ [27]. These two pigments represent a structural link between Phe a and BPhe a, since they differ from these original two pigments only in one position. With these modifications it is possible to affect selectively either the absorption characteristics or the redox properties of the native BPhe a. Pronounced changes in the primary electron transfer reactions can directly be related to the redox potential of the different chromophores and the corresponding changes in free energy of the first intermediates. The reaction dynamics do not correlate with the changes in the absorption spectra.

The values of free energies of the intermediates, the reaction rates and the step-wise reaction scheme can be understood considering optimization of the primary ET in wild type RCs. The intermediate $P^+B_A^-$ allows a high speed of the primary ET step. The fast decay of the state $P^+B_A^-$ accelerates forward ET preventing direct recombination to P. Back reactions are slowed down via the large energy gap between $P^+B_A^-$ and $P^+H_A^-$. All these design features are required to obtain the high quantum yield of photosynthesis close to unity.

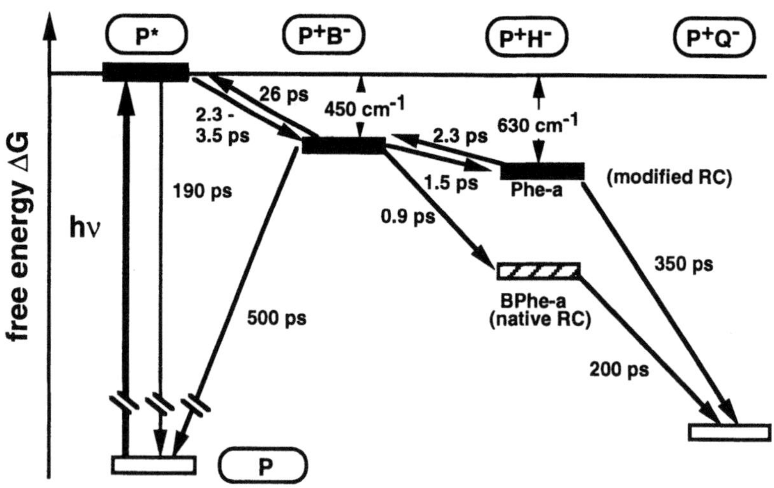


**Fig. 6.** Reaction and energy scheme used to explain primary photosynthesis in bacterial reaction centers of *Rb. sphaeroides*. It is based on a step-wise ET along the pigment chain of the reaction center starting at the special pair P via $B_A$, $H_A$ to $Q_A$.

### 3.4 Femtosecond Infrared Spectroscopy of Photosynthetic Reaction Centers

A complete data set of the transient absorbance changes of the RCs from the purple bacterium *Rb. sphaeroides* covering the spectral range from 1000-1800 $cm^{-1}$ and a temporal range of 0-1 ns was measured with a spectral resolution of 4-7 $cm^{-1}$ and a temporal resolution of 100 fs. In the following, this data set is analyzed under two different aspects: (i) temporal evolution at certain spectral positions and (ii) transient difference spectra taken at certain delay time positions.

#### 3.4.1 Temporal Evolution

Almost throughout the entire investigated spectral range (1800 $cm^{-1}$ - 1000 $cm^{-1}$), an initial absorption increase occurs. However, a detailed analysis shows that at most wavelengths, the signal rise is slower than the temporal response function of the experimental setup [28]. This delay is pronounced in certain spectral regions around 1460 $cm^{-1}$ (see Fig. 7.) and 1240 $cm^{-1}$. The delay can be fit well assuming that the instantaneous absorption increase is weak and that most of the absorption increase occurs with a time constant of 200±100 fs. Recent dynamic hole burning experiments in the near infrared have shown transient features with a similar time constant (250fs) [29].

The subsequent temporal evolution may be well described by two well known ET kinetic components: The dominant feature shows the same time constant (3.5 ps) as the decay of the initial state P* while the $P^+H_A^- \rightarrow P^+Q_A^-$-ET step (≈220 ps) is observed with small amplitudes at certain spectral positions especially in the high frequency range (>1600 $cm^{-1}$, C=O region). The fast ET step from $P^+B_A^- \rightarrow P^+H_A^-$ (in 0.9 ps) which was observed in VIS/NIR measurements is not resolvable in the IR experiments. It can be concluded that the IR absorbance

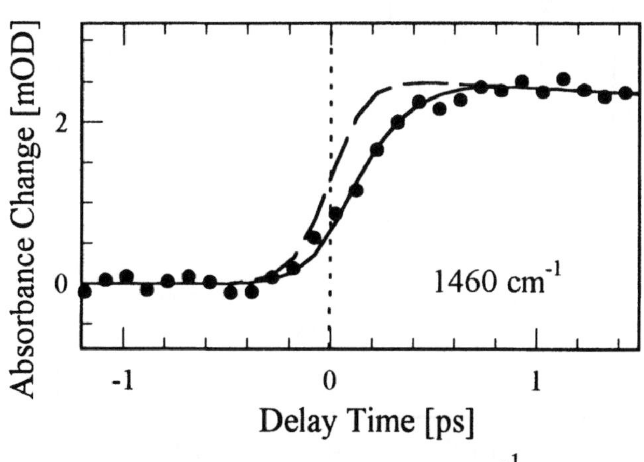


**Fig. 7.** Transient absorption signal measured at 1460 $cm^{-1}$ at the peak of an intense IR-absorption band. The absorption rise (solid line and dots) is delayed with respect to instrumental response function (dashed line). The delay can be modelled by a exponential function with a time constant of 200±100 fs.

changes are mainly caused by modifications of the special pair P (or of the protein surrounding P) and not by the other chromophores. Within the dynamic range of the experiments, there is no evidence for further kinetic components which might be assigned to a slow protein relaxation due to a response of the protein on the changed charge distribution.

### 3.4.2 Transient Difference Spectra

In Fig. 8, transient difference spectra recorded at distinct delay times at $t_D = 1$ ps, 10 ps and 1000 ps are shown. They are related to the intermediate states P*, $P^+H_A^-$ and $P^+Q_A^-$, respectively. The discussion of these spectra can be roughly divided into two spectral regions:

**(i) The spectral range of the C=O vibrations:** In the spectral region >1600 $cm^{-1}$ absorption bands of the C=O groups of the chromophores and of the protein (Amide I band) are expected. Here, a detailed assignment of distinct absorption bands can be given [19, 39]: In P* (1 ps), a red shift of the 9-keto carbonyl group is observed and the 10a-ester carbonyl group is not affected. However, already in the state $P^+H_A^-$, the 9-keto carbonyl group (1706 $cm^{-1}$/1684 $cm^{-1}$) and

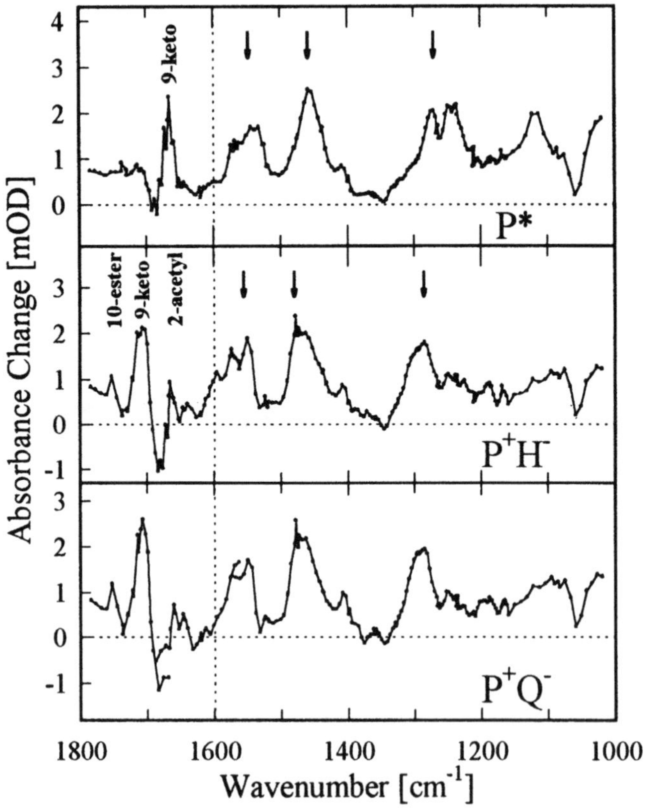


**Fig. 8.** Time resolved difference spectra at delay times of 1 ps, 10 ps and 1000 ps where the RCs are predominantly in the states P*, $P^+H_A^-$, and $P^+Q_A^-$.

the 10a-ester carbonyl group (1752 $cm^{-1}$ /1736 $cm^{-1}$) of the special pair are blue shifted just as observed in steady state FTIR spectra [19]. Also some weak bands corresponding to the 2-acetyl group can be identified around 1650 $cm^{-1}$. Only weak changes occur between $P^+H_A^-$ and $P^+Q_A^-$ again showing that the difference spectra are dominated by the special pair. Within the experimental error, the 1 ns difference spectrum and the steady state FTIR difference spectrum (according to a mean delay time of several milliseconds) are identical indicating that no further slow reactions occur.

**(ii) The low frequency range:** In the spectral region <1600 $cm^{-1}$, the C=C, C-C, C-N stretching and C-H bending modes of the chlorophyll tetrapyrolle ring system are expected. Due to the lack of exact model calculations, a detailed assignment of difference bands to distinct modes is not yet possible. However, a striking feature is found especially in this part of the spectrum: a pronounced positive absorption background occurs in connection with three strong broad positive bands which are observed in P* and $P^+$ at slightly different peak positions (see arrows in Fig. 8). In addition, as seen in Fig. 7, these positive bands do not appear instantaneously after electronic excitation but within a delay time of 200 fs. Qualitatively, these observations can be explained as follows: As a consequence of excitonic coupling between two chromophores to a dimer (the special pair P), low lying electronic transitions appear when the RCs populate P* or $P^+$ [28,31]. Due to the low excitation frequency of these electronic transitions (2000-3000 $cm^{-1}$), additional coupling between electronic and some distinct vibronic transitions is expected leading to a transfer of oscillator strength from the (in general strong) electronic transition to these (in general weak) vibronic transitions. Therefore the absorption cross section in the vibrational part of the spectrum of $P^*$ and $P^+$ is increased. This explanation is confirmed by the observation, that a similar absorption increase is not observed in the $P^+$/P-difference spectrum of heterodimer mutants, where no effective excitonic coupling is possible in the dimer [28]. Within this model, many of the observations can be explained qualitatively: (i) the positive P*/P- and $P^+$/P-difference spectra, (ii) the fact that the absorbance changes of the special pair dominate the difference spectra since the other chromophores (B, H and Q) have to be treated predominantly as monomers where no increase of vibronic oscillator strength is expected; (iii) and the fact that the $P^+B_A^- \rightarrow P^+H_A^-$-ET-step in 1 ps can not be observed: The small population of $P^+B_A^-$ (16%) combined with the small $B^-$/B-contribution to the difference spectra (as small as that of $H^-$/H or $Q^-$/Q) indicate that the corresponding kinetic component is so small that it can not be observed within the present signal to noise level.

However, as shown before, the positive bands are not present immediately after electronic excitation but rise within an ultrafast 200 fs relaxation process. The amplitude of the 200 fs kinetic component is too large to be understood as simple vibrational relaxation. Therefore, in order to understand this process, the electronic structure of the special pair (or the electronic-vibronic-coupling mechanism) must change considerably during that relaxation. It is known, that the excitonic coupling of both BChl-molecules in the special pair depends critically on the arrangement of both molecules. In addition, one can conclude qualitatively from quantum mechanical arguments that in a symmetric special pair P, coupling between electronic and vibronic transition should be weak, since no intramolecular charge separation within P is possible. In this framework, we explain the strong 200 fs kinetic component as follows: Immediately after electronic excitation, the RCs populate an almost symmetric initial excited state with uniform charge distribution (small charge transfer character). With the 200 fs process, a motion along a low frequency mode (for example a motion of both BChl molecules against each other) occurs, destroying the symmetry of the dimer and leading to an increased net charge separation in the dimer. Therefore, this reaction may facilitate ET along the active branch and may be one reason for the "unidirectionality of ET".

## References


1. Deisenhofer, J.,and Michel, H., (1989) EMBO J. 8, 2149.
2. Ermler, U., Fritzsch, G., Buchanan, S.K., Michel, H., (1994) Structure 2, 925.
3. For a recent review see: The photosynthetic reaction center *Vol.I and II*, ed. Deisenhofer, J. and Norris, J.R., (1993) Academic Press, San Diego.
4. Budil, D.E., Gast, P., Chang, C.H., Schiffer, M., Norris, J.R., (1987) Annual Review of Physical Chemistry 38, 561.
5. Bixon, M., Michel-Beyerle, M.E. and Jortner, J., (1988) Israel J. Chem. 28, 155.
6. Bixon, M., Jortner, J., and Michel-Beyerle, M.E., (1991) Biochim. Biophys. Acta 1056, 301.
7. Parson, W.W., Warshel, A., (1993) in: The photosynthetic reaction center *Vol. II*, ed. Deisenhofer, J., and Norris, J.R., (1993) (Academic Press, San Diego) p. 23.
8. Kuhn, H., (1986) Phys. Rev. A34, 3409.
9. For a review see: Femtosecond Reaction Dynamics ed.Wiersma, D.A., (1994) North-Holland, Amsterdam.
10. Vos, M.H., Rappaport, F., Lambry, J.-C., Breton, J., and Martin, J.L., (1993) Nature 363, 320.
11. Vos, M.H., Jones, M.R., McGlynn, P., Hunter, C.N., Breton, J., and Martin, J.L., (1994) Biochim. Biophys. Acta 1186, 117.
12. Martin, J.L., Breton, J., Hoff, A.J., Migus, A., and Antonetti, A., (1986) Proc. Natl. Acad. Sci. USA 83, 957.
13. Breton, J., Martin, J.L., Migus, A., Antonetti, A., Orzag, A., (1986) Proc. Natl. Acad. Sci. USA 83, 5121.
14. Holzapfel, W., Finkele, U., Kaiser, W., Oesterhelt, D., Scheer, H., Stilz, H.U., and Zinth, W., (1989) Chem. Phys. Letters 160, 1.

15. Holzapfel, W., Finkele, U., Kaiser, W., Oesterhelt, D., Scheer, H., Stilz, H.U., and Zinth, W., (1990) Proc. Natl. Acad. Sci. USA 87, 5168.
16. Dressler, K., Umlauf, E., Schmidt, S. Hamm, P., Zinth, W., Buchanan, S., and Michel, H., (1991) Chem. Phys. Letters 183, 270.
17. Marcus, R.A., (1987) Chem. Phys. Letters 133, 471.
18. Arlt, T., Schmidt, S., Kaiser, W., Lauterwasser, C., Meyer, M., Scheer, H., and Zinth, W., (1993) Proc. Natl. Acad. Sci. USA 90, 11757.
19. Hamm, P., Zurek, M., Mäntele, W., Meyer, M., Scheer, H., and Zinth, W., (1995) Proc. Natl. Acad. Sci. USA 92, 1826.
20. Hamm, P., Lauterwasser, C., and Zinth, W., (1993) Opt. Lett. 18, 1943
21. Hamm, P., Wiemann, S., Zurek, M., and Zinth, W., (1994) Opt. Lett. 19, 1642.
22. Hamm, P., Gray, K.A., Oesterhelt, D., Feick, R., Scheer, H., and Zinth, W., (1993) Biochim. Biophys. Acta 1142, 99.
23. Murchison, H.A., Alden, R.G., Allen, J.P., Peloquin, J.M., Taguchi, A.K.W., Woodbury, N.W., Williams, J.C.; (1993) Biochemistry 32, 3498.
24. Williams, J.C., Alden, R.G., Murchison, H.A., Peloquin, J.M., Woodbury, N.W., Allen, J.P., (1992) Biochemistry 31, 11029.
25. Schmidt, S., Arlt, T., Hamm, P., Huber, H., Nägele, T., Wachtveitl, J., Meyer, M., Scheer, H., and Zinth, W., (1994) Chem. Phys. Lett. 223, 116.
26. Schmidt, S., Arlt, T., Hamm, P., Huber, H., Nägele, T., Wachtveitl, J., Meyer, M., Scheer, H., and Zinth, W., (1995) Spectrochim. Acta, in press.
27. Huber, H., Meyer, M., Nägele, T., Hartl, I., Scheer, H., Zinth, W., and Wachtveitl, J., (1995) Chem. Phys. in press.
28. Hamm, P., and Zinth, W., (1995), J. Phys. Chem., in press.
29. Streltsov, A.M., Yakovlev, A.G., Shkuropatov, A.Ya., Shuvalov, V.A., (1995) FEBS Letters 357, 239.
30. Maiti, S., Walker, G. C., Cowen, B. R., Pippenger, R., Moser, C. C., Dutton, P. L., and Hochstrasser, R. M., (1994) Proc. Natl. Acad. Sci. USA 91, 10360.
31. Walker, G. C., Maiti, S., Cowen, B. R., Moser, C. C., Dutton, P. L., and Hochstrasser, R. M., (1994) J. Phys. Chem. 98, 5778.

# FEMTOSECOND KINETICS OF SPECIAL PAIR BLEACHING AND ELECTRON TRANSFER IN *RHODOBACTER SPHAEROIDES* REACTION CENTERS

V.A.Shuvalov[1,2,#], A.M. Streltsov[1], A.G. Yakovlev[1] and A.Ya. Shkuropatov[2]

[1]Laboratory of Photobiophysics, Moscow State University, Moscow, 119899, Russia; [2]Institute of Soil Science and Photosynthesis, Russian Academy of Sciences, Pushchino, Moscow reg., 142292, Russia

**Abstract.** The absorbance spectrum of reaction centers of *Rhodobacter sphaeroides* at room temperature consists of relatively narrow spectral components which are moving in femtosecond time domain and can be bleached by 90-fs laser pulses in the short wavelength region with a subsequent broadening and red shift of the bleaching (time constant ~250 fs). This effect is explained by a formation of vibronic wave packet in the ground state by the interaction with phonons at 293K. The formation of the vibronic wavepacket by fs excitation induces nuclear motions accompanied by oscillation of the stimulated emission from excited primary electron donor P* (described by M.Vos et al.) and by wavepacket motions leading to electron transfer from P* to bacteriochlorophyll (B) and then to bacteriopheophytin (H). This motions have low frequency (about 15 $cm^{-1}$) and is related to protein nuclear motions which are along the reaction coordinate. When the wavepacket approaches at first time (~1.5 ps delay) the intersection of the reactant potential energy surface and the product surface ~60% of excited state is converted to $P^+B^-$ state. The $P^+H^-$ state formation is delayed by ~2 ps with respect to that of $P^+B^-$ probably due to the transfer of wavepacket from P* to $P^+B^-$ surface. This wavepacket moves also slowly and approaches the intersection of the surfaces of $P^+B^-$ and $P^+H^-$ within ~2 ps (~8 $cm^{-1}$) inducing the electron transfer to H.

**Keywords:** Femtosecond spectroscopy; reaction center; electron transfer

**Abbreviations**: ΔA, light-minus-dark absorbance changes; P, primary electron donor; B and H, bacteriochlorophyll and bacteriopheophytin monomers, respectively, located in L protein subunit, primary electron acceptors; ET, electron transfer; RC, reaction center; S, Pekar-Huang-Rhys factor

[#] Corresponding author

# 1. INTRODUCTION

Molecular dynamic simulation [1-3] have predicted that low-frequency (1-100 cm-1) global protein motions play a key role in biochemical reactions. The femtosecond spectroscopy is a new tool for an investigation of light-induced reactions of chromophores accompanied by nuclear motions and electron transfer in pigment protein complexes. The excitation of the pigments by fs-pulses with broad spectral width creates simultaneous population of several vibronic levels with a formation of superposition of corresponding wavefunctions. The formed wavepacket has semiclasical behavior of motions [4] on the potential energy surface of the excited state. The wavepacket formation and motions play a key role in fs-, ps-photochemistry if the motions is along the reaction coordinate and the wavepacket is transferred to the product potential energy surface [5].

The electron transfer (ET) reaction in bacterial hexachromic reaction center protein (RC) occurs from the excited state of the bacteriochlorophyll special pair, primary electron donor P. This reaction takes place within ~3 ps at 300K and is accompanied by the reduction of bacteriopheophytin (HL) via intermediary electron acceptor, monomeric bacteriochlorophyll (BL) (where L indicates chromophore location in the L protein subunit) (for a review see [6,7]).

The absorption and hole burning measurements at 2-60K on bacterial RCs have revealed the progression of the enhanced (S=1.2 - 2) around 150 $cm^{-1}$ (73, 110 and 148 $cm^{-1}$ ) frequency modes which are accompanied by a ~30 $cm^{-1}$ protein mode seen clearly in hole burning experiments on RCs with prereduced HL [8] The frequencies found in hole burning experiments in the range of 30-150 $cm^{-1}$ were established in Raman spectra as well [9].

The important contribution to the understanding of the primary step was made by a visualization of coherent nuclear motion in RCs by femtosecond spectroscopy in a wide range of temperatures (10K-290K) [10]. The presence of the phase coherence of low frequency motions on the ps time scale of electron transfer was demonstrated. The vibrational motions of the excited state reveals the fundamental frequencies around 15 $cm^{-1}$ , and at 69, 92, 122 and 153 $cm^{-1}$ in native *Rhodobacter sphaeroides* RCs [10] and at 15 and 77 $cm^{-1}$ in DLL mutant of *Rhodobacter capsulatus* [11]. The important finding would be a clarification what motion is along the ET coordinate.

The ps study of ET in borohydrate treated Rcs in which BM is modified and partially removed and the 800-nm band reflects mostly BL molecule has shown that ET among pigments has the same ps kinetics as for native Rcs [12]. The bleaching of the H band is clearly delayed by a couple of ps with respect to that of the B band [13]. This finding was interpreted as an indication of a formation of the P+B- state before ET to H. The lack of information about ET at shorter time prevented us from proper conclusion about quantum mixing of P* and $P^+B^-$ [13]. More proper conclusion was done by fs-studies [14,15] . It was found [15] that the step-wise ET model has time constants of 2.3 ps, 0.9 ps and 200 ps for ET from P* to BL, from $BL^-$ to HL and from $HL^-$ to quinone (QA), respectively. It is of interest to study the wavepacket motions along ET coordinate in Rcs in which BM is modified.

Another interesting point is an obscure temperature dependence of the P absorbance spectrum. The spectral peak is blue shifted from 900 nm at 2K to 860 nm at 293K [6] with suppression of the Stokes components in contrast to the calculated overlapping of the vibration wave functions of the ground and excited states and to low temperature measurements (7-60K) which showed the coincidence with calculations (see [16]).

This work shows that the absorbance spectrum of RCs at room temperature (but probably not at 10K [11]) consists of the relatively narrow spectral components which are moving in femtosecond time scale and can be bleached by femtosecond laser pulses in the short wavelength region with a subsequent broadening and red shift with a time constant of ~250 fs. These data are discussed in terms of the population of the vibronic wave packets in the ground state by the interaction with phonons at 293K. The motion of these packets is probably responsible for the absorbance spectrum of P at 293K with enhanced short wave length components and suppressed Stokes components [17].

This work also shows that in borohydrate modified Rcs the low-frequency oscillation (about 15 $cm^{-1}$) related to protein nuclear motions is along the reaction coordinate. When the wavepacket formed by fs excitation approaches at first time (~1.5 ps) the intersection of the reactant potential energy surface and the product surface ~60% of excited state is converted to $P^+B^-$ state. The $P^+H^-$ state formation is delayed by ~2 ps with respect to that of $P^+B^-$ probably due to the transfer of wavepacket from P* to $P^+B^-$ state and to the motion of that until an approaching the intersection of the surfaces of $P^+B^-$ and $P^+H^-$ states.

## 2. MATERIALS AND METHODS

Rcs from *R. sphaeroides* (R-26) were isolated as described earlier [12]. The borohydrate treatment to modify a bacteriochlorophyll monomer located in M subunit was performed as described earlier [12]. When the A800/A860 ratio reached 1.4-1.5 Rcs were dialyzed against 100 mM Tris-HCl (pH 8.0) buffer containing 0.1% lauryldimethyl-amine-N-oxide and were purified by using DEAE-cellulose chromatography. The quantum yield of photooxidation of P on ms-time scale was close to 100%. The 2 mM ascorbate was added to keep P in the neutral state before the arrival of each pump pulse (1 Hz repetition rate). The concentration of the sample was adjusted to an optical density of 0.5 at 860 nm and at 293K (optical path length of 1 mm).

The femtosecond spectrometer described earlier [17] consisted of cpm dye laser (~50 fs pulse duration) with Rhodamine 590 (Exciton) jet pumped by a continuous argon laser (Spectra Physics) and DODCI (Exciton) jet. Fs pulses were amplified in a 6 pass dye jet amplifier and in a dye cell both pumped by 10 ns pulses at 538 nm. The amplified pulses were focused in a water cell ($H_2O$ + $D_2O$) to produce a fs continuum. Its near-IR part (around 840 nm) was amplified in a Styryl 9M (LDS 821, Exciton) cell. The spectrally filtered emission centered at 800-845 nm was used as excitation and a small fraction (~4%) of the whole NIR emission - as measuring and reference pulses. The measuring and exciting pulses propagated through delay lines and the sample. Then the measuring and reference pulses were passed through a polychromator and directed by the three-lens system to photodiode array of OMA-2 (PARC, Princeton). Cross-correlation function of the exciting and measuring pulses showed ~90 fs pulse duration. A spectral width of excitation pulse at 845 nm was 450 $cm^{-1}$. The measuring pulse was 90%-depolarized. The relative position of the zero time delay within a 700-950 nm range differed by less than 100 fs.

## 3. RESULTS

Fig. 1 shows the 293K transient difference absorbance spectra within Q (-) transition of P in *R. sphaeroides* RCs excited at 840 nm at different delays [17]. At 70-100 fs delay using stronger (curve 2) and weaker (curve 4) excitation the $\Delta A$ spectra centered at 847 nm are relatively narrow and blue shifted with respect to the absorbance spectrum (curve 5). In contrast to that the $\Delta A$ spectra measured at 760 fs (P ) (curve 1) or 17 ps (P HL ) (curve 3) coincide very well with the absorbance spectrum of P in its major part (820-890 nm) at 293K. The increase of the delay from 50 fs to 1 ps shows the gradual red shift of the center of the bleaching and its broadening for both intensities of the excitation.

Fig. 2 shows the kinetics of the center of the $\Delta A$ spectra upon stronger (o) and weaker (o) excitation. The center is shifted from 847 to 860 nm by monoexponential kinetics with a time constant of 250 $^{+}_{-}$50 fs in both cases.

The described effects with some longer excitation pulses were also observed and studied in details by Woodbury group [18]. They studied fs kinetics and $\Delta A$ spectra with excitation at short and long wavelength sides of the P band. In both cases a hole was burned and centered near the excitation wavelength and broadens within 1 ps. These results support the idea that observed effects reflect a dynamic hole burning in the P band.

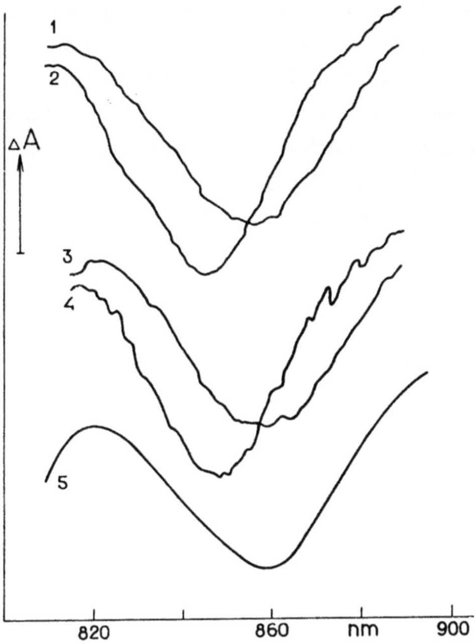


Fig.1 The spectra of absorbance changes (ΔA) of *R. sphaeroides* RCs at 293K excited by 90 fs pulses at 840 nm. The ΔA spectra measured at 760 fs (1) and 100 fs (2) delays using the excitation which bleaches ~30% of RCs. The ΔA spectra measured at 17 ps (3) and 70 fs (4) delays using the excitation which bleaches ~12% of RCs. The ΔA bar corresponds to 0.1 for curves 1 and 2 and to 0.04 for curves 3 and 4. Curve 5 shows the reversed absorption spectrum of RCs.

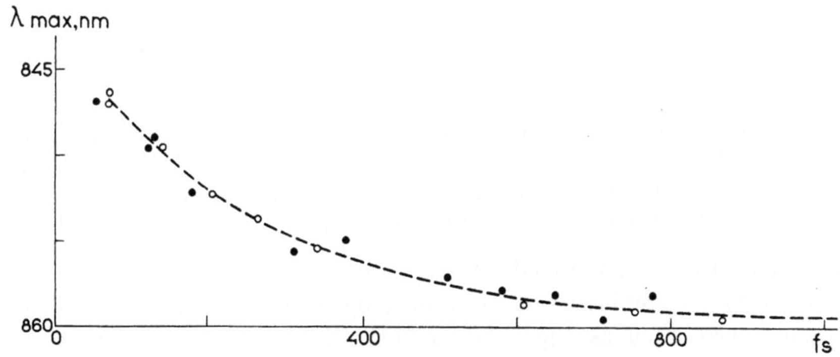


Fig.2 Kinetics of the shift of the ΔA peak for *R. sphaeroides* RCs at 293K excited by 90 fs pulses at 840 nm. The data set for 30% bleaching (o) and 12% bleaching (o). Dashed curve shows the exponential decay with a time constant of 250 fs.

Fig.3 shows the difference (light-dark) absorbance spectra in the range of 700-950 nm for modified RCs at different delays after excitation by 90-fs pulses at

845 nm. The absorbance changes in the P band were about 13% of the absorbance. At 0.2 ps delay the spectrum of ΔA reflects the excited state P* and is characterized by a bleaching of the P band at 865 nm and by a small bleaching at 813 nm which reflects the second excitonic component of the P transition. At latter delay a new bleaching centered at 805 nm is developed and at 2.2 ps approaches ~60% of maximal bleaching . At this delay no bleaching around 760 nm (H band) is observed. After 3 ps delay the bleaching at 760 nm is developed and at 4.5 ps approaches ~70% of maximal bleaching.

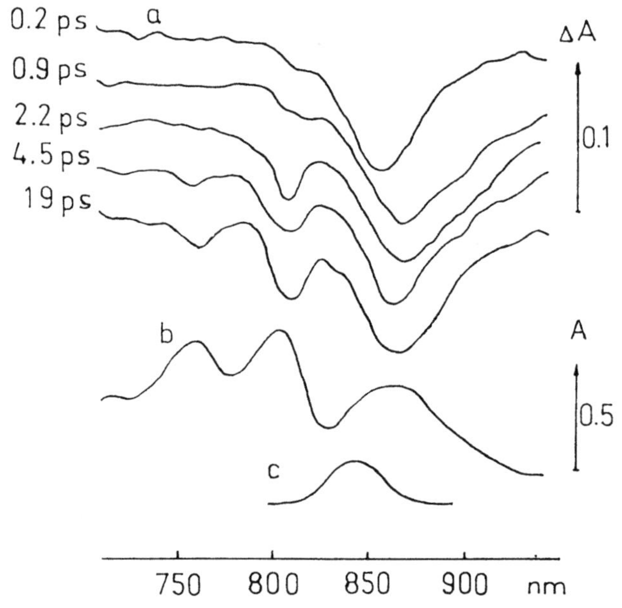


Fig. 3. (a) The difference (light-dark) spectra of borohydrate modified *Rhodobacter sphaeroides* (R-26) RCs excited by 90-fs pulses centered at 845 nm and measured at different delays (indicated by numbers in ps). (b) Absorbance spectrum of the sample measured by the same fs setup. (c) The spectrum of the fs excitation pulses.

For kinetic measurements around H band (760 nm), B band (805 nm) and P band (865 nm) the amplitudes of corresponding spectral troughs on the broad background were used as suggested in [13]. Fig 4 shows kinetics of the bleaching troughs at 760 nm and 805 nm and kinetics at 905 nm (stimulated emission from P*) normalized to the bleaching trough at 865 nm. The bleaching of the P band is instantaneous with excitation pulse (not shown). The first part of the bleaching of the B band is delayed by ~1.5 ps with respect to that of the P band probably reflecting the formation of the P+B- state. After 2.5-ps delay some relaxation of the B band is observed. The first part of the bleaching of the H band is delayed by ~3.5 ps with respect to that of the P band and ~2 ps to that of the B band. After 5 ps delay the H band bleaching is gradually increased up to ~9 ps. This bleaching is accompanied by the additional bleaching of the B band with approx. same kinetics. This behavior might be related to the relaxation of $P^+B^-$ energy level down to the level of $P^+H^-$ probably due to the polarization of protein medium.The thermal equilibrium between these two states becomes possible under these conditions [19]. The 905-nm stimulated emission kinetics are roughly described by 2.8 ps exponential curve (not shown) in agreement with earlier measurments [24].

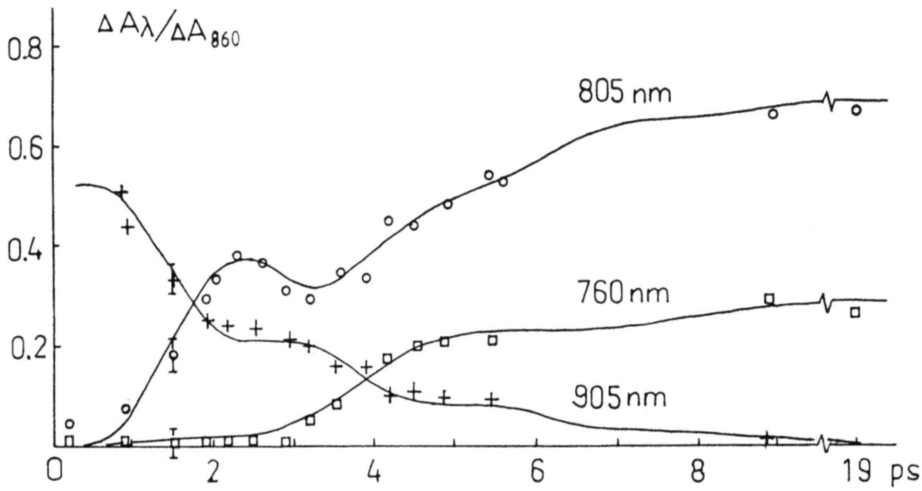


Fig.4. Kinetics of ΔA at 905 nm (simulated emission from P*) and the bleaching troughs around 805 nm (B band) and 760 nm (H band) normalized to the bleaching troughs at 860 nm (for determination of the spectral troughs see [13]). Solid curves are the modeling ones based on the wavepacket motions on the potential surfaces of the P* , $P^+B^-$ and $P^+H^-$ states (see Fig. 3 and text for details).

## 4. DISCUSSION

The data on the dynamic fs bleaching of the P band can be interpreted in two different ways. The bleaching of the short wavelength side of the P band and subsequent red shift and broadening of the bleaching measured from 50 fs to 1 ps in the P band excited at 840 nm can be related to (1) the Stokes shift of the stimulated emission from 840 to 900 nm during the measured kinetics (Fig.2) or to (2) the real red shift and broadening of the bleached P band. According to the measurements at 10K [10] the amplitude of the stimulated emission is not more than ~0.5 of that of the band bleaching. At 293K this emission amplitude is even less [20]. It means that the small amplitude of the bleaching centered at 865-875 nm at 70-100 fs delay is not due to the presence of the stimulated emission at 840 nm. Then the spectrum of ΔA measured at 17 ps (Fig.1) when no stimulated emission is present, shows that the bleaching spectrum coincides with the absorbance spectrum. Thus the observed dynamic broadening and red shift of the P band bleaching reflect the dynamic hole burning. Furthermore the direct measurements of the stimulated emission excited at short wavelength demonstrated that the stimulated emission appears at 50 fs at 910 nm and its Stokes shift approaches the maximal value (960 nm) at 150 fs at 10K and at <100 fs at 290K [11]. This time scale is much shorter than the kinetics of the band shift presented in Fig.2. Therefore we believe that the major part of the kinetics presented in Fig.2 reflects the dynamic hole burning of the P band. The data presented by Woodbury group [18] support this idea.

Excitation at short wavelength side of the P band at 10K shows the simultaneous (<30 fs) bleaching of the total band centered at 895 nm [11]. It means that at 10K the P band is bleached as a whole and there is no dynamic hole burning of the P band in fs time domain. One can suggest that at 10K the optical transition takes place from the zero vibronic level of the ground state. In contrast to that at 293K the optical transition probably occurs from the vibronic wave packet of the ground state formed due to the thermal activation by phonons of vibronic levels with n>1.

Since the vibronic wave packet ($\Psi(x,t)$) is a result of the superposition of the wave functions for several (ν+1) vibronic levels (see [4]) we have

$$\Psi_n(x) = 1/(1.77 * 2^n * 2\, x_o)^{0.5} \exp((-0.5\, x/x_o))\, H_n(x/x_o) \qquad (1)$$

where

$x^2 = h/(m\, \omega)$,

$H_n(x/xo) = (-1)^n \exp((x/x_o)^2)\, d^n \exp((-x/x_o)^2)/d(x/x_o)^n$ ,

$\omega$ is a frequency of the oscillation. Then

$$\Psi(x,t) = \Sigma^{\nu}{}_{j=0}\, C_{n+j}\, \Psi_{n+j} \exp(-i\omega(n+j+0.5)t) \qquad (2)$$

From eq (2) one can find the time dependence of the optical transition from the ground state wavepacket which is proportional to the overlapping of the vibronic wave functions of the ground and excited states. Since the optical transition should occur with the conservation of the nuclear oscillation momentum the overlapping is high for the left part of the potential curve (short wavelength transitions) and small for its right part (long wavelength transitions) due to displacement of the potential curves (Fig.5). This explains enhancement of short wavelength components and suppression of Stokes components in the P absorbance spectrum at 293K. This is not the case in the temperature range of 7-60K where the phonon energy is probably not enough to produce a vibronic wave packet and where the Stokes components are comparable with anti-Stokes ones [16].

Thus the data presented in Fig.1 and 2 can be interpreted in the following way. The fs excitation at 840 nm bleaches RCs which have the wavepacket at the left side of the potential curve. These RCs are not recovered for several ms. Then in an assembly of unbleached RCs the redistribution of the spectral components occurs over whole range of the P band and the $\Delta A$ spectrum becomes close to the P absorbance spectrum. The computer simulation of this effect shows that the time constant for the band shift roughly corresponds to half of the period of oscillation of the wave packet if this shift directly corresponds to the motion of the wave packet in unbleached RCs. Since the time constant is equal to ~250 fs (Fig.2) the period of 500 fs would correspond to the frequency oscillation of 67 $cm^{-1}$. Similar frequency was found in hole burning experiments [8], Raman spectra [9] and oscillation dynamics of the wave packet of the excited state [10].

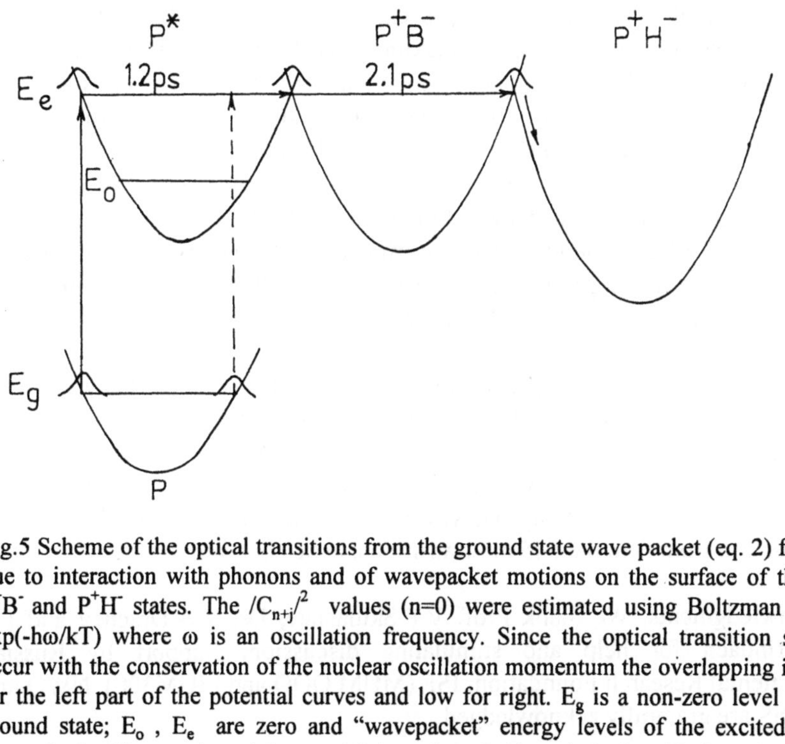


Fig.5 Scheme of the optical transitions from the ground state wave packet (eq. 2) formed due to interaction with phonons and of wavepacket motions on the surface of the P*, $P^+B^-$ and $P^+H^-$ states. The $/C_{n+j}/^2$ values (n=0) were estimated using Boltzman factor exp(-hω/kT) where ω is an oscillation frequency. Since the optical transition should occur with the conservation of the nuclear oscillation momentum the overlapping is high for the left part of the potential curves and low for right. $E_g$ is a non-zero level of the ground state; $E_o$ , $E_e$ are zero and "wavepacket" energy levels of the excited state, respectively. The numbers 1.2 ps and 2.1 ps show half period of oscillations (14 and 8 $cm^{-1}$, respectively) of the wavepackets on the P* and $P^+B^-$ potential surfaces, respectively, corresponding to the time of the wavepacket approach of the surfaces intersections.

The model of ET in Rcs is based on the assumption that after 90-fs excitation of P and relaxation of P* within ~300 fs the formation occurs of low frequency (~14 cm-1) wavepacket which moves along the ET coordinate (see Fig.5). After the half period of the wavepacket motion (~1.2 ps) on the potential surface of P*B state the intersection of P*B and P+B- occurs which is accompanied by a partial ET from P* to B and by a decrease of the stimulated emission by a factor 2.5. (In Fig. 4 the modeling curves are shown by solid lines). It means that 40% of the wavepacket are returned to the P*B surface and 60% are transferred to the P+B- surface. At this delay the B band bleaching is ~40% of the P band bleaching. It means that the difference extinction coefficient ($\Delta\varepsilon$) for active B at 805 nm is ~0.67 of the P band at 863 nm. Taking into account the value for $\Delta\varepsilon$ for P at 860 nm equal to 108 $mM^{-1}cm^{-1}$ [21] one can obtain that $\Delta\varepsilon$ for B is 72 $mM^{-1}$ $cm^{-1}$. This value is close to that for bacteriochlorophyll in solution [22].

It is also suggested that the wavepacket continues the motion on the $P^+B^-$ surface approaching the intersection with the $P^+H^-$ surface. The time for this motion is 2.1 ps as a half of 4.2- ps period (~8 $cm^{-1}$). The crossing the $P^+B^-$ and $P^+H^-$ surfaces produces about 65% of the maximal bleaching of the H band (Fig. 4). It means that almost all produced $B^-$ is converted to $H^-$. After ET from $B^-$ to H the B band should relax to the original state. This is not the case since this band is bleached at further delay showing only a small relaxation at 3 ps (Fig.4). To explain this it was suggested that this behavior is due to the protein relaxation around B molecule (induced by a formation of the $P^+H^-$ dipole) with a time constant of 0.83 ps producing the lowering the $P^+B^-$ energy which becomes close to the $P^+H^-$energy. It was shown [19] that this effect is temperature dependent and the B band is not bleached at 77K.

The fs data presented in Figs. 3 and 4 beginning from ~2-ps delay are very similar to the ps data described earlier [13] for modified Rcs supporting the fact that the electron transfer from P* to $B^-$ and from $B^-$ to H are separated in time. In earlier paper [13] it was shown that the H band at 545 nm is partially (~20%) bleached even at ~2 ps delay. However there is an indication [23] that some contribution from B to the 545 nm transition can be observed.

Acknowledgments. We thank Prof. V.P.Skulachev, Dr. V.A.Drachev and Dr. D.P.Krindach for help and stimulating discussion. Support by Russian Fundamental Research Foundation, ISF (MUM OOO and MUM300), NWO and INTASS are gratefully acknowledged.

## 5. REFERENCES

1 Karplus, M. and McCammon, J.A. (1983) A. Rev. Biochem. 53, 263-300.

2. Go, N., and Nishikawa, T. (1983) Proc. Natl. Acad. Sci. 80, 3696 - 3700.

3. Schulten, K. and Tesh, M. (1991) Chem. Phys. 158, 421-446

4. Sokolov A.A., Loskutov Yu.M., Ternov I.M. (1962) Quantum Mechanics. Moscow: State Education Publisher.

5. Peteanu, L.A., Schoenlein, R.W., Wang, Q., Mathies, R. and Shank, C.V. (1993) Proc. Natl. Acad. Sci. USA 90, 11762-11766.

6. Shuvalov V.A. (1990) Primary Light Energy Conversion at Photosynthesis. Moscow: Nauka

7. Arlt T., Schmidt S., Kaiser W., Lauterwasser C., Meyer M., Scheer H. and Zinth W. (1993) Proc. Natl. Acad. Sci. USA 90, 11757-11760

8. Shuvalov V.A., Klevanik A.V., Ganago A.O., Shkuropatov A.Ya. and Gubanov V.S. (1988) FEBS Lett. 237, 57-60

9. Cherepy N.J., Shreve A.P., Moore L.J., Franzen S., Boxer S.G. and Mathies R.A. (1994) J. Phys. Chem. 98, 6023-6029

10. Vos M.H., Jones M.R., Hunter C.N., Breton J., Lambry J.-C., Martin J.-L. (1994) Biochem. 33, 6750-6757

11. Vos M.H., Rappaport F., Lambry J.-C., Breton J. and Martin J.-L. (1993) Nature 363, 320-325

12. Shuvalov V.A., Shkuropatov A.Ya., Kulakova S.M., Ismailov M.A. (1986) Biochim. Biophys. Acta 849, 337-348

13. Shuvalov, V.A. and Duysens, L.N.M. (1986) Proc. Natl. Acad. Sci. USA 83, 1690-1694

14. Chekalin, S.N., Matveetz, Yu.A., Shkuropatov, A.Ya.,Shuvalov, V.A., Yartzev, A. (1987) FEBS Lett. 216, 245-248

15. Schmidt, S., Arlt, T., Hamm, P., Huber, H., Nagele, T., Wachtveitl, J., Meyer, M., Scher, H. and Zinth, W. (1994) Chem. Phys. Lett. 223, 116-120

16. Klevanik A.V., Ganago A.O., Shkuropatov A.Ya. and Shuvalov V.A. (1988) FEBS Lett. 237, 61-64

17. Streltsov A.M., Yakovlev, A.G., Shkuropatov A.Ya. and Shuvalov V.A. (1995) FEBS Lett. 357, 239-241

18. Peloquin J.M., Lin S., Taguchi A.K.W., and Woodbury N.W. (1995) J.Phys. Chem. 99, 1349-1356

19. Shuvalov, V.A. and Parson, W.W. (1981) Proc. Natl. Acad. Sci. USA 78, 957-961

20. Woodbury N.W., Becker M. Middendorf D. and Parson W.W. (1985) Biochem. 24, 7516-7521

21. Straley, S.C., Parson, W.W., Mauzerall, D.C., Clayton, R.K. (1973) Biochim. Biophys. Acta 305, 597-609

22. Fajer, J., Brune, D.C., Davis, M.S., Forman, A. and Spaulding, L.D. (1975) Proc. Natl. Acad. Sci. USA 72, 4956-4960

23. Shkuropatov, A.Ya. and Shuvalov, V.A. (1993) FEBS Lett.322, 168-171

24. Martin, J.-L., Breton, J., Hoff, A.J. and Antonetti, A. (1986) Proc. Natl. Acad. Sci. USA 83, 957-961

# Ultrafast Energy Transfer Within the Bacterial Photosynthetic Reaction Center

D.M. Jonas, M.J. Lang, Y. Nagasawa, S.E. Bradforth, S.N. Dikshit, R. Jimenez, T. Joo, and G.R. Fleming

Department of Chemistry and James Franck Institute
The University of Chicago, Chicago, IL 60637 USA

**Abstract.** Energy transfer from the accessory bacteriochlorophylls to the special pair within the reaction center requires ~120 fs.

**Keywords.** energy transfer, reaction center, transient absorption

## 1 Introduction

Prerequisites for an understanding of the primary processes in photosynthesis -- energy and electron transfer -- are: (1) knowledge of the structures of the antenna and reaction center proteins, (2) knowledge of the energetics of the species involved, (3) understanding of the strength and range of the electronic interactions between the prosthetic groups (chromophores), (4) knowledge of the strength, distributions, and timescales of the chromophore-protein coupling, and (5) understanding how distributions in parameters influencing the energetics and electronic coupling strengths affect experimental observations.

The energetics are of great interest for determination of the mechanism of primary charge separation[1-8], but a major difficulty is that electron transfer in the reaction center is sensitive to the relative energies of orbitals on different chromophores (donor LUMO and acceptor LUMO). Neither the strength of the electronic interactions nor the energetics are sufficiently known experimentally to resolve the controversies about the electron transfer mechanism. Unfortunately, there seems to be no direct method of determining the relative orbital energies on different chromophores in the reaction center if the protein precludes electrochemical measurements[9]. By contrast, energy transfer[10-14] is sensitive to the spectroscopically observable orbital energy differences on the donor compared to those on the acceptor (there may be practical difficulty in determining which electronic states are involved). For this reason, we have studied accessory bacteriochlorophyll (B) to special pair (P) energy transfer in reaction centers with the aim of shedding light on the coupling between the pigments of the reaction center[15].

## 2 Energy Transfer in the Reaction Center

Figure 1 shows the room temperature absorption spectra of normal and oxidized reaction centers from *Rhodobacter sphaeroides* (*Rb. sphaeroides)* R26. Oxidation was achieved chemically by means of 100 mM potassium ferricyanide [16]. Note the complete removal of the band at 870 nm assigned to the lower exciton component of the special pair and the small blue shift of the band(s) around 800 nm, which in normal and oxidized reaction centers is expected to be dominated by absorption from the accessory bacteriochlorophylls [9].

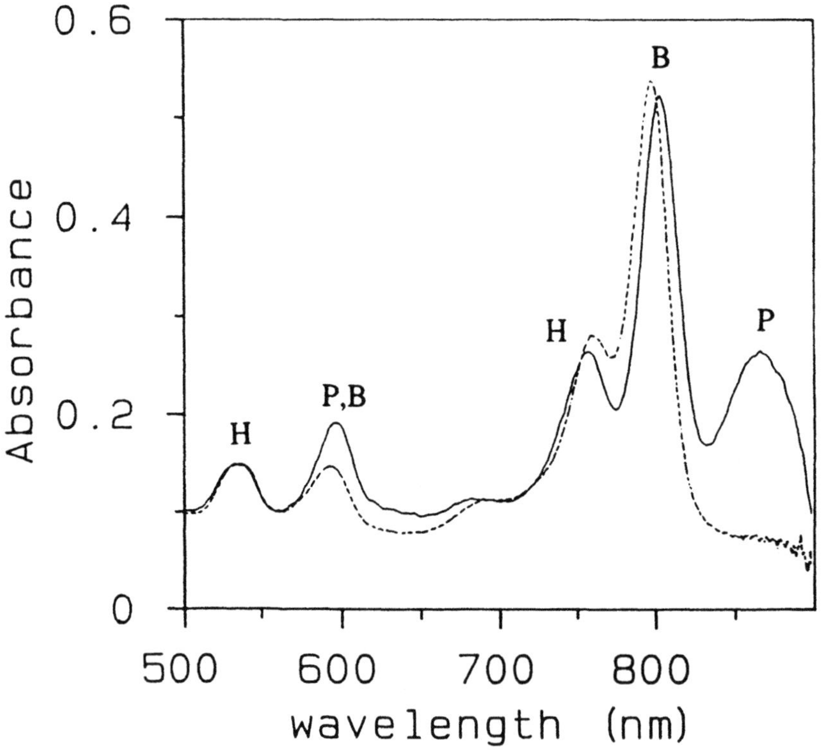


**Fig. 1.** Absorption spectra of normal ( —— ) and oxidized ( ------- ) reaction centers from *Rb. sphaeroides* R26 recorded at 293K. The bands are labeled P (special pair), B (accessory bacteriochlorophylls), H (bacteriopheophytins) according to the dominant absorbing species in that region. Note the absence of the special pair absorption band at 870 nm in oxidized reaction centers.

Figure 2 shows fluorescence up-conversion data from R26 reaction centers in BOG detergent which had been reduced with 30 mM sodium ascorbate[17]. B was excited using 90 fs pulses with an 11 nm bandwidth centered at 797 nm from a Ti:Sapphire laser (Mira 900F, Coherent) and the P* fluorescence polarized perpendicular to the excitation pulse was detected over an ~10 nm bandpass around 930 nm. The pulse energy was 0.4 nJ in a 50 μm spot diameter, the repetition rate was 76 MHz, and the sample was stirred with an electric toothbrush in a static cell. The fluorescence decays bi-exponentially in accord with the charge separation kinetics[17]. No fluorescence was observed when the reaction centers were chemically oxidized with 100 mM potassium ferricyanide, suggesting that all detected fluorescence comes from functional reaction centers which undergo charge separation after energy transfer.

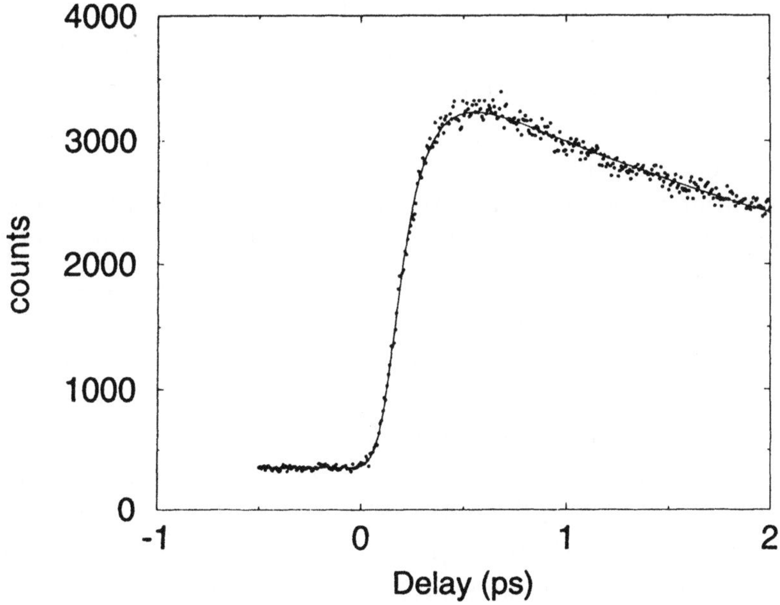


**Fig. 2.** Fluorescence up-conversion trace from reduced *Rb. sphaeroides* R26 reaction centers excited using 90 fs pulses with 11 nm bandwidth centered at 797 nm. Fluorescence at 930 nm was time resolved by sum frequency mixing with a gate pulse (a replica of the 797 nm pump) and detecting an 8 nm bandpass of upconverted light at 429 nm. The asymmetry between the lower and upper portions of the initial rise (which is not observed when exciting $P_-$ directly at 860 nm) indicates that a rise time is required to fit the data. The fitted curve gives a rise time of 120 fs.

A rise time is clearly required to fit the fluorescence upconversion data in Figure 2, and the best fit using convolution with the 127 fs FWHM instrument response gives a rise time of 120 fs. The result is consistent with the statement of Stanley and Boxer [18], that a rise time was required to fit their P* emission at 955 nm following excitation of B at 803 nm. A fluorescence upconversion study of B*→P* energy transfer is also reported by Michel-Beyerle and coworkers in this volume. The rise time in Figure 2 cannot be interpreted directly as an energy transfer rate for a number of reasons. First, spectral evolution of the stimulated emission from P*, after direct excitation with 45 fs pulses has been observed [19]. The timescale of the relaxation was estimated to be ninety to a few hundred fs and is thus comparable to the rate of energy transfer. Second, energy transfer may well proceed through the upper exciton component of P ($P_+$) and subsequent relaxation of $P_+$ to $P_-$ will enter the fluorescence rise time. (Subscripts + and - denote excitonically coupled states which are symmetric and antisymmetric under exchange of the two identical chromophores.) Finally, the B*→P* process may involve electronic or vibrational coherences which are not describable at the level of rate constants[20, 21].

In order to begin to address these issues, we have carried out pump-probe studies of B with much higher time resolution[15]. A cavity-dumped Kerr-lens mode-locked Ti:Sapphire laser [22] generating 20-27 fs pulses at repetition rates between 5 and 152 kHz at about 800 nm center wavelength was used as the pump and probe source. Excitation pulse energies were kept below 1 nJ (in a 50 μm spot diameter) to minimize saturation effects, the repetition rate was 10 kHz and the sample (*Rb. sphaeroides* R26 in LDAO detergent; OD(800 nm) = 0.05 in the 100 μm pathlength cell) was flowed at a velocity of 100 cm/s by a peristaltic pump. This flow rate is sufficient to ensure a fresh sample volume for each pulse at 10 kHz repetition rate. From a measurement of the pump beam absorption by the sample (5%), we calculate that about 15% of the reaction centers have been electronically excited by each pump pulse. The total volume of ~3 mL was cooled by ice water.

Figure 3 shows a magic angle detected pump-probe transient for *Rb. sphaeroides* R26. The initial bleaching decays very rapidly and is followed by a rising signal which has a time constant in accord with the charge separation kinetics - ~3 ps [13]. This latter component presumably arises from the well-known band shift of B, resulting from the formation of $P^+ H_A^-$ [23]. Careful inspection of higher signal to noise data obtained previously at 152 kHz [15] reveals oscillatory behavior in the rapidly decaying portion, but the amplitude of the oscillations is sufficiently weak and the signal decay sufficiently rapid that we have not been able to specify frequency(ies) with any precision.

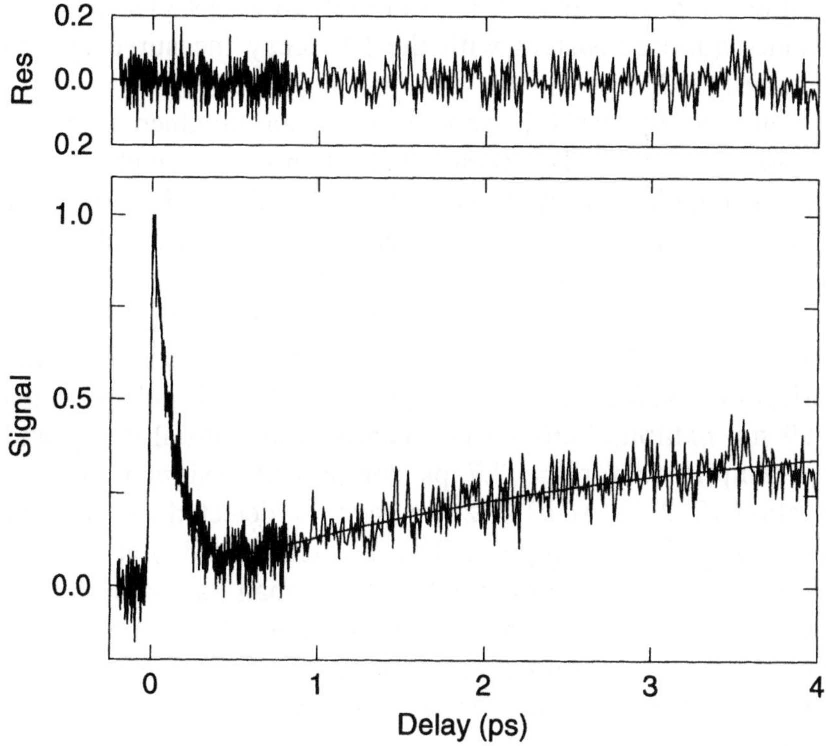


**Fig. 3.** Transient absorption of *Rb. sphaeroides* R26 reaction centers recorded with 24 fs pump and probe pulses of 49 nm bandwidth (FWHM) centered at 799 nm. The signal was recorded with the probe polarization at the magic angle to the pump polarization. The pulse energy was 0.9 nJ, the spot size was 50 μm, the repetition rate was 10 kHz, and the flow rate was 100 cm/s, which is sufficiently rapid to ensure that each sample volume was refreshed between pump pulses. The fit shown is the convolution of $0.718\times\exp(-t/120\text{ fs}) + 0.282\times(1 - \exp(-t/2.7\text{ ps}))$ with the autocorrelation. The residual errors between experiment and fit are shown in the top panel.

The small amplitude of the oscillations results in part from the broad band width (40-50 nm) of the pump and probe pulses [24, 25] and in part from the small displacement of vibrational modes between ground and excited B [26]. Finding a unique fit for the data in Figure 3 is not straightforward. Best fits to several magic angle data sets give a decay time for the rapid drop of 120-150 fs. The fitting is made difficult by the presence of a coherence spike of unknown amplitude, by the weak oscillations in the data and by the fact that the long time constant (~3 ps) is poorly defined by fits over this time range (0-4 ps). Fits going out to 20 ps give long-time constants quite consistent with the P* decay measured by other techniques[17, 27].

In an earlier paper[15], we presented data (obtained at 152 kHz repetition rate with pulses centered at 805 nm) for which the pump and probe polarizations were parallel to each other. The experiments at 152 kHz have a higher signal to noise ratio than data obtained at 10 kHz, but each sample volume was excited by many (~20-80) laser pulses at 152 kHz and the pump-probe signal was not zero when the probe arrived at the sample before the pump. The data for parallel polarization obtained at 10 kHz repetition rate with pulses centered at 799 nm exhibits both a significantly faster initial decay and a larger amplitude for the 2.7 ps component assigned to charge separation[17, 27]. At 10 kHz, we have not detected a pump-probe signal at negative pump-probe delays. Data recorded at the magic angle (Fig. 3) have a slower initial decay than data recorded with parallel pump and probe polarizations under identical conditions. Pump-probe data obtained with perpendicular pump and probe polarizations are also consistent with a rapid decrease in absorption anisotropy.

A change in the anisotropy might be expected because energy transfer from B to P (either $P_+$ or $P_-$) will replace the bleach due to excitation of B with a bleach of the ground state of P, thus reducing the $P \rightarrow P_+$ absorption band of the reaction center near 815 nm[20, 28]. The apparent absence of a $P \rightarrow P_+$ residual bleach after the initial decay and before the 3 ps rise suggests the pump-probe signal also contains contributions from absorption by an excited state of P. Quantitative simulations will be needed to determine whether energy transfer proceeds through $P_+$ or goes directly to $P_-$. The issue of whether B is monomeric or excitonically coupled (via P) must also be considered, however. If the two B's are excitonically coupled, as suggested by Won and Friesner [1], then possible contributions of $B_- \rightarrow B_+$ relaxation to the signal must also be considered.

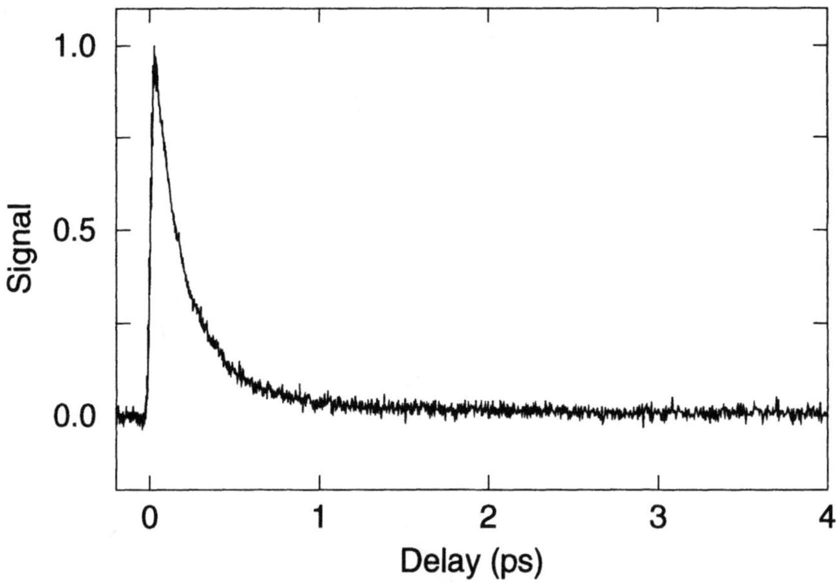


**Fig. 4.** Transient absorption of *Rb. sphaeroides* R26 reaction centers, chemically oxidized by 100 mM ferricyanide, recorded with parallel pump and probe polarizations. The pulses had a duration of 27 fs, were centered at 803 nm with a bandwidth of 38 nm, and had a pulse energy of 6 nJ. The flow rate was 80 cm/s, the spot size was 70 μm, and the repetition rate was 152 kHz, indicating that each sample volume was excited many times in succession. Despite the high repetition rate, the pump-probe signal is zero to within the noise when the probe arrives before the pump.

Addition of ferricyanide to the reaction centers oxidizes P and eliminates the 870 nm absorption band attributed to $P_-$. Data from a parallel polarization pump and probe experiment on an oxidized sample at 152 kHz are shown in Fig. 4. For oxidized reaction centers, the pump-probe signal is zero at negative delays, even at 152 kHz repetition rate. Immediately after the pump-probe measurements on the chemically oxidized reaction centers, it was verified that reduction with sodium ascorbate restored the spectrum of the normal reaction centers, including the P→P_ absorption band at 870 nm. This shows that the oxidation was reversible and presumably had not yet damaged the reaction center. Oxidation slightly lengthens the initial rapid decay, which is bi-exponential with ~120 fs and ~400 fs decay components, and eliminates the ~3 ps rise (Fig. 4), strengthening the assignment of this rise to charge separation in normal reaction centers.

## 3 Mechanism of the Energy Transfer

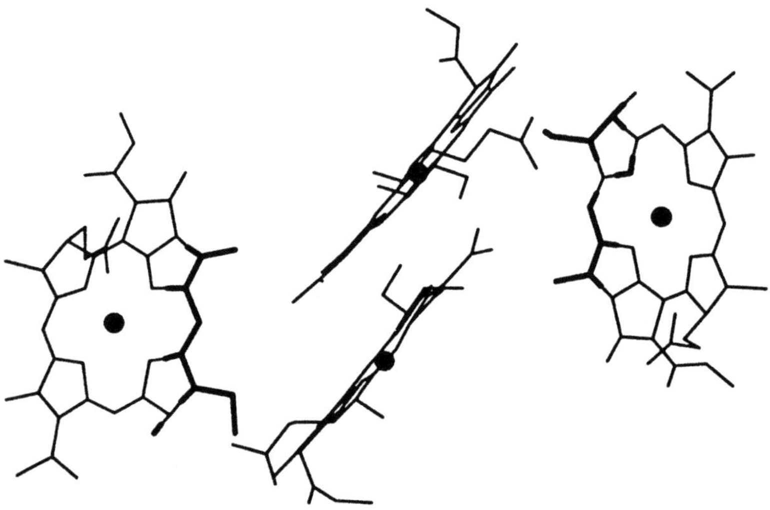

**Fig. 5.** Partial structure of the *Rb. sphaeroides* reaction center showing the proximity between the special pair (central pair, P) and the accessory bacteriochlorophylls (outer pair, B). The view is from the periplasmic face with the approximate $C_2$ axis of the reaction center pointing out of the plane of the figure. Half of each bond extending from B nuclei that lie within 5Å of the special pair is shown as a thick line. Reproduced from ref. [15], © 1995, American Chemical Society.

Figure 5 shows a view of the locations of P and B within the reaction center complex of *Rb. sphaeroides* [29]. Significant portions of both B molecules lie within 5Å of the special pair (thick lines in the figure) and weak coupling models for the energy transfer process are likely to be inadequate. Conventional (weak coupling) Förster theory [10] gives a rate proportional to the overlap integral :

$$J_{Forster} = \int F(\nu)\, \varepsilon(\nu)\, \nu^{-4} d\nu \tag{1}$$

where $F$ is the fluorescence quantum spectrum of the donor and $\varepsilon$ is the molar extinction coefficient of the acceptor as a function of frequency, $\nu$. In normal reaction centers, the observed energy transfer is an order of magnitude faster than the maximum Förster rate ($\tau$ = 2 ps) calculated for B$\rightarrow$P energy transfer [20]. The rate calculated by Dexter type exchange transfer [11] is proportional to:

$$J_{Dexter} = \frac{\int F(\nu)\, \varepsilon(\nu)\, \nu^{-4} d\nu}{\int F(\nu)\nu^{-3}\, d\nu \quad \int \varepsilon(\nu)\nu^{-1}\, d\nu} \tag{2}$$

which is just a product of Franck-Condon factors. In comparing expressions (1) and (2) it is clear that, while strong absorption and emission is required for rapid Förster energy transfer, in the exchange case, although spectrally overlapping Franck-Condon regions are required, the optical transitions between upper and lower states on the donor and acceptor need not be strong.

The rapid bleach recovery in oxidized reaction centers seems qualitatively inconsistent with Förster dipole-dipole energy transfer. Recently, Reimers and Hush [30] have calculated the electronic spectrum of the special pair cation $P^+$ using the INDO/S-CI method. They find four excited states of $P^+$ below B* and one slightly above. In particular, the transition designated as $Q_y$ ($L^+$) (i.e. a $Q_y$ transition largely localized on the $P_L$ member of the special pair) is calculated to be at 826 nm and assigned (by Gaussian fitting) to lie at 847 nm in the experimental spectrum of Reed [31]. This upper state of the special pair cation appears, based on Franck-Condon requirements, to be a likely acceptor in the oxidized reaction centers despite the small transition dipole.

## 4 Conclusion

After excitation of B at 800 nm, absorption by B recovers in ~130 fs and the 930 nm emission from $P_-$ rises in ~120 fs. The bandwidths and pulse durations differ between the two experiments, but the similarity between these two timescales suggests either direct energy transfer from B to $P_-$ or an extremely short lived intermediate, possibly $P_+$. The ~130 fs energy transfer from B to P within the photosynthetic reaction center is accompanied by a change in the pump-probe anisotropy. The rapid bleach recovery attributed to energy transfer is followed by a bleach increase with a time constant of about 2.7 ps which we attribute to the electrochromic shift of B upon electron transfer from P to H and the bleach of $P \rightarrow P_+$ absorption at 815 nm. Reaction centers which have been chemically oxidized with potassium ferricyanide also exhibit a rapid bleach decay in less than 150 fs which is not followed by the 2.7 ps rise. Our results are both quantitatively (in the case of normal reaction centers) [20] and qualitatively (in the case of oxidized reaction centers) inconsistent with weak coupling Förster theory. In normal reaction centers, one possible explanation which seems to be consistent with Franck-Condon requirements is that energy transfer proceeds via $P_+$ and is followed by rapid internal conversion to $P_-$. This hypothesis is qualitatively consistent with our polarization data, but quantitative simulations are necessary to distinguish between energy transfer through $P_+$ and direct energy transfer to $P_-$. A detailed description of energy transfer within the normal and oxidized reaction centers probably requires inclusion of short range electronic interactions, such as those considered by Scholes and Ghiggino [14].

## Acknowledgment

We would like to thank James R. Norris, David Tiede, Julia Popov, and Maxim Popov for the reaction center samples and helpful advice. This work was supported by the National Science Foundation. Y.N. thanks the JSPS for a postdoctoral fellowship for research abroad. R.J. acknowledges the NPSC for a graduate fellowship.

## References

1. R.A. Friesner and Y. Won, *Biochim. Biophys. Acta*, **77**, 99 (1989).

2. M. Bixon, J. Jortner, and M.E. Michel-Beyerle, *Biochim. Biophys. Acta*, **1056**, 301 (1991).

3. M. Marchi, J.N. Gehlen, D. Chandler, and M.J. Newton, *J. Am. Chem. Soc.*, **115**, 4178 (1993).
4. J.N. Gehlen, M. Marchi, and D. Chandler, *Science*, **263**, 499 (1994).
5. A. Warshel and W.W. Parson, *Ann. Rev. Phys. Chem.*, **42**, 279 (1991).
6. A. Warshel, Z.T. Chu, and W.W. Parson, *Journal of Photochemistry and Photobiology A - Chemistry*, **82**, 123 (1994).
7. R. Egger and C.H. Mak, *J. Phys. Chem.*, **98**, 9903 (1994).
8. S. Schmidt, T. Arlt, P. Hamm, H. Huber, T. Nagele, J. Wachtveitl, M. Meyer, H. Scheer, and W. Zinth, *Chem. Phys. Lett.*, **223**, 116 (1994).
9. R.K. Clayton, *Photosynthesis: physical mechanisms and chemical patterns,* Cambridge University Press, New York (1980).
10. T. Förster, *Delocalized excitation and excitation transfer,* in *Modern Quantum Chemistry,* O. Sinanoglu, Editor. Academic Press, Inc., New York (1965) p. 93.
11. D.L. Dexter, *J. Chem. Phys.*, **21**, 834 (1953).
12. R. van Grondelle, J.P. Dekker, T. Gillbro, and V. Sundstrom, *Biochim. Biophys. Acta*, **1187**, 1 (1994).
13. G.R. Fleming and R. van Grondelle, *Physics Today*, February, 1994, p. 48.
14. G.D. Scholes and K.P. Ghiggino, *J. Phys. Chem.*, **98**, 4580 (1994).
15. Y. Jia, D.M. Jonas, T. Joo, Y. Nagasawa, M.J. Lang, and G.R. Fleming, *J. Phys. Chem.*, **99**, 6263 (1995).
16. V. Nagarajan, W.W. Parson, D. Davis, and C.C. Schenck, *Biochem.*, **32**, 12324 (1993).
17. M. Du, S.J. Rosenthal, X. Xie, T.J. DiMagno, M. Schmidt, D.K. Hanson, M. Schiffer, J.R. Norris, and G.R. Fleming, *Proc. Natl. Acad. Sci. USA*, **89**, 8517 (1992).
18. R.J. Stanley and S.G. Boxer, *J. Phys. Chem.*, **99**, 859 (1995).
19. M.H. Vos, J.-C. Lambry, S.J. Robles, D.C. Youvan, J. Breton, and J.-L. Martin, *Proc. Natl. Acad. Sci. USA*, **89**, 613 (1992).
20. J.M. Jean, C.-K. Chan, and G.R. Fleming, *Israel Journal of Chemistry*, **28**, 169 (1988).
21. M.H. Vos, M.R. Jones, C.N. Hunter, J. Breton, J.-C. Lambry, and J.-L. Martin, *Biochem.*, **33**, 6750 (1994).

22. M.S. Pshenichnikov, W.P. de Boeij, and D.A. Wiersma, *Opt. Lett.*, **19**, 572 (1994).
23. J. Breton, J.-L. Martin, G.R. Fleming, and J.-C. Lambry, *Biochem.*, **27**, 8276 (1988).
24. D.M. Jonas, S.E. Bradforth, S.A. Passino, and G.R. Fleming, *J. Phys. Chem.*, **99**, 2594 (1995).
25. D.M. Jonas and G.R. Fleming, *Vibrationally Abrupt Pulses in Pump-Probe Spectroscopy,* in *Ultrafast processes in Chemistry and Photobiology,* M.A. El-Sayed, I. Tanaka, and Y.N. Molin, Editor. Blackwell Scientific, London (1995) p. 225.
26. N.R. Cherepy, A.P. Schreve, L.J. Moore, S. Franzen, S.G. Boxer, and R.A. Mathies, *J. Phys. Chem.*, **98**, 6023 (1994).
27. M.H. Vos, J.-C. Lambry, S.J. Robles, D.C. Youvan, J. Breton, and J.-L. Martin, *Proc. Natl. Acad. Sci. USA*, **88**, 8885 (1991).
28. A. Warshel and W.W. Parson, *J. Am. Chem. Soc.*, **109**, 6152 (1987).
29. C.-H. Chang, O. El-Kabbani, D. Tiede, J. Norris, and M. Schiffer, *Biochem.*, **30**, 5352 (1991).
30. J.R. Reimers and N.S. Hush, *J. Am. Chem. Soc.*, **117**, 1302 (1995).
31. D.W. Reed, *J. Biol. Chem.*, **244**, 4936 (1969).

# Slow Charge Separation in a Minority of Reaction Centers Correlated with a Blueshift of the $P_{860}$-Band in *Rb. Sphaeroides.*

G. Hartwich, H. Lossau, A. Ogrodnik and M. E. Michel-Beyerle
Institut für Physikalische und Theoretische Chemie, TU München
Lichtenbergstr. 4, D-85747 Garching

**Abstract.** The fluorescence decay pattern of the primary donor state $^1P^*$ in *Rb. sphaeroides* R26 as a function of quinone ($Q_A$) content and excitation (or detection) wavelength has been investigated with a 40ps instrumental response function. The open question, whether the fluorescence components between 40ps and 1ns originate from slow charge separation or from recombination of energetically unrelaxed $P^+H_A^-$-states ($H_A$ denoting bacteriopheophytin) is addressed by comparing $Q_A$-containing reaction centers (RCs) with $Q_A$-depleted ones. The removal of $Q_A$ increases the recombination fluorescence lifetime of $P^+H_A^-$ from about 100ps to tens of nanoseconds. The key observation reported in this paper is that the fluorescence decay components in the range of 100ps to 1ns are not affected by the $P^+H_A^-$ lifetime. Thus, these components reflect slow charge separation of a minority (<4% of the RCs). Moreover, this minority manifests itself in a blue-shifted $Q_y$-absorption band of the primary donor ($P_{860}$-band) detected in the time resolved fluorescence excitation spectra of the 100ps-1ns components, and also affects the steady state fluorescence excitation spectrum. The fluorescence components faster than 40ps representing the sample majority as well as the delayed recombination fluorescence (>10ns) exhibit no shift. So far, all RC preparations investigated show these features of a blue shifted $P_{860}$-band and charge separation in the 100ps time-scale, indicating that the RCs of *Rb. sphaeroides* are intrinsically heterogeneous.

**Keywords.** *Rb. sphaeroides*, fluorescence decay, slow charge separation, superexchange interaction

**Abbreviations.** RC, reaction center; *Rb. sphaeroides, Rhodobacter sphaeroides;* P, primary donor; $H_{A,B}$, $B_{A,B}$, and $Q_{A,B}$, bacteriopheophytin, monomeric bacteriochlorophyll and ubiquinone at their binding sites in the reaction center; OD, optical density; EET, excitation energy transfer; IRF, instrumental response function.

## 1. Introduction

Recent absorption and stimulated emission studies with improved S/N ratio [1] as well as recent spontaneous emission studies [2-5] reveal that the primary donor ($^1P^*$) decay in reaction centers (RCs) of *Rb. sphaeroides* is multi-exponential within the first 25ps. The decay can be characterized by two decay times (~ 2ps and ~ 10ps) with relative amplitudes of about 70% and 30%. The origin of non exponential decay patterns has been discussed in terms of a parking state at the inactive B-branch [5-7] and inhomogeneity of charge separation [2,8-13]. Excited state fluorescence measurements on a longer time-scale additionally show lifetime components in the time range of 40ps-1ns (amplitudes <4%) [3,10,12,14] and a minor component with tens of nanoseconds (amplitude $<10^{-4}$) [10,15-18,]. Since in quinone ($Q_A$) depleted RCs the lifetime of the charge separated state $P^+H_A^-$ ($H_A$ denoting bacteriopheophytin) is 13-40ns, depending on temperature and magnetic field, the latter fluorescence component can be attributed to the $P^+H_A^-$ recombination [27].

So far, the origin of the 40ps to 1ns fluorescence components has been discussed in terms of two models: (i) In the dynamic solvation model the fluorescence originates from recombination of an initially energetically non-relaxed $P^+H_A^-$ state. The amplitude of this fluorescence decreases as a function of time due to an energetic stabilisation of $P^+H_A^-$ by the reorganisation of the surrounding medium [12-15]. (ii) In the inhomogeneous charge separation model a distribution of charge separation rates is reflected in these fluorescence components with increasing time constants and decreasing amplitudes [2,10].

In the presence of $Q_A$ the $P^+H_A^-$-lifetime is reduced from ≈10ns to 100-200ps [14,26]. In this paper we utilize the possibility of manipulating the $P^+H_A^-$-lifetime by $Q_A$ extraction and reconstitution to determine the origin of the 40ps to 1ns fluorescence. The central goal is to discriminate between the two models (i) and (ii). Due to the short lifetime of $P^+H_A^-$ in the $Q_A$-containing preparation these fluorescence components are expected to vanish in model (i), while they are expected to be insensitive in model (ii). Furthermore, a spectral characterisation of these components will be presented.

## 2. Materials and Methods

### 2.1 Preparation

RCs from *Rb. sphaeroides* R26 were prepared by standard methods [19], quinone extraction was achieved according to [20]. Residual $Q_A$-content was determined according to [21] to be less than 1%. In order to optimize comparability of data from $Q_A$-free and $Q_A$-containing RCs, the latter samples were prepared from the same stock of Q-free RCs by quinone reconstitution. This was achieved by incubation with a 50fold molar excess of ubiquinone dissolved in ethanol/dimethylsulfoxide (1:2, 5mg/ml) for 4h at 0°C. The $Q_A$-content was determined to be higher than 95%.

A ~ 150 μm thin sample of the RCs ($OD_{860}$ ~ 0.1) in TL-buffer (aqueous buffer at pH = 8 containing 10mM Tris, trishydroxymethylaminemethane and 0.1 Vol% LDAO, lauryldimethylaminoxide) is sandwiched between the long sides of two rectangular prisms forming a rectangular cube. Two orthogonal faces of the cube were positioned perpendicular to the direction of excitation and fluorescence detection, orienting the film in 45° to both directions while thus minimising refraction effects.

### 2.2 Steady State Fluorescence

The fluorescence excitation spectra were recorded on a photon counting spectrometer [22,23]. Because of the perpendicular configuration of the excitation and detection axis in this apparatus the detected fluorescence is dichroic. Thus, the fluorescence intensity $F(\lambda_e, \lambda_d)$ was recorded as a function of excitation wavelength $\lambda_e$ both with the polarisation of emission parallel, $F_{\|}(\lambda_e, \lambda_d)$ and perpendicular, $F_{\perp}(\lambda_e, \lambda_d)$ to that of excitation. The unpolarised spectra were calculated according to $F = F_{\|} + 2F_{\perp}$. The emission was detected at $\lambda_d$ = 930nm, approximately being the emission maximum of the special pair $^1P^*$.

The fluorescence intensity $F(\lambda_e, \lambda_d)$ is proportional to the amount of photons absorbed per second, $[I_0(\lambda_e)-I(\lambda_e)]\cdot\lambda_e/hc$, to the quantum yield of the excitation energy transfer (EET) from the excited pigments to $^1P^*$, $\Phi_{EET}(\lambda_e, \lambda_d)$, and to the fluorescence quantum yield $\Phi_F(\lambda_e, \lambda_d)$. One obtains [23]:

$$\frac{F(\lambda_e,\lambda_d)}{I_0(\lambda_e)}\cdot\frac{hc}{\lambda_e} \propto \Phi_{EET}(\lambda_e,\lambda_d)\cdot\Phi_F(\lambda_e,\lambda_d)\cdot\left(1-10^{-OD(\lambda_e)}\right) \qquad (1)$$

We define the fluorescence excitation spectrum $Fl(\lambda_e,\lambda_d) := F(\lambda_e,\lambda_d)/I_0(\lambda_e)\cdot hc/\lambda_e$, which by comparison with the absorption $(1\text{-}10^{-OD(\lambda_e)})$ allows the determination of $\Phi_{EET}(\lambda_e, \lambda_d)$ without determination of $\Phi_F(\lambda_e, \lambda_d)$, when spectra are normalized at the $P_{860}$-band, where $\Phi_{EET}(\lambda_e, \lambda_d) = 1$. This applies for homogeneous emission. For a sample with inhomogeneous decay kinetics of the excited state, the fluorescence intensity $F(\lambda_e, \lambda_d)$ is the sum of fluorescence intensities of the individual fluorescent species i with concentration $\rho_i$ in a distribution of RCs. In absence of delayed fluorescence the quantum yield $\Phi_{i,F}(\lambda_e, \lambda_d)$ can be expressed by the corresponding lifetime $\tau'_i(\lambda_e, \lambda_d)$ and its radiative transition rate $k_{i,F}(\lambda_d)$:

$$\begin{aligned}\frac{F(\lambda_e,\lambda_d)}{I_0(\lambda_e)}\cdot\frac{hc}{\lambda_e} &= \sum_i \rho_i\cdot\Phi_{i,EET}(\lambda_e,\lambda_d)\cdot\left(1-10^{-OD_i(\lambda_e)}\right)\cdot k_{i,F}(\lambda_e,\lambda_d)\cdot\tau'_i(\lambda_e,\lambda_d)\\ &= \sum_j a_j(\lambda_e,\lambda_d)\cdot\tau_j(\lambda_e,\lambda_d)\end{aligned} \qquad (2)$$

*In practice* the time resolved fluorescence will be characterised by a limited amount of *representative* components (section 3.3). Thus, the lifetimes obtained in a fitting procedure are average lifetimes $\tau_j$ depending on the amount of admitted components, the inspected time window and on the constraints of a global fitting procedure. Assuming identical radiative rates $k_{i,F}(\lambda_d)$, the fit-amplitudes $a_j$ will essentially be a corresponding average of the absorption characteristics $1\text{-}10^{-OD_i(\lambda_e)}$, the quantum yield of energy transfer $\Phi_{i,EET}(\lambda_e,\lambda_d)$ and the concentration $\rho_i$ of each species i.

### 2.3 Time resolved fluorescence

For comparison of $Q_A$-free and $Q_A$-containing RCs the samples were excited at 864nm with 40ps pulses from a Hamamatsu PLP-01 laser diode at 10 MHz repetition rate. The fluorescence decay was detected at the magic angle with time correlated single-photon-counting-technique [10]. The dependence of the fluorescence decay on the excitation wavelength (770nm to 900nm) was determined by using a modelocked Ti:$Al_2O_3$-laser (Coherent Mira Basic) pumped by an $Ar^+$-laser (Coherent Innova 425) exciting the sample at a repetition rate of 76MHz . The instrumental response function (IRF) was ~ 40ps in both systems (full width at half maximum).

Time constants are extracted from the fluorescence decay pattern by fitting a convolution of the instrumental response function and exponential decay functions to the data using the Levenberg-Marquardt method [24]. Deconvolution of $Q_A$-free and $Q_A$-containing RCs at different excitation wavelengths was done simultaneously in a global analysis with identical, but free lifetimes and with individual amplitudes as free running parameters. Quality of the fit is judged by observing the residuals and values of the reduced $X_v^2$.

## 3. Results and Discussion

### 3.1 Influence of $Q_A$-content on the fluorescence decay pattern

The lifetime of $P^+H_A^-$ increases from 200ps in presence of $Q_A$ to ~12ns in absence of $Q_A$ at room temperature and from 130ps [14,25] to ~23 ns [21,26] at 85K, respectively. For discussing the influence of the $P^+H_A^-$-lifetime on the fluorescence decay the fluorescence decay is crudely divided into three regimes (i-iii). i) The fast fluorescence which is below our time resolution and reflects primary charge separation of the majority of RCs, ii) the intermediate components in the subnanosecond region, the origin of which is subject of this paper, and iii) the nanosecond components, which can be attributed to delayed fluorescence due to recombination from the state $P^+H_A^-$ to $^1P^*$ [10,15,17,27].

The quantification of the fluorescence decay data (Table 1) has been achieved by a global fitting procedure, assuming an identical set of time constants for both $Q_A$-free and $Q_A$-containing RCs, but allowing for different amplitudes. Individual fits of the various decay traces do not yield a unique set of time constants for the intermediate components, so that they apparently do not represent kinetically distinct states of the system. Therefore, these time constants should rather be regarded as a description of a continuous distribution of lifetimes. The fast primary charge separation kinetics (fast prompt fluorescence) is taken to be independent of $Q_A$-content as subpicosecond studies have shown [2], and the lifetime (not resolved in our experiments) was taken from the literature.

The manipulation of the $P^+H_A^-$-lifetime by the presence or absence of $Q_A$ leads to marked differences in the fluorescence decay pattern of $^1P^*$ (Figure 1). These differences concentrate on the amplitude of the ns-delayed fluorescence, which is substantially higher in $Q_A$-free RCs than in reconstituted ones. For the amplitudes of the intermediate components only minor differences can be seen in the fluorescence decay of $Q_A$-free and $Q_A$-reconstituted RCs (Figure 1 and Table 1).

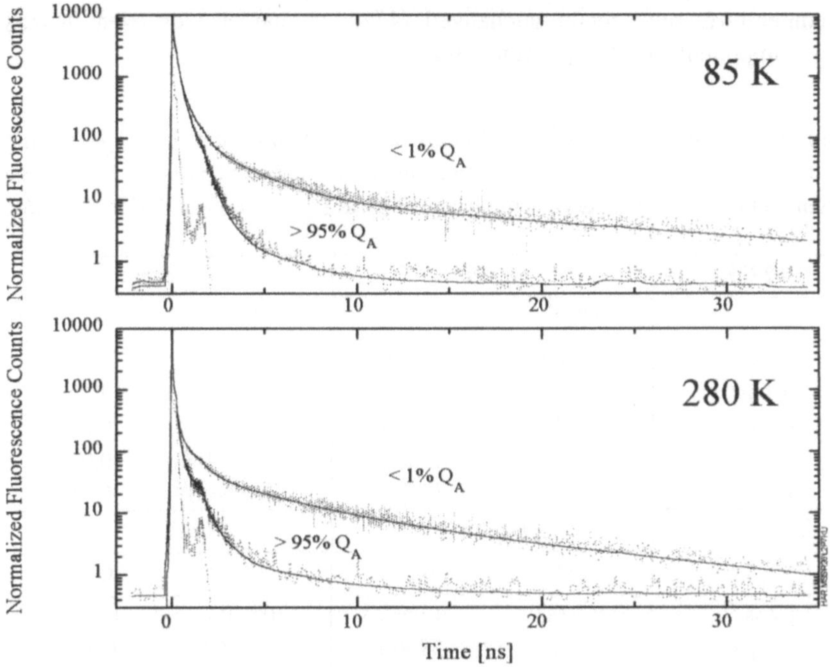


**Fig. 1.** Decay of the 920nm fluorescence of *Rb. sphaeroides* RCs excited at 864nm in absence and presence of $Q_A$.

At 280K the slowest component with a time constant of ~ 13ns originates from delayed fluorescence of $P^+H_A^-$-recombination in $Q_A$-depleted RCs. Its amplitude becomes negligible when $Q_A$ is reconstituted. This finding is in good agreement with the $Q_A$-content determined to be less than 1% for the $Q_A$-depleted sample and >95% for the reconstituted one. The small contribution of a component with a 3.7ns time constant is much more pronounced in the $Q_A$-depleted sample than in the reconstituted one, thus behaving similarly to the 13ns component. The intermediate fluorescence is represented by components with time constants of 103ps and 648ps. Their amplitudes are almost independent of the $Q_A$-content.

At 85K the lifetime of the component attributed to recombination from $P^+H_A^-$ in $Q_A$-free RCs is increased to ~ 17ns. The amplitude of this recombination fluorescence component shows a much weaker temperature dependence as one would expect from the temperature dependence of the Boltzmann-factor with $\Delta G = 0.21\text{eV}$. This unusually weak temperature dependence has been shown to be due to a significant energetic heterogeneity of the state $P^+H_A^-$ [10]. As for the 280K data this amplitude nearly vanishes on $Q_A$-reconstitution. The amplitude of the 2.3ns component again behaves in a similar way as the 17ns component upon $Q_A$-reconstitution. Going to lower temperatures the amplitudes of the intermediate components in $Q_A$-free RCs increase by a factor of 3-4 on lowering the temperature, but remain unchanged (472ps-component) or even increase (125ps-component) upon $Q_A$-reconstitution.

**Table 1.** Multiexponential global fit of fluorescence decay of $Q_A$-free and $Q_A$-containing RCs of *Rb. sphaeroides* with lifetimes $\tau$, amplitudes $a(\tau)$ and relative fluorescence quantum yield $\Phi_F^{rel}$ for each decay component at 280K ($\chi^2 = 1.02$) and 85K ($\chi^2 = 1.06$).

| T = 280K | $Q_A$-free | | $Q_A$-cont. | |
|---|---|---|---|---|
| $\tau$ | $a^{rel}$ c) | $\Phi_F^{rel}$ | $a^{rel}$ | $\Phi_F^{rel}$ |
| 3.5ps a) | 100,000 | 49.9% | 100,000 | 72.2% |
| 103ps | 695 | 10.2% | 779 | 16.6% |
| 648ps | 137 | 12.6% | 75 | 10.0% |
| 3.5ns | 27 | 13.5% | 1.3 | 0.9% |
| 13.3ns b) | 7.3 | 13.8% | <0.1 | 0.3% |
| T = 85K | $Q_A$-free | | $Q_A$-cont. | |
| $\tau$ | $a^{rel}$ | $\Phi_F^{rel}$ | $a^{rel}$ | $\Phi_F^{rel}$ |
| 1.2ps a) | 100,000 | 13.5% | 100,000 | 15.1% |
| 125ps | 2,055 | 28.9% | 2,921 | 46.0% |
| 472ps | 647 | 34.4% | 635 | 37.8% |
| 2.3ns | 50 | 13.0% | 3 | 0.9% |
| 16.8ns b) | 5.4 | 10.2% | <0.1 | 0.2% |

a) Literature data [2-5,25,28], for T = 280K an average value of 3.5ps for $\tau_1$ ~2.3ps (79%) and $\tau_2$ ~ 10ps (21%) was taken.

b) The longest ns time constant has been obtained from a global fit of the fluorescence data collected in time window expanded to 80ns.

c) All amplitudes are normalized relative to the amplitude of the shortest lifetime, $a_1$=100000.

The intermediate fluorescence components constitute an integral part of functional RCs after isolation from the membrane. This conclusion is supported by (i) time resolved fluorescence excitation spectra (section 3.3) as well as by the (approximate) independence of these fluorescence amplitudes of (ii) sample preparation, (iii) detection wavelength in the range of 900nm to 950nm and (iv) excitation power. At high average excitation power, where a steady state concentration of more than 2/3 of the RCs is pumped into the state $P^+Q_A^-$, the amplitude of components which do not originate from functional RCs relative to the amplitude of the fastest component should increase by a factor of at least 3.

For the intermediate components only a slight increase of 20-40% is observed, as it is expected due to the minor contribution of internal conversion (see below).

This 3ns component, however, increases by a factor of 3. This allows us to identify this component in $Q_A$-reconstituted RC to be mainly due to free pigment. Indeed this lifetime matches with that of bacteriochlorophyll in solution [36]. RC preparations investigated prior to $Q_A$-extraction may show a significantly higher amount of this component. Apparently the procedure of $Q_A$-extraction serves as an additional cleaning step. For this reason we preferred $Q_A$-reconstituted samples as a reference for our comparative study. It should be noted, that the 3ns component is much larger in $Q_A$-free RCs prior to reconstitution. In $Q_A$-free RC it is dominated by a process related to $P^+H_A^-$-recombination, supposedly reflecting dynamic solvation.

Other groups have also shown that the main part of the intermediate components does not originate from sample impurities like free pigments or inactive RCs [12]. Furthermore these components are also found in membrane bound RCs, which have suffered less preparational strain [37].

*Origin of the intermediate fluorescence components.*

In a dynamic solvation model employed for $Q_A$-reduced RCs the intermediate components are assigned to solvation processes in the state $P^+H_A^-$. In the simplest form of this model all fluorescence arises from a single population of reaction centers. The large number of degrees of freedom inherent in the protein matrix gradually stabilizes the charge separated state $P^+H_A^-$, which initially lies close to $^1P^*$. During the lifetime of $P^+H_A^-$ this state runs through several substates with decreasing energy. From these substates delayed recombination fluorescence is expected, the amplitudes of which decrease according to the energetic relaxation.

In $Q_A$-containing RCs the lifetime of $P^+H_A^-$ decreases to 100-200ps and thus all delayed recombination fluorescence components with a longer lifetime should vanish. This is not the case: at 280K the intermediate component with 648ps time constant is only slightly decreased. Moreover, at 85 K, where the overall contribution of the intermediate fluorescence components is much larger, the amplitude of this intermediate component is invariant to $Q_A$-reconstitution. This independence of $Q_A$-content (and therefore on the lifetime of $P^+H_A^-$) reveals that most of the intermediate fluorescence components do not originate from recombination of the state $P^+H_A^-$ They are due to prompt fluorescence reflecting slow charge separation processes. The amplitudes of these fluorescence components therefore represent the concentration of a small subpopulation (< 4%, Table 1) and the time constants are characteristic for the long $^1P^*$-lifetimes of this minority resulting from a distribution of the primary charge separation rates.

The occurrence of very slow charge separation on the time scale of several 100ps of course is limited by other decay processes of the $^1P^*$-state. The intrinsic lifetime in absence of charge separation cannot be observed directly, but a crude estimate is obtained from RC-mutants, which are not able to undergo primary charge separation (e. g. the $D_{LL}$- and $D_{LL\text{-}2a}$-mutant of *R. capsulatus* [29]). There, a considerable contribution of fluorescence components with lifetimes up to ~1ns was found (Appendix). Assuming a similar intrinsic $^1P^*$ lifetime for the RCs under investigation, 35% of the RCs represented by the 648ps component and only 10% of the RCs represented by the 100ps component will decay by processes other than charge separation (most likely by internal conversion).

The predominant involvement of charge separated states in the intermediate components is supported by the electrical field effect observed on the fluorescence, since it is largest on these components. According to [22, 30] the intermediate fluorescence increases by a factor of ~2 at $1 \cdot 10^6$ V/cm. Such a strong electrical field effect is expected only, if a charge separated state is involved and cannot be due to processes involving solely neutral states, such as internal conversion. Indeed in the $D_{LL}$-mutant, i. e. in absence of charge separation, the lifetime of $^1P^*$ was found to change only by 1.5% at $5.4 \cdot 10^5$ V/cm [31].

This large field effect observed on the intermediate components dominates the effect on steady state emission. Its anisotropy with respect to the electric field carries important information with respect to the route of the slow charge separation process reflected in the intermediate fluorescence components. Steady state DELFY-experiments (dichroic excitation spectra of electric field modulated fluorescence yield) [32,33] revealed the orientation of the electric dipole moment responsible for the electrical field effect. This has been determined to be identical with that of $P^+H_A^-$. Thus, the process occurring on the time scale of the intermediate fluorescence components is direct charge separation from $^1P^*$ to $P^+H_A^-$. We conclude that a minority of less than 4% of the RCs performs superexchange mediated charge separation from $^1P^*$ to $P^+H_A^-$ via high lying states $P^+B_A^-$. Since the majority performs a two step charge separation with an activationless primary electron transfer from $^1P^*$ to $P^+B_A^-$ [4,5,34,35] a strong energetic inhomogeneity of the $P^+B_A^-$ states is demanded. Due to a significant activation energy electron transfer from $^1P^*$ to $P^+B_A^-$ cannot compete with the superexchange mechanism in this minority.

Important features of the intermediate components is their spectral fingerprint represented in the sections 3.2-3.3 as well as their strong increase upon lowering the temperature. This increase upon cooling can easily be understood from such an energetic inhomogeneity induced by conformational substates of the protein. At lower temperatures the effect of this inhomogeneity on the dispersion of the charge separation dynamics to $P^+B_A^-$ becomes significantly stronger [9], thus more RCs are forced to carry out activationless superexchange mediated charge separation to $P^+H_A^-$. A minor part of the increase of the intermediate fluorescence amplitudes components at 85K originates from the temperature dependent redshift of the primary donor $Q_y$-band ($P_{860}$-band). Since the various components have different spectral characteristics (section 3.3), the ratio of intermediate relative to prompt fluorescence components is increased because the excitation wavelength is held constant at 864nm.

Spectral characterization of this minority of RCs performing slow charge separation we acquired steady state and time resolved fluorescence excitation spectra.

### 3.2 Steady State Fluorescence Excitation Spectrum

In Figure 2 the steady state fluorescence excitation spectrum of $Q_A$-containing RCs at 85K is compared to the simultaneously recorded absorption spectrum. The fluorescence excitation spectrum is corrected for dichroic effects (Materials and Methods, section 2.2) and the maxima of the $P_{860}$-band in both spectra have been normalized to the same amplitude.

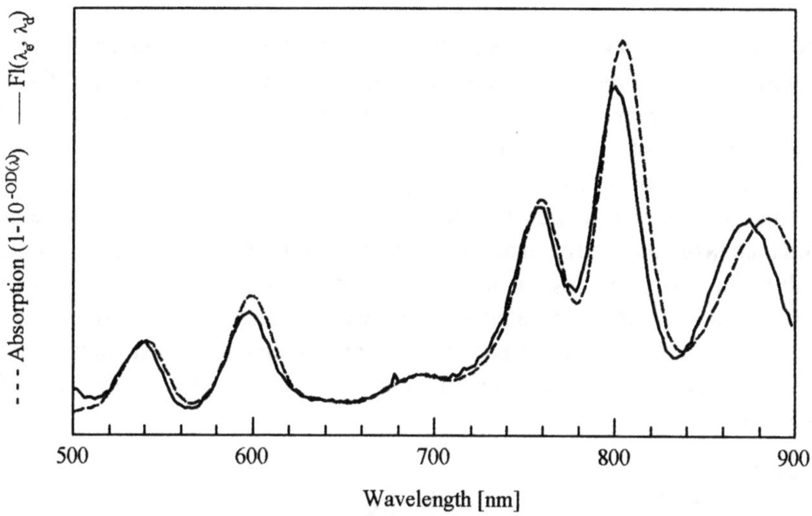


**Fig. 2.** Simultaneously recorded 85K fluorescence excitation $Fl(\lambda_e)$ (---) and absorption $1-10^{-OD(\lambda_e)}$ (—) of $Q_A$-containing *Rb. sphaeroides* RCs. $Fl(\lambda_e)$ is detected at 930nm and normalized to P-absorption $(1-10^{-OD(878\,nm)})$.

The two spectra differ in essentially two features: (i) the $P_{860}$-band is blue-shifted by 230 $cm^{-1}$ and (ii) the 800nm band at its red wing is depressed in the fluorescence excitation spectrum as compared to the absorption spectrum.

Energy transfer plays no role when excitation is done directly into the $P_{860}$-band. Thus, the only possible explanation for the blue-shift is that the fluorescence quantum yield $\Phi_F(\lambda_e)$ is inhomogeneous within the $P_{860}$-band (eq. 1). According to Table 1 the steady state fluorescence excitation spectrum is dominated by the intermediate fluorescence components and therefore represents the absorption characteristics of the sample subspecies from which these components originate (eq.2). This dominance results from the long lifetimes of these components inspite of their small amplitudes. The small amplitudes indicate that the corresponding subspecies constitute a sample minority, thus making a negligible contribution to the absorption signal. Apparently the blue shift of the fluorescence excitation spectrum reveals a blueshifted absorption of this sample minority, which in terms of dispersive charge separation dynamics carries out slow charge separation. This finding thus manifests an inhomogenous $P_{860}$-band and a correlation between energetically high lying states of $^1P^*$ and slow charge separation.

The explanation for the depression at the red wing of the 800nm band may involve decreased excitation energy transfer and/or altered absorption characteristics for some of the lifetime components.

The conjecture that the intermediate fluorescence components should predominate in the blue wing of the $P_{860}$-band will be put to test in section 3.3.

### 3.3 Influence of the excitation wavelength on timeresolved fluorescence

In Figure 3 the fluorescence decay at 85K of the $Q_A$-reconstituted RCs ($Q_A$-content >95%) is represented by the fit curves derived from a global analysis of the data for different excitation wavelengths in the range of 770nm to 900nm.

The parameters of a global fit of all decay curves are given in Table 2. Time constants emerging from this global fit differ from those of Table 1 being in accordance with the notion that they are merely representative time constants discribing a more or less continuous distribution of decay times with a significant wavelength dependence. Here we are dealing with time constants presenting an average for all wavelengths. Any wavelength dependence therefore is forced to show up in wavelength dependent amplitudes of the individual components.

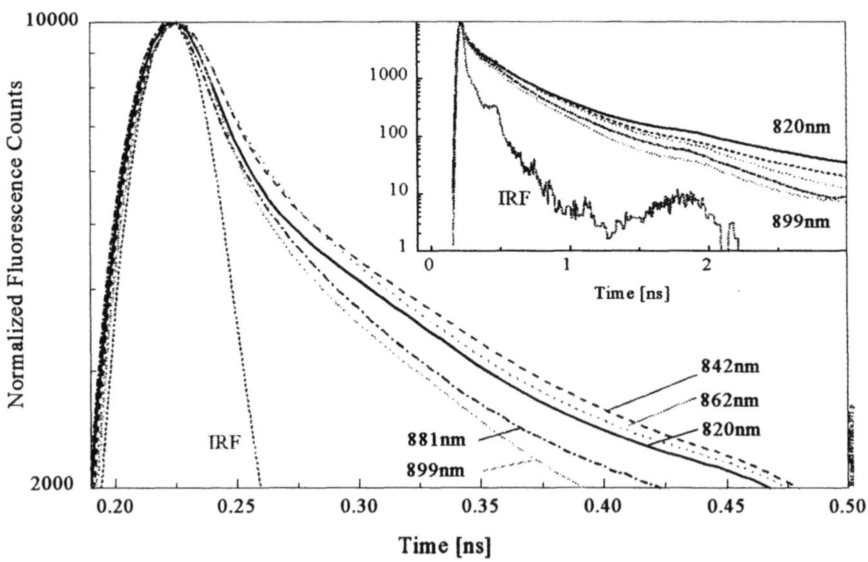


**Fig. 3.** Fit curves of the 920 nm fluorescence decay of $Q_A$-containing *Rb. sphaeroides* RCs at 85K at different excitation wavelengths. Data are collected in a 11ns time window, displayed is the region of the intermediate components. The curves are normalized to the amplitude of the 1.2ps component (set to 10,000). Inset: display of the decay within the first 3ns.

**Table 2.** Multiexponential fit of the 85K fluorescence decay of $Q_A$-containing RCs of *Rb. sphaeroides* after excitation at wavelengths $\lambda_e$ with lifetimes $\tau$ and amplitudes $a(\tau)$ for each decay component (global fit together with the data for $Q_A$-free RCs; $\chi^2 = 1.4$).

| T = 85 K | a(1.2ps)[a)] | a(66ps) | a(342ps) | a(1.23ns) | a(17ns)[b)] |
|---|---|---|---|---|---|
| 779 nm | 100000 | 3685 | 1291 | 306 | 84.3 |
| 791 nm | 100000 | 3500 | 1228 | 132 | 24.3 |
| 800 nm | 100000 | 4437 | 1487 | 111 | 13.6 |
| 810 nm | 100000 | 4077 | 1168 | 109 | 7.5 |
| 820 nm | 100000 | 4395 | 1392 | 181 | 6.8 |
| 831 nm | 100000 | 4756 | 1590 | 172 | 1.9 |
| 842 nm | 100000 | 5589 | 1784 | 112 | 2.2 |
| 851 nm | 100000 | 6552 | 1918 | 98 | 3.3 |
| 862 nm | 100000 | 5638 | 1887 | 66 | 3.0 |
| 871 nm | 100000 | 5598 | 1661 | 43 | 3.9 |
| 881 nm | 100000 | 5192 | 1566 | 34 | 6.7 |
| 891 nm | 100000 | 5073 | 1253 | 28 | 6.4 |
| 899 nm | 100000 | 4952 | 1111 | 25 | 6.0 |

a) Literature data [3-5,25,28] for time constant below our resolution are used. Amplitude of the 1.2ps component is set to 100,000; the other amplitudes are normalized to this value.
b) Data of $Q_A$-free RCs. Since the time window is limited to 11ns the 17ns time constant was taken from Table 1 (excitation at 864nm) and held fixed.

The 66ps component with a relative amplitude a(66ps) = 3.7% at 779nm shows a weakly pronounced maximum around 851nm (with a relative amplitude of 6.6%, Table 2). The 665ps component with a relative amplitude of less than 2% essentially shows the same wavelength dependent behaviour as the 66ps component. In contrast, the 1.23ns component steadily increases to shorter wavelengths. This component is mainly responsible for the different ordering of the decay curves displayed in a larger time window in the inset of Figure 3. The contribution of $P^+H_A^-$ recombination fluorescence originating only from max. 5% of the RCs which happened to be not $Q_A$-reconstituted (time constant of 17ns) is to small to allow the detection of a significant excitation wavelength dependence. Its amplitude relative to the fastest component amounts to less than $3 \cdot 10^{-6}$ in the range from 820nm to 900nm. Instead, in Table 2 the 17ns components observed in $Q_A$-free RCs are displayed. However, due to the high repetition rate of the Ti-sapphire laser only a time window limited to 11ns is accessible. Thus, the obtained amplitudes are considerably spoiled by contributions from components in the 1-3ns regime, in particular when going to shorter wavelengths.

For a better comparison of the influence of the excitation wavelength on the individual time components we constructed the excitation spectra of the yield $\tau_j \cdot a_j$ of the individual components by normalizing them to the steady state excitation spectrum $Fl(\lambda_e)$ according to equation 2, see Figure 4.

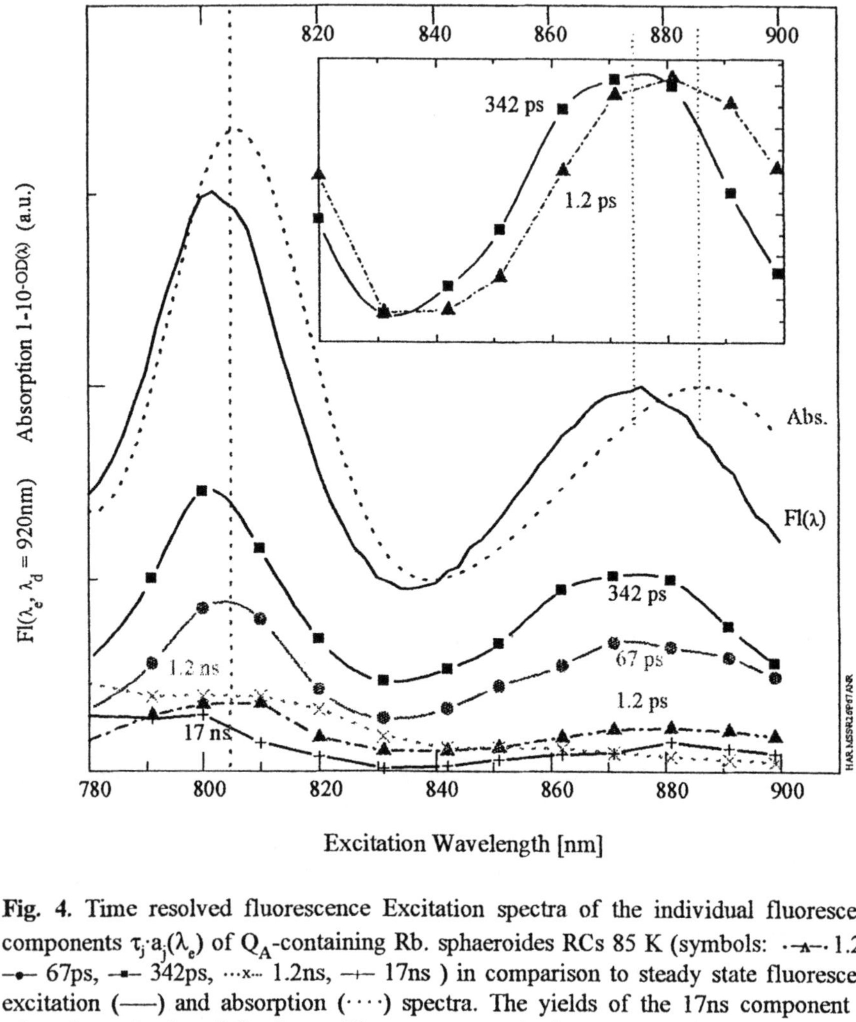


**Fig. 4.** Time resolved fluorescence Excitation spectra of the individual fluorescence components $\tau_j \cdot a_j(\lambda_e)$ of $Q_A$-containing Rb. sphaeroides RCs 85 K (symbols: ·-▲-· 1.2ps, -●- 67ps, -■- 342ps, ···x··· 1.2ns, -+- 17ns ) in comparison to steady state fluorescence excitation (—) and absorption (····) spectra. The yields of the 17ns component are taken from $Q_A$-free RCs. Inset: Fluorescence excitation spectra of the 1.2ps and the 305ps component normalized to identical maximum intensity in the primary donor band.

The following discussion of the time resolved fluorescence excitation spectra will focus on (a) the monomer absorption region around 800nm which is depenent on excitation energy transfer and (b) the primary donor region which is not.

*(a) The 800nm absorption region:* Normalizing the excitation spectra of the individual time components in Figure 3 to equal intensity at the maximum of the

$P_{860}$-band (not shown) reveals the origin of the depression of the 800nm-band in the steady state fluorescence excitation spectrum: The 1.2ps component shows a spectrum that is almost identical to the 800nm absorption band. For this component the intensity ratio of the 800nm-band relative to the $P_{860}$-band is the same as in the absorption spectrum indicating an efficiency of excitation energy transfer from B to P of ~100%. The depression in the steady state fluorescence excitation spectrum is caused by the 67ps and 342ps component. These components show an overall depression which is more pronounced at the red side of the 800nm-band as compared to the absorption spectrum. The 17ns curve is almost identical to the 1.2ps curve, but with an increased intensity below 800nm. In the $Q_A$-reconstituted sample its contribution is to small to be reflected in the steady state fluorescence excitation spectrum. The 1ns component has almost no $P_{860}$-band and steadily increases up to 779nm thus slightly contributing to the depression in the red wing of the 800nm band and shifting this band to lower wavelengths.

In conclusion the spectral changes in the 800nm region of the steady state excitation spectrum essentially result from the intermediate fluorescence components representing a sample minority. A small contribution to the band shift is due to the 1ns component originating from free pigment.

*(b) The primary donor absorption region:* The 1.2ps component representing the great majority of RCs shows a maximum that is close to the absorption maximum of the $P_{860}$-band (Figure 4, inset). The slight blue shift of the 1.2ps excitation spectrum displayed may be due to the fact that additional components with time constants up to 40ps (not resolved here) are represented in this component. The 342ps component dominating the steady state excitation spectrum is responsible for the blue-shift of the $P_{860}$-band (Figure 4). A similar spectral behaviour results for the 66ps component. The amplitude of the 1.23ns component shows almost no band in this region, but rather steadily increases to shorter wavelength. On the basis of the excitation spectrum (Figure 4) and its lifetime, we conclude that this component results from the far red wing of free pigment emission.

As already pointed out above, this component is much stronger prior to $Q_A$-reconstitution and thus reflects in $Q_A$-free RCs mainly delayed recombination fluorescence originating from high lying $P^+H_A^-$-states. In agreement the fluorescence excitation spectrum of this component in $Q_A$-free RCs indeed shows a pronounced $P_{860}$-band (not shown).

The 17ns component closely follows the absorption spectrum and therefore reflects the behaviour of the majority of the RCs (note that in fig. 4 the amplitudes of the 17ns component of a $Q_A$-free preparation are displayed). It was found previously that this delayed fluorescence from $P^+H_A^-$-recombination is strongly affected by an energetic inhomogeneity of this radical pair state, favouring the emission of the high lying states [10]. The missing blue shift on this fluorescence component indicates that it does not mainly originate from the blue shifted minority of RCs with slow charge separation. Thus slow charge separation is not correlated to energetically high lying states of $P^+H_A^-$. This is expected, if the speed of charge separation is determined by the energetic location of the state $P^+B_A^-$. Thus the finding that $P^+B_A^-$ is the primary acceptor state of charge separation in the majority of reaction centers [5,34, Häberle *et al.*, this volume] is corroborated by the missing spectral correlation between the 17ns component and the intermediate components. On the other hand there has to be a correlation between the energetic distribution of $^1P^*$- and $P^+B_A^-$-states in the sense, that the activation energy for the $^1P^* \rightarrow P^+B_A^-$ reaction is larger for the minority with blue shifted $^1P^*$ states.

In summary, we have shown that the blue shift of the $P_{860}$-band in the fluorescence excitation spectrum is due to a blue shifted $P_{860}$-absorption of a minority of RC responsible for the intermediate components. In consequence the $P_{860}$-band is inhomogeneous with energetically high lying states of $^1P^*$ performing slow superexchange mediated charge separation as discussed in section 3.1.

## 4. Conclusions

(1) In RCs of *Rb. sphaeroides* R26 intermediate (40ps - 1ns) fluorescence components from $^1P^*$ are almost independent of the lifetime of $P^+H_A^-$, proving that they do not originate from recombination of energetically unrelaxed $P^+H_A^-$-states.

(2) Due to their small amplitudes the intermediate components originate from a minority of RCs with slow, superexchange mediated charge separation to $P^+H_A^-$. This is corroborated by electric field effects being most pronounced on these components and by DELFY-experiments revealing that this process is governed by a dipole moment being parallel to that of $P^+H_A^-$.

(3) The unpolarized steady state fluorescence excitation spectrum at 80K deviates from the RC absorption, exhibiting a blue-shifted $P_{860}$-band. Time resolved fluorescence excitation spectra assign this blue-shift to the absorption properties of the intermediate fluorescence components. This proves that the $P_{860}$-band is inhomogenous with the energetically high lying states of $^1P^*$ performing slow superexchange mediated charge separation to $P^+H_A^-$. Charge separation of the majority of RCs forming $P^+B_A^-$ in the first step [4,34,35,38-40] is corroborated and an energetic correlation between the energetically dispersed $^1P^*$ states with an even larger distribution of $P^+B_A^-$-states is suggested to achieve such a mechanistic change.

**Acknowledgements**
This work was supported by the Deutsche Forschungsgemeinschaft (SFB 143). Tilman Häberle is gratefully acknowledged for providing his data processing software for the global fit procedure.

**Appendix**

*Fluorescence decay of RCs of the $D_{LL}$-mutant and the $D_{LL\text{-}2a}$-mutant of R. capsulatus* :

RCs of the $D_{LL}$-mutant of *R. capsulatus* are not able to undergo primary charge separation [29, 41]. The fluorescence lifetime of these mutants reveals the intrinsic lifetime of $^1P^*$- in the absence of quenching due to charge separation and may serve as an estimate for this time constant in *Rb. sphaeroides.* Since the oxidation potential of P in the $D_{LL\text{-}2a}$-mutant is more similar to that of *Rb. sphaeroides* we compare its fluorescence decay pattern with that of the $D_{LL}$-mutant. In both mutants fluorescence components faster than 200ps are missing. Thus we conclude that also in the $D_{LL\text{-}2a}$-mutant no charge separation occurs. Instead considerable fluorescence components with time constants of ~800ps at room temperature as well as at 85K were found (Table 3). Furthermore the excitation wavelength dependence of the relative amplitude of the 800ps component in the $D_{LL}$-mutant is similar to that of the intermediate fluorescence components of RCs of *Rb. sphaeroides* R26, but less pronounced.

**Table 3.** Biexponential global fit of fluorescence decay of the $D_{Ll\text{-}2a}$- and the $D_{LL}$-mutant after excitation at 864nm at 300K and 85K. The fraction of the amplitudes $a(\tau)$ for each decay component at 280K and 85K is given in brackets.

| $D_{LL}$ | | $D_{LL\text{-}2a}$ | |
|---|---|---|---|
| 300 K | 85 K | 300 K | 85 K |
| 230ps (80%) | 400ps (70%) | 210ps (55%) | 460ps (51%) |
| 650ps (20%) | 830ps (30%) | 845ps (45%) | 890ps (49%) |

**References**

[1] M.H. Vos, J.-C. Lambry, S. Robles, D.G. Youvan, J. Breton and J.-L. Martin **(1991)** Proc. Nat. Acad. Sci., USA, *88*: 8885-8889.

[2] M. Du, S.J. Rosenthal, X. Xie, T.J. DiMagno, M. Schmidt, D.K. Hanson, M. Schiffer, J.R. Norris and G.R. Fleming **(1992)** Proc. Nat. Acad. Sci., USA, *89*: 8517-8521.

[3] M.C. Müller, K. Griebenow and A.R. Holzwarth **(1992)** Chem.Phys.Letters, *199*: 465–469.

[4] Y. Jia, T.J. DiMagno, C.-K. Chan, Z. Wang, M. Du, D.K. Hanson, M. Schiffer, J.R. Norris, G.R. Fleming and M.S. Popov **(1993)** J. Phys. Chem., *97*: 13180–13191.
[5] T. Arlt, S. Schmidt, W. Kaiser, C. Lauterwasser, M. Meyer, H. Scheer and W. Zinth **(1993)** Proc. Natl. Acad. Sci., USA, *90*: 11757–11761.
[6] J.K.H. Hörber, W. Göbel, A. Ogrodnik and M.E. Michel-Beyerle **(1985)** in ref. [42], pp. 292–297.
[7] A. Ogrodnik, U. Eberl, R. Heckmann, M. Kappl, R. Feick and M.E. Michel-Beyerle **(1990)** in ref. [43], pp. 157–168.
[8] A.R. Holzwarth, M.G. Müller and K. Griebenow **(1992)** in ref. [44], pp. 219–226.
[9] Z. Wang, R.M. Pearlstein, Y. Jia, G.R. Fleming and J.R. Norris **(1993)** Chem. Phys., *176*: 421–425.
[10] A. Ogrodnik, W. Keupp, M. Volk, G. Aumeier and M.E. Michel-Beyerle **(1994)** J. Phys. Chem., *98*: 3432–3439.
[11] R. Jankowiak and G.J. Small **(1993)** in ref. [45], pp. 133–178.
[12] J.M. Peloquin, J.C. Williams, X. Lin, R.G. Alden, A.K.W. Taguchi, J.P. Allen and N.W. Woodbury **(1994)** Biochemistry, *33*: 8098–8100.
[13] N.C. Woodbury, J.M. Peloquin, R.G. Alden, X. Lin, S. Lin, A.K.W. Taguchi, J.C. Williams and J.P. Allen **(1994)** *33*: 8101–8112.
[14] C. Kirmaier and D. Holton **(1993)** in ref. [45], pp. 49–70.
[15] N.W. Woodbury and W.W. Pason **(1984)** Biochim. Biophys. Acta, *767*: 345–361.
[16] J.K.H. Hörber, W. Göbel, A. Ogrodnik, M.E. Michel-Beyerle and R.J. Cogdell **(1986)** FEBS Letters, *198*: 273–278.
[17] J.K.H. Hörber, W. Göbel, A. Ogrodnik, M.E. Michel-Beyerle and R.J. Cogdell **(1986)** FEBS Letters, *198*: 268–272.
[18] A. Ogrodnik **(1990)** Biochim. Biophys. Acta, *1020*: 65–71.
[19] G. Feher and M.Y. Okamura **(1978)** in ref. [46], pp. 349–386.
[20] M.R. Gunner, D.E. Robertson and P.L. Dutton **(1986)** J. Phys. Chem., *90*: 3783–3795.
[21] M. Volk, G. Aumeier, T. Häberle, A. Orgodnik, R. Feick and M.E. Michel-Beyerle **(1992)** Biochim. Biophys. Acta, *1102*: 253–259.
[22] A. Ogrodnik **(1993)** Mol. Cryst. Liq. Cryst., *230*: 35–56.
[23] G. Hartwich, M. Friese, H. Scheer, A. Ogrodnik and M.E. Michel-Beyerle **(1995)** Chem. Phys., *197*: 423-434.
[24] P.R. Bevington and D.K. Robinson, "Data Reduction and Error Analysis for the Physical Science" 2nd Edition; McGraw-Hill, International **(1992)**.
[25] C. Kirmaier and D. Holten **(1990)** Proc. Natl. Acad. Sci., USA, *87*: 3552–3556.
[26] M. Volk, A. Ogrodnik and M.E. Michel-Beyerle **(1995)** in ref. [47].
[27] A. Ogrodnik and M.E. Michel-Beyerle **(1989)** Z. Naturforsch., *44a*: 763–764.
[28] P. Hamm, K.A. Gray, D. Oesterhelt, R. Feick, H. Scheer and W. Zinth **(1993)** Biochim. Biophys. Acta, *1142*: 99–105.
[29] S.J. Robles, J. Breton and D.C. Youvan **(1990)** Science, *248*: 1402–1405.
[30] A. Ogrodnik and M.E. Michel-Beyerle **(1992)** in ref. [48], pp. 349–373.
[31] U. Eberl, M. Gilbert, W. Keupp, T. Langenbacher, J. Siegl, I. Sinning, A. Ogrodnik, S.J. Robles, J. Breton, D.C. Youvan and M.E. Michel-Beyerle **(1992)** in ref. [44], pp. 253–260.
[32] U. Eberl, A. Ogrodnik and M.E. Michel-Beyerle **(1990)** Z. Naturforsch., *45a*: 763–770.

[33] U. Eberl **(1992)** Dissertation, TU München.
[34] W. Holzapfel, U. Finkele, W. Kaiser, D. Oesterhelt, H. Scheer, H.U. Stilz and W. Zinth **(1990)** Proc. Natl. Acad. Sci., USA, *87*: 5168–5172.
[35] R. Marcus and R. Almeida **(1990)** J. Phys. Chem., *94*: 2973–2977.
[36] Connolly, Samuel and Janzen **(1982)** Photochem. Photobiol., *36*: 565–574.
[37] N.W. Woodbury and W.W. Parson **(1986)** Biochim. Biophys. Acta, *850*: 197–210.
[38] M. Bixon, J. Jortner and M.E. Michel-Beyerle **(1991)** Biochim. Biophys.Acta, *1056*: 301–315.
[39] M.H. Vos, J.-C. Lambry, S.J. Robles, D.C. Youvan, J. Breton and J.-L. Martin **(1992)** Proc. Natl. Acad. Sci., USA, *89*: 613–617.
[40] C.K. Chan, T.J. DiMagno, L.X.Q. Chen, J.R. Norris and G.R. Fleming **(1991)** Proc. Natl. Acad. Sci., USA, *88*: 11202–11206.
[41] S.J. Robles, J. Breton and D.C. Youvan in ref. [43].
[42] "Antennas and Reaction Centers of Photosynthetic Bacteria: Structure, Interaction and Dynamics", (ed. M.E. Michel-Beyerle), Springer, Berlin **(1985)**.
[43] "Reaction Centers of Photosynthetic Bacteria", (ed. M.E. Michel-Beyerle), Springer, Berlin **(1990)**.
[44] "The Photosynthetic Bacterial Reaction Center II", (ed. J. Breton and A. Vermeglio), Plenum Press, New York **(1992)**.
[45] "The Photosynthetic Reaction Center", (ed. J.R. Norris and R. Deisenhofer), Academic Press, New York **(1993)**.
[46] "The Photosynthetic Bacteria", (ed. R.K. Clayton and W.R. Sistrom), Plenum Press, New York **(1978)**.
[47] "Anoxygenic Photosynthetic Bacteria", (ed. R.E. Blankenship, M.T. Madigan and C.E. Bauer), Kluwer Acad. Publ., Netherlands **(1995)**.
[48] "Photoprocesses in Transition Metal Complexes, Biosystems and Other Molecules. Experiment and Theory", (ed. E. Kochanski), Kluwer Acad. Publ., Netherlands **(1992)**.

# Reaction Center Heterogeneity Probed by Multipulse Photoselection Experiments With Picosecond Time Resolution

Su Lin, Xiaomei Lin, JoAnn C. Williams, Aileen K.W. Taguchi, James P. Allen and Neal W. Woodbury

Arizona State University, Department of Chemistry and Biochemistry, and the Center for the Study of Early Events in Photosynthesis, Tempe, AZ 85287-1604, USA

**Abstract.** Using double pulse excitation of an electron-transfer-impaired reaction center mutant from *Rhodobacter sphaeroides*, it has been possible to provide direct evidence for reaction center conformational heterogeneity which affects the charge separation yield. A saturating prepulse was used to photoselect a subpopulation of reaction centers which returned rapidly to the ground state instead of forming the long-lived $P^+Q_A^-$ state. Femtosecond transient absorbance measurements were then performed on this photoselected subpopulation. It was found that the yield of electron transfer in this subpopulation was substantially lower than that in the bulk reaction center population. Though these measurements are preliminary, they represent the first direct demonstration of a reaction center conformational heterogeneity that clearly affects the efficiency of the initial electron transfer reaction.

**Key words:** Population Heterogeneity, Quantum Yield, Charge Separation, *Rhodobacter sphaeroides*, Photosynthesis

## 1 Introduction

As instrumentation for monitoring absorbance changes and emission signals on ultrafast time scales has improved in sensitivity as well as in spectral and temporal resolution, greater and greater kinetic and spectral complexity has been found to be associated with the early electron transfer reactions of photosynthesis. This has been most carefully studied in the reaction centers of purple nonsulfur bacteria [see reviews 1-7]. It is now clear that the kinetics of the absorbance changes and the spontaneous emission decay in the bacterial reaction center cannot be adequately described by simple models which involve a series of first order electron transfer reactions between only two or three homogeneous reaction intermediates (i.e. $P^*$, $P^+B_A^-$, $P^+H_A^-$). In wild type reaction centers at room temperature, absorbance changes have been reported with exponential lifetimes near 1 ps, 3 ps, and 15 ps and spontaneous emission has been shown to occur with exponential lifetimes of 2-3 ps, 10-15 ps, and hundreds of picoseconds [8-14]. In addition, oscillatory kinetic behavior has been observed upon excitation with very short pulses at both room temperature and low temperature [15-20]. The spectral evolution is also complex at room temperature, involving at least four spectrally distinct states as determined by singular value decomposition analysis [14]. As the temperature is lowered, the spectral evolution becomes less complex [21]; only two spectrally distinct components are required to describe absorbance change data over a broad spectral range at 20 K [14], but several exponential decay components are still required for an adequate description of the kinetics [e.g. refs. 14,22-24]. The presence of this kinetic complexity is even more pronounced in certain reaction center mutants that show a decreased driving force for the initial electron donor [8,14,25].

One possible explanation for kinetic and spectral complexity is conformational heterogeneity of the reaction centers which affects either the driving force or relevant couplings for electron transfer. This heterogeneity could either take the form of static, ground state heterogeneity which is inherent at all times in the sample, or conformational dynamics which is induced by the charge separation reaction itself [8,10,13,14,26-29]. The distinction between static and dynamic heterogeneity is two-fold. First, static conformational heterogeneity must exist prior to light absorption in the ground state of the system while conformational dynamics, if it is to explain the kinetic complexity of the reaction, must involve conformational changes which are induced or at least favored by the charge separation reaction. Second, static heterogeneity must involve slow conformational interconversions on the time scale of electron transfer while conformational dynamics must involve nuclear movement that takes place on the time scale of the observed kinetic components.

Though there have been a number of kinetic experiments performed on fast time scales which have been interpreted in terms of static or dynamic heterogeneity in the reaction center population [8,10,13,14,26-29], to our knowledge, no one has succeeded in performing an experiment to directly select a specific subpopulation of reaction centers and show that that subpopulation actually has a different yield or rate of electron transfer than the bulk population. Note that both conventional and dynamic hole burning measurements select populations of reaction centers [e.g. refs. 30,31], but these populations are selected based on the energy of the P to $P^*$ transition, rather than on the basis of electron transfer rate or yield.

## 2 Materials and Methods

The construction and characterization of the triple mutant, LH(L131) + LH(M160) + FH(M197), have been discussed previously [32,33]. Transient absorbance spectroscopy was performed using the apparatus described in past publications [31,33] with the exception that in some cases the excitation was performed by two consecutive pulses. The first pulse was a 90 ps, 1 mJ, saturating pulse at 532 nm which resulted in about 75% conversion of the reaction centers to $P^+Q_A^-$. This was followed after 1 ns by a 590 nm, 150 fs duration, 10 μJ pulse which reexcited the reaction centers that returned to the ground state following the stronger 532 nm pulse. The transient absorbance changes due to the second excitation pulse were then probed using a weak continuum pulse as described for other transient absorbance measurements [31,33].

## 3 Results and Discussion

Fig. 1 shows the results of a transient absorbance experiment using 590 nm excitation pulses which can either be preceded or not by a strong 532 nm preflash roughly 1 ns before the 590 nm pulse. The measurements were performed at 850 nm where one can observe the absorbance decrease in the $Q_Y$ band of P without interference from the stimulated emission at longer wavelengths. Thus, a time dependent recovery of the initial absorbance decrease in this region corresponds to a recovery of the ground state reaction center population. The sample used is the triple hydrogen bond mutant, LH(L131) + LH(M160) + FH(M197), which has about a 50% quantum yield of overall charge separation at room temperature [33]. As can be seen in the top panel, the strong preflash converts about 75% of the reaction centers to the state $P^+Q_A^-$ (the high conversion efficiency is because the preflash is about 90 ps long, allowing each reaction center two or more opportunities to absorb a photon). If there were substantial heterogeneity in the reaction center population that was static on the nanosecond time scale, then one would expect that those reaction centers that returned to the ground state after the strong preflash would be preferentially those with the lowest yield of charge separation. After exciting this subpopulation of reaction centers with a second pulse at 590 nm, the resulting transient absorbance changes should be indicative of reaction centers with a lower yield than the bulk population. If, on the other hand, the reaction centers were homogeneous in their rate and yield of charge separation, the photoselected population should have the same yield and kinetics as the bulk population without a preflash. As can be seen from Fig. 1B, roughly half of the initial absorbance decrease due to excitation of P recovered after a few hundred picoseconds in the experiment without a prepulse while at least 80% of the initial absorbance decrease due to excitation of P recovered in the photoselected reaction center population. This implies that there was a

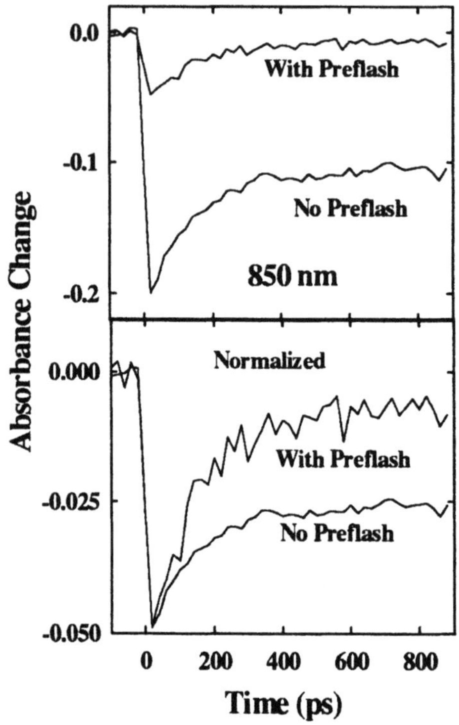


**Fig. 1** Kinetics of transient absorbance changes in triple mutant reaction centers at 850 nm measured either with or without a preflash. The top panel shows the raw data; the bottom panel shows the two traces normalized at the peak.

substantial effect of the preflash on the yield of electron transfer, in agreement with the concept that different reaction centers in the population undergo electron transfer with different yields. A preliminary analysis of this data indicates that the difference in rates between the 10% of the reaction centers in the population that were slowest and the 10% that were fastest may be as much as a factor of 10. (The reason why there is not a very large rate change between the traces with and without a preflash in Fig. 1 comes, at least in part, from the fact that with a yield as low as 50% in the bulk population, one would never see more than a factor of two change in *observed* decay rate for any subpopulation since the observed decay rate is the sum of the charge separation rate and the intrinsic rate of $P^*$ decay. The intrinsic $P^*$ decay time is apparently about 200 ps in this mutant.)

The results of Fig. 1, though significant, are subject to qualifications. A more literal interpretation of the experiment in Fig. 1 is that those reaction centers which returned to the ground state after a prepulse have a different quantum yield of charge separation than the bulk population did before any excitation. There is no guarantee that these reaction centers are in the same configuration a nanosecond after having been excited and returning to the ground state than they were prior to excitation. In fact, one could interpret these results as a time-dependent conformational change which is induced by the first flash but which has not yet had time to recover before the second flash. Thus, these results do not rule out a substantial contribution of dynamic conformational heterogeneity to the complex kinetics of charge separation and fluorescence decay [14, 22].

In addition, the reaction centers used for these measurements were mutant reaction centers with seriously impaired electron transfer capabilities. The reason for using a low quantum yield mutant is that in wild type reaction centers, which undergo electron transfer with a quantum yield near unity, it is very difficult to select for a substantial subpopulation of low yield reaction centers with a prepulse. In fact, these measurements were repeated on R-26 reaction centers, a carotenoidless mutant that otherwise behaves as wild type, and no discernible transient absorbance signals were observed in the 850 nm region after the saturating prepulse, due to the very high yield of $P^+Q_A^-$ (data not shown). The reliance of the experiment on the use of one particular mutant lends uncertainty to the relevance of these results to the wild type system.

Work is in progress to extend these preliminary results. Similar experiments are being performed on wild type reaction centers in which electron transfer to the quinone is blocked. In this case, there is some return to the ground state on the several nanosecond time scale, and one can ask if these reaction centers have a particularly rapid $P^+H_A^-$ recombination rate or whether they have a slower forward charge transfer rate than the rest of the population, as might be expected if the standard free energy difference between $P^*$ and $P^+H_A^-$ was smaller in magnitude than normal in the reaction centers that first returned to the ground state after the prepulse. Also, results from other low yield mutants are being compared to the triple mutant to determine if the observed heterogeneity is unique to this particular mutant.

## Acknowledgments

The authors acknowledge Drs. M. Bixon, J. Jortner, and M. Michel-Beyerle for making manuscripts available before publication. This work was supported by grants DMB91-58251, MCB 9219378 and MCB 9404925 from the National Science Foundation as well as grant GM45902 from the National Institutes of Health. Instrumentation was purchased with funds from NSF grant DIR-8804992 and Department of Energy grants DE-FG-05-88-ER75443 and DE-FG-05-87-ER75361. This is publication No. 243 from the Arizona State University Center for the Study of Early Events in Photosynthesis.

## References


1 Kirmaier, C. and Holten, D. (1987) Photosynth. Res. 13, 225-260

2 Feher, G., Allen, J.P., Okamura, M.Y. and Rees, D.C. (1989) Nature 339, 111-116

3 Parson, W.W. (1991) in Chlorophylls (Scheer, H., ed.), pp. 1153-1180, CRC Press, Boca Raton

4 Martin, J.-L. and Vos, M.H. (1992) Ann. Rev. Biophys. Struct. 21, 199-222

5 Kirmaier, C. and Holten, D. (1993) in The Photosynthetic Reaction Center (Deisenhofer, J. and Norris, J.R., eds.), Vol. II, pp. 49-70, Academic Press, San Diego

6 Zinth, W. and Kaiser, W. (1993) in The Photosynthetic Reaction Center (Deisenhofer, J. and Norris, J.R., eds), Vol. II, pp 71-88, Academic Press, San Diego

7 Woodbury, N.W. and Allen, J.P. (1995) in Anoxygenic Photosynthetic Bacteria (Blankenship, R.E., Madigan, M.T. and Bauer, C.E., eds.), in the series Advances in Photosynthesis, Kluwer Academic Publishing, Netherlands, pp. 527-557

8 Jia, Y., DiMagno, T.J., Chan, C.-K., Wang, Z., Du, M., Hanson, D.K., Schiffer, M., Norris, J.R., Fleming, G.R. and Popov, M.S. (1993) J. Phys. Chem. 97, 13180-13191

9 Du, M., Rosenthal, S.J., Xie, X., DiMagno, T.J., Schmidt, M., Hanson, D.K., Schiffer, M., Norris, J.R. and Fleming, G.R. (1992) Proc. Natl. Acad. Sci. USA 89, 8517-8521

10 Muller, M.G., Griebenow, K. and Holzwarth, A.R. (1992) Chem. Phys. Lett. 199, 465-469

11 Hamm, P., Gray, K.A., Oesterhelt, D., Feick, R., Scheer, H. and Zinth, W. (1993) Biochim. Biophys. Acta 1142, 99-105

12 Holzapfel, W., Finkele, U., Kaiser, W., Oesterhelt, D., Scheer, H., Stilz, H.U. and Zinth, W. (1989) Chem. Phys. Lett. 160, 1-7

13 Kirmaier, C. and Holten, D. (1990) Proc. Natl. Acad. Sci. USA 87, 3552-3556

14 Woodbury, N.W., Peloquin, J.M., Alden, R.G., Lin, X., Lin, S., Taguchi, A.K.W., Williams, J.C. and Allen, J.P. (1994) Biochemistry 33, 8101–8112

15 Vos, M.H., Lambry, J.-C., Robles, S.J., Youvan, D.C., Breton, J. and Martin, J.-L. (1991) Proc. Natl. Acad. Sci. USA 88, 8885-8889

16 Vos, M.H., Rappaport, F., Lambry, J.-C., Breton, J. and Martin, J.-L. (1993) Nature 363, 320-325

17 Vos, M.H., Jones, M.R., Hunter, C.N., Breton, J., Lambry, J.-C. and Martin, J.-L. (1994a) Biochemistry 33, 6750-6757

18 Vos, M.H., Jones, M.R., Hunter, C.N., Breton, J. and Martin, J.-L. (1994b) Proc. Natl. Acad. Sci. USA 91, 12701-12705

19 Vos, M.H., Jones, M.R., McGlynn, P., Hunter, C.N., Breton, J. and Martin, J.-L. (1994c) Biochim. Biophys. Acta 1186, 117-122
20 Stanley, R.J. and Boxer, S.G. (1994) J. Phys. Chem. 99, 859-863.
21 Kirmaier, C. and Holten, D. (1988) FEBS Lett. 239, 211-218
22 Peloquin, J.M., Williams, J.C., Lin, X., Alden, R.G., Murchison, H.A., Taguchi, A.K.W., Allen, J.P. and Woodbury, N.W. (1994) Biochemistry 33, 8089–8100
23 Vos, M.H., Lambry, J.-C., Robles, S.J., Youvan, D.C., Breton, J. and Martin, J.-L. (1992) Proc. Natl. Acad. Sci. USA 89, 613-617
24 Lauterwasser, C., Finkele, U., Scheer, H. and Zinth, W. (1991) Chem. Phys. Lett. 183, 471-477
25 Nagarajan, V., Parson, W.W., Gaul, D. and Schenck, C.C. (1990) Proc. Natl. Acad. Sci. USA 87, 7888-7892
26 Small, G.J., Hayes, J.M. and Silbey, R.J. (1992) J. Phys. Chem. 96, 7499-7501
27 Wang, Z., Pearlstein, R.M., Jia, Y., Fleming, G.R. and Norris, J.R. (1993) Chem. Phys. 176, 421-425
28 Ogrodnik, A., Keupp W., Volk, M., Aumeier, G. and Michel-Beyerle, M.E. (1994) J. Phys. Chem. 98, 3432-3439
29 Bixon, M., Jortner, J. and Michel-Beyerle, M.E. (1995) Chem. Phys., in press
30 Reddy, N.R.S., Lyle, P.A. and Small, G.J. (1992) Photosynth. Res. 31, 167-194
31 Peloquin, J.M., Lin, S., Taguchi, A.K.W. and Woodbury, N.W. (1995) J. Phys. Chem. 99, 1349-1356
32 Lin, X., Murchison, H.A., Nagarajan, V., Parson, W.W., Williams, J.C. and Allen, J.P. (1994) Proc. Natl. Acad. Sci. USA 91, 10265-10269
33 Woodbury, N.W., Lin, S., Lin, X., Peloquin, J.M., Taguchi, A.K.W., Williams, J.C. and Allen, J.P. (1995) Chem. Phys., in press

# Characteristics of the electron transfer reactions in the M210W reaction centre only mutant of *Rhodobacter sphaeroides*.

**M. E. van Brederode[1], L. M. P. Beekman[1], D. Kuciauskas[1], M. R. Jones[2], I. H. M. van Stokkum[1] and R. van Grondelle[1].**

[1]Faculty of Physics and Astronomy, Vrije Universiteit,, Amsterdam.

[2]Department of Molecular Biology and Biotechnology, University of Sheffield, Sheffield, United Kingdom.

**Abstract.**

Charge separation and recombination is studied in membrane bound reaction centres (RCs) of the M210W mutant of *Rhodobacter sphaeroides* in which the tyrosine at the M210 position has been altered into a tryptophan. Many of the observed properties can be explained by assuming that in the M210W RC the first charge separated state is approximately degenerate with $P^*$, while with $Q_A$ reduced the charge separation is to a state higher in energy than $P^*$. We observe a pronounced wavelength dependence of the $P^+Q_A^-$ recombination, with 'red' RC being faster than 'blue' RCs. It is suggested that a variation of the electronic coupling element over the near infrared absorption band of P may explain this effect. In the 5 K excitation spectrum we observe enhanced emission from the red wing of the P absorption band, due to RCs with either a very slow rate of charge separation or RCs with an increased probability of charge recombination. Finally, the $B_A$ absorption appears to be absent in the 5 K fluorescence excitation spectrum.

**Introduction.**

In the bacterial RC, excitation of a bacteriochlorophyll dimer (P) triggers a sequence of electron transfer reactions along one of two symmetrically aligned pigment chains. Recent experiments by Schmidt et al. have shown that the first electron acceptor is the monomeric bacteriochlorophyll on the active branch, $B_A$, located between the primary donor P and the bacteriopheophytin, $H_A$ [1]. The primary charge separation is multi-exponential and in WT RCs the $P^*$ kinetics can be described with a bi-exponential decay with time constants between 2 ps and 12 ps as resolved in spontaneous emission and transient absorption measurements [2,3,4,5]. From $B_A^-$ the electron moves to $H_A$ with a lifetime of about 1 ps [1,6, Beekman *et al.* unpublished results]. The lifetime of the excited state measured in time resolved fluorescence measurements contains components up to several nanoseconds [7,8,9]. These delayed fluorescence components are attributed to recombination from $P^+H_A^-$ to P*. From $P^+H_A^-$ the electron moves with a time constant of about 200 ps to a quinone, $Q_A$. Blocking further electron transfer from $Q_A^-$ to $Q_B$

results in recombination from the state $P^+Q_A^-$ to the ground state on a millisecond timescale.

The mechanistic and energetic description of the sequence of electron transfer steps in the bacterial reaction centre are still under debate. In particular, the multi-exponential behaviour and temperature dependence of the primary charge separation, in the wildtype (WT) and mutants of *Rhodobacter (Rb.) sphaeroides* and *Rb. capsulatus* have been extensively investigated and rather diverse explanations have been presented. One interpretation involves a "parking state" in which transiently an electron is transferred to one of the pigments in the inactive branch [4,11]. A dynamic solvation model in which the charge separated state is stabilized in time after its initial formation, leading to a time dependent driving force is described in [8] and [7]. Finally, a static or dynamic distribution of the driving force of the primary charge separation has been proposed, which could also lead to the observed multi-exponential behaviour [2,3,5].

In the detergent isolated M210W mutant the redox potential ($E_m$) of $P/P^+$ has been reported to be increased by 50 meV relative to the WT RC [10]. A similar increase was observed for the membrane bound M210W RC (Beekman *et al.*, unpublished results). For the detergent solubilised M210W RC this change in $E_m$ was 16 meV smaller than the change in $\Delta G$ between P* and $P^+H_A^-$ as determined from delayed fluorescence measurements [10]. Since in WT RCs the free energy difference between $P^*$ and $P^+B_A^-$ is estimated to be about 50-60 meV [1], this increase in the $E_m$ of $P/P^+$ for M210W could give rise to a situation in which $P^*$ and $P^+B_A^-$ are degenerate [12]. In combination with a free energy distribution in both P* and $P^+B_A^-$, this could result in a fraction of RCs in which $P^+B_A^-$ is located above the energy level of $P^*$ and a fraction in which the reverse is true. An energetic distribution with $2\sigma$=100 meV for $P^+B_A^-$ similar as was estimated for the $P^+H_A^-$ state [13], would be sufficient to create this situation.

In detergent isolated open M210W RCs at room temperature, the decay of P* was observed to proceed with a time constant of 33 ± 4 ps; alternatively, a biphasic decay of 5 ps and 30 ps was obtained [14]. In membrane bound M210W RCs P* decay was found to be biphasic with time constants of about 30 ps and 90 ps having an amplitude ratio of 0.35 to 0.65, respectively (Beekman *et al.*, unpublished results). At 80 K the P* decay is slowed down and the spectral changes in the 850-950 nm region can be described by two spectrally distinct components with time constants of 13 and 155 ps, where the 13 ps component describes an apparent blue shift of the P* stimulated emission. Van Noort *et al.* observed a decrease of the rate of P* decay to about 300 ps at 80 K, with hardly any additional increase in decay time between 80 K and 20 K [11].

The rate of recombination from the $P^+Q_A^-$ state in the M210W mutant has been found to increase relative to the WT. At room temperature the lifetime for $P^+Q_A^-$ recombination is 45 ms compared to 100 ms for the WT. As in the WT the reaction rate increases with decreasing temperature in the M210W mutant [10].

In this work we have studied the M210W mutant RC expressed in the RC-only strain of *Rb. sphaeroides* [15]. We present low temperature transient absorption measurements of the primary charge separation reaction, an analysis of the $P^+Q_A^-$ recombination and the decay of the triplet state of P, $^3P$, which we think is directly formed from P* due to the slow charge separation at low temperatures. Also a 5 K fluorescence excitation spectrum is shown. The results are discussed in terms of models proposed for the kinetics and energetics of charge separation in bacterial RCs.

**Materials & Methods.**

*Sample preparation and mutant construction.*

Construction of the RC-only strain was carried out as described in [16], using the codon change TAC->TGG. Preparation of intracytoplasmic membranes was also as described in [16]. For spectroscopy, aliquots of concentrated membranes were suspended in 50 mM potassium phosphate buffer. The $P^+Q_A^-$ recombination measurements were performed in 20 mM 1,10 o-phenanthroline. For the picosecond transient absorbance measurements $Q_A$ was reduced by 20 mM dithionite.

*$P^+Q_A^-$ recombination and quantum yield.*

Transient absorbance measurements on a μs and ms timescale after a 8 ns 590 nm laserflash were performed on a set-up essentially as described in [17]. An extra monochromator, that was scanned together with the monochromator before the detector, was placed in the measuring beam before the sample to minimize the actinic effect of the measuring light.

The $P^+Q_A^-$ quantum yield was determined from a saturation curve of the ΔOD signal integrated over the first 500 μs after the excitation flash. The saturation curves were analysed according to the formula $\Delta OD = \Delta OD_{max}(1\text{-}10^{-\sigma\Phi E})$ [18], in which σ is the absorbance cross-section at the excitation wavelength, Φ is the quantum yield and E is the excitation energy. By comparison of the apparent decay constant of the saturation curve, σΦ, of the WT RC and the M210W RC the quantum yield of the mutant relative to the wildtype RC was calculated. It was checked that at 77 K the extinction coefficient of the wildtype and the mutant were the same at the wavelength of excitation in the $Q_x$ region. We did not find any shifting of the $Q_x$ band as reported by others for detergent solubilised RCs [14], and we assumed that the extinction coefficients were the same.

*Fluorescence excitation spectrum.*

For the steady state fluorescence excitation spectrum the sample was selectively excited by scanning a Ti-Sapphire laser (Coherent), pumped by an Ar-ion laser (Coherent), between 750 and 914 nm. The fluorescence was detected by measuring the fluorescence after a 940-960 nm interference filter with a Chromex Charge Coupled Device (CCD) set-up. Excitation energy was measured with a powermeter (Delta Developments), which is calibrated for wavelength sensitivity. Excitation energy was between 80 and 500 μW on an excitation surface of 1 $cm^2$. The optical density of the sample was 0.3 at the wavelength of maximal OD at 5 K.

*Picosecond transient absorption measurements.*

The picosecond transient absorption measurements were performed on an amplified high repetition rate Ti-Sapphire set-up (Mira-Rega combination Coherent), which is described elsewhere [19,20]. Briefly, the central wavelength (803 nm) coming from the Rega was split in two. One part was used to excite the sample, the other part was used to create a white light continuum by focusing on a Ti-Sapphire crystal. Using filters, the detection wavelengths were selected. The excitation beam was modulated with 1 kHz and put through a variable optical delay, both excitation and detection beam were focused and overlapped in the cuvette. The detection beam was measured after a monochromator with a 6-12 nm resolution using a photodiode. The signal was filtered out by a lockin amplifier. At room temperature and 77 K the excitation energy was typically 500 $\mu$W, 50 $\mu$W and the detection energy 100 $\mu$W and 30 $\mu$W, respectively. The room temperature measurements were performed with a repetition frequency of 250 kHz. To avoid accumulation of triplets at 77 K, these measurements were performed with a repetition rate of 10 kHz. From the OD of the sample, the excitation frequency and the illuminated volume of excitation, it was calculated that at 77 K approximately 13% of the RCs were excited per excitation flash and at RT 5%.

*Analysis of the data.*

Kinetics and decay associated spectra were determined from a global analysis of the kinetic traces as described in [21].

**Results.**

*$P^*$ decay and $P^+H_A^-$ formation at room temperature and 77 K with $Q_A$ reduced..*

At 77 K transient absorbance changes of the M210W mutant with $Q_A$ reduced were measured on a 900 ps timescale at 780, 860, 880, 905, 918, 930, 950, 960 and 970 nm. In Figure 1A traces recorded at 77 K at 780, 918 and 940 nm are shown.

Global analysis of the data revealed two spectrally distinct components with lifetimes of 400 ± 10 ps and 1.7 ns ± 0.3 ns at 77 K. In the stimulated emission region the decay is dominated by the 400 ps component. The 1.7 ns component shows the characteristics of a $P/P^+$ difference spectrum, i.e. a bleach between 860-930 nm and an absorption increase from 930 to 970 nm and is attributed to direct decay of $P^+H_A^-$.

At 780 nm the pump-probe signal is composed of two absorption increases: the first due to the pulse-limited formation of $P^*$ and the second due to the electrochromic bandshift of $B_A$ due to the formation of $P^+H_A^-$ in 400 ps. On a longer time scale the signal decreases with a 1.7 ns lifetime. At 860 nm the relative contribution of the 400 ps component is approximately 75%. At this wavelength the signal is dominated by the ground state bleaching and we can not distinguish between $P^*$ or $P^+$. Since 75% of the P ground state recovery shows a 400 ps phase, whereas only 25% recovers with 1.7 ns, we conclude that only approximately 25% of the RCs actually performs charge separation under these conditions. From the quantum yield of charge separation and the 400 ps lifetime of $P^*$ we estimate a true lifetime of 1.6 ns for the charge separation process and a lifetime of all other loss channels of approximately 530 ps.

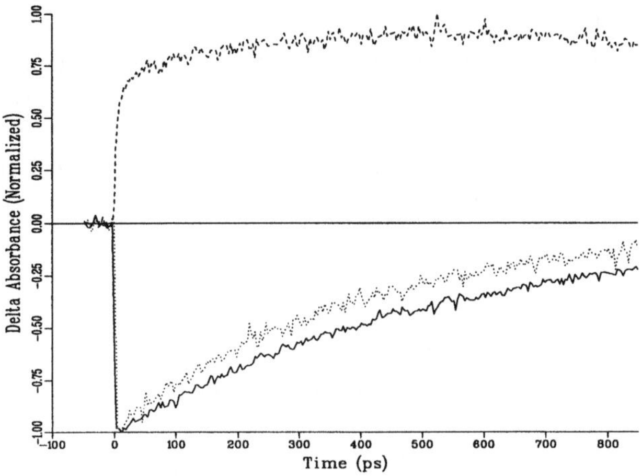


*Figure 1. Transient absorbance changes of the M210W mutant at 77 K recorded at 780 nm (dashed), 860 nm (solid) and 940 nm (dotted).*

Global analysis of a similar set of kinetic measurements at 293 K (see Fig. 2 for a few characteristic traces) also revealed two decay components of 152 ± 4 ps and 1.0 ± 0.1 ns.

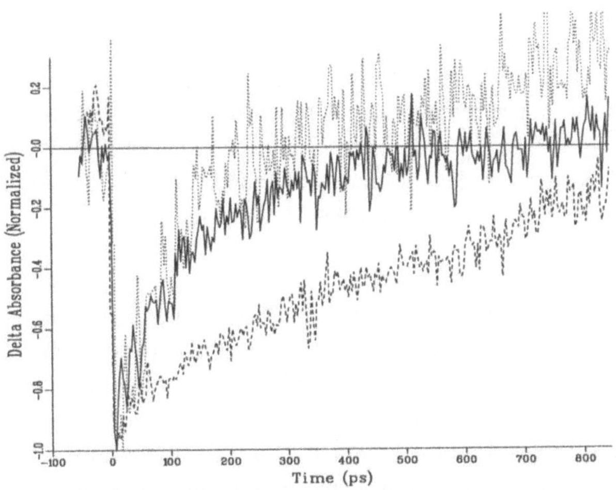


*Figure 2. Transient absorbance changes of the M210W mutant with $Q_A$ reduced at room temperature, recorded at 960 nm (dotted), 940 nm (solid) and 840 nm (dashed).*

The spectral characteristics of both decays follow the fast and slow component of the 77 K data set. However, the quantum yield of charge separation as determined from the kinetic trace at 840 nm, has now increased to approximately 75%, corresponding to a 200 ps charge separation time and a 600 ps time constant for other losses from P*.

Assuming that the $P^+H_A^-$ decay time, $\tau_{rg} = 1.7$ ns, is independent of temperature, the room temperature data can be fitted to the simple model as shown in Figure 3, in which $\Delta G = 67$ meV, $\tau_{ic} = 600$ ps, $\tau_{rg} = 1.7$ ns, $\tau_{cs} = 200$ ps and $\tau_{cr} = 2.4$ ns. Note the 6-fold faster decay of $P^+H_A^-$ to the ground state for the M210W RC as compared to the WT RC(about 10 ns).

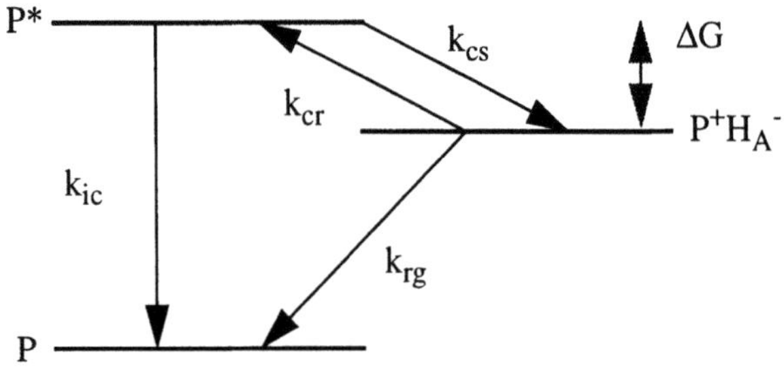


*Figure 3. Simplified scheme for the primary processes in the RC. $k_{cs}$, $k_{cr}$, $k_{ic}$, $k_{rg}$ are the rate constants for charge separation, charge recombination via P*, internal conversion and direct charge recombination from $P^+H_A^-$ to the groundstate, respectively.*

At room temperature the increase of the measured $P^+H_A^-$ decay to 1 ns is largely due to recombination to $P^*$. Note further that within the context of this model and the accuracy of the measurements, $k_{ic}$ is more or less independent of temperature between 77 K and 300 K ($\tau_{ic}$=533 ps at 77 K, $\tau_{ic}$=600 ps at RT). We finally remark that in the low temperature data no trace was found for a 10-20 ps process as found earlier by others for the detergent isolated M210W RC [10].

*$P^+Q_A^-$ recombination and $^3P$ decay.*

Transient absorbance changes of the M210W mutant, after excitation with a 5 ns 590 nm light flash, were recorded at 59 wavelengths between 700 and 950 nm, on time scales from 0-500 μs, 0-20 ms and 0-200 ms. The time evolution of the difference spectra is shown in Figure 4.

From global analysis of all the kinetic traces at the three time-scales, it followed that three independently decaying spectral components were needed to satisfactorily fit the data. The lifetimes of the spectral components were 100 μs, 4.7 ms and 13.9 ms.The corresponding decay associated spectra (DAS) are shown in Figure 5.

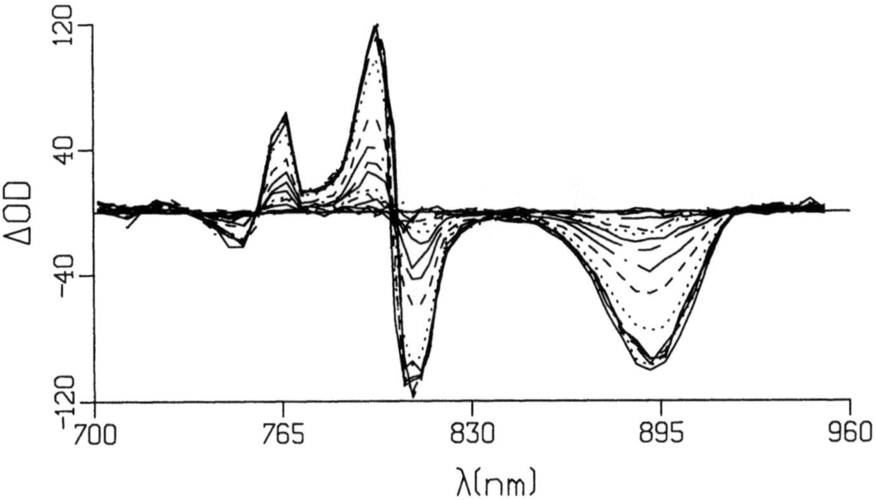


*Figure 4. Time evolution of the absorbance difference spectra between 20 μs (largest amplitude spectrum) and 200 ms of the M210W mutant measured at a temperature of 5 K.*

The 100 μs DAS is mainly composed of a bleaching of P and shows only some minor features around 800 nm. In view of the characteristic lifetime and spectrum [14], this component is attributed to the transition $^3P \rightarrow P$. Most likely, these triplets are directly formed from P*, since the rate of primary charge separation in the M210W mutant at cryogenic temperatures is slowed down to several hundreds of picoseconds in open RC [10, 11]. As a result, the formation of triplets by intersystem crossing directly from P* can efficiently compete with the charge separation process. Furthermore, the electron transfer time from $P^+H_A^-$ to $Q_A$ is not strongly affected in this mutant [10,14], which makes the formation of triplets via the radical pair mechanism from $P^+H_A^-$ equally unlikely as in open WT RCs. The two spectra with the lifetimes of 4.7 ms and 13. 9 ms are associated to a multi-exponential decay of $P^+Q_A^-$. Two remarkable differences between these two decay associated spectra should be noted.

Firstly, the DAS show a large difference in the P band with the 4.7 ms DAS clearly red shifted relative to the 13.9 ms DAS. Secondly, the electrochromic bandshift around 800 nm has a significantly larger amplitude for the red 'fast' RCs than for the 'blue' slow RCs. At 77 K the $P^+Q_A^-$ recombination spectrum of the M210W mutant shows the same characteristics in the P band region as at 5 K, with lifetimes of 6 and 20 ms (results not shown). The triplet spectrum has a lower amplitude, most likely due to the fact that above 50 K the carotonoids start to quench $^3P$.

Global analysis of the transient absorbance changes of the $P^+Q_A^-$ recombination

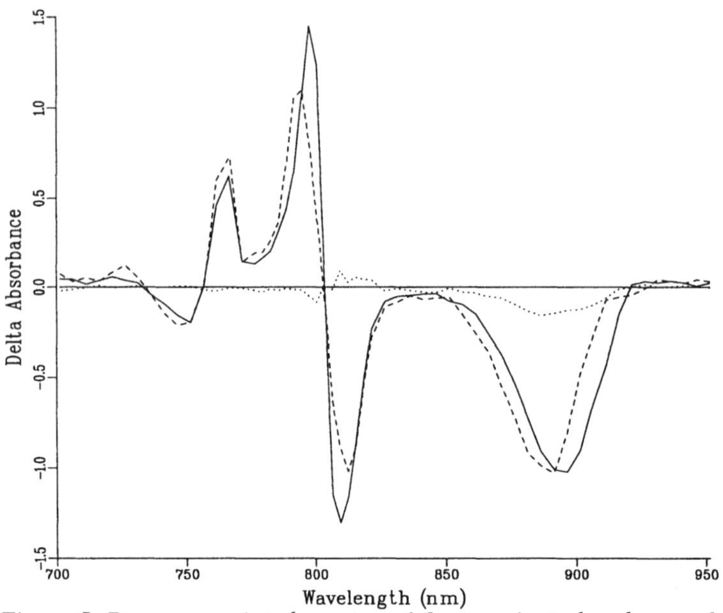

*Figure 5. Decay associated spectra of the transient absorbance changes of the M210W mutant measured on a μs and ms timescale at 5 K. The accompaning lifetimes of the spectra are 100 μs, 4.7 ms and 13.9 ms for the dotted, solid and dashed spectrum, respectively. For comparison in the P band region the dashed spectrum (13.9 ms) was multiplied by 2.3.*

in the M210W mutant at RT, using two spectral components yielded identical DAS in the P band, with lifetimes of 24 ms and 105 ms (results not shown). Also for $P^+Q_A^-$ recombination measured at 77 K in membrane bound WT and M210L RCs of *Rb. sphaeroides,* no spectral shift in the DAS in the P band was observed, although these reactions are clearly bi-exponential at that temperature (results not shown).

The temperature dependence of the $P^+Q_A^-$ recombination kinetics was probed at 885 nm and fitted to a bi-exponential decay (Figure 6). For comparison the temperature dependence of the $P^+Q_A^-$ recombination in the membrane bound WT RC of *Rb. sphaeroides* is also shown. We note that at higher temperatures it was not always clear that within the signal to noise a bi-exponential fit was required.

The results agree well with the earlier measurements of Nagarajan et al. [10], who analysed the same process assuming a mono-exponential decay. At 80 K they obtained a lifetime of 12 ms, whereas we find lifetimes of 6 and 20 ms with relative contributions of 72% and 28%, respectively. At all temperatures in the M210W RC the $P^+Q_A^-$ recombination rate has increased relative to the WT RC. Upon cooling from 293 K to temperatures lower than 40 K, the lifetime of the fast component in the $P^+Q_A^-$ recombination in the M210W mutant decreases from 24 ms to 4-5 ms, whereas the lifetime of the slow component decreases from 105 ms to 13-16 ms. The latter is about the same as the lifetime measured for the fast phase in $P^+Q_A^-$ recombination in WT RCs at low temperatures.

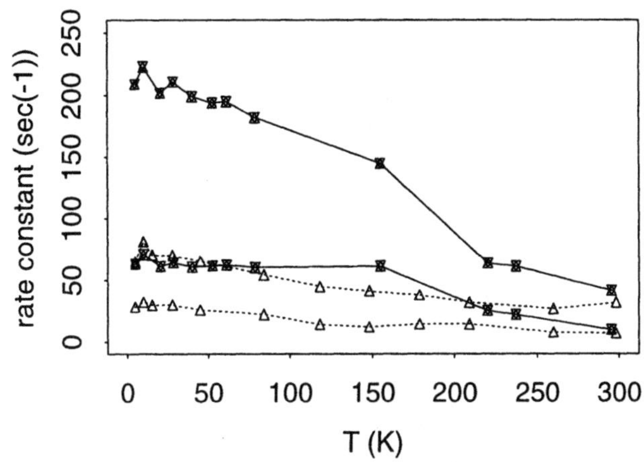


*Figure 6. Temperature dependance of the $P^+Q_A^-$ recombination kinetics in the WT (dashed lines) and M210W mutant (solid lines).*

*The quantum yield of $P^+Q_A^-$ formation and $^3P$ formation in M210W RCs.*

The quantum yield of $P^+Q_A^-$ formation in M210W RCs was determined to be 0.8 at RT and 0.6 at 5 K relative to the WT RC, this compares well to results presented in [10, 14]. From the integral of the different DAS in the P band, it is estimated that at 5 K 60% of the RCs decayed with 4 ms and 30% with 14 ms. The integral of the triplet spectrum in the P band was 10% of the total, which results in a quantum yield of triplet formation of 6% at 5 K. At 77 K 68% decayed with the 6 ms component, 26% with the 20 ms component while the triplet amplitude was 6% of the total.

*Wavelength dependence of the steady state fluorescence.*

The excitation spectrum of the steady state fluorescence detected between 940 and 960 nm is shown in Figure 7 and compared with a 1-T spectrum. Two remarkable differences between the 1-T and excitation spectra can be observed. Firstly, the 800 nm monomeric bacteriochlorophyll band is strongly suppressed and shifted to the red in the excitation spectrum. A similar suppression of the 800 nm band in the excitation spectrum was seen for the M210L mutant of *Rb. sphaeroides* (results not shown). Secondly, the fluorescence yield increases upon excitation in the red side of the P band relative to excitation in the blue. It was checked that the shift in the P band was not due to shifting of the fluorescence spectra by comparing the total fluorescence spectra in the detection region, when exciting with different excitation wavelengths in the P band.

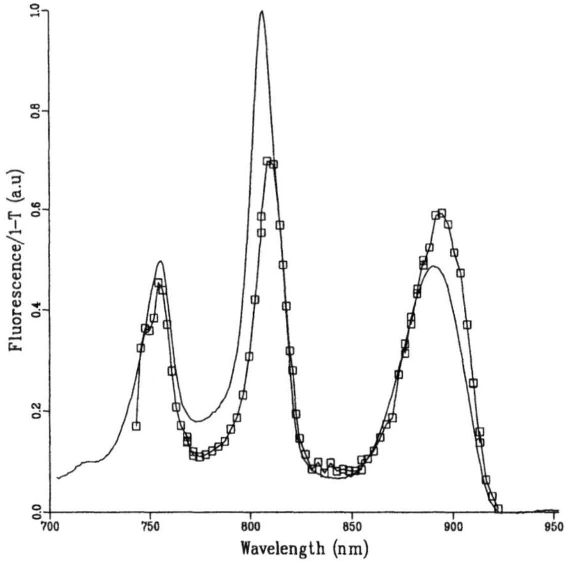


*Figure 7. Excitation spectrum of the steady state fluorescence (markers) of the M210W mutant compared with the 1-T spectrum at 5 K.*

**Discussion.**

In this work we have studied the wavelength and temperature dependence of charge separation, fluorescence emission, and charge recombination in membrane bound RCs of the M210W mutant of *Rb. sphaeroides*. We observe remarkable temperature and wavelength dependencies for some of the reactions, some of which can be explained by a static distribution of energy levels relative to the position of the energy levels of the early charge separated states. Other, which may be explained by a distribution in the electronic coupling between the pigments.

*$P^*$ decay and $P^+H_A^-$ formation at room temperature and 77 K.*

Our experiments with M210W RCs with $Q_A$ reduced show that at 293 K the quantum yield of charge separation is about 75%. The extrapolated rate of charge separation is about 200 ps, while the rate of $P^*$ deactivation is about 600 ps. The latter is slower than the value of 170 ps estimated earlier by [10] for open M210W RCs at RT.

The rate of charge separation drops to 1.6 ns at 77 K, suggesting that the process has become strongly activated. A simple model in which the free energy difference between $P^*$ and $P^+H_A^-$ is 67 meV explains the acceleration of the recombination kinetics to 1 ns at 293 K. Assuming that in M210W $P^+B_A^-$ is still about 200 meV above $P^+H_A^-$ indeed places $P^+B_A^-$ above $P^*$, in agreement with the observed decrease of the rate of charge separation.

Within the error of the experiment we do not observe a wavelength dependence of the decay time of P* in closed M210W RCs. This in contrast to measurements on closed WT RCs at RT, where a 20 % faster decay is observed in the red wing of the stimulated emission (Beekman *et al.*, unpublished results). One possible explanation is that at 77 K in closed M210W RCs the decay is largely determined by the sum of the loss processes from $P^*$. Furthermore, we do not observe a time dependent blueshift of the stimulated emission with an apparent time constant of 13 ps as reported for open solubilised M210W RCs by Nagarajan *et al.* [10], although our time resolution was certainly sufficient. We are currently repeating these experiments for open M210W RCs to investigate in detail whether or not this is related to the presence of the membrane, or in fact reflects wavelength dependent kinetics.

Finally, in reduced M210W RCs the rate of $P^+H_A^-$ decay at 77 K is estimated to be only 1.7 ns, at least a 6-fold acceleration in comparison with WT RCs. We do not have an explanation, but this effect might be related to the presence of the Trp residue at the position M210. In this respect we note that also the rate of $P^+Q_A^-$ recombination has increased by about a factor of 5.

*The wavelength and temperature dependence of $P^+Q_A^-$ recombination.*

The rate of $P^+Q_A^-$ recombination at 5 K and 77 K shows a pronounced wavelength dependence, with the faster rates observed in the red wing of the P band. A similar wavelength dependence of the stimulated emission decay has recently been observed for the membrane bound WT RC of *Rb. sphaeroides* [22,19]. Also the apparent blue shift of the stimulated emission observed for open M210W RCs may be related to fast 'red' RCs. One possibility to explain these effects is that a variation in electronic couplings exists, with the 'red' RCs having the stronger coupling. We note that this correlation between red shift and charge separation may also underlie the increase of the electron transfer rate with decreasing temperature in WT RCs. The difference in the amplitude of the electrochromic bandshift over the monomeric bacteriochlorophylls, with the fast 'red' RCs having the larger amplitude, supports the idea that a difference in electronic coupling underlies the difference in ET rate. However, for the WT and the M210L RCs the $P^+Q_A^-$ recombination at 77 K needed only a single spectral component, although the reactions were multi-exponential. Apparently, the distribution of electronic couplings is not as manifest in the other RCs with mutations at the M210 positions, for reasons which are not clear yet.

*The wavelength dependence of the steady state fluorescence.*

The steady-state fluorescence not only probes the relatively short lived charge separation components, but is especially sensitive to the long lifetimes that are largely attributable to recombination. It is even possible that a fraction of the M210W RCs is incapable of charge separation at 5 K. Previous experiments have demonstrated that the $P \rightarrow P^*$ absorption band is inhomogeneously broadened [12] and the same is true for the $P^+H_A^-$ and $P^+B_A^-$ states [13]. If we assume that the changes in the $P/P^+$ redox potential are also reflected in the free energy of the $P^+B_A^-$ state, then primary electron transfer in the M210W mutant may easily occur between two degenerate states or even be activated. Activated electron transfer will in particular involve RCs with a red P→P* transition, where the energy barrier between P* and $P^+B_A^-$ will be greatest. Since the charge separation quantum yield was only 0.6 at 5 K, this red fraction of RCs may indeed be present. Furthermore, the RCs that are still able to perform charge separation, due to a favourable value of the free energy of $P^+B_A^-$, will have an enhanced level of charge recombination. Assuming a time dependent driving force for the charge separation [7,8], this recombination could even proceed at cryogenic temperatures. Therefore, both hypothetical models will lead to an enhanced fluorescence excitation in the red wing of the absorption. The model above also suggests an enlarged probability of triplet formation for the 'red' RCs. Our analysis of the transient absorption data at 5 K and 77 K indicate that the DAS corresponding to the $^3P$→P transition overlaps with the P→P* absorption spectrum. There are two possible interpretations for this. Firstly, we might not be able to resolve the shift in the $^3P$→P transition due to the small amplitude of the 100 μs decay. An overlap of the obtained $^3P$–P spectrum and the fluorescence excitation spectrum shows that in principle they could be the same. Alternatively, the $^3P$→P transition may have lost the selectivity, but that seems to be in contrast with the observed wavelength dependence of the $P^+Q_A^-$ recombination rate.

Finally, a remarkable feature is observed in the 800 nm band of the excitation spectrum. About half the band is missing and the peak is red shifted relative to the major absorption band. In fact the maximum in the excitation spectrum coincides with the presumed position of the $B_B$ absorption maximum, which is seen as a shoulder in the low temperature absorption spectrum. It is noteworthy that the same phenomena can be observed in the excitation spectrum of the M210W mutant of *Rb. sphaeroides* as measured by Shochat et al.[14], although this was not noticed by the authors. A similar phenomenon was observed by us for the fluorescence excitation spectrum of the M210L RC. Apparently, excitation of $B_A$ does not, or with a much lower probability, result in formation of P* and consequently is not observed in the fluorescence excitation spectrum. One possible explanation for this is, that $B_A$ participates in a charge transfer state and its excitation results in a direct charge separation. Alternatively, the $B_A$ excited state may have a fast rate of internal conversion, resulting in an effective loss of absorbed energy. This latter possibility may be clarified by measuring the $P^+Q_A^-$ excitation spectrum, which we are currently doing.

**Acknowledgements.**

This research was supported by the Dutch Foundation for Fundamental Research (NWO) through the Foundation for Life Sciences (SLW) and the EC contract CT930278. MRJ is a BBSRC Senior Research Fellow.

**References.**

1) Schmidt, S., Arlt, T., Hamm, H., Huber, H., Nägele, T., Wachtveitl, J., Meyer, M., Scheer, H. and Zinth, W. (1994) Chem. Phys. Lett. 223, 116-120.

2) Müller, M.C., Griebenow K. and Holzwarth, A.R. (1992) Chem. Phys. Lett. 199, 465-469.

3) Jia, Y., DiMagno, T.J., Chan, C.-K., Wang, Z., Du, M., Hanson, D.K., Schiffer, M., Norris, J.R., Fleming, G.R. and Popov, M.S. (1993) J. Phys. Chem. 97, 13180-13191
4) Hamm, P., Gray, K.A., Oesterhelt, D., Feick, R., Scheer, H. and Zinth, W. (1993) Biochim. Biophys. Acta 1142, 99-105.
5) Kirmaier, C., Holten, D. and Parson W. W. (1990), Proc. Natl. Acad. Sci. U.S.A., 87, 3552.
6) Arlt, C., Schmidt, S., Kaiser, W., Lauterwasser, C., Scheer, H. and Zinth, W. (1993) Proc. Natl. Acad. Sci. USA 90, 11757-11761.
7) Peloquin, J.M., Williams, J.C., Lin, X., Alden, R.G., Taguchi, A.K.W., Allen, J.P. and Woodbury, N.W. (1994) Biochemistry 33, 8089-8100.
8) Woodbury, N.W., Peloquin, J.M., Alden, R.G., Lin, X., Lin, S., Taguchi, A.K.W., Williams, J.C. and Allen J.P. (1994) Biochemistry, 33, 8101-8112
9) Woodbury, N.W. and Parson, W.W. (1984) Biochim. Biophys. Acta 767, 345-361.
10) Nagarajan, V., Parson, W.W., Davis, D. and Schenck, C.C. (1993) Biochemistry 32, 12324-12336.
11) Van Noort, P.I.(1994), PhD thesis, State University of Leiden.
12) Reddy, N.R.S., Lyle, P.A. and Small, G.J. (1992), Photosyn. Res. 31, 167
13) Ogrodnik A. Keupp W., Volk M., Aumeier G. and Michel-Beyerle M.E. (1994) J. Phys. Chem. 98, 3432-3439.
14) Shochat, S., Arlt, T., Francke, C., Gast, P., Van Noort, P.I., Otte, S.C.M., Schelvis, H.P.M., Schmidt, S., Vijgeboom, E., Vrieze, J., Zinth, W. and Hoff, A.J. (1994) Photosynth. Res. 40, 55-66.
15) Jones, M.R., Visschers, R. W., Van Grondelle, R. and Hunter, C.N. (1992) Biochemistry 31, 4458-4465.
16) Jones, M.R., Heer-Dawson, M., Mattioli, T.A., Hunter, C. N. and Robert, B. (1994), FEBS Lett 339, 18-24.
17) Groot, M.-L., Peterman, E.J.G., Van Stokkum, I.H.M., Dekker, J.P. and Van Grondelle, R. (1995) Biophys. J. 68, 281-290.
18) Cho, H.M., Mancino, L.J. and Blankenship, R.E. (1984). Biophys. J. 45, 455-461.
19) Beekman, L.M.P., Visschers, R.W., Monshouwer, R., Van Mourik, F., Heer-Dawson, M., McGlynn, P., Hunter, C.N., Robert, B., Van Stokkum, I.H.M., Van Grondelle, R. and Jones, M.R. (1995) submitted to Biochemistry.
20) Monshouwer, R., Ortiz De Zarate, I., Van Mourik, F. and Van Grondelle, R. (1995) submitted to Chem. Phys. Lett.
21) Van Stokkum, I.H.M., Brouwer, A.M., Van Ramesdonk, H.J. and Scherer, T. (1993) Kon. Ned. Acad. Wet. 96, 43-68.
22) Vos, M.H., Jones, M.R., Hunter, C.N., Breton J. and Lambry, J.-C. (1994) Biochemistry 33, 6750-6757.

# Ultrafast Electron and Excitation Energy Transfer in Modified Photosynthetic Reaction Centers from *Rhodobacter sphaeroides*

T. Häberle[1], H. Lossau[1], M. Friese[1], G. Hartwich[1], A. Ogrodnik[1], H. Scheer[2] and M.E. Michel-Beyerle[1]

[1] Institut für Physikalische und Theoretische Chemie
TU München, Lichtenbergstr. 4, D-85747 Garching

[2] Botanisches Institut der LMU München, Menzinger Str. 67, D-80638 München

**Abstract.** Femtosecond time-resolved fluorescence measurements have been performed on RCs with the monomeric bacteriochlorophylls ($B_A$ and $B_B$) replaced by $13^2$-OH-Ni-BChl. Most relevant with respect to the primary electron transfer (ET) mechanism is the lowering of the energy of the state $P^+B_A^-$ 1500 $cm^{-1}$ below the primary donor $^1P^*$. On the basis of such energetics $P^+B_A^-$ has to be a real intermediate in the Ni-RC. Nevertheless, at 275 K the primary ET rate is found to be the same as in the native RC. This invariance indicates the sequential mechanism ($^1P^* \rightarrow P^+B_A^- \rightarrow P^+H_A^-$) to be operative also in the native RC. Another important feature of the Ni-RC is related to EET. After excitation of Ni-$^1B^*$ the rise time of the dimer fluorescence is observed to be shorter than in the native RC. Taking into consideration the low steady state quantum yield of EET (Ni-$^1B^* \rightarrow {}^1P^*$) this might be due to either (i) an enhancement of EET in the modified RC by a factor of two or (ii) an ultrashort lifetime of 40 fs for Ni-$^1B^*$. In the later case the short risetime would reflect the rate determining deactivation of a bottleneck state in the EET pathway from Ni-$^1B^*$ to the emitting $^1P^*$-state. The bottleneck may either be a vibronically excited or an upper excitonic state of $^1P^*$. The wavelength dependence of the observed rise time gives support in favour of model (ii).

**Keywords.** *Rhodobacter sphaeroides* R26, nickel-bacteriochlorophyll, femtosecond time-resolved fluorescence, primary charge separation, excitation energy transfer.

**Abbrevations.** TCSPC, time correlated single photon counting; BChl, Bacteriochlorophyll; RC, reaction center; ET, electron transfer; EET, excitation energy transfer; pigments bound to the RC: P, primary donor; B, bacteriochlorphyll; H, bacteriopheophytin.

## 1. Introduction

The simple question of the driving force for electron transfer (ET) has shown to be most important for the mechanistic interpretation of the primary charge separation in the photosynthetic reaction center (RC). It is closely related to the role of the bacteriochlorophyll monomer ($B_A$), located between the primary donor (P) and the bacteriopheophytin ($H_A$) in the X-ray structure [1-6]. The first fs- and ps-time-resolved measurements were interpreted with a single time constant for the spectral transients [7-10]. The absence of a transient intermediate in spectral regions characteristic of $B_A$ would be consistent with $B_A$ being a superexchange mediator for a direct ET-process from $^1P^*$ to $P^+H_A^-$, and thus with a positive driving force between $^1P^*$ and $P^+B_A^-$ [11-16]. Improved time resolution and sensitivity in the work of ZINTH et al. [17-19] revealed an additional, fast time constant supporting $P^+B_A^-$ to be a kinetic intermediate in the charge separation process. Since the maximum concentration of this state amounts to only 10%, it might easily escape experimental observation. Moreover, the investigation of primary kinetics is complicated by a variety of additional factors, e.g. congested difference spectra including strong electrochromic and excitonic shifts, dispersive kinetics due to inhomogeneity of energetics [20] and/or electronic couplings [21], and coherence phenomena [22-24].

One of the key features of primary ET dynamics is the free energy difference $\Delta G_1$ between $^1P^*$ and $P^+B_A^-$. Depending on its sign and value this may lead either to a sequential pathway of charge separation ($^1P^* \rightarrow P^+B_A^- \rightarrow P^+H_A^-$) or to direct formation of $P^+H_A^-$ via superexchange [11,25-27]. Therefore, it is of special interest to examine the consequences of changing the free energy of the state $P^+B_A^-$. An attractive method consists in the chemical exchange of pigments [28]. Unlike mutagenetic changes this procedure does not alter amino acid residues which might contribute to the medium reorganization parameters.

In SCHEER's group methods for selective chemical exchange of the monomeric bacteriochlorophylls against derivatives thereof have been developed [28]. These rest on thermal expansion of the protein matrix in the presence of an excess of the modified pigments to be implanted. By repetitive exchange and purification more than 95% of the RC-preparation can be modified.

By transmetallation of the central $Mg^{++}$ ion of BChl against other divalent metal ions the redox potential BChl/BChl$^-$ can be changed. For example, Ni-BChl shows the most significant lowering of the redox potential by ≈0.29 V, measured by cyclic voltametry in the solvent tetrahydrofuran (THF) [29]. Synthesis of a series of transmetallated BChls including Ni-BChl and Zn-BChl as well as their implantation into the RC have been accomplished recently [30,31].

After thermal exchange of native BChl against derivatives of the type $13^2$-OH-BChl ENDOR- and CD-spectra, as well as time-resolved fluorescence and absorption measurements show no differences to the native RC [32]. For Ni- and Zn-RCs EXAFS-measurements exhibit the same fivefold nitrogen coordination for the central metal ion in the porphin ring as for Mg of BChl in the native case [33]. Thus, the method of thermal exchange is not expected to affect grossly the position of the pigments in their binding pockets. On this basis we assume, that the electronic matrix element for ET remains unaffected.

*What is known so far on ET dynamics and energetics in the Ni-RC?*
The Ni-RC is functional [34], the $P^+Q_A^-$ state being formed on the timescale of 110 ps and recombining with almost the same kinetics as in the native RC. However, the quantum yield of $P^+Q_A^-$ formation is lower, amounting to ≈80% at 90 K and 60% at 280 K [35]. The temperature dependent loss of quantum yield is accompanied by the respective increase of the yield of triplet $^3P^*$.

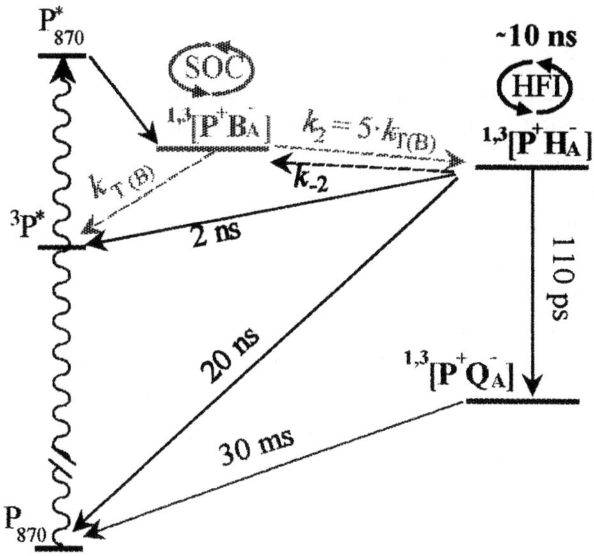


**Figure** 1: Rate scheme for ET-processes at in the Ni-RC at 90 K. SOC = spin orbit coupling, HFI = hyperfine interaction.

The approximate energies of $P^+H_A^-$ and $P^+NiB_A^-$ as well as the rates measured so far are compiled in the kinetic scheme of fig. 1. The proximity of the energy levels of $P^+NiB_A^-$ and $P^+H_A^-$ is borne out by a variety of observations, including temperature dependent triplet formation in quinone containing RCs. From the thermally activated participation of $P^+NiB_A^-$ in the recombination dynamics of the radical pair $P^+H_A^-$ an energy gap between the two states of less than ≈ 500 $cm^{-1}$ can be estimated. On the basis that the energy difference $\Delta G(^1P^* - P^+H_A^-) \approx 2000\ cm^{-1}$ is not altered by the modification, this is equivalent to $\Delta G(^1P^* - P^+NiB_A^-) > 1500\ cm^{-1}$. Apparently, only part of the *in vitro* lowering of the redox potential is realized in the RC. This might reflect the difference in the coordination number of $Ni^{2+}$, being five in the RC and four in THF.

Consistent with the primary formation of $P^+NiB_A^-$ caused by the large energy gap $\Delta G(^1P^* - P^+NiB_A^-)$ is the observation that the triplet state of the primary donor $^3P^*$ develops within 50 ps. This is attributed to fast singlet-triplet transitions in the radical pair $P^+NiB_A^-$ induced by the paramagnetic Ni-ion followed by fast recombination prior to the formation of the slowly recombining $P^+H_A^-$ states [T. Langenbacher, G. Bieser, G. Rousseau, unpublished results].

Since the sequential mechanism in the Ni-RC prevails, we can put the charge separation mechanism in the native RC to test via the primary ET rate in the Ni-RC. Thereby we evade all the difficulties of spectral characterization of minor concentrations of $P^+B_A^-$. Knowing ET in the native RC to proceed in the activationless regime, what do we expect?

(i) If in the native RC primary charge separation would lead directly to the state $P^+H_A^-$ with $P^+B_A^-$ acting as *superexchange mediator*, lowering the energy of $P^+B_A^-$ is expected to *increase* the rate either by lowering the energy denominator in the expression for the superexchange matrix element, or by transition to the sequential mechanism.

(ii) If, however, in the native RC primary charge separation forms $P^+B_A^-$ as a real kinetic intermediate, lowering its energy would put the primary charge separation in the inverted Marcus-regime. Depending on the value of the reorganisation parameter $\lambda$, the rate would either *remain constant* or *slightly decrease* [see Bixon *et al.*, this volume].

In this paper we present femtosecond time-resolved measurements of the primary ET dynamics in the $Ni_{AB}$-RC. The method of upconversion of the fluorescence of the primary donor $^1P^*$ will be also implied to the mechanism of excitation energy transfer from $^1B^*$ to $^1P^*$ in the Ni-RC.

## 2. Experimental

*Materials.* Photosynthetic RCs of *Rb. sphaeroides* R26 were isolated by standard methods [36]. The transmetallated bacteriochlorophylls were obtained by the procedure described in HARTWICH *et al.* [31]. Pigments were exchanged thermally following STRUCK AND SCHEER [28]. Efficiency of the exchange was better than 95%. Quinone was removed following GUNNER et al. [37]. Samples were solved in Tris(tris-hydroxymethylamine-methane)-LDAO(lauryl-dimethyl-aminoxide) buffer. Measurements were performed in a 1 mm quartz cell cooled to 275 K. The excitation wavelengths used are marked with arrows in the absorption spectrum in fig 2.

In quinone-depleted RCs the slowest time constant for recovery of the ground state (P) is determined by the triplet ($^3P^*$) lifetime which is shorter than 1 μs [38] in Ni-RCs. This allows measurements with relatively high turnover rates, favoring fs-fluorescence measurements in the upconversion technique. Fluorescence measurements using single photon timing (TCSPC) with a time resolution of 50ps were performed on RC-preparations containing quinone. To allow for low temperature studies the RC-solution contained 60% glycerol.

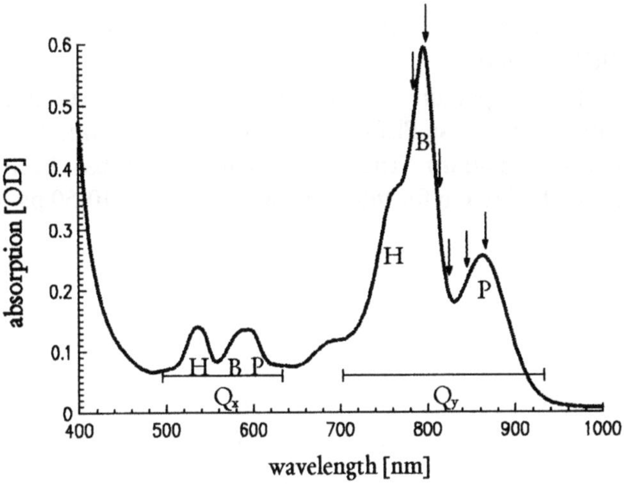


**Fig** 2: Absorption spectrum of $Ni_{AB}$-RC at room temperature with band assignment.

*Methods.* Femtosecond time-resolved fluorescence was measured with the upconversion technique. The laser source was a Kerr lens modelocked $Ti^{3+}$:$Al_2O_3$-laser (COHERENT MIRA BASIC), pumped by 8 W of the output of an $Ar^+$-laser (COHERENT INNOVA 425). The laser delivered 450 mW of average power at the desired excitation wavelengths. Pulse repetition rate was 76 MHz.

The laser beam was split in two parts with equal power, one focusing onto the sample (pulse energy 0.7 nJ, spot diameter 50 μm), the other one (gate beam) passes a stepping motor driven delay stage (resolution 16.7 fs, range 1.3 ns). The latter is combined with the fluorescence from the sample in a β-barium-borate (BBO) crystal (thickness 1 mm) under an angle of 11°, thus allowing non-critical type I phase matching. The geometry was chosen to upconvert fluorescence being polarized parallel to the excitation. The sample cell was rotating with up to 3000 rpm, thus reducing sample degradation by distributing the excitation energy over a larger sample volume. The intensity of sum frequency light is detected with a photomultiplier (HAMAMATSU R4220 P) after a suitable bandpass filter. The sum frequency photons are separated from the remaining background photons by counting them synchronously to the phase of light choppers in the excitation and gate pathways. In order to reduce the influence of long term drifts of the laser and degradation of the samples upon the decay curves, we collect up to 16 individual time traces and alter the direction of acquisition after each run. The fluorescence signal did not decrease significantly during a measurement cycle, which took up to several hours. The time window was at least 16 ps.

The instrument response function (IRF) is taken before each fluorescence measurement using scattered light from the sample. The alignment is not changed except for adjusting the phase matching angle of the crystal and removing a long pass filter after the sample. The full width at half maximum (fwhm) of the IRF was 200 fs.

Time correlated single photon counting (TCSPC) results were obtained using 40 ps pulses from a Hamamatsu PLP-01 laser diode for excitation (repetition rate 10 MHz, wavelength 864nm). The experimental setup has been described elsewhere [20]. The IRF fwhm for those measurements was 40-50 ps.

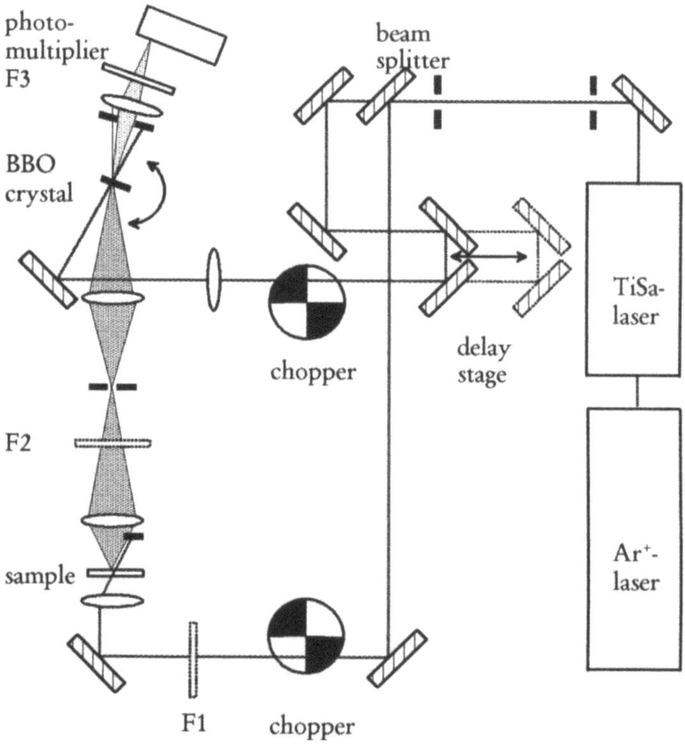


**Fig 3:** Optical setup of the upconversion apparatus. F = filter.

Time constants are extracted by fitting a convolution of the IRF and exponential decay functions [39] to the data using the Levenberg-Marquardt method [40,41]. Quality of fit is judged by observing the residuals and values of the reduced $\chi_\nu^2$. We obtained $\chi_\nu^2$ values between 1.0 and 1.5. This deconvolution method, together with the high quality of our data, allows to resolve lifetimes down to a quarter of the IRF fwhm.

## 3. Results and Discussion

### 3.1. Fluorescence Decay of the Dimer excited at 866nm. Comparative Study of the ET Mechanism in Native and Modified Reaction Centers.

*(i) Femtosecond results. Fluorescence upconversion.* The fluorescence decay of $^1P^*$ in $Ni_{AB}$-RCs excited at 866nm and detected at 930nm is shown in fig 4 together with the instrumental response function (IRF). The time constants are 2.4 ps (79% amplitude) and 8 ps (21%) when using a biexponential fit. Within error these time constants and the relative amplitudes for the spontaneous emission in the $Ni_{AB}$-RCs at 275 K are identical with the ones reported for native RCs from *Rb. sphaeroides* R26 [42-44]. The residuals show some structure below 1 ps, which can be removed by including a third, short time component in the fit; this leads to a shift of the other time constants to slightly larger values yielding 0.6 ps (9%), 2.8 ps (78%), and 10 ps (13%). No longer components with amplitudes larger than 1% were observed in time windows up to 100 ps. The results of these measurements are summarized in table 1, together with data on excitation energy transfer (discussed in section 3.2).

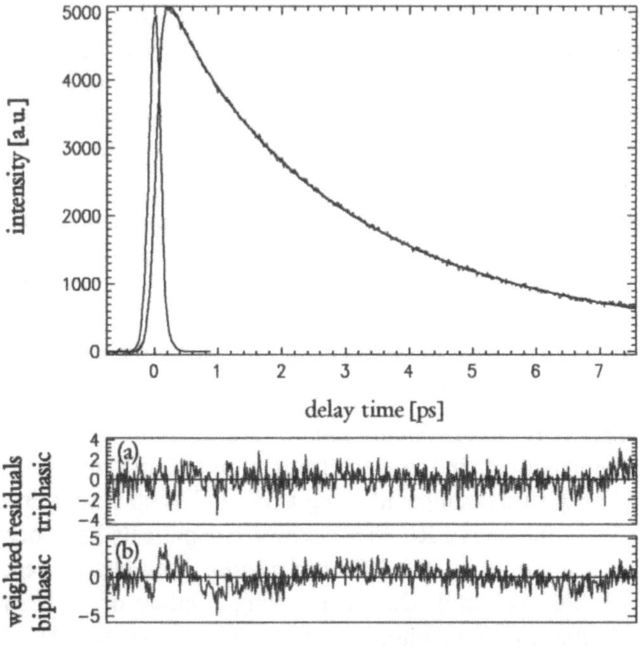


**Fig** 4: Decay of the dimer fluorescence in $Ni_{AB}$-RCs from *Rb. sphaeroides* R26. Excitation wavelength 866nm, detection wavelength 930nm, $T$ = 275 K. (a) residuals for triphasic fit, (b) for biphasic fit.

**Table** 1: Decay times and amplitudes from fits of multi-exponential decay patterns of measured fluorescence traces from $Ni_{AB}$-RCs. The amplitude corresponding to $\tau_{rise}$ was fixed to $\alpha_{rise} = -\Sigma\alpha_i$. [a]Measurement with large time window (>32 ps); long time constant $\tau_4$ has been fitted, but is prone to a large error. [b]For experimental reasons unreliable value. [c]No long-lived fluorescence was observed.

| excitation wavelength [nm] | # exp. terms | $\tau_{rise}$ [ps] | $\alpha_1$ [%] | $\tau_1$ [ps] | $\alpha_2$ [%] | $\tau_2$ [ps] | $\alpha_3$ [%] | $\tau_3$ [ps] | $\alpha_4$ [%] | $\tau_4$ [ps] |
|---|---|---|---|---|---|---|---|---|---|---|
| 785 | 2[a] | (0.058)[b] | | | 66 | 2.4 | 32 | 7 | 2.2 | 94[a] |
| 800 | 2[a] | 0.075 | | | 83 | 2.7 | 16 | 10 | 0.5 | 288[a] |
| 814 | 2 | 0.073 | | | 68 | 2.2 | 32 | 8 | | |
| 826 | 2 | 0.060 | | | 68 | 2.2 | 32 | 8 | | |
| 846 | 2[a] | 0.010 | | | 67 | 2.3 | 32 | 8 | 0.8 | 83[a] |
| 866 | 2[a,c] | | | | 79 | 2.4 | 21 | 8 | 0[c] | >50 |
| 866 | 3[a,c] | | 9 | 0.6 | 78 | 2.8 | 13 | 10 | 0[c] | >50 |

A central result of this paper is the invariance of the rate $k_1$ of the primary charge separation upon lowering the free enthalpy of $P^+B_A^-$ to $\Delta G_1 \approx 1500$ cm$^{-1}$ below $^1P^*$ in the Ni-RC. This invariance of $k_1$ on $\Delta G_1$ can be explained within the frame of multimode nonadiabatic ET theory. The rate $k_1$ in the Ni-RC can be described with the identical set of parameters used to model the dependence of this rate on $\Delta G_1$ in a series of native and mutagenetically altered RCs [Bixon *et al.*, this volume].

Our conclusion on sequential charge separation to be operative in native RCs of *Rb. sphaeroides* is complementing the findings on RCs, where the energy $\Delta G_2$ of the second acceptor $H_A$ has been diminished by replacing bacteriopheophytin by pheophytin [45].

The substantial lowering of the energy of $P^+B_A^-$ is consistent with fluorescence measurements on the 100ps timescale.

*(ii) Picosecond results. Single photon timing.* The quantum yield of fluorescence related to time components in the range from 50 ps to 2 ns ('middle components') of native and Ni-RCs was compared using a TCSPC setup with a time resolution of 50 ps (fig 5). In the native RC these slower components have been shown to originate from a minority of the RCs undergoing slow charge separation (see Hartwich *et al.*, this volume). Due to the long lifetime of such subsets, their relative contribution to the quantum yield $\phi = \alpha \cdot \tau$ ($\alpha$ amplitude, $\tau$ lifetime) is much higher than their relative concentration, determining the amplitude $\alpha$. While the relative quantum yield of fluorescence related to these components in

the native RC varies from 57% at 280 K up to 78% at 85 K (table 2), they are almost negligible in the Nickel-RC.

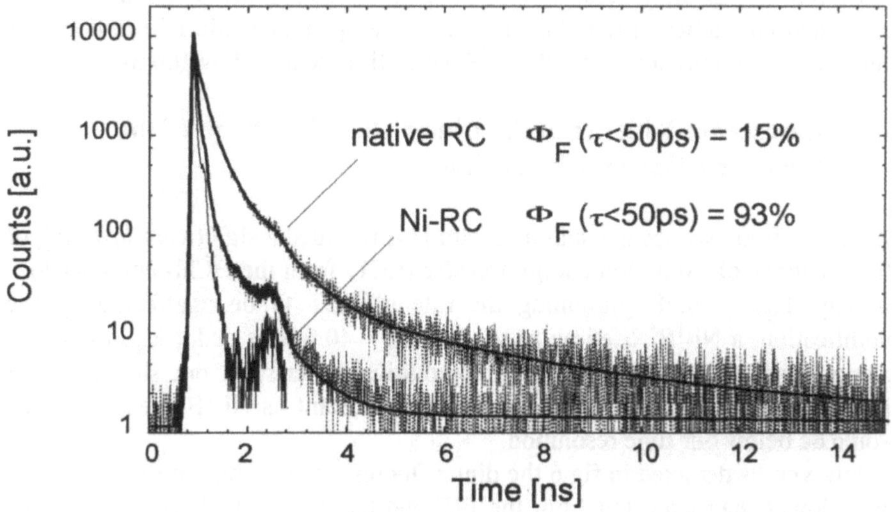


**Fig** 5: Fluorescence decay of the native and Ni-RC in the presence of $Q_A$ at 85 K. $\lambda_{exc}$=864nm; $\lambda_{det}$=920nm

**Table** 2: Relative fluorescence quantum yield of subpopulations of RCs with different ET rates, obtained from TCSPC. The assignment of the components to different processes is: (a) primary charge separation of the majority of RCs, (b) primary charge separation of slow RCs, (c) delayed fluorescence reflecting the lifetime of $P^+H_A^-$.

| | | | 85 K | 180 K | 280 K |
|---|---|---|---|---|---|
| Ni-RC | (a) | $\tau \leq 50$ ps | 93% | 92% | 96% |
| | (b) | 50 ps $\leq \tau \leq$ 2 ns | 7% | 8% | 4% |
| | (c) | 2 ns $\leq \tau$ | 0% | 0% | 0% |
| R26-RC | (a) | $\tau \leq 50$ ps | 15% | 30% | 72% |
| | (b) | 50 ps $\leq \tau \leq$ 2 ns | 84% | 69% | 27% |
| | (c) | 2 ns $\leq \tau$ | 1% | 1% | 1% |

Accepting that in the native RC slow charge separation results from an energetic distribution of the state $P^+B_A^-$ relative to $^1P^*$ (Hartwich *et al.*, this volume), we conclude that even the energetically highest Ni-RCs are deep enough in the inverted region in order to keep their primary rate rather independent of the exact value of their energy. Since there is no reason to assume that the width $2\sigma$ of the energetic distribution of radical ion pairs is *smaller* in the Ni-RC than in the native one, the free energy of $P^+B_A^-$ in the Ni-RCs must be lower than that of $^1P^*$ by at least half of this width ($\sigma = 400\ cm^{-1}$, [20]).

Nevertheless, explaining the absence of 'middle' components in the ps-fluorescence with the insensitivity of the primary rate in the inverted region to distributed energetics only holds, as long as the paramagnetic Ni-ion does not enhance intersystem crossing ($^1P^* \rightarrow {}^3P^*$) on a time scale below 100 ps.

## 3.2. Fluorescence Kinetics of the Dimer after Excitation of the Monomeric Bacteriochlorophyll.

Stationary fluorescence excitation spectra [34] revealed a significant lowering of the efficiency of excitation energy transfer (EET) from the BChl-monomer to P to only $\Phi_{EET} = 0.28$. Assuming the rate of EET to be unaffected by the modification, a Ni-$^1B^*$ lifetime as short as $\tau_B = 40$ fs was deduced. In case the energy transfer proceeds directly to the emitting dimer state, one should expect the fluorescence to rise with the same time constant as Ni-$^1B^*$ decays, which would be below our time resolution.

However, as depicted in fig 6 the dimer fluorescence excited in the Ni-B band rises slower than expected from the IRF and has to be fitted with a rise time component of 75±20 fs (fig 7). This rise time holds for excitation wavelengths of 785, 800, 814, and 826nm. From excitation wavelengths of 846nm on, however, no rise time component is required.

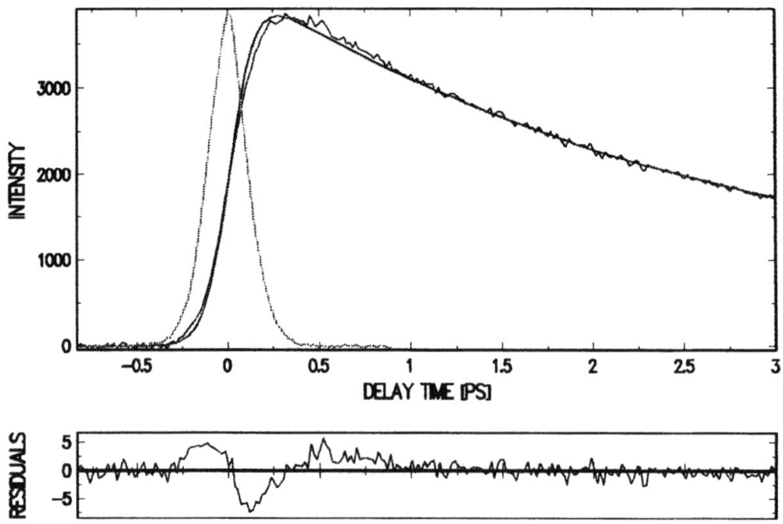


**Fig** 6: Decay of the fluorescence (930nm) of $^1P^*$ in the $Ni_{AB}$-RC after excitation in the monomer band (800nm). Fit without rise time component. Only a small time window is shown to illustrate the need for a rise time component.

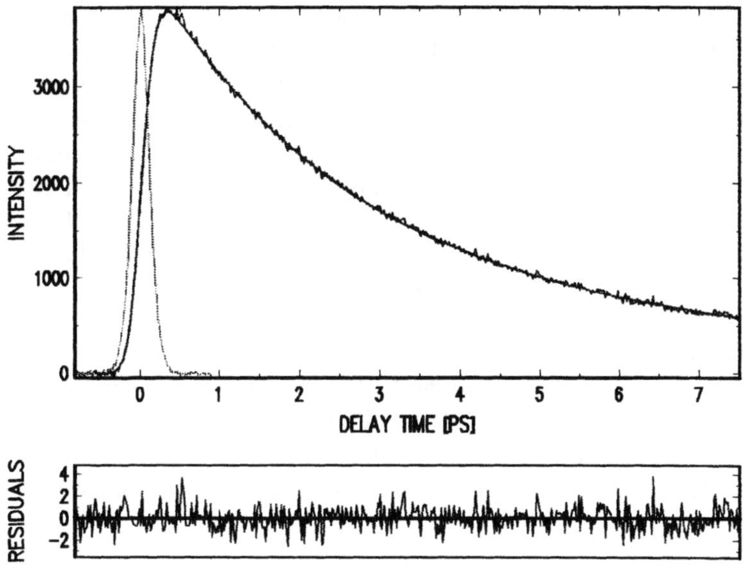


**Fig** 7: Fluorescence kinetics $^1P^*$ of the $Ni_{AB}$-RC after excitation in the monomer band (800nm). Fit with two decay times and one rise time component.

Adding one more decay time constant in the 1 ps range improves the fit by some extent. The dominating 2.5 ps component is split in two nearby time components between 1.4 ps and 3.8 ps with comparable amplitudes. No significance as individual components can be attributed to them, but the resulting trace can be considered as an improved empirical description of the inhomogeneous decay, so that a more precise rise time component can be obtained. It results in rise time constants that are 10 to 15 fs longer.

In the upconversion data longer time components (>50 ps) only contributed to the decay curve at $\lambda_{exc} \leq 846$nm with amplitudes <2% (table 1). They could originate from extreme red emission of a minor contamination with bacteriochlorophyll molecules, which are less excited when using 866nm as excitation wavelength (compare Hartwich *et al.*, this volume).

*Comparison with the native RC.* The efficiency of EET from the $^1B^*$ to P in the native RC was determined to be 0.93 [46]. Early measurements of the fluorescence rise time [7,47] gave only upper limits of about 100 fs. More recently, the rise time was determined in fluorescence upconversion measurements to be 120 fs (Jonas *et al.*, this volume) or 136 fs (stimulated emission, Wynne et al., this volume). A complementary approach is to observe the decay of $^1B^*$ in transient absorption. This has been performed by Jia *et al.* [48] and Jonas *et al.* (this volume), resulting in a fast recovery of the B absorption (130 fs).

*Simple kinetic model* (Fig 8 a). In the simplest model, EET proceeds directly from $^1B^*$ to the emitting state $^1P^*$. The rise time of the dimer fluorescence is then equal to the decay time of $^1B^*$, which is determined by the energy transfer rate $k_{EET}$ and the loss channels. The latter ones do not lead to $^1P^*$ formation and are summarized here under $k_{IC}$: $\tau_B = 1/(k_{EET}+k_{IC})$. The quantum efficiency $\phi_{EET}$ for the energy transfer from $^1B^*$ to $^1P^*$ is governed by the competition of the EET rate with the loss channels, $\phi_{EET} = k_{EET}/(k_{EET} + k_{IC})$. With $\phi_{EET} = 0.93$ and $\tau_B$ = 120 fs for native RCs one obtains a rate $k_{EET} = (130\text{ fs})^{-1}$. Within this model one obtains for the Ni-RC from $\phi_{EET} = 0.28$ and $\tau_B = 75$ fs a rate $k_{EET} = (270\text{ fs})^{-1}$, i.e. the energy transfer rate in the Ni-RC would be smaller by a factor two compared to the native RC.

However, if we assume $k_{EET}$ in the Ni-RC to be the same as in the native RC, the low $\phi_{EET} = 0.28$ would indicate an ultrafast internal conversion rate of $k_{IC} = (55\text{ fs})^{-1}$, and a lifetime of Ni-$^1B^*$ of only 40 fs [34]. The measured 75 fs rise time therefore is unexpectedly slow and requires an extension of the model.

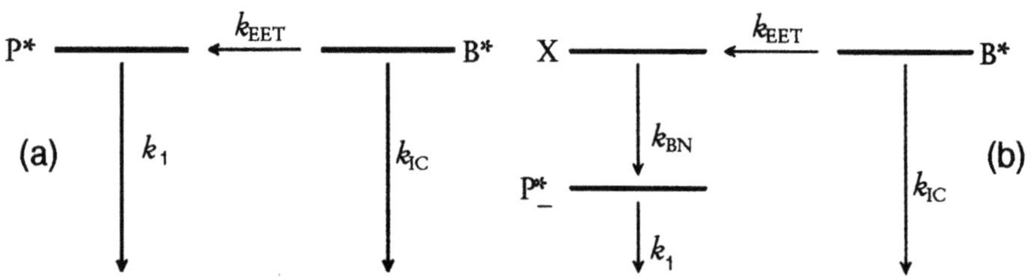


**Fig 8:** Rate schemes for EET from $^1B^*$ to P. (a) Simple model, (b) extended model involving a bottleneck state X

*Extended kinetic model involving a bottleneck state* (Fig 8 b). The slower rise time of the dimer fluorescence can be explained by a model, where the energy transfer first proceeds to an intermediate state X before a subsequent slower process populates the emitting $^1P^*$state. X could be either the upper excitonic state of the dimer $P_+^*$ or vibronically excited states of $^1P^*_-$

From the solution of the kinetic scheme (b) we obtained for the population of the vibronically relaxed state $^1P^*_-$

$$[P^*_-] = \alpha_1 \exp(-t/\tau_B) + \alpha_2 \exp(-k_{BN}t) + \alpha_3 \exp(-k_1 t) \tag{1}$$

Here $\tau_B$ denotes the lifetime of $^1B^*$, $\tau_B = 1/(k_{EET}+k_{IC})$, $k_{BN}$ the bottleneck rate and $k_1$ the primary ET rate (neglecting inhomogeneities). Assuming $k_1$ to be much slower than $k_{BN}$ and $1/\tau_B$, one obtains for the amplitudes:

$$\alpha_1 = \frac{\tau_B}{(\tau_{BN} - \tau_B)}, \quad \alpha_2 = \frac{-\tau_{BN}}{(\tau_{BN} - \tau_B)}, \quad \alpha_3 = 1 \tag{2}$$

with $\tau_{BN} = 1/k_{BN}$. The precursor states to $^1P^*$ are reflected in two additional time components. In the case $\tau_B < \tau_{BN}$, $\alpha_1$ is positive (decay) and $\alpha_2$ negative (rise). Phenomenologically, the fast decay term sits on top of the rising slope of the time trace, so that this rise time appears to be slowed down. Thus, if both time constants are below the width of our instrument response function, the fitted rise time is an average value dominated by the slower of the two components (which always has a negative amplitude). Taking $\tau_B = 40$ fs, we obtain from the fit a bottleneck time constant of $\tau_{BN} \cong 70$ fs, forming a limit for the fluorescence rise time. This bottleneck does not show up in the recovery of $^1B^*$ [48] and in the fluorescence rise time (Jonas et al., this volume) of native RCs, because here the rate limiting step for $^1B^*$-decay is the EET from $^1B^*$ to P within 130 fs.

Additional evidence for the bottleneck mechanism comes from the fact that the delayed rise of the $^1P^*$ emission pertains to excitation wavelengths as long as 826nm. According to [34], in the Ni-RC the upper exciton band $P_+$ of the $Q_y$ absorption of the dimer predominates at this wavelength in the excitation spectrum of the $^1P^*$ emission at 90K. This is a special feature of the Ni-RC caused by the low efficiency of 0.28 for EET from $^1B^*$ to P.

At 275K, the temperature of the fluorescence upconversion presented here, excitation at 826nm occurs in the trough of the absorption spectrum between the 800nm- and the 865nm-band. Assuming that $\phi_{EET}$(Ni-$^1B^* \rightarrow$ P) is temperature independent, less than 10% of the emission originates from RCs excited in the 800nm-band, while more than 90% of the signal are caused by RCs excited either in the upper exciton component of the dimer ($P_+$) or in the vibrational manifold of the lower exciton component ($P_-$). If this holds, the delayed rise of the $^1P^*$ emission in the Ni-RC is not reflecting the decay characteristics of $^1B^*$, but rather the bottleneck rate from excited intermediate states.

Excitation wavelength dependence of the delayed rise of the emission at low temperature is in progress, which would relief us from the assumptions made above. Further discrimination between the models might follow from the direct measurement of the Ni-$^1B^*$ lifetime in transient absorption (or emission) experiments analogous to the ones performed on $^1B^*$ in the native RC [48].

## 4. Conclusions

Electron transfer and excitation energy transfer studies have been performed with femtosecond and picosecond time-resolved fluorescence measurements on RCs with the monomeric bacteriochlorophylls replaced by $13^2$-OH-Ni-BChl.

(1) Most interesting for the investigation of primary charge separation is the implantation of the Ni-BChl at the monomer sites in the RC, shifting the energy of the state $P^+B_A^-$ 1500 $cm^{-1}$ below $^1P^*$. Nevertheless, at 275 K the primary ET rate is found to be the same as in the native RC. Knowing from independent pump-probe experiments, that $P^+B_A^-$ has to be a real intermediate in the Ni-RC, this invariance indicates the sequential mechanism ($^1P^* \rightarrow P^+B_A^- \rightarrow P^+H_A^-$) to be

operative also in the native RC. This result is also consistent with the almost complete vanishing of the slower fluorescence components (between 50 ps and 2 ns). For additional confirmation of the 'inverted' behaviour of the rate $k_1$ in the Ni-RC, low temperature measurements are in progress.

(2) The rise time of the dimer ($^1P^*$) fluorescence after excitation of the monomer ($^1B^*$) is 75 fs. In the absence of direct measurements of the lifetime of $^1B^*$, this 'slow' rise time can be due to two effects: (a) A slowed down (270 fs) EET ($^1B^* \rightarrow P$) in Ni-RC as compared to native RCs, or (b) the population of an intermediate state between $^1B^*$ and the emitting lowest excitonic component of $^1P^*$ assuming that the rate of EET is unaffected by the modification of B. If this holds, the decay of the intermediate state within ≈70 fs limits the rise time of the dimer fluorescence. The prevalence of the slow rise time up to excitation wavelengths of 826nm is strongly in favour of mechanism (b). It is the unique ultrashort lifetime (40 fs) of $^1B^*$, which gives access to the relaxation of either a higher vibronic or the upper excitonic state of $^1P^*$, while it is not accessible in native RCs with intrinsically long $^1B^*$-lifetimes.

**Acknowledgement.** The work has been supported by the Deutsche Forschungsgemeinschaft (SFB143).

**References.**

[1] J. Deisenhofer, O. Epp, K. Miki, R. Huber and H. Michel, **(1984)** J. Mol. Biol. *180:* 385-398.

[2] J.P. Allen, G. Feher, T.O. Yeates, D.C. Rees, J. Deisenhofer, H. Michel and R. Huber, **(1986)** Proc. Nat. Acad. Sci. USA, *83:* 8589-8593.

[3] C.H. Chang, J. Norris, M. Schiffer, U. Smith, J. Tang and D. Tiede, **(1986)** FEBS Lett. *205:* 82-86.

[4] U. Ermler, H. Michel and M. Schiffer, **(1994)** J. Bioenerg. Biomembrane, *26:* 5-15.

[5] U. Ermler, G. Fritzsch, S.K. Buchanan and H. Michel, **(1994)** Structure, *2:* 925-936.

[6] J. Deisenhofer, O. Epp, I. Sinning and H. Michel, **(1995)** J. Mol. Biol. *246:* 429-457.

[7] J. Breton, J.L. Martin, A. Migus, A. Antonetti and A. Orszag, **(1986)** Proc. Nat. Acad. Sci. USA, *83:* 5121-5125.

[8] J. Breton, A. Antonetti, J.L. Martin, A. Migus and J. Petrich, **(1986)** FEBS Lett. *209:* 37-43.

[9] J.L. Martin, J. Breton, A.J. Hoff, A. Migus and A. Antonetti, **(1986)** Proc. Nat. Acad. Sci. USA, *83:* 957-961.

[10] G.R. Fleming, J. Breton and J.L. Martin, **(1988)** Nature, *333:* 190-192.

[11] R.A. Marcus, **(1987)** Chem. Phys. Lett. *133:* 471-477

[12] M.E. Michel-Beyerle, M. Bixon and J. Jortner, **(1988)** Chem. Phys. Lett. *151:* 188-194.

[13] Y. Won and R.A. Friesner, **(1988)** Biochim. Biophys. Acta, *935:* 9-18.

[14] M. Bixon, J. Jortner, M.E. Michel-Beyerle and A. Ogrodnik, **(1989)** Biochim. Biophys. Acta, *977:* 273-286.
[15] Y.M. Hu and S. Mukamel, **(1989)** Chem. Phys. Lett. *160:* 410-416.
[16] M. Bixon, J. Jortner and M.E. Michel-Beyerle, **(1990)** in Perspectives in Photosynthesis, (J. Jortner and B. Pullman, Eds) Kluwer Acad. Publ. Netherlands, pp. 325-336.
[17] W. Holzapfel, U. Finkele, W. Kaiser, D. Oesterhelt, H. Scheer, H.U. Stilz and W. Zinth, **(1989)** Chem. Phys. Lett. *160:* 1-7.
[18] W. Holzapfel, U. Finkele, W. Kaiser, D. Oesterhelt, H. Scheer, H.U. Stilz and W. Zinth, **(1990)** Proc. Nat. Acad. Sci. USA, *87:* 5168-5172.
[19] T. Arlt, S. Schmidt, W. Kaiser, C. Lauterwasser, M. Meyer, H. Scheer and W. Zinth, **(1993)** Proc. Nat. Acad. Sci. USA, *90:* 11757-11761.
[20] A. Ogrodnik, W. Keupp, M. Volk, G. Aumeier and M.E. Michel-Beyerle, **(1994)** J. Phys. Chem. *98:* 3432-3439.
[21] G.J. Small, J.M. Hayes and R.M. Silbey, **(1992)** J. Phys. Chem. *96:* 7499-7501.
[22] M.H. Vos, J.C. Lambry, S.J. Robles, D.C. Youvan, J. Breton and J.L. Martin, **(1991)** Proc. Nat. Acad. Sci. USA, *88:* 8885-8889.
[23] M.H. Vos, F. Rappaport, J. Lambry, J. Breton and J. Martin, **(1993)** Nature, *363:* 320-325.
[24] M.H. Vos, M.R. Jones, C.N. Hunter, J. Breton, J.C. Lambry and J.L. Martin, **(1994)** Biochem. *33:* 6750-6757.
[25] M. Bixon, J. Jortner, M. Plato and M.E. Michel-Beyerle, pp. 399-419 in "The Photosynthetic Bacterial Reaction Center: Structure and Dynamics" (ed. J. Breton and A. Verméglio), Plenum Press, New York **(1988)**.
[26] M. Bixon, J. Jortner and M.E. Michel-Beyerle, **(1991)** Biochim. Biophys. Acta, *1056:* 301-315.
[27] M. Bixon, J. Jortner and M.E. Michel-Beyerle, **(1995)** J. Phys. Chem. *197:* 389-404
[28] A. Struck and H. Scheer, **(1990)** FEBS Lett. *261:* 385-388.
[29] C. Geskes, G. Hartwich, H. Scheer, W. Mäntele and J. Heinze, **(1995)** J. Amer. Chem. Soc. *117:* 7776-7783.
[30] G. Hartwich, **(1994)** Ph.D. Thesis, TU München.
[31] G. Hartwich, E. Cmiel, I. Katheder, W. Schäfer, A. Scherz and H. Scheer, **(1996)** J. Amer. Chem. Soc., in press
[32] H. Scheer and A. Struck, pp. 157-193 in "The Photosynthetic Reaction Center", (ed. J. Deisenhofer and J.R. Norris), Academic Press, New York **(1993)**.
[33] L. Chen, Z. Wang, G. Hartwich, I. Katheder, H. Scheer, D.M. Tiede, A. Scherz, P.A. Montano and J.R. Norris, **(1995)** Chem. Phys. Lett. *234:* 437-444
[34] G. Hartwich, M. Friese, H. Scheer, A. Ogrodnik and M.E. Michel-Beyerle, **(1995)** Chem. Phys. *197:* 423-434.
[35] P. Müller, **(1995)** Ph.D. Thesis, TU München.

[36] G. Feher and M.Y. Okamura, pp. 349-386 in "The Photosynthetic Bacteria", (ed. R.K. Clayton and W.R. Sistrom), Plenum Press, New York, **(1978)**.

[37] M.R. Gunner and P.L. Dutton, pp. 259-270 in "The Photosynthetic Bacterial Reaction Center - Structure and Dynamics", (ed. J. Breton and A. Verméglio), Plenum Press, New York, **(1988)**.

[38] G. Buchmann, **(1994)** Diplomarbeit, TU München.

[39] A. Grinvald and I.Z. Steinberg, **(1976)** Anal. Biochem. *75:* 260-280.

[40] Bevington, P.R. and D.K. Robinson, "Data Reduction and Error Analysis for the Physical Sciences", McGraw-Hill, International, **(1992)**.

[41] Press, W.H., S.A. Teukolsky, W.T. Vetterling and B.P. Flannery, "Numerical Recipes in C. The Art of Scientific Computing", University Press, Cambridge, **(1992)**.

[42] M. Du, S.J. Rosenthal, X. Xie, T.J. DiMagno, M. Schmidt, D.K. Hanson, M. Schiffer, J.R. Norris and G.R. Fleming, **(1992)** Proc. Nat. Acad. Sci. USA, *89:* 8517-8521.

[43] P. Hamm, K.A. Gray, D. Oesterhelt, R. Feick, H. Scheer and W. Zinth, **(1993)** Biochim. Biophys. Acta, *1142:* 99-105.

[44] R.J. Stanley and S.G. Boxer, **(1995)** J. Phys. Chem. *99:* 859-863.

[45] S. Schmidt, T. Arlt, P. Hamm, H. Huber, T. Nägele, J. Wachtveitl, M. Meyer, H. Scheer and W. Zinth, **(1994)** Chem. Phys. Lett. *223:* 116-120.

[46] C.A. Wraight and R.K. Clayton, **(1973)** Biochim. Biophys. Acta, *333:* 246-260.

[47] J. Breton, G.R. Fleming, J.L. Martin and J.C. Lambry, **(1988)** Biochem. *27:* 8276-8284.

[48] Y. Jia, D.M. Jonas, T. Joo, Y. Nagasawa, M.J. Lang and G.R. Fleming, **(1995)** J. Phys. Chem. *99:* 6263-6266

# Pressure Dependence of Energy and Electron transfer in Photosynthetic Complexes

N. R. S. Reddy, H.-C. Chang, H.-M. Wu, R. Jankowiak, and G. J. Small
*Ames Laboratory-USDOE and Department of Chemistry*
Iowa State University, Ames, IA 50011

**Abstract.** Effect of high pressure on the primary charge separation kinetics of the PS II RC has been investigated using spectral hole burning. These results show that pressure has a pronounced effect on primary electron transfer. This effect is plastic (irreversible) in nature in contrast with the elastic effects of pressure on the low temperature $Q_y$- absorption and non-line narrowed hole spectra of P680. The electron-phonon coupling shows only a weak dependence on pressure. Based on these data and using the theory of hole profiles, the energy of the acceptor state in the primary charge transfer is proposed to have pressure shift rate of 1.0 $cm^{-1}$/MPa. In *Rb. sphaeroides* LH II antenna complex, the pressure dependence of the B800 and B850 bands as well as the B800→B850 energy transfer rate have been measured. A possible explanation for the pressure independence of the energy transfer rate is provided.

**Keywords.** Electron and energy transfer, spectral hole burning, high pressure

## 1 INTRODUCTION

Excitation energy transfer and primary charge separation in photosynthetic units are problems of long-standing interest [1-11]. In recent years there has been a burgeoning of activity because of the increasing availability X-ray structures for protein complexes. The bacterial RC of *Rhodobacter sphaeroides* and *Rhodopseudomonas viridis* have received the most attention since their crystal structures are known and their $Q_y$-absorption spectra are particularly well resolved [5-11]. However, primary charge separation in the D1-D2-cyt $b_{559}$ RC of photosystem II has received increased attention [12-13] since its isolation in 1987 by Nanba and Satoh [14]. Both ultrafast and spectral hole burning spectroscopies have been extensively applied to the study of the bacterial and the PS II RC as well as several antenna complexes. Despite a significant number of advances, there remain important questions on transport mechanisms whose solutions depend on a firm understanding of the excited electronic state structure and energetics of coupled pigments, the nuclear response function associated with the motion of excitation energy or electron transfer, vibrational dephasing and the intrinsic structural heterogeneity of proteins. From the viewpoint of experiment, improvements in the sensitivity and resolution of time and frequency domain spectroscopies will continue to be important as well as the introduction of new approaches. Pressure, as temperature, should prove useful in testing transport properties.

Recently, we initiated a program to study the effects of pressure on the excited electronic state structure and transport dynamics of photosynthetic protein complexes. Windsor and coworkers [15-17] had already demonstrated that pressures of ~350 MPa have a significant effect on the kinetics of secondary charge separation in the RC of *Rb. sphaeroides* and *Rps. viridis*. These works and those on the $Q_y$-absorption spectra of the *Rps. viridis* [18] and *Rb. sphaeroides* [19,20] RC and of the FMO antenna complex of *Chlorobium tepidum* [21] have clearly demonstrated that the linear pressure shifts (to the red) of $Q_y$-states can be large (~0.5 $cm^{-1}$/MPa for P870 and P960 of *Rb. sphaeroides* and *Rps. viridis*). In addition, the pressure shifts for different $Q_y$-states of a given complex are generally not the same. This is important since one can alter the energy gap(s) associated with energy and electron transfer. Site directed mutagenesis has been used in bacterial RC to investigate the effects of specific residues on the driving force of charge separation [22]. We state in this context, using pressure as a variable, that we effectively obtain a different mutant at each pressure. In addition, however, pressure may affect other parameters important to transfer such as the reorganization energy, the electronic coupling and the structural heterogeneity which leads, for example, to a distribution of values for donor-acceptor energy gaps. Fortunately, spectral hole burning can be used to examine the pressure dependencies of these quantities.

In this paper we present a sampling of our results of high pressure-hole burning studies of the primary charge separation kinetics of the PS II RC and energy transfer dynamics in LH II antenna complex from *Rb. sphaeroides*. Full reports of these studies will appear elsewhere [23,24]. To the best of our knowledge our data are the first to show that pressure can have a significant effect on primary charge separation; at 4.2 K the P680* lifetime of 2.0 ps at 0.1 MPa lengthens by a factor of 3.5 as the pressure is increased to 267 MPa. This pronounced effect is irreversible (plastic) in contrast with the elastic pressure dependence of the low temperature (4.2, 77 K) absorption and non-line narrowed P680 hole spectra. This contrasting behavior, together with other observations, allow for a quite straightforward theoretical analysis of the pressure dependence of the primary charge separation kinetics.

Despite similarities with bacterial RCs in pigment composition, the low temperature (4.2 K) $Q_y$- region absorption spectrum of the PS II RC consists of only two bands (at 681 and 670 nm). This spectral congestion in the PS II RC absorption spectra has made the understanding of excited state electronic structure and primary charge separation kinetics more difficult. Both the special pair, P680 and the active pheophytin molecule contribute to the 680 nm band [25-27]. Fortunately, spectral hole burning can isolate P680* and the Pheo *a* $Q_y$-state. The latter gives rise to persistent nonphotochemical hole burning while P680* does not

[25]. Additional Chl *a* of differing amounts is present in typical PS II RC preparations in functional and dysfunctional forms [27-29]. Despite the apparent variability in the amount of the additional Chl *a* contributing to PS II RC, hole burning of many different preparations has shown that the P680* lifetime at 4.2 K (1.9 ± 0.2 ps) is invariant to the amount of additional (beyond 4) Chl *a* [30]. Indeed, the P680* lifetime has also been found to be the same for the PS II RC-CP47 complex, where CP47 is one of the two (CP43 being the other) proximal antenna complexes.

Recent structural data for the LH II complex from *Rb. acidophila* [31] show that within a structural subunit the eighteen B850 BChl *a* are arranged as dimers with a ninefold symmetry. The Mg....Mg distance within a dimer is 8.7 Å and between adjacent dimers is 9.7 Å. This results in strong exciton interactions between BChls [32,33]. The nearest $Mg_{B800}...Mg_{B800}$ and $Mg_{B800}...Mg_{B850}$ distances are 21 Å and 17.6 Å respectively. These data have confirmed the results of earlier hole burning studies [32] that the B850 BChl *a* are strongly exciton coupled and that their coupling with B800 BChl *a* is weak. Narrow (4.5 ± 0.5 $cm^{-1}$) zero phonon holes (ZPH) burned into the B800 band established the B800→B850 energy transfer time to be 2.4 ± 0.2 ps at 4.2 K [32,34]. Hole burning has also established that the 750 $cm^{-1}$ BChl *a* vibrations promote the B800 to B850 Forster type energy transfer. The low temperature (4.2 K) homogeneous width of 220 $cm^{-1}$ for the B850 band further explains the weak temperature dependence of the B800→B850 energy transfer. Novel nonphotochemical action spectroscopy has been used to identify the weakly absorbing B870 and B896 states [33]. These states have been identified as the lowest excitonic levels of the B850 and B875 bands. These states have further been implicated as "shuttle" states for energy transfer from B850 to B875 and from B875 to P870 respectively. Recent X-ray structure data does *not* identify any 'special' BChls among the 18 (B850) BChls supporting the interpretation of B870 and B896 states based on exciton bands. Hole burning at high pressures has shown that the B800→B850 energy transfer rate remains unchanged for pressures of upto 800 MPa even though the energy gap is increased by 190 $cm^{-1}$.

## 2 EXPERIMENTAL

Both hole burning and high pressure experimental setups have been described in detail earlier [23]. Briefly, high pressures (upto 1.5 GPa) were generated by a three-stage compressor (model U11, Unipressequipment Division, Polish Academy of Sciences, Warsaw, Poland). The high pressure cell located in the sample chamber of a liquid helium cryostat contains the sample under study and is connected to the compressor by a flexible capillary tube. Helium gas is used as the pressure transmitting medium. Absorption and hole burned spectra were recorded with a Bruker HR120 Fourier transform spectrometer. Typically, the spectrometer was used at a resolution of 4 $cm^{-1}$ except for high resolution scans of the zero phonon holes where a resolution of 0.2 $cm^{-1}$ was used. A Coherent CR699-21 ring dye laser (linewidth = 0.07 $cm^{-1}$), pumped by a 6 W Coherent

Innova argon ion laser, was used for hole burning. For experiments on LH II, a Coherent CR899-21 Ti:sapphire laser (linewidth = 0.07 $cm^{-1}$) pumped by a 15W Innova 200 argon ion laser was used.

Samples of LH II antenna complex from *Rb. Sphaeroides* and the D1-D2-cyt $b_{559}$ RC complex from spinach leaves were generously provided by M. Seibert and R. Picorel. Details of the isolation procedures are given in refs.23 and 32.

## 3 RESULTS AND DISCUSSION

### 3.1 Pressure effects on the PS II RC.

Figure 3.1 shows the $Q_y$-absorption spectra (at 77 K) of the PS II RC obtained at 0.1 (a), 141 (b) and 308 MPa (c), respectively. Spectrum (d) obtained after pressure cycling is coincident with spectrum (a) obtained at the start of the pressure cycle. Similar spectra (not shown) were obtained at 4.2 K. Using the data from Figure 3.1, we estimate the linear pressure shift rates for the 680 and 670 nm bands to be -0.08 and -0.04 $cm^{-1}$/MPa respectively. Linear shift rates for P680 and pheophytin, both absorbing at 680 nm (cf. Introduction) can be separated by employing triplet bottleneck holeburning and nonphotochemical holeburning respectively. These values are -0.09 $cm^{-1}$/MPa for P680 and -0.04 $cm^{-1}$/MPa for

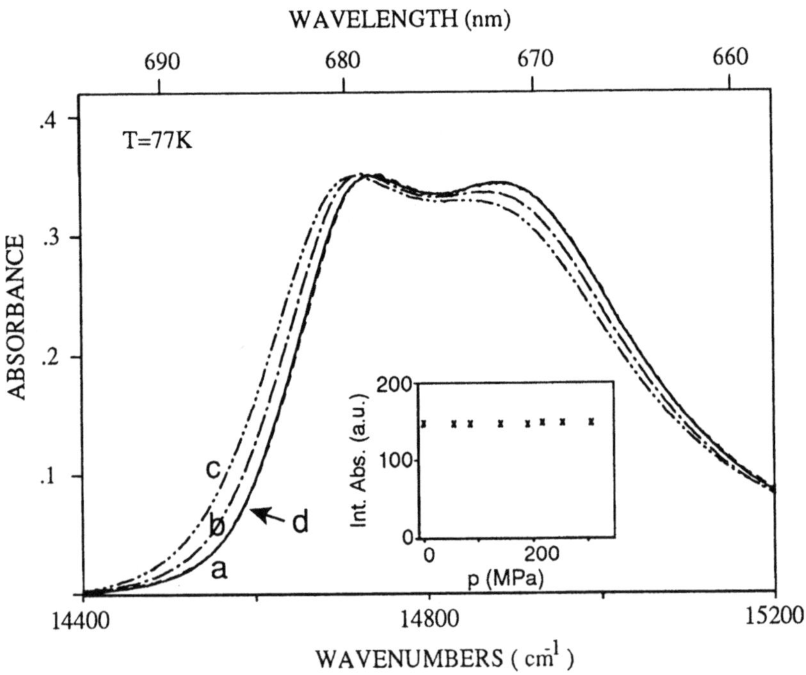


Figure 3.1. Absorption ($Q_y$-region) of PS II-RC obtained at T = 77 K for p = 0.1 MPa (a), 141 MPa (b), and 308 MPa (c), respectively. Spectrum d is the spectrum upon pressure release. The inset shows the absence of a pressure dependence of the integrated absorption intensity.

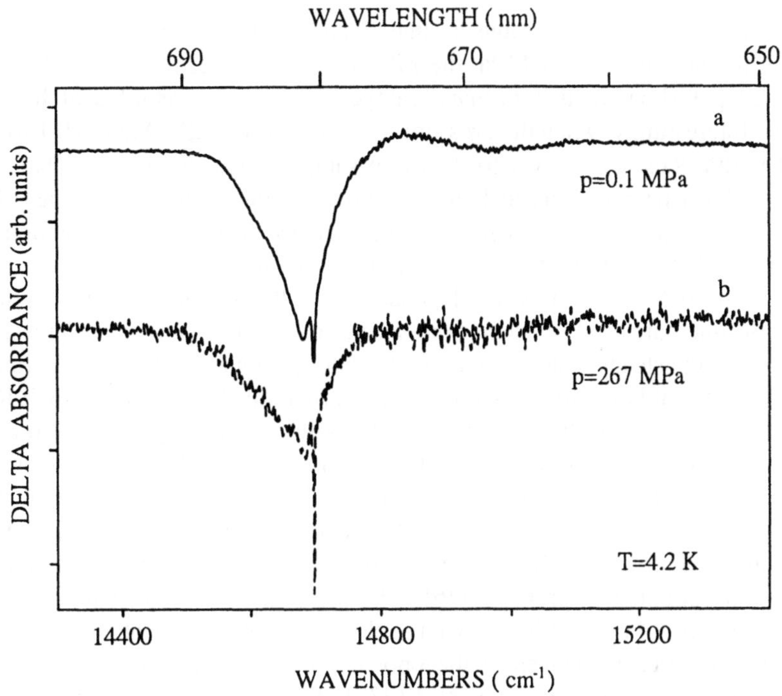


Figure 3.2. Line-narrowed transient triplet bottleneck hole spectra (4.2 K) obtained at two different pressures: p = 0.1 MPa (a) and p = 267 MPa (b), respectively; $\lambda_B$ = 680.4 nm. The ZPH widths are 5.2 cm$^{-1}$ at p = 0.1 MPa and 1.5 cm$^{-1}$ at p = 267 MPa. Burn intensity at the sample ~100 mW/cm$^2$.

pheophytin [23]. These experiments were performed at 77 K to eliminate interference from the 684 nm Chl *a* [35]. Further since the shift rates are estimated from the hole maxima hole burning needs to be done under non-line narrowing conditions. Gaussian fits to transient hole burned spectra at 4.2 K yields a shift rate of -0.07 cm$^{-1}$/MPa. The upper dimer component of the special pair shifts at the same rate as P680. The P680 transient hole is interfered with by the ground state absorption of the Chl *a* monomer associated with $^3$P680$^*$. Correction for the interference leads to a value of 0.16 cm$^{-1}$/MPa for the linear broadening of P680 [23].

It should be emphasized at this stage that the absorption profile, the transient hole profile as well as hole burning efficiency are all unaffected by pressure cycling. i.e. after pressure cycling they are identical to those of virgin samples. The pressure dependence of the primary charge separation kinetics was determined using line narrowed triplet bottleneck holeburning. Two of the hole burned spectra obtained at 0.1 MPa and 267 MPa are shown in Figure 3.2. Spectrum a, obtained at 0.1 MPa, shows the ZPH at the burn laser frequency and the associated phonon side band structure. The ZPH width of 5.2 ± 0.3 cm$^{-1}$ yields a P680* lifetime of

1.9 ± 0.2 ps. This value is same as that obtained in earlier studies. Upon increasing the pressure to 267 MPa, the ZPH width decreases to 1.5 $cm^{-1}$ (P680* lifetime 7.0 ps) showing that primary charge separation is retarded at higher pressures. Interestingly, when the pressure is decreased from 267 MPa to 0.1 MPa the ZPH width does *not* recover to its virgin sample value (5.2 $cm^{-1}$). That we continue to burn narrow ZPH at high pressures together with the fact that the transient hole also broadens with pressure shows that the aforementioned broadening of P680 band is due to increase in inhomogeneous broadening.

On the basis only of the pressure dependences of the 4.2 and 77 K absorption and the non-line narrowed transient hole spectra of P680 as well as the pressure dependence of the steady state population of $^3$P680* (estimated from the integrated area of the transient hole), the pressure induced structural changes of the PS II RC would be classified as elastic (reversible). The observation, at high resolution, of an irreversible decrease in the width of the ZPH however, establishes that there is a plastic component to the structural changes. It is well established that the complexity of proteins can result in both elastic and conformational (plastic) pressure induced structural changes. Conformational drift has been observed in several small oligomeric proteins [36]. It has also been demonstrated that, although pressures of up to 100-200 MPa at physiological or ambient temperatures produce no significant changes in the optical spectra of heme proteins, subtle structural changes can be detected by NMR [37]. Windsor and coworkers [15-17] have reported that in *Rb. sphaeroides* and *Rps. viridis* RC at room temperature, the secondary charge separation process ($P^+H_L^-Q_A \rightarrow P^+H_LQ_A^-$) is increasingly inhibited at higher pressures. The elastic nature of the low temperature absorption spectrum and non-line narrowed P680 hole spectrum upon pressure cycling suggests that dark charge transfer states are important in understanding the pressure effects on primary charge separation. Combined with the weak pressure dependence of the electron-phonon coupling the just mentioned elastic behavior rules out the energy of P680* and the electronic coupling V as sources for the plastic pressure dependence of the primary charge separation [23].

A theory of hole profiles in the low temperature limit valid for arbitrarily strong electron-phonon coupling has been developed in 1988 [38]. This has been recently generalized for arbitrary temperature and used to accurately account for the temperature and burn frequency dependence of P680 hole spectra at 1 atm. Use of this theory to simulate the absorption and non-line narrowed hole burned spectra at elevated pressures, including those shown in Figure 3.2, yields the following information [23]. The mean protein phonon frequency, $\omega_m$ (=20 $cm^{-1}$) and the associated Huang-Rhys factor, S (=1.9) are independent of pressure. In contrast with bacterial RC spectra, the special pair marker mode is only of minor importance as evidenced by $S \leq 0.3$ and increase in pressure does not alter the situation. The change in the intensity of the ZPH can be well accounted for by a decrease in the homogeneous width of the zero phonon line for the P680*←P680 transition. This however does not explain the broadening of the structure accompanying the ZPH. Simulations that include an increase in the inhomogeneous broadening, $\Gamma_{inh}$, with pressure ($d\Gamma_{inh}/dp$=0.16 $cm^{-1}$/MPa) account well the high pressure spectra [23].

Small *et al* [39] have developed a theory for non adiabatic electron-transfer and obtained the following expression for the electron transfer rate. Denoting the donor state, P680$^*$ by D and the acceptor state, $P^+Chl^-$ (two-step model) or $P^+Pheo^-$ (superexchange model) by A, the rate expression for the superexchange model is

$$\langle k_{DA}\rangle = 2\pi V^2(2FC_{loc})\{2\pi[\Gamma^2 + \Sigma(T)^2]\}^{-1/2}\exp\{-[(\Omega_0-\omega_{loc})-S\omega_m]^2/2\{\Gamma^2 + \Sigma(T)^2\}$$

where $\Omega_0$ is the mean donor-acceptor energy gap, V is the electronic coupling matrix element, $\Sigma(T)$ is the homogeneous width of the nuclear factor associated with the Golden rule and $\Gamma^2$ is the variance in $\Omega$ values [39]. $\Sigma(T)^2 = S(\sigma^2 + \omega_m^2)$ where $2\sigma$ is the width of the one-phonon profile and $S(T) = S\ \mathrm{ctnh}(\hbar\omega_m/2kT)$. $\omega_{loc}$ is the frequency of a localized intramolecular Chl/Pheo mode with a Franck-Condon factor $FC_{loc}$. For calculations $\omega_{loc}$ was set to 710 $cm^{-1}$ and $FC_{loc}$ was taken to be 0.2 [23]. These values are justifiable when Chl *a* modes ($Q_y \leftarrow S_0$ transition) in this region are considered. In the case of a two-step model $\omega_{loc}$ is set to zero and $FC_{loc}$ = 1 since no modes other than the phonon modes are considered [40]. The values of $\sigma$=13.8 and $\omega_m$= 20 $cm^{-1}$ were obtained from hole burning results. The room temperature free energy gap between P680$^*$ and $P680^+Pheo^-$ has been reported to be 890 $cm^{-1}$ [41]. For uncorrelated distributions of donor and acceptor levels, we obtain $2.35\Gamma \approx (\Gamma_{inh,D}^2 + \Gamma_{inh,A}^2)^{1/2}$. Assuming $\Gamma_{inh,D} = \Gamma_{inh,A}$ = 100 $cm^{-1}$, we obtain $\Gamma$=60 $cm^{-1}$. The value of $\Gamma$ was set to either 100 or 200 $cm^{-1}$. The latter value is due to recent reports that $\Gamma_{inh}$ for charge separated states could be significantly larger than 100 $cm^{-1}$ [42].

The pressure dependence of primary charge separation kinetics can be modeled as due to structural changes (irreversible) affecting the acceptor state. These structural changes affect both the inhomogeneous broadening and energy of the acceptor state. For the two step model, assuming that the $\Gamma_{inh,A}$ increases by a factor of 1.5 for pressure increase from 0.1 MPa to 267 MPa, calculations [23] show that $\Omega_0$ increases by 259 and 457 $cm^{-1}$ for $\Gamma$(0.1 MPa)= 100 and 200 $cm^{-1}$ respectively. For the superexchange model the corresponding increases are 231 and 410 $cm^{-1}$ respectively. i.e. The observed charge separation kinetics (reduction in ZPH width) require a pressure shift of the acceptor state energy of about 1 $cm^{-1}$/MPa depending on the value of $\Gamma$. For pressures ≤400 MPa, Rollinson *et al* [43] report the linear pressure shift rates (at room temperature) for intramolecular charge transfer states of several compounds to be in the vicinity of 1 $cm^{-1}$/MPa. Thus, the pressure shift value suggested by our calculations does not appear to be unreasonable.

### 3.2 Pressure effects on the LH II antenna complex.

In this section, we give an example of application of hole burning spectroscopy at high pressure to LH II antenna complex from *Rb. sphaeroides*. We will then compare these results with those from B875 LH I (*Rb. sphaeroides*), B1015 (*Rps.*

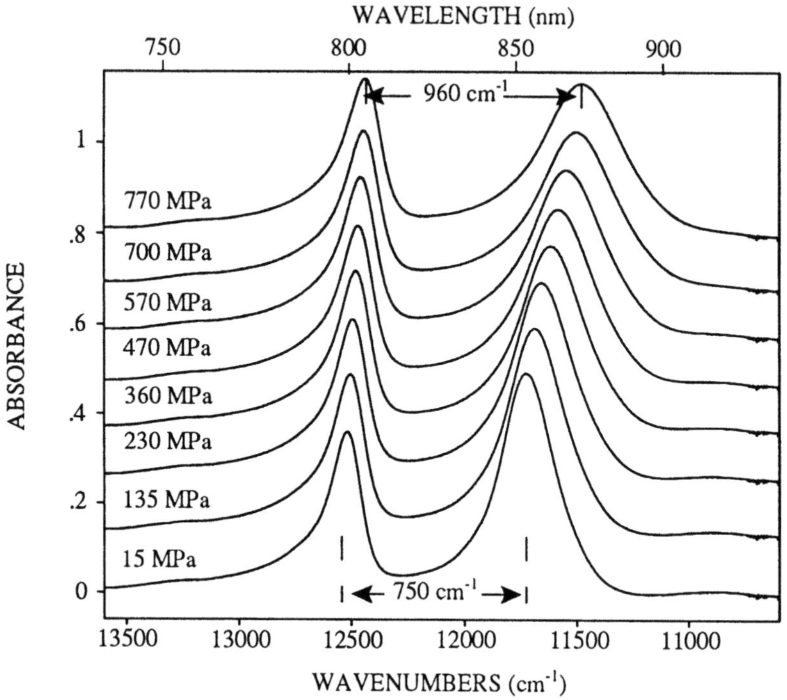


Figure 3.3. Absorption spectra (at 85 K) of the B800-B850 antenna complex of Rb. sphaeroides at various high pressures. The separation of the B800 and B850 bands increases from 750 $cm^{-1}$ at ambient pressure to 960 $cm^{-1}$ at 770 MPa. The pressure shift rates are -0.084 $cm^{-1}$/MPa (B800) and -0.28 $cm^{-1}$(B850)

*viridis*) and the FMO antenna complex of *Chlorobium tepidum* [21].

The pressure dependence of low temperature (77 K) $Q_y$-absorption spectra of the LH II antenna complex is shown in Figure 3.3. Both the B800 and B850 bands shift to the red as the pressure is increased from 0.1 MPa to 770 MPa. Similar data (not shown) were obtained at 4.2 K. The linear pressure shift rates (4.2 K) for the two bands are -0.10 and -0.28 $cm^{-1}$/MPa respectively. Owing to the difference in shift rates the separation between the B800 and B850 bands increases from 750 $cm^{-1}$ to 960 $cm^{-1}$ at 770 MPa. We examine later in this section what, if any, effect this increase in the donor-acceptor energy gap has on the B800→B850 electronic energy transfer.

Reddy *et al* [32] have established that the widths of the narrow ZPHs burned in the B800 band are determined by the B800→B850 energy transfer time. i.e. the observed ZPH width (at 0.1 MPa) of 5.8 ± 0.3 $cm^{-1}$ reflects the B800* lifetime 1.8 ± 0.1 ps . The B800 bandwidth of 170 $cm^{-1}$ is then due to inhomogeneous broadening. The burn wavelength independence [32] (except for $\omega_B$ < 858 nm) of the broad hole in the B850 band and the accompanying vibronic hole structure in B800 region (Figure 3.4 (b)) establish that the B850 band is predominantly

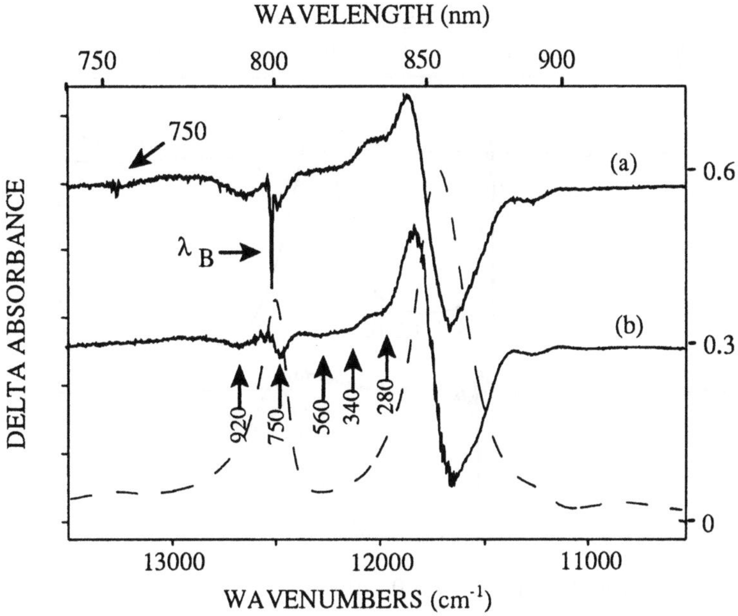


Figure 3.4 Nonphotochemical hole burned spectra of B800-B850 antenna complex of *Rb.sphaeroides.* Top spectrum shows a narrow zero phonon hole (ZPH) at the burn frequency (12517 $cm^{-1}$) and a broad hole in the B850 band. A weak 750 $cm^{-1}$ vibronic hole building on the ZPH at 12517 $cm^{-1}$ is also shown. Spectrum (b), obtained by burning at 11640 $cm^{-1}$, shows clearly the vbronic holes (vertical arrows) generated by various intramolecular vibrational modes (frequencies as labelled) of BChl *a*. Burn conditions are 200 $mW/cm^2$ for 30 min.

homogeneously broadened. In the hole burned spectrum shown in Figure 3.4 (b) the dominant intermolecular modes [32] in the B800→B850 Forster energy transport viz. the 750 $cm^{-1}$ mode and the ~920 $cm^{-1}$ modes are apparent. The 750 $cm^{-1}$ vibronic hole building on the narrow ZPH burned into the B800 band is shown in Figure 3.4 (a). The vibronic hole at 920 $cm^{-1}$ (Figure 3.4 (b)) is composed of three closely spaced holes whose Franck-Condon factors sum to the same value as that for the 750 $cm^{-1}$ hole (i.e. 0.05). Recent X-ray structural data by McDermott *et al* [31] (cf. Introduction) proves the conclusions of Reddy *et al* [32] on the nature of B800 and B850 bands. With increasing pressure while the width of B800 band is, within experimental error, constant, the B850 band broadens at the rate of 0.2 $cm^{-1}$/MPa. The integrated intensity however, does not change either with pressure or pressure cycling.

To study the pressure effects on energy transfer dynamics, the low temperature (4.2 K) hole burned spectra were obtained at various pressures by burning in the B800 band. Figure 3.5 shows one such spectrum obtained at a pressure of 680 MPa. An ambient pressure hole burned spectrum is also shown. High resolution

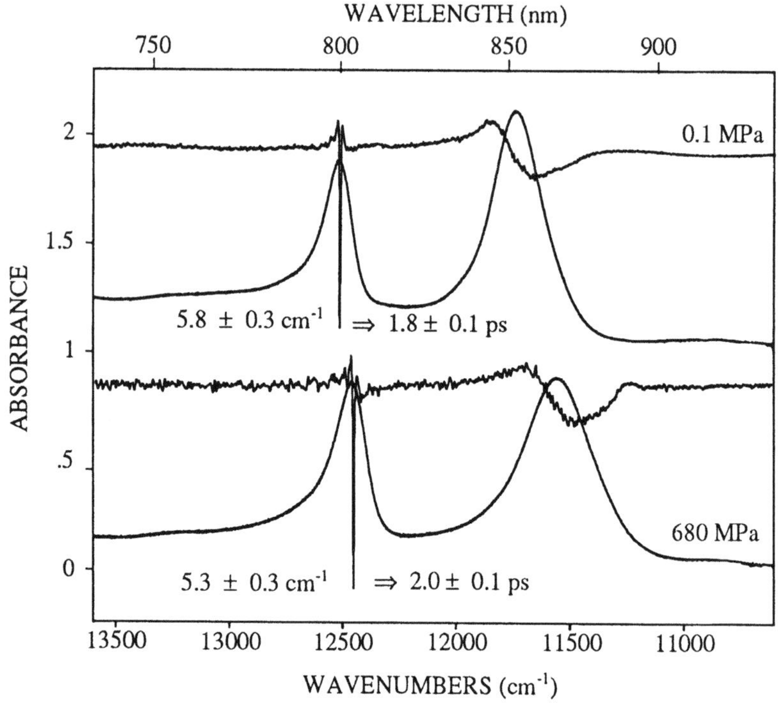


Figure 3.5 Low temperature (4.2 K) hole burned spectra obtained by burning in the B800 band show efficient energy transfer to the B850 band both at ambient (top spectrum) and high (bottom) pressures (p=680 MPa). The ZPH width of 5.8 $cm^{-1}$ at ambient pressure is essentially unchanged at 680 MPa. Hole burned spectra are shown in arbitrary absorbance units.

scans show that the ZPH width (5.3 ± 0.3 $cm^{-1}$) at 680 MPa differs very little from that at ambient pressure (5.8 ± 0.3 $cm^{-1}$) obtained from the same sample. These hole widths translate to B800* lifetimes of 2.0 and 1.8 ± 0.1 ps respectively. This result would at first sight appear to be very surprising since the applied pressure increases the separation between the B800 and B850 bands by 190 $cm^{-1}$. For a theoretical explanation of this result we refer again to the expression for average rate constant for electron/energy transfer. Before considering that equation however, we describe an important result obtained in our studies on pressure dependence of the energy transfer dynamics in the FMO complex from *Chlorobium tepidum* [21]. The $Q_y$-absorption spectra for this complex shows three prominent peaks at 804, 814 and 825 nm. The linear pressure shift rates for these bands are -0.08, -0.11 and -0.11 $cm^{-1}$/MPa respectively. The 814 and 825 nm bands are predominantly contributed to by BChls 6 and 7 (numbering scheme of Fenna *et al* [44]). That the shift rates for these two bands are identical can be understood if one assumes that the coupling (V) between BChls 6 and 7 is pressure independent. The shift of the bands is then mainly due to lowering of the diagonal energies of the individual BChls. Returning to the LH II complex, given the weak pressure dependence of electronic coupling matrix element, V and Γ and Σ(T), we note (from the rate equation) that the energy transfer rate is dependent on the overlap provided by an active intramolecular mode(s). (Further details on the pressure dependence of V, Γ and Σ(T) are given elsewhere [24].) The 750 $cm^{-1}$ mode has been identified as the dominant mode for energy transfer at ambient pressure. Pressure independence of B800→B850 energy transfer now follows from the expression for $\langle k_{DA} \rangle$ when it is realized that at high pressures the 920 $cm^{-1}$ modes instead of the 750 $cm^{-1}$ mode become active in the energy transfer. Note also that Franck-Condon factor for the 920 $cm^{-1}$ modes is same as for 750 $cm^{-1}$ mode. Knowing the pressure shifts rates for B800 and B850, we predict that pressures of about 1 GPa would slow down the energy rate by a factor of 2 or 3. Using site directed mutagenesis van der Laan *et al* [45] reported that the B800→B850 energy transfer in mutated antenna complexes of *Rb. sphaeroides* and *Rb. acidophila.* They observed at most a factor of 1.5 variation in the energy transfer rate when the B800-B850 band separation was varied over 400 $cm^{-1}$.

Currently, there exists no theory for the effects of pressure on the absorption and hole burned spectra of excitonically coupled pigment molecules. The experimental and theoretical study on the pressure dependence of the ZPH of isolated (non-interacting) chromophores in polymer films by Sesselman *et al* [46], led Laird and Skinner to develop a microscopic theory [47] which has been applied with considerable success to isolated chromophores in proteins [48-50]. Aside from the assumption that the pigments are non-interacting, the theory assumes that the positions of host molecules relative to the pigment molecule are uncorrelated, the host medium is homogeneous and isotropic and local compressibilities can be replaced by the bulk compressibility. For systems which would appear to conform to these restrictions, the agreement between the κ-values determined by pressure dependent hole burning and the bulk values has been good. A key equation which is obtained from the Laird and Skinner theory is

$$\Delta\nu(\nu,\Delta p)=n\kappa 3^{-1}(\nu-\nu_v)\Delta p$$

where $\nu_v$ is the vacuum frequency, $(\nu-\nu_v)$ the solvent shift at ambient pressure and n is the power of the attractive solute-solvent pair-wise potential ($\alpha R^{n}$) which is assumed to dominate the repulsive interaction for pressures which are not too high. Using a BChl *a* vacuum frequency of 13340 $cm^{-1}$ and κ=0.19/GPa [51], the above equation predicts that pressure shift rates of B800 and B850 differ at most by a factor of 2. But experimental values differ by almost a factor of three. Similarly, the observed shift rates for P960 and B1015 bands are the same even though they both are BChl *b* complexes absorbing at different wavelengths. Application of this theory to the FMO complex [21] has been discussed earlier. In view of the above stated assumptions, it would thus seem unjustifiable to apply the theory to complexes with excitonically coupled pigment molecules.

We now compare the pressure broadening of B850 band with that of P960 band. While the B850 broadens with pressure at the rate of 0.2 $cm^{-1}$/MPa, the width of P960 band with experimental error remains unchanged. We propose that this difference is due to the differences in structural arrangement of BChl *a* in the

Table 3.1 Linear pressure shift rates and Electron phonon coupling strengths for some photosynthetic complexes.

| Pigment Band (nm) | $R_{Mg...Mg}$(Å) | Interplanar Separation (Å) | Linear Pressure shift ($cm^{-1}$/MPa) | Electron Phonon Coupling |
|---|---|---|---|---|
| B800 (*Rb. Sph.*) | 21 | ~ 21 | -0.10 | weak |
| B850 (*Rb. Sph.*) | 8.7 | 4-5 | -0.28 | weak |
| B875 (*Rb. Sph.*) | ≤8.7 | 4-5 | -0.41 | weak |
| B1015 (*Rps. viridis*) | ≤8.7 | 4-5 | -0.42 | weak |
| B805<br>B814 FMO<br>B825 | 11-14 | ~ 11-14 | -0.08<br>-0.11<br>-0.11 | weak<br>weak<br>weak |
| P960 (*Rps. viridis*) | 7 | 3.4 | -0.42 | strong |
| P870 (*Rb. Sph.*) | 7 | 3.4 | ~-0.5 | strong |
| P680 (PS II) | ~10-11 | ~11-14 | -0.07 | medium |

two systems. The eighteen B850 BChl *a* within a structural subunit [31] are arranged as dimers with a ninefold symmetry. The Mg...Mg distance within a dimer is 8.7 Å which is comparable to the BChl *a* separation (7 Å) of the P960 special pair [52]. However, while the B850 dimers are still closely spaced (9.7 Å) and strongly interacting, the P960 special pair BChls from different reaction centers are not. In the case of P960 special pair BChls the two dimer components are widely separated. Any change in BChl interaction due to pressure would only result in a change in the dimer component separation. A decrease in the B850 inter/intra dimer separation on the other hand would result in an increase in the homogeneous bandwidth. That the B850 band broadening is due increase in homogeneous width has been checked by burn wavelength dependence of hole spectra at high pressure. This increase in the homogeneous bandwidth explains the B850 broadening with pressure. Similar band broadening has been observed for B875 which is now reported to have dimers arranged in a 16-fold symmetry (32 BChls/subunit) [53]. Further details of our high pressure studies on Rb. sphaeroides antenna complexes will appear elsewhere [24].

In Table 3.1 we summarize the results of our high pressure experiments on some photosynthetic complexes. The pressure shift for various from antenna and reaction center complexes are listed along with the nearest neighbor Mg...Mg distances between the Chls contributing to the band. These data indicate that pressure is a useful probe of coupling between Chl molecules. Earlier work has shown that the linear pressure shifts for monomer molecules isolated in glasses, polymers and proteins fall in the range of -0.05 to -0.15 $cm^{-1}/MPa$ at 4.2K. Evidently, the Chl-Chl coupling energies associated with the three FMO bands, B800 and P680 are too weak to yield pressure shifts greater than expected for an isolated Chl molecule. These results also provide some support for the suggestion by Small [54] that the standard weak coupling models for primary charge separation in the bacterial RC may not be valid.

To summarize, we have shown that pressure can have a pronounced effect on the primary charge separation of a reaction center. In PS II RC, knowing the pressure dependence of inhomogeneous broadening $\Gamma_{inh}$ it was possible to estimate the pressure dependence of $\Gamma^2$, the variance of the donor-acceptor energy gap value distribution. Using the theory of hole profiles it was then possible to reasonably estimate the pressure dependence of the acceptor state energy (~ 1 $cm^{-1}/MPa$). In *Rb. sphaeroides* LH II antenna complex, the pressure dependence of the B800 and B850 bands as well as the B800→B850 energy transfer rate have been measured. A possible explanation for the pressure independence of the energy transfer rate is provided. Finally, we have summarized the pressure shift coefficients for some photosynthetic complexes studied recently in our laboratory.

**4 ACKNOWLEDGEMENTS**

Research at the Ames Laboratory was supported by the Division of Chemical Sciences, Office of Basic Energy Sciences, U.S. Department of Energy. Ames Laboratory is operated for USDOE by Iowa State University under contract W-7405-Eng-82. We would like to thank M. Seibert and R. Picorel of the National Renewable Energy Laboratory and S. Kolaczkowskii of St. Thomas University for providing us with the RC and antenna complex preparations. H-CC is indebted to Dow Chemical for support through a Research Fellowship.

**REFERENCES**

1 Geacintov, N.E. and Breton, J. (1987) in Critical Reviews in Plant Science; Vol. 5 (Congor, B.V. ed.) pp 1-44, CRC Press, Boca Raton.
2 Pearlstein, R.M. (1982) in Photosynthesis: Energy Conversion by Plants and Bacteria; (Govindjee ed.) pp 293-330, Academic, New York.
3 van Grondelle, R. (1985) Biochim. Biophys. Acta 811, 147-195.
4 Knox, R.S. (1986) in Encyclopedia of Plant Physiology; Vol. 19 (Staehelin, L.A. and Arntzen, C.J. eds.) pp 86, Springer-Verlag, Berlin.
5 Reddy, N.R.S., Lyle, P.A. and Small, G.J. (1992) Photosyn. Res. 311, 167-194.
6 Kirmaier, C. and Holten, D. (1987) Photosyn. Res. 13, 225-260.
7 Feher, G., Allen, J.P., Okamura, M.Y. and Rees, D.C. (1989) Nature 339, 111-116.
8 Parson W.W. (1991) in Chlorophylls; (Scheer, H. ed.) pp 1153-1180, CRC Press, Boca Raton.

9 Kirmaier, C. and Holten, D. (1993) in The Photosynthetic Reaction Center, Vol. II; (Deisenhofer, J. and Norris, J.R. eds.) pp 49-70, Academic Press, New York.
10 Shuvalov, V.A. *ibid.* pp 71-88.
11 Friesner, R.A. and Won, Y. (1989) Biochim. Biophys. Acta 977, 99-122.
12 Seibert, M. (1993) in The Photosynthetic Reaction Center, Vol I; (Deisenhofer, J. and Norris, J.R. ed.) pp 319-356, Academic Press, New York.
13 Renger, G. (1992) in Topics in Photosynthesis, Vol. 11, (Barber, J., ed.) pp. 45, Elsevier.
14 Nanba, O. and Satoh. K. (1987) Proc. Natl. Acad. Sci. USA. 84, 109-112.
15 Redline, N.L. and Windsor, M.W. (1991) Chem. Phys. Lett. 186, 204-209.
16 Redline, N.L. and Windsor, M.W. (1992) Chem. Phys. Lett. 198, 334-340.
17 Hoganson, C.W., Windsor, M.W., Farkas, D.I. and Parson, W.W. (1987) Biochim. Biophys. Acta 892, 275-283.
18 Reddy, N.R.S. and Small, G.J. (1995) Advances in Photosynthesis (in press).
19 Clayton, R.K. and DeVault, D. (1972) Photochem. Photobiol. 15, 165-175.
20 Freiberg, A. Ellervee, A., Kukk, P., Laisaar, A., Tars, M. and Timpmann, K. (1993) Chem. Phys. Lett. 214, 10-16.
21 Reddy, N. R. S., Jankowiak, R. and Small, G.J. (accepted) J. Phys. Chem.
22 Woodbury, N.W., Peloquin, J.M., Alden, R.G., Lin, X., Lin, S., Taguchi, A.K.W., Williams, J.C. and Allen, J.P. (1994) Biochemistry 33, 8101-8112.
23 Chang, H.-C., Jankowiak, R., Reddy, N.R.S. and Small, G.J. (1995) Chem. Phys. (in press).
24 Reddy, N.R.S., Wu, H.-M., Jankowiak, R. and G.J. Small, (submitted) J. Phys. Chem.
25 Jankowiak, R., Tang, D., Small, G.J. and Seibert, M. (1989) J. Phys. Chem. 93, 1649-1654.
26 Breton, J. (1990) in Perspectives in Photosynthesis (Jortner, J. and Pullman, B. eds.) Kluwer Academic: Dordrecht, pp 23-28.
27 van der Vos, R., van Leeuwen, P.J., Braun, P., Hoff, A.J. (1992) Biochim. Biophys. Acta 1140, 184-198.
28 Jankowiak, R. and Small, G.J. (1993) in The Photosynthetic Reaction Center, Vol 2; (Deisenhofer, J. and Norris, J.R. eds.) pp 133-177, Academic Press, New York.
29 Otte, S.C.M., van der Vos, R. and van Gorkom, H.J. (1990) J. Photochem. Photobiol. B15, 5-14.
30 Chang, H.-C., Jankowiak, R., Reddy, N.R.S., Yocum, C.F., Picorel, R., Seibert, M. and Small, G.J. (1994) J. Phys. Chem. 98, 7725-7735.
31 McDermott, G., Prince, S.M., Freer, A.A., Hawthornthwaite-Lawless, A.M., Papiz, M.Z., Cogdell, R.J. and Isaacs, N.W. (1995) Nature, 374, 517-521.
32 Reddy, N.R.S., Small, G.J., Seibert, M. and Picorel, R. (1991) Chem. Phys. Lett. 181, 391-399.
33 Reddy, N.R.S., Seibert, M., Picorel, R., and Small, G.J. (1992) J. Phys. Chem. 86, 6458-6464.

34 van der Laan, H., Schmidt, Th., Visschers, R.W., Visscher, K.J., van Grondelle, R. and Völker, S. (1990) Chem. Phys. Lett. 170, 231-238.
35 Chang, H.-C., Small, G.J. and Jankowiak, R. (1995) Chem. Phys. 194, 323-333.
36 Protein Interactions (1992) Chapman-Hall: New York.
37 Jonas, J. and Jonas, A. (1994) Ann. Rev. Biophys. Biomol. Struct. 23, 287-318.
38 Hayes, J.M., Gillie, J.K., Tang, D. and Small, G.J. (1988) Biochim. Biophys. Acta 932, 287-305.
39 Small, G.J., Hayes, J.M. and Silbey, R.J.(1992) J. Phys. Chem. 96, 7499-7501.
40 Gillie, J.K., Small, G.J. and Golbeck, J.H.(1989) J. Phys. Chem. 93, 1620-1627.
41 Booth, P.J., Crystall, B., Ahmad, I., Barber, J., Porter, G. and Klug, D.R. (1991) Biochemistry 30, 7573-7586.
42 Bixon, M., Jortner, J. and Michel-Beyerle, M.E. (1995) Chem. Phys. (in press).
43 Rollinson, A.M. and Drickamer, H.G. (1980) J. Chem. Phys. 73, 5981-5996.
44 Fenna, R.E., Matthews, B.W. (1975) Nature, 258, 573-577.
45 van der Laan, H., De Caro, C., Schmidt, Th., Visschers, R.W., van Grondelle, R., Fowler, G.J.S., Hunter, C.N. and Volker, S. (1993) Chem. Phys. Lett. 212, 569-580.
46 Sesselmann, Th., Richter, W., Haarer, D. and Morawitz, H. (1987) Phys. Rev. B, 14, 7601-7611.
47 Laird, B.B. and Skinner, J.L. (1989) J. Chem. Phys., 90, 3274-3281.
48 Freidrich, J., Gafert, J., Zollfrank, J., Vanderkooi, J.M. and Fidy, J. (1994) Proc. Natl. Acad. Sci. 91, 1029-1033
49 Zollfrank, J., Freidrich, J. and Parak, F. (1992) Biophys. J. 61, 716-724
50 Zollfrank, J., Freidrich, J., Fidy, J. and Vanderkooi, J.M. (1991) J. Chem. Phys. 94, 8600-8603
51 Tars, M., Ellervee, A., Kukk, P., Saarnak, A. and Freiberg, A. (1995) Lith. J. Phys. 34, 320-328.
52 Deisenhofer, J., Epp, O., Miki, K., Huber, R. and Michel, H. (1985) Nature, 318, 618-624.
53 Karrasch, S., Bullogh, P. and Ghosh, R. (1995) EMBO J. 14, 631-638.
54 Small, G.J. (1995) Chem. Phys. (in press).

# Femtosecond spectroscopy and vibrational coherence of membrane-bound RCs of *Rhodobacter sphaeroides* genetically modified at positions M210 and L181.

Marten H. Vos[1], Michael R. Jones[2], C. Neil Hunter[2], Jacques Breton[3], Jean-Christophe Lambry[1] and Jean-Louis Martin[1]

[1] Laboratoire d'Optique Appliquée, INSERM U275, CNRS URA 1406, Ecole Polytechnique-ENSTA, 91120 Palaiseau, France

[2] Krebs Institute for Biomolecular Research and Robert Hill Institute for Photosynthesis, Department of Molecular Biology and Biotechnology, University of Sheffield, Western Bank, Sheffield, S10 2UH, United Kingdom

[3] SBE/DBCM, CEN de Saclay, 91191 Gif-sur-Yvette Cedex, France

## 1 Introduction

The mechanism of primary charge separation in bacterial reaction centers is still subject to debate [1]. Amongst other questions, the functional role of coherent motions from non-thermalized vibrational states has to be evaluated. The potential energy surface of the excited donor state $P^*$ is strongly displaced with respect to the ground state for a number of low-frequency vibrational modes. This has been extensively documented by photochemical holeburning [2-4] and resonance Raman [5,6] studies. Using femtosecond spectroscopy to monitor oscillations in the stimulated emission band shape, we have previously shown that phase correlations for those modes are maintained on the timescale of electron transfer [7,8]. In the usual assumption that vibrational dephasing precedes or at most is concomitant with the processes of energy exchange, these observations imply that the entire vibrational manifold of $P^*$ is not thermalized on the timescale of ET.

Femtosecond spectroscopy thus offers a unique opportunity to characterize the modes that are activated by the $P \rightarrow P^*$ transition, both in their frequencies and in their dephasing properties. In an attempt to provide a molecular description of the modes, and their relation to ET, we have started to study mutant RCs.. A first series of experiments concerns *R. sphaeroides* antenna-less chromatophores containing single-site mutations at positions M210 and L181. In this class of mutants, the electronic properties of the cofactor system are known to be modified [9,10] and ET can be both slower and faster than in WT [9-15] as also illustrated in the four mutants studied here: Tyr to Leu, Trp and His in position M210 and Phe to Tyr in position L181.

Using these single-site mutants the vibrational dynamics in $P^*$ will be measured under conditions where $P^*$ is long and short-lived. For long-lived mutants, for example, spectral characteristics of the vibrational motions can be disentangled from those related to leak from the $P^*$ potential energy surface to the $P^+$ product state.

In this paper we will concentrate on the spectroscopic properties of these mutants, at 10K, where both vibrational and electronic spectra are better resolved. A full account of the vibrational dynamics is in progress.

## 2 Materials and Methods

The construction of the mutant reaction centers bearing changes at the M210 position to Leu and His have been described previously [16]. The changes Tyr M210→Trp and Phe L181→Tyr were made in a similar manner, using in the case of the latter change a 0.85 kb *Xba*I-*Sal*I restriction fragment encompassing the *pufL* gene as the template for mutagenesis. Codon changes were TTC→TAC (Phe L181→Tyr) and TAC→TTC (Tyr M210→Trp). The deletion and expression systems used to construct mutants with a reaction center-only phenotype are described in detail in Ref. 17. The growth of mutant strains of *R. sphaeroides* and the preparation of intracytoplasmic membranes for spectroscopy was as described in Refs. 8,16. The measurements were performed at 10K in a convection cryostat.

The arrangement of the femtosecond pump-probe spectrometer running at 30 Hz was essentially as described previously [8]. The ~30 fs pump pulses were centered at 880 nm. The infrared probe beam was generated by a continuum cut-off below ~800 nm and was compressed to be essentially chirp-free (20 fs) over the wavelength range 880-960 nm. Improvements were made in the detection system. After dispersion in the monochromator, the test and reference probe beams were focussed on an EG&G OMA4 CCD camera. The camera was read out in a dual array configuration. The readout time of the camera was fast enough for shot-to-shot normalization at 30 Hz, of entire spectra with a resolution of ~1 nm. This system allowed relatively fast data collection covering a large spectral range and with a high signal to noise ratio. The ground state spectra were also measured with the same set-up, in a configuration where the pump beam was blocked. In this way there is a slight uncertainty in the baseline correction, which is small compared to the intensity of the absorption band.

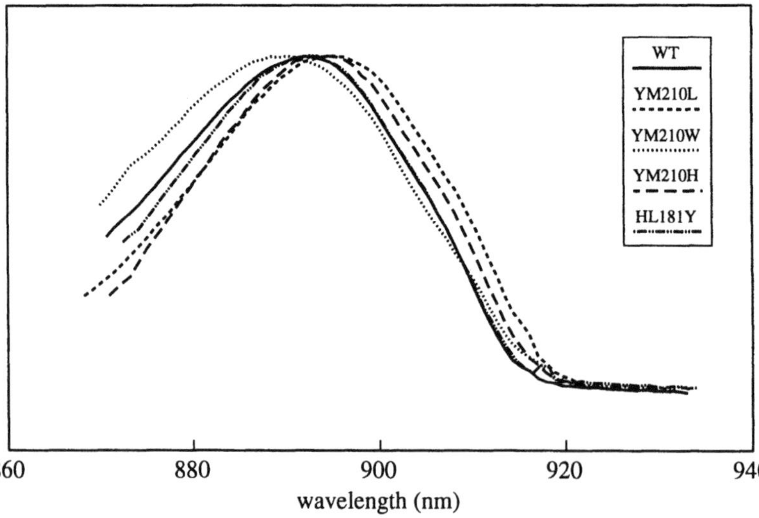


**Fig. 1** *Ground state absorption spectra of the P absorption band of WT and mutant membrane-bound RCs at 10K. The spectra were obtained with the femtosecond continuum in the absence of a pump pulse.*

## 3 Results and Discussion

### 3.1 Ground state spectra

Ground state spectra at 10K of the P absorption region, obtained directly with the femtosecond spectrometer, are shown in Fig. 1. The spectra are similar for the different mutants and the absorption maxima are all in the 890-895 nm range. This is in general agreement with earlier reports on low temperature electronic spectra of isolated RCs modified at position M210 [15,16,18] and it indicates that the cofactor system is not much modified in these mutants. In particular the steepness of the low-energy edge, where the 0-0 transition can be accessed, is very similar for WT and Y(M210)→L,H and F(L181)→Y (10-100% rise 21±1 nm). By contrast, the red edge of the Y(M210)→W mutant is slightly less steep (10-100% rise 25±1 nm). This is possibly related to structural inhomogeneities introduced by replacing Tyr by the bulkier residue Trp and/or to small differences in the vibronic progression underlying the absorption band (see below).

### 3.2 Transient absorption spectra

Transient absorption spectra in the P band absorption and $P^*$ emission region are shown in Fig. 2 for the different mutants and WT at delay times of 500 fs and 3 ps. The spectra are superpositions of the bleaching of the P band in the blue-most part of the spectra and stimulated emission, which is red-shifted with respect to the P-band. In addition, after charge separation, a minor $P^+$ band centered at ~955 nm contributes to the transient spectra.

The time evolution of the stimulated emission spectra is governed by charge separation and by nuclear motions on the $P^*$ potential energy surface. The former gives rise to decay of the stimulated emission band and the latter to modifications in its band shape, which have an oscillatory character reflecting coherent nuclear motions. In WT, Y(M210)→H and F(L181)→Y reaction centers charge separation occurs on the timescale of ~1 ps and hence the time evolution of the stimulated emission band between 500 fs and 3 ps is mainly determined by $P^*$ decay. The oscillatory component of the stimulated emission signal is better characterized in kinetics at single wavelengths (see Fig. 5). WT and Y(M210)→H have similar decay characteristics. For the F(L181)→Y mutant the 3-ps spectrum is dominated by the bleaching of the P ground state band and the appearance of the $P^+$ band at 955 nm. At this time delay, $P^*$ is virtually depleted. This is consistent with earlier work reporting faster electron transfer in isolated RCs of this mutant in *R. sphaeroides* and in *R. capsulatus* [9,14]. We found that at the center of the stimulated emission band, at 920 nm, the overall kinetics decays mainly in approximately 0.7 ps (not shown).

For the Y(M210)→L,W mutants, $P^*$ mainly decays on the hundreds of picoseconds timescale, in agreement with the observed substantially slower electron transfer reaction at higher temperatures in isolated RCs from these mutants [10,11,15]. Therefore, on the timescale of a few picoseconds, the spectral changes of these mutants shown in Fig. 2, reflect mainly the vibrational dynamics. In addition to studying the $P^*$ spectrum without complications due to $P^*$ decay, these mutants can also be used to estimate the intrinsic vibrational dephasing of the activated modes in the hypothesis that a single mutation does not alter the vibrational pattern.

Fig. 3 shows more extensively the transient spectra of the Y(M210)→L mutant displayed as $-\Delta A$ to facilitate comparison both with the ground state absorption

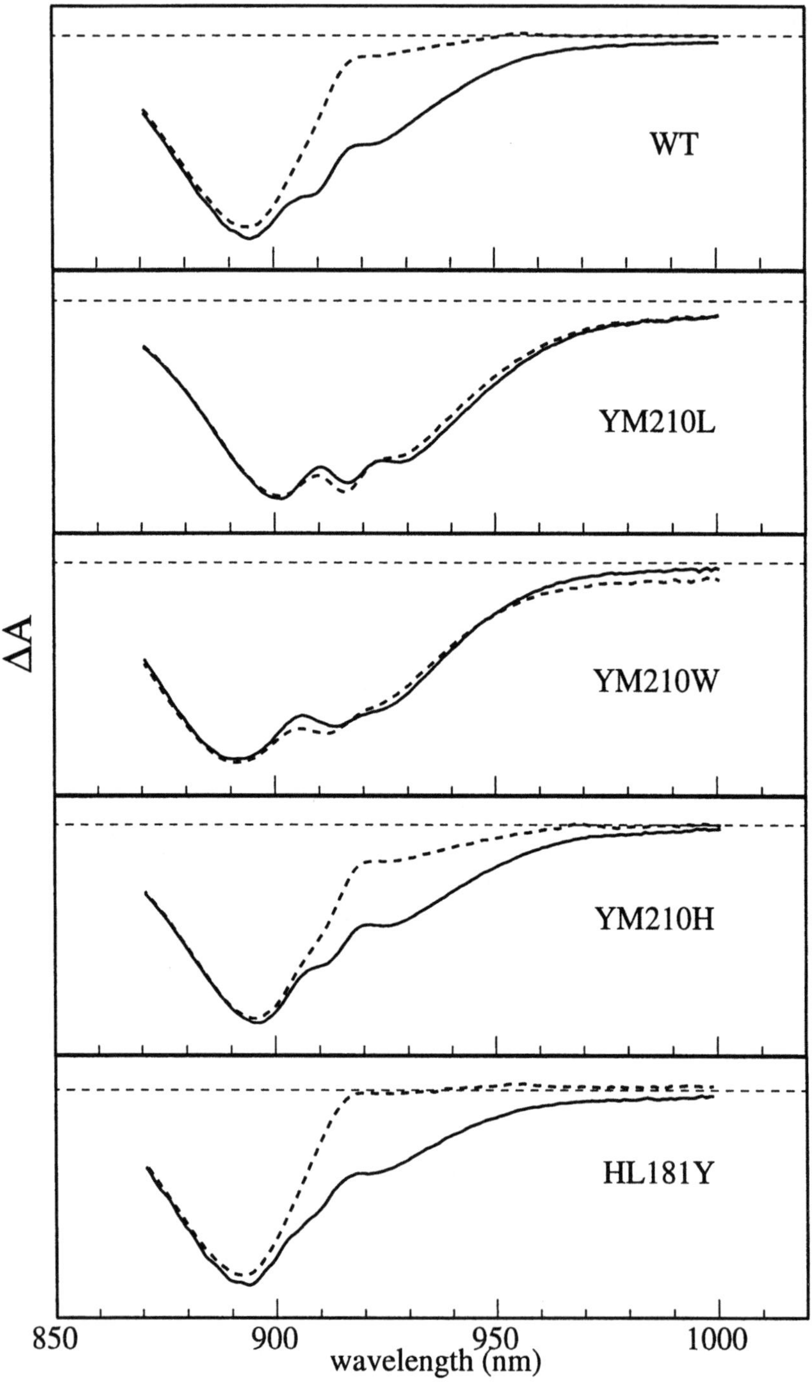


**Fig. 2** *Transient absorption spectra at delay times of 500 fs (solid lines) and 3 ps (dashed lines).*

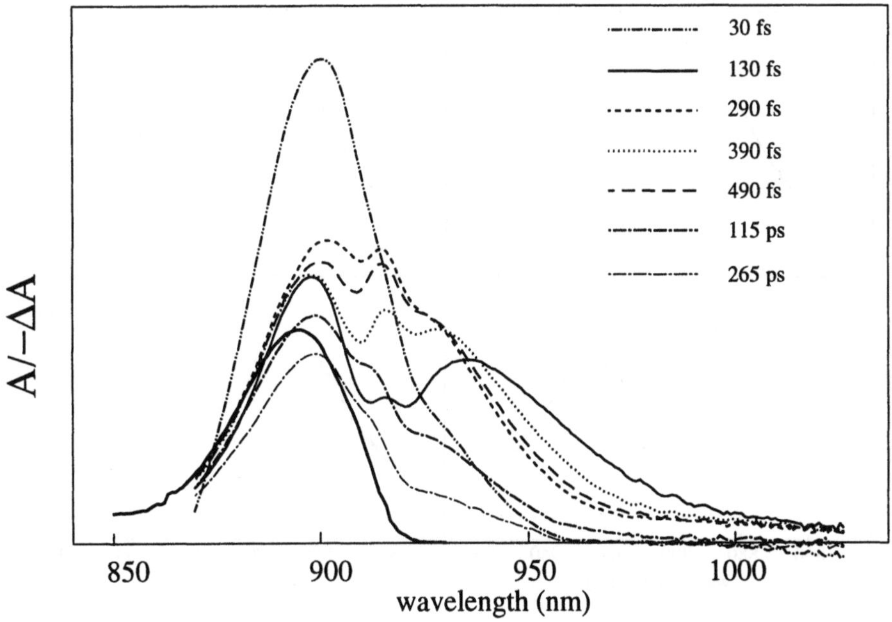


**Fig. 3** *Ground state absorption (thick solid line) and transient absorption spectra of the Y(M210)→L mutant at various delay times. The transient spectra are depicted as −ΔA for comparison.*

spectrum and with Fig. 4. During the first tens of femtoseconds the spectra are dominated by ground state bleaching and stimulated emission superimposed in the ground state absorption region. The redshift on the 100-fs timescale corresponds to the motion of the initially created wave packet out of the Franck-Condon region, as discussed previously for WT [8]. After ~150 fs the emission band shifts back to the blue and subsequently keeps oscillating for several picoseconds, with the two sides of the band in counterphase, as illustrated for the first 500 fs in Fig. 3. This spectral behaviour in the absence of electron transfer is what would be expected on the basis of our interpretation of the corresponding WT spectra [8].

At long delay times, in the hundreds of picoseconds time range, stimulated emission decreases as for WT, Y(M210)→H and F(L181)→Y on the picosecond timescale (Fig. 2). However, in contrast to the latter cases, the signal simultaneously decreases somewhat at the blue side of the ground state absorption. This indicates that the quantum yield of charge separation is not 100% and that $P^*$ decays in part to the ground state. From a kinetic analysis we find that the overall decay of $P^*$ takes ~190 ps, and that this decay is for approximately 30% to the ground state. For Y(M210)→W a qualitatively similar spectral evolution on the picosecond and hundreds of picoseconds timescale is observed. For this case the overall decay of $P^*$ takes ~320 ps, and for approximately 50% to the ground state. These data are in qualitative agreement with the previous finding of $P^*$ decay to the ground state in ~200 ps in the $D_{LL}$ mutant of *R. capsulatus,* where $H_L$ is absent [19].

As reported earlier for WT [8], the transient spectra associated with $P^*$ display a characteristic shape with three maxima at ~895, 913 and 925 nm. The exact position and the relative strength of these features depends on the delay time. Such features are observed for WT and all mutants, albeit that they are considerably smoother for

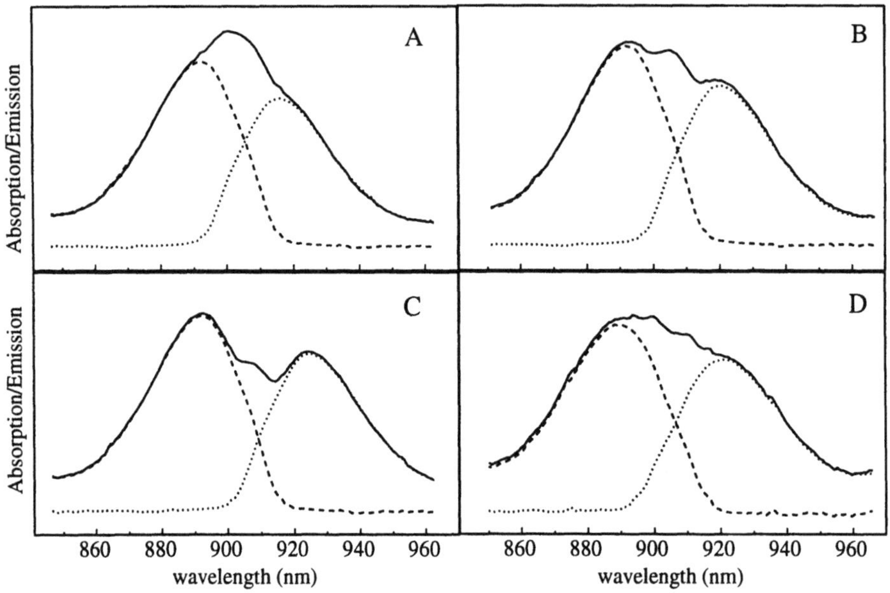


**Fig. 4** *Simulation of transient absorption spectra in the spectral region of P absorption and emission. Dashed: measured ground state absorption spectra. Dotted: 80% of mirror-image of absorption spectra, with the maxima put at 916 nm (A), 920.5 nm (B,D) and 925 nm (C). Solid: dashed line + dotted line. A-C: WT absorption spectrum. D: Y(M210)→W mutant absorption spectrum.*

Y(M210)→W (Fig. 2). From our Y(M210)→L (Fig. 3) and Y(M210)→W (not shown) data it is clear that the features extend up to at least 265 ps. We previously offered a tentative explanation for these features in WT, where they disappear with $P^*$ in the picosecond time range, in terms of a vibrationally unrelaxed excited state. This explanation now seems unlikely, in view of the clear presence of these features even hundreds of picoseconds after the pulse, when vibrational relaxation is strongly expected to have been completed, even at 10K.

In fact a very simple spectroscopic explanation of the observed features can be given, which relates to the very steep rising edge of the absorption spectra (Fig. 1). In Fig. 4 we have simulated the transient $-\Delta A$ signal by adding to the (bleached) ground state spectrum an emission spectrum shaped as the mirror image of the absorption spectrum, with various "Stokes" shifts. This is a much simplified picture, as the emission profile from a non-stationary wavepacket is not quantitatively calculated and no holeburning effects are taken into account. However, for the WT-absorption spectrum used in Fig. 4A-C, the superposition of the two sharply rising edges gives rise to a third maximum, with its intensity strongly depending on the shift between the two spectra and qualitatively this superposition corresponds reasonably well with the transient spectra for WT and the various mutants (Fig. 2). Also the profile development when the wavepacket initially moves out of the Franck-Condon region is qualitatively reproduced (cf. Fig. 3).

Results similar as in Fig. 4A-C are obtained when simulations are conducted for the spectra of the mutants (not shown), with the exception of the spectrum of

Y(M210)→W (Fig. 4C). In the latter simulated spectrum from an absorption profile where the low-energy rising edge is somewhat less steep (Fig. 1) the central hump is considerably smoothed. This corresponds well with the observed transient spectra for Y(M210)→W (Fig. 2).

The spectral profile thus appears very sensitive to the characteristics of the ground-state spectrum. Correspondingly, the appearance of the three maxima gradually decreases above 40K, in concert with the slight broadening of the ground state absorption band (not shown).

### 3.3 Vibrational dynamics

The fundamental frequencies of the activated vibrational modes modulate the kinetics at both sides of the emission spectrum. Fig. 5 shows the kinetics at 940 nm, at the red side of the emission spectrum, along with Fourier transform (FT) spectra of the oscillatory parts. The most important features are briefly discussed here. A more extensive discussion of the vibrational dynamics in the various mutants will be presented elsewhere.

The main low-frequency bands of the FT spectra are found at the same positions for all mutants and WT. This indicates that the single mutations do not strongly affect the frequencies and the coupling of the activated modes. The observed small differences for the relative amplitudes of the bands, which are most apparent for Y(M210)→W, may be due to differences in the distribution of the electron-phonon couplings amongst the modes. Consequently, the observed small difference in the shape of the P absorption band for Y(M210)→W compared to the WT and the other mutants (Fig. 1) may in part be due to a difference in the vibronic progression underlying this band.

The most important result concerns the damping of the oscillations in the various mutants. For WT, F(L181)→Y, and Y(M210)→H the oscillatory features disappear with the decay of $P^*$. For Y(M210)→L,W, where $P^*$ decays on the hundreds of picoseconds timescale, the oscillations are observed much longer than for WT and they are not completely damped after ~4 ps. These results are consistent with the fact that the vibrational motions underlying the oscillations take place on the excited state potential energy surface. Furthermore, they demonstrate that for WT, and for the mutants with similar or faster electron transfer than WT, the observed damping is due to depletion of $P^*$ rather than to intrinsic vibrational dephasing processes.

## 4 Summary and concluding remarks

Near-infrared femtosecond spectroscopy has been used to study the $P^*$ excited state, at low temperature, on a series of single-site mutants at positions M210 and L181. The vibrational dynamics are much the same in these mutants compared to WT. The clear relation between $P^*$ decay and damping of oscillations in the transient spectra, which reflect coherent motions, confirms the attribution of the oscillations to $P^*$ and set a longer *lower limit* for the intrinsic vibrational dephasing time than previously inferred from measurements on WT [8].

The separation of timescales between $P^*$ decay and vibrational motion in some mutants allows a characterization of the spectroscopy related to vibrational motion only, in a system with WT-like vibrational dynamics. The superposition of the ground-state bleaching and stimulated emission is shown to give strongly modulated transient spectra, with characteristics depending on the time-dependant nuclear configuration.

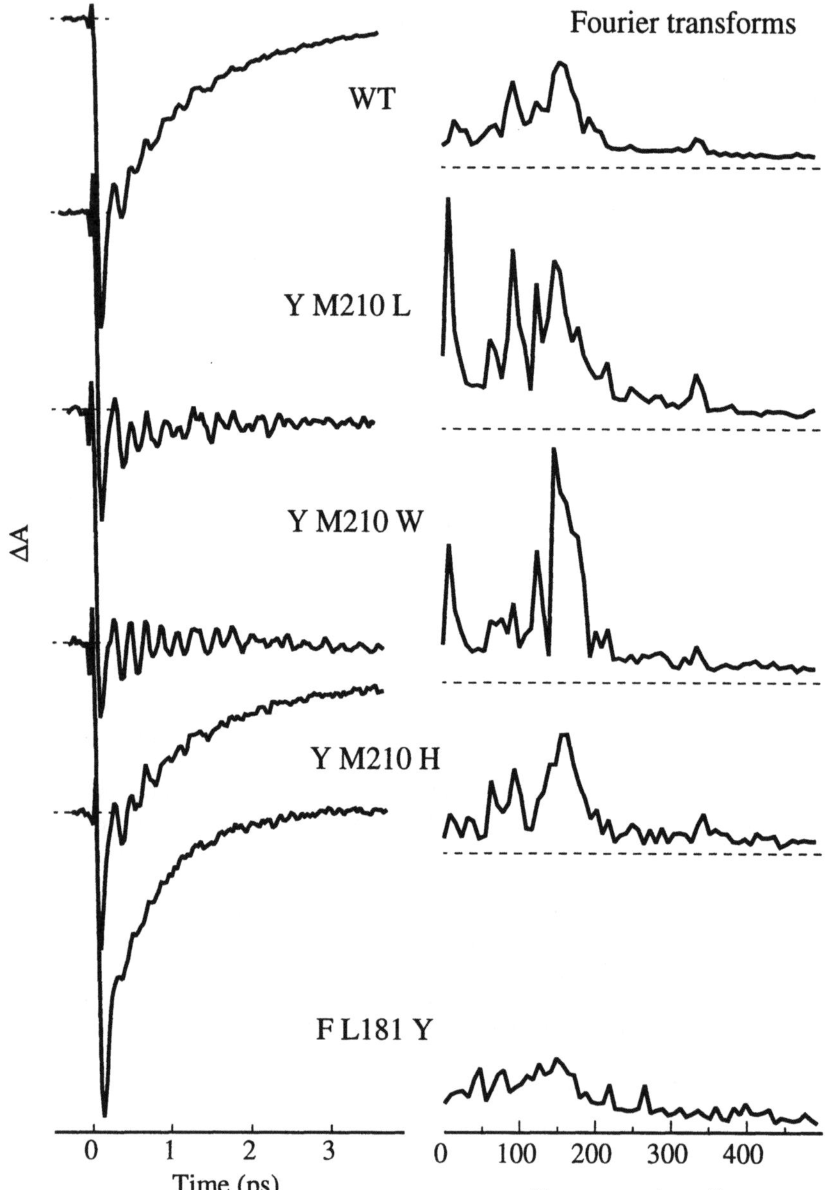


**Fig. 5** *Left: kinetics at the red side of the stimulated emission band, at 940 nm. Right: Fourier transforms of the oscillatory parts.*

The question of the functional role of the light-activated vibrational modes in assisting the electron transfer reaction cannot be directly addressed from this study, as the energy gaps and probably the electronic couplings are modified in these mutants. However, the mutants studied here can be used to separate contributions from vibrational dynamics and population decays in future studies.

## 5 Acknowledgments

M.R.J. and C.N.H. acknowledge support from the Wellcome Trust and the European Community. M.R.J. is an AFRC Senior Research Fellow.

## 6 References

1. The Photosynthetic Reaction Center, Deisenhofer, J. & Norris, J.R., eds. Academic Press, New York, 1993.
2. Johnson, S.G., Tang, D., Jankowiak, R., Hayes, J.M., Small, G.J. & Tiede, D.M. (1990) J. Phys. Chem. 94, 5849-5855.
3. Lyle, P.A., Kolaczkowski, S.V. & Small, G.R. (1993) J. Phys. Chem. 97, 6924-6933.
4. Middendorf, T.R., Mazzola, L.T., Gaul, D.F., Schenck, C.C. & Boxer, S.G. (1991) J. Phys. Chem. 95, 10142-10151.
5. Shreve, A.P., Cherepy, N.J., Franzen, S., Boxer, S.G. & Mathies, R.A. (1991) Proc. Natl. Acad. Sci. USA 88, 11207-11211.
6. Palanappian, V., Aldema, M.A., Frank, H.A. & Bocian, D.F. (1992) Biochemistry 31, 11050-11058.
7. Vos, M.H., Rappaport, F., Lambry, J.-C., Breton, J. & Martin, J.-L. (1993) Nature 363, 320-325.
8. Vos, M.H., Jones, M.R., Hunter, C.N., Breton, J., Lambry, J.-C. & Martin, J.-L. (1994) Biochemistry 33, 6750-6757.
9. Jia, Y., DiMagno, T.J., Chan, C.-K., Wang, Z., Du, M., Hanson, D.K., Schiffer, M., Norris, J.R., Fleming, G.R. & Popov, M.S. (1993), J. Phys. Chem. 97, 13180-13191.
10. Nagarajan, V., Parson, W.W., Davis, D. & Schenck, C. (1993) Biochemistry 32, 12324-12336.
11. Finkele, U., Lauterwasser, C., Zinth, W., Gray, K.A. & Oesterhelt, D. (1990) Biochemistry 29, 8517-8521.
12. Nagarajan, V., Parson, W.W., Gaul, D. & Schenck, C. (1990) Proc. Natl. Acad. Sci. USA 87, 7888-7892.
13. Chan, C.-K., Chen, L.X-Q., DiMagno T.J., Hanson D.K., Nance S.L., Schiffer M., Norris, J.R. & Fleming, G.R. (1991) Chem. Phys. Lett. 176, 366-372.
14. Hamm, P., Gray, K.A., Oesterhelt, D., Feick, R., Scheer, H. & Zinth, W. (1993) Biochim. Biophys. Acta 1142, 99-105.
15. Schochat, S., Arlt, T., Francke, C., Gast, P., van Noort, P.I., Otte, S.C.M., Schelvis, H.P.M., Schmidt, S., Vijgenboom, E., Vrieze, J., Zinth, W. & Hoff, A.J. (1994) Photosynth. Res. 40, 55-66.
16. Jones, M.R., Heer-Dawson, M., Mattioli, T.A., Hunter, C.N. & Robert, B. (1994) FEBS Lett. 339, 18-24.
17. Jones, M.R., Visschers, R.W., van Grondelle, R. & Hunter, C.N. (1992) Biochemistry 31, 4458-4465.
18. Gray, K.A., Farchaus, J.W., Wachtveitl, J., Breton, J. & Oesterhelt, D.A. (1990) EMBO J. 9, 2061-2070.

19. Breton, J., Martin, J.-L., Lambry, J.-C., Robles, S.J. & Youvan, D.C. (1990) in Reaction Centers of Photosynthetic Bacteria (Michel-Beyerle, M.E., ed.) pp. 293-302, Springer, Berlin.

# Femtosecond Infrared Spectroscopy on Reaction Centers of *Rb. Sphaeroides*

*Klaas Wynne, Gilad Haran, Gavin D. Reid, Chris C. Moser, Gilbert C. Walker, Sudipta Maiti, P. Leslie Dutton and Robin M. Hochstrasser*

Department of Chemistry and Johnson Foundation, University of Pennsylvania, Philadelphia, PA19104-6323, USA

**Abstract.** In this paper we will introduce some of the theory of ultrafast visible-pump IR-probe spectroscopy which is employed to understand recent experimental studies on chromophore vibrations in the RC. New methods to generate femtosecond probe pulses tunable in the IR are presented. These pulses have been used to study energy transfer in the RC as well as low lying electronic states in the special pair.

**Keywords.** Ultrafast IR spectroscopy, vibrations, electronic states, energy transfer, *Rb. Sphaeroides*

## 1.Introduction

In the past few years we have performed a series of experiments that employ infrared (IR) light to probe the dynamics of the photosynthetic bacterial reaction center (RC) on an ultrafast timescale. The reasons these studies were undertaken are twofold: First studies in the mid- and far-IR can reveal the dynamics of vibrational modes, both on cofactors as well as in the protein, responding to the electron transfer reactions. In the second place near- and mid-IR pulses can probe transient absorption and emission changes due to the transient population of electronic states of the cofactors. In addition, since the visible transient absorption spectrum is congested, new states or transients may be observed by studying the near-IR.

In this paper we will present some of the theory of ultrafast visible-pump IR-probe spectroscopy, outlining the differences between using a continuous-wave (CW) IR probe in combination with a short visible gate pulse versus using an ultrashort IR probe pulse. This theory is employed to understand recent experimental studies on the RC in the range 1700-2000 $cm^{-1}$. A brief description will be given of our efforts to generate femtosecond tunable pulses in the near- and mid-IR. Finally two sets of experiments on electronic states of the RC will be discussed: an investigation of energy transfer from the accessory bacteriochlorophyll (B) to the special pair (P) and observations of low lying electronically excited states of the special pair.

## 2. Visible-Pump/IR-Probe Spectroscopy

Two approaches are currently in use for visible-pump IR-probe spectroscopy: CW probing and pulsed probing. In both cases a sample is excited with an ultrashort visible pulse. In CW-probing the transient changes in sample absorbance are probed by a CW IR field and the transients are recorded on top of the CW IR beam. These transients on top of the CW beam can be read out by sending this beam with a visible gate pulse of variable delay into a nonlinear upconversion crystal and detecting the generated sum-frequency radiation. In pulsed-probing a short IR pulse detects the change of sample absorbance and either the total probe transmission is registered or the probe-pulse is sent into a monochromator to select a frequency band of the probe spectrum to be detected [1].

Electronic transitions in condensed phases typically have absorption bands with a full width at half maximum (FWHM) of 100-1000 $cm^{-1}$ and therefore a corresponding dephasing time of $T_2 = 2 - 100$ fs. Vibrational transitions, on the other hand, have absorption bands with a FWHM of 5-20 $cm^{-1}$ and therefore a dephasing time of $T_2 > 0.5$ ps. Since the time resolution in current visible-pump IR-probe experiments is on the order of 40-400 fs, coherent transients will occur at short delay times. Since these coherent transients appear on a time scale that may be similar to the kinetics of the studied system, they have to be fully understood in order to extract the dynamics from the data. Here we will give a very brief summary of our theoretical results [1] regarding coherent transients in the IR. In the case that the IR-probe monitors the transient bleaching of a ground state absorption, the bleaching signal will rise instantaneously if it is probed by a CW-probe. However, if a pulsed probe is used, the bleaching signal will rise with the dephasing time of the ground state vibrational transition. If the IR-probe monitors a transient absorption (for example a vibrational mode that has a changed frequency or an enhanced transition strength in an excited electronic state) the signal will rise instantaneously if a pulsed IR-probe is used whereas the signal with a CW-probe will rise with the dephasing time of the excited-state vibrational transition. A much more detailed description of the theory of visible-pump IR-probe spectroscopy is given in Ref. 1.

## 3.Femtosecond Experiments on Vibrational Transitions in the RC

Our experiments in the range 1700-2000 $cm^{-1}$ were published last year,[2] therefore we will only give a brief summary of the main results. Samples of *Rb. Sphaeroides R-26* were excited at 870 nm by a Ti:sapphire based regenerative amplification system, using pulses of ~200 nJ and a width of approximately 350 fs. The transient absorbance changes were monitored by the output of a CW CO-laser which was upconverted after the sample in a $AgGaS_2$ crystal by an 870 nm gate pulse. An overall increase of absorbance after excitation was noted in the range of IR frequencies used; There is no sign of large scale protein relaxation in the signals, that is, all the IR transients follow the well known electron transfer kinetics (3 ps or 200 ps); There may be a small signal due to protein relaxation at 1665 $cm^{-1}$ possibly due to a Stark effect enhanced transition strength of amide I carbonyls; Anisotropy measurements indicate that a transient absorption at 1702 $cm^{-1}$ is due to $P_M$; It seems that the excited state surface of $P^*$ is very similar to surface of P. At 1682 $cm^{-1}$ a transient bleaching is observed that shows a slow (3 ps) and not an instantaneous rise as expected (see above). This will occur if the 9-keto C=O vibrations are very similar in P and $P^*$ (frequency shifted by less than the linewidth of ~9 $cm^{-1}$) and the bleach at 1682 $cm^{-1}$ only develops after $P^*$ decays into $P^+H^-$ of $P^+B^-$. There may be a signal due to the formation of the intermediate state $P^+B^-$. The data at 1665 $cm^{-1}$ shows a rise of the absorbance consistent with the formation of $B^-$. However, given the current signal-to-noise ratio in the experiment, the transient is equally consistent with a model involving only $P^*$. Future studies that carefully compare the signals with CW- and pulsed-probing can be expected to resolve this issue.

## 4.Recent Developments in the Generation of Ultrashort IR Pulses

After developing new ultrashort pulse, dispersion compensated, Ti:sapphire regenerative amplifiers [3] these new light sources were employed to generate tunable femtosecond pulses in the near- and mid-IR through parametric generation and amplification. In a scheme similar to that published recently [4] the nonlinear material BBO was used to generate pulses tunable from 1.2 to 2.4 μm with a pulse width ranging from 48 to 120 fs. KTP was used to improve the tuning range to 1 to 4.3 μm with a pulse width on the order of 100 fs. $AgGaS_2$ and GaSe crystals are now being used to extend the tuning range into the mid-IR (5-20 μm).

## 5.Electronic Transitions

The ultrashort near-IR pulses generated through parametric processes were employed to study various electronic transitions in the RC [5]. The bands due to the electronic transitions are wide, have correspondingly short dephasing times and therefore coherent effects at short delay times will be absent. Below results will be presented on energy transfer from $B^*$ to $P^*$ and on the observation of excited electronic states of $P^*$ and $P^+H^-$.

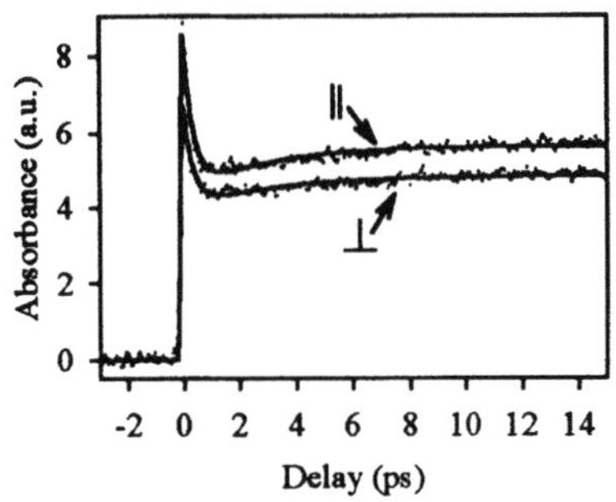


**Figure 1**. *Transient absorption signal observed at 1.2 μm after pumping RCs of* Rb. Sphaeroides R-26 *at 800 nm.*

## 5.1. Results Pumping B

In the following experiments samples of the RC of *Rb. Sphaeroides R-26* are excited at 800 nm and therefore it may be expected that Förster energy transfer from excited B to the special pair takes place. The experiments are done with two different Ti:sapphire regenerative amplifier laser systems, both with approximately 30-40 fs time resolution and employing BBO and KTP parametric devices, respectively, for generating IR. Also white light continuum generation in sapphire is used to create near-IR pulses. Initial experiments at 1.2 μm (see ***Figure 1***) show a rise of $P^+H^-$ transient absorption as seen previously in experiments where the RC sample was excited at 610 nm [6]. In addition a fast transient is observable at short times that decays with 200 fs; This transient appears to reflect the rate of energy transfer from B* to P*. Also the rise of the stimulated emission from $P^*$ at 950 nm was measured (see ***Figure 2***); the rise is single exponential and has a characteristic time of 136±5 fs.

There may be many interpretations for the fast 200 fs decay observed at 1.2 μm: energy transfer or "internal conversion" from $B^*$ to $P^*$ or electron transfer from $B^*$ to $P^+B^-$ or from $B^*$ to $B^+H^-$. There are several reasons to assign this transient to relaxation from $B^*$ to $P^*$. First of all the rise of the gain signal at 950 nm is on a similar time scale as the decay observed at 1.2 μm. Furthermore a rise of the $P^+H^-$ signal with the "regular" 3 ps is observed and a signal from P* is observed that decays with ~3 ps. There are reasons, however, to expect that $P^*$ and $B^*$ are strongly mixed. For example it can be calculated that the theoretical Förster energy transfer time can be as fast as 25 fs. It is therefore possible that the 100-200 fs decay is an internal conversion process rather than Förster-like energy transfer.

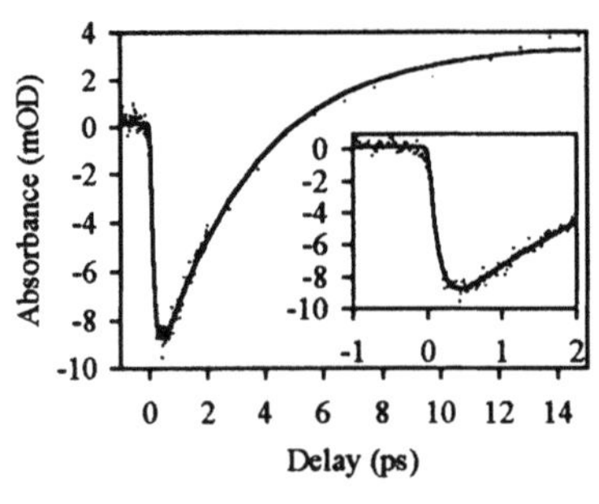


**Figure 2.** *Transient gain observed at 950 nm. The inset shows the same signal at short delay times.*

These issues will be discussed in more detail in a forthcoming publication [7].

## 5.2. Electronic Transitions in $P^*$

The RC has approximate $C_2$ symmetry which allows one to construct stationary states and estimate transition dipoles. Calculations of the excited state ($P^*$) absorption were performed, based on the electronic structure calculations of Warshel and Fischer [8-10] and as described before in Ref. 11 Essentially excitonic states are constructed from individual chlorophyll excitations as: $|P_Y^{+/-}\rangle \approx |P_M^*P_L\rangle \pm |P_MP_L^*\rangle$ and charge resonance states as: $|C^{+/-}\rangle \approx |P_M^+P_L^-\rangle \pm |P_M^-P_L^+\rangle$. The actual mixing coefficients are obtained from the electronic structure calculations. Using the experimental transition and permanent dipole moments of bacteriochlorophyll-a, it is predicted that transitions from $P_Y^-$ to higher lying states that have mainly $C^{+/-}$ character, should be observed in the near-IR.

The experiments were performed with the same ultrafast laser setups and on the same type of RC samples as described in the preceding section. In the region 1.4-2.2 μm an instrument limited rise is followed by a decay with 3 ps to a constant background. The constant background can be assigned to absorption from $P^+H^-$ on the basis of a previous study of the RCs of *Rb. Sphaeroides* [12]. Around 5 μm another band due to $P^*$ can be observed; The long time signal is due to the 2600 $cm^{-1}$ band of $P^+H^-$ (hole exchange transition). The first band is centered at 6000 $cm^{-1}$ with a transition dipole moment of approximately 5 D. The second band is centered at approximately 1900 $cm^{-1}$ with a transition dipole moment >1 D (See ***Figure 3***). The latter band can be assigned to the $P_Y^+ \leftarrow P_Y^-$ transition based on anisotropy measurements.

The result that the $P_Y^+ \leftarrow P_Y^-$ transition is at approximately 1900 $cm^{-1}$ may appear inconsistent with previous hole burning results [13] which indicate that the $P_Y^+ \leftarrow P_Y^-$ transition is at 1300 $cm^{-1}$. It should be noted, however, that the electronic absorption may extend *further into the IR* and that the band observed at 1900 $cm^{-1}$ may be a Herzberg-Teller transition based on a vibration of ca. 600 $cm^{-1}$. Although the transition from the ground state to $P_Y^+$ may have a strong 0-0 component, the $P_Y^+ \leftarrow P_Y^-$ transition may have a weak origin and be mainly vibrationally induced.

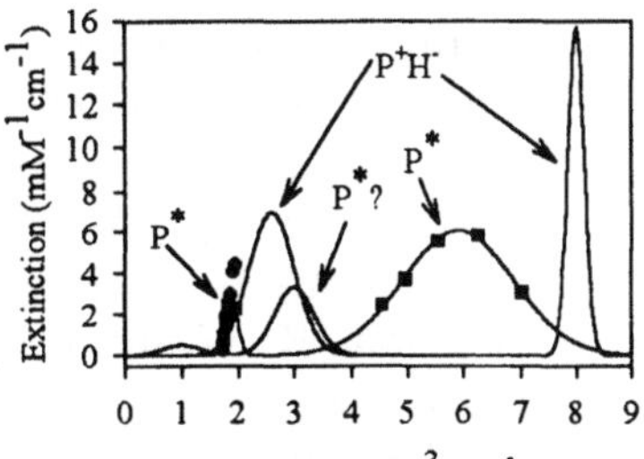


**Figure 3**. *Spectrum of the transient absorption from $P^*$ and $P^+H^-$ observed in the RC.*

## 6.Conclusions

In summary, ultrafast vibrational and electronic spectroscopy has been performed on the RC and energy transfer from $B^*$ to $P^*$, or internal conversion between states of similar origin, was measured to occur in 100-200 fs. In addition new excited electronic states of the accessory bacteriochlorophyll and the special pair have been observed. A summary of the excited electronic states that have been observed in our experiments is shown in *Figure 4*. The spectrum of $P^*$ between 2.4 and 6 μm is currently being investigated.

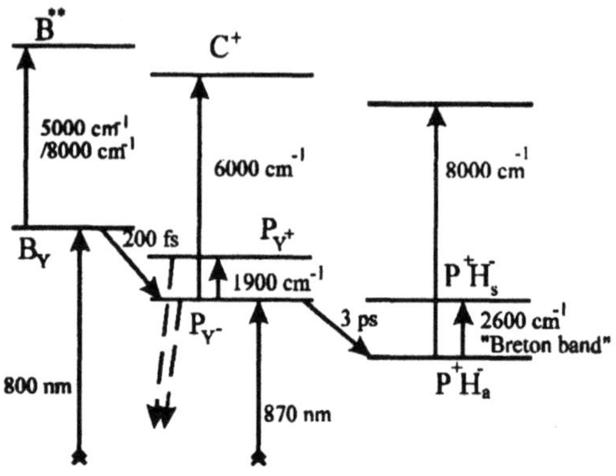


**Figure 4**. *Level scheme for the RC*

## 7.References


1 K. Wynne, R.M. Hochstrasser (1995) Chem. Phys. 193, 211-236.

2 S. Maiti, G.C. Walker, B.R. Cowen, R. Pippenger, C.C. Moser, P.L. Dutton, R.M. Hochstrasser (1994) Proc. Natl. Acad. Sci. USA 91, 10360-10364.

3 K. Wynne, G.D. Reid, R.M. Hochstrasser (1994) Opt. Lett. 19, 895-897.

4 V.V. Yakovlev, B. Kohler, K.R. Wilson (1994) Opt. Lett. 19, 2000-2002.

5 G.C. Walker, S. Maiti, G.D. Reid, K. Wynne, C.C. Moser, R.S. Pippenger, B.R. Cowen, P.L. Dutton, R.M. Hochstrasser (1994) in *Ultrafast Phenomena IX*, Eds. P.F. Barbara, W.H. Knox, G.A. Mourou, A.H. Zewail, Springer, Berlin, pp.439-440.

6 N.W. Woodbury , M. Becker, D. Middendorf, W.W. Parson (1985) Biochemistry 24, 7516-7521.

7 G. Haran, K. Wynne, C.C. Moser, P.L. Dutton, R.M. Hochstrasser, in preparation.

8 W.W. Parson, A. Warshel (1987) J. Am. Chem. Soc. 109, 6152-6163.

9 P.O.J. Scherer, S.F. Fischer (1989) Chem. Phys. 131, 115-127.

10 P.O.J. Scherer, S.F. Fischer (1991) in: *Chlorophylls*, Ed. H. Scheer, CRC Press, Boca Raton.

11 G.C. Walker, S. Maiti, B.R. Cowen, C.C. Moser, P.L. Dutton, R.M. Hochstrasser (1994) J. Phys. Chem. 98, 5778-5783.

12 J. Breton, E. Nabedryk, W.W. Parson (1992) Biochem. 31, 7503-7510.

13 D. Tang, S.G. Johnson, R. Jankowiak, J.M. Hayes, G.J. Small, D.M. Tiede (1990) Jerusalem Symp. Quantum Chem. and Biochem. 22 (Perspect. Photosynth.) 99-120.

# ENERGETICS OF THE PRIMARY CHARGE SEPARATION IN BACTERIAL PHOTOSYNTHESIS

M. Bixon,[1] Joshua Jortner,[1] and M.E. Michel-Beyerle[2]
[1]School of Chemistry, Tel Aviv University, Tel Aviv 69978, Israel.
[2]Institut für Physikalische und Theoretische Chemie,
Technische Universität München,
Lichtenbergstr. 4, D-85748 Garching, Germany.

**Abstract.** We consider the free energy relationship for the primary electron transfer (ET) (at T = 295 K) from the electronically excited singlet state of the bacteriochlorophyll dimer in the bacterial photosynthetic native reaction center (RC), some of its single-site mutants and chemically engineered RCs containing accessory $13^2$-OH-Ni-bacteriochlorophyll (Ni-B). This analysis resulted in the reasonable value $\lambda_1 = 800\pm250\ cm^{-1}$ for the medium reorganization energy and $\Delta G_1(N) = -480\pm180\ cm^{-1}$ for the energy gap of the native RC. These energetic parameters imply that ET in the native RC corresponds to (nearly) optimal activationless ET. The quantum free energy relation predicts very fast ET time for the Ni-B substituted center ($\Delta G_1 \simeq -1500\ cm^{-1}$) which reflects prosthetic group(s) vibrational excitation induced by ET in the inverted region. The low negative value of $\Delta G_1(N)$ implies that the dominant room temperature ET mechanism for the native RC involves sequential ET.

Key Words: energetics, primary charge separation, Ni-RC, RC mutants.

## 1. Introduction

All the mechanisms proposed for the primary electron transfer (ET) in the bacterial photosynthetic reaction center (RC) [1-5] attribute a special role to the accessory bacteriochlorophyll (B). The nature of the primary process, i.e., the superexchange [5-15], the sequential [16-27] or the parallel sequential-superexchange [28-30] mechanism, is dominated by the (free) energy gap $\Delta G_1$ of the ion pair state $P^+B^-H$ relative to $^1P^*$. Several sources of information on $\Delta G_1$ in the native RC were advanced: (1) Molecular dynamics simulations [31-33]. These energetic data are intrinsically limited due to incomplete input information, e.g., the location of accessory water molecules and the long-range interactions. (2) Recombination of the ion pair $P^+BH^-$. Experimental magnetic data [34] resulted in the lower limit $\Delta G_1 > -600\ cm^{-1}$. (3) Kinetic data on ET in the chemically modified pheophytin → bacteriopheophytin RC resulted in $\Delta G_1 = -450\ cm^{-1}$ [27], which is presumably applicable also for the native RC. In what follows we shall construct a free-energy relationship for the decay rates of $^1P^*$ in the native RC [25,35], in a series of some single-site mutants [25,35-39] and in the chemically modified $13^2$-OH-Ni-bacteriochlorophyll (Ni-B) → bacteriochlorophyll (B) RC [40,41]. From this phenomenological analysis we shall extract the energy gap $\Delta G_1$ for the native RC and the medium reorganization energy for the primary ET. The energetics will provide a clue for the mechanism of the primary process.

## 2. Energetic Parameters for the Native RC and for 'Good' Single-Site Mutants

We have constructed a free-energy relationship for the primary ET in single-site mutants [25,35,36-39] using the following information.
(i) The kinetic data for the decay of $^1P^*$. The kinetic decay of $^1P^*$ is nonexponential, reflecting the effects of static heterogeneity, which are manifested by long time tails [25,30,35,42,43]. On the basis of the representation of the decay of $^1P^*$ in a heterogeneous system in terms of two exponentials [25,35,42,43], as was previously done in the analysis of the experimental data, we showed [30] that the fast decay component provides a reasonable (within 10%-30%) representation of the decay of $P^*$ in the corresponding homogeneous system. Accordingly, we use the experimental fast decay component for the representation of the primary ET rate, which will be denoted by $k_{ET}$.
(ii) The oxidation potentials $E_R(P^+/P)$ of P in the RCs [25,35,36-39]. It is assumed that for these mutants

$$\Delta G_1 = E_R(P^+/P) + B \quad , \tag{1}$$

where B is a constant. Eq. (1) implies that equal changes are induced by mutations in the redox potentials of $^1P^*/P^+$ and of $P/P^+$ and that the mutation perturbs only P and not $B_A$ of $H_A$.

Single-site local L181 or M208 mutations of *R.capsulatus* [35] and of *R.sphaeroides* do not necessarily obey relation (1), as site mutagenesis may also change the energetics of the prosthetic groups $B_A$ and/or $H_A$ and their ionic states, and also modify the electronic coupling V. A valid free-energy relationship between $k_{ET}$ and $E_R(P^+/P)$ is expected to be obeyed for 'good' single-site mutants, which obey the following constraints regarding the geometry and the interactions:
(A) Minimization of geometrical changes, which may modify V. Such changes will be induced by the replacement of the tyrosine (Y) M210 in *R.sphaeroides* (M208 in *R.capsulatus*) residue by a smaller (i.e., threonine, leucine or isoleucine) or by a bulkier (i.e., tryptophane (W)) residue. On the other hand, phenylalanine (F) and histidine (H) are

closer in size to Y, and H may even hydrogen-bond like Y.
(B) Minimization of the perturbation of the prosthetic groups $B_A$ and $H_A$ by the mutations. All the site mutants of Y M210 (M208) [35,38] are excluded, and so is the GM203 → D (glycine (G) to aspartic acid (D)) mutant, which was specifically designed to affect $B_A$ [36].

Guided by these empirical rules we have taken for the analysis of the free energy relation for 'good' mutants all the L181 mutants of the Chicago-Argonne group [35], the Y-Y mutant from the Munich group [25] and the relevant mutants from the Arizona group [36,37]. For the native RC we have taken the data for *R.capsulatus* [35], and *R.sphaeroides* [25,38]. These data are scattered (due to different samples and interrogation methods) by ± 35%, providing a lower limit for the uncertainty of the experimental data. In Figure 1 we present the experimental $k_{ET}$ data at T = 295 K for the native RC, together with those for the 'good' mutants. These data fit nicely the theoretical free energy dependence

$$k_{ET} = (2\pi V^2/\hbar)\ F(\lambda_1, S_c, \hbar\omega_c, \Delta G_1, T) \ , \tag{2}$$

where V is the electronic coupling and F is the nuclear Franck Condon factor

$$F = (2\pi k_B T\lambda_1)^{-1/2}\exp(-S_c)\sum_{n=0}^{\infty}\frac{(S_c)^n}{n!}\exp[-(\Delta G_1+n\hbar\omega_c+\lambda_1/4\lambda_1 k_B T)]. \tag{3}$$

F is characterized by low frequency modes with the medium reorganization energy $\lambda_1$ and by the high frequency mode $\omega_c$ with a coupling strength $S_c$. The lifetime for primary ET is given by

$$\tau_{ET} = 1/k_{ET} \ . \tag{2a}$$

For the 'good' mutants we implicitly assume that V and $\lambda_1$ are invariant and $\Delta G_1$ is given by Eq. (1). At this stage we do not commit ourselves to the nature of the charge separation from $^1P^*$, i.e., sequential or superexchange mechanism, although the simple form of (2) implies that one of these mechanisms dominates. The parameters $\lambda_1$ (taken to be mutant invariant) and $\Delta G_1$ (mutant dependent according to Eq. (1)) will be determined from the analysis of the experimental data [25,35-37]. Good account of the free energy relationship for $\tau_{ET}$ (or $k_{ET}$) can be accomplished (Fig. 1) with the following parameters: (i) The medium reorganization energy $\lambda_1$ = 800 cm$^{-1}$. (ii) The uniform shift of the (free) energy scale, Eq. (1), is B = -4500 cm$^{-1}$. (iii) For the high frequency mode we have chosen either the traditional values

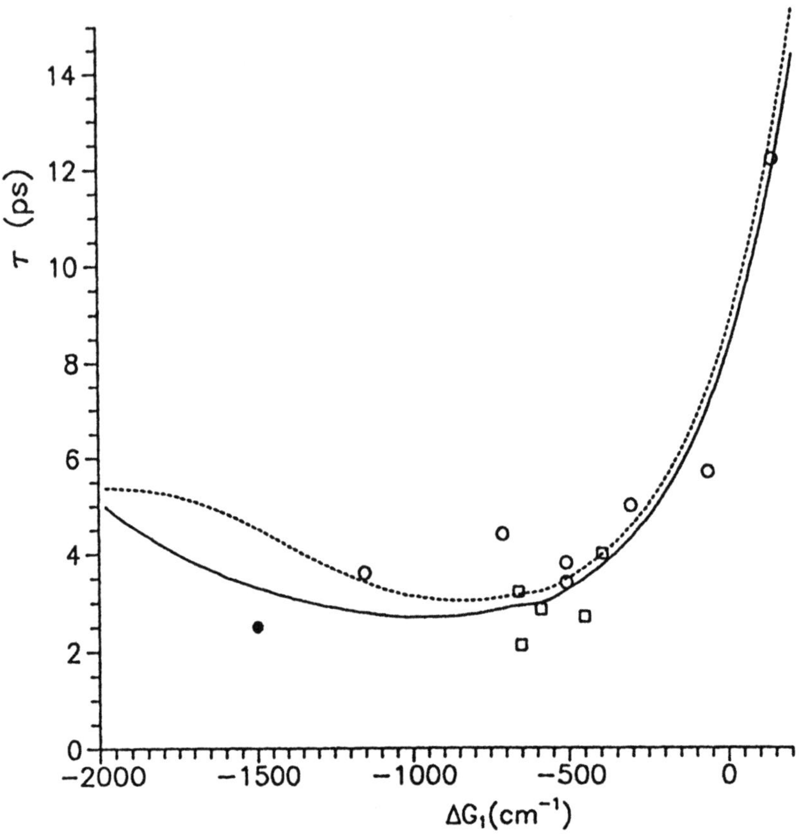


Figure 1. The free-energy relationship for the experimental data of $\tau_{ET}$ vs $\Delta G_1$ for the native RC (data marked by a vertical arrow), for 'good' single site mutants and for the chemically modified Ni-B RC at room temperature (T = 295 K). The electronic coupling is V = 20 $cm^{-1}$. The low frequency 'medium' reorganization energy is $\lambda_1$ = 800 $cm^{-1}$. The high frequency vibration is characterized by $S_c$ = 0.5 and $\omega_c$ = 1500 $cm^{-1}$ or $\omega_c$ = 1000 $cm^{-1}$. The experimental data for the single site mutants are taken from: o - references 36,37, ▯ - reference 35 and • - reference 45. The experimental data point, with the corresponding $\Delta G_1$ values taken from the $E_R(P^+/P)$ redox potentials (references 25,35-37) for the chemically modified Ni-B are taken from reference 45, with $\Delta G_1 \simeq$ -1500 $cm^{-1}$ (see text). The theoretical curves are calculated at T = 300 K from Eqs. (2), (2a) and (3) with the parameters given above. Solid curve $\omega_c$ = 1000 $cm^{-1}$, dashed curve $\omega_c$ = 1500 $cm^{-1}$. Note that the free energy relationship is presented by $\tau_{ET}$ vs $\Delta G_1$, rather than by the conventional form of $\ell n k_{ET}$ vs $\Delta G_1$.

[28,44] $\hbar\omega_c$ = 1500 $cm^{-1}$ and $S_c$ = 0.5, or a somewhat lower frequency $\omega_c$ = 1000 $cm^{-1}$ again with $S_c$ = 0.5. At room temperature (T = 295 K) the calculated free energy relationship is independent of $\omega_c$ for $\Delta G_1 >$ -1000 $cm^{-1}$ and depends only weakly on $\omega_c$ for lower values of $\Delta G_1$ (-1000 $cm^{-1} > \Delta G_1 >$ -2000 $cm^{-1}$) (Fig. 1). (iv) The electronic coupling is taken to be V = 20 $cm^{-1}$. From this analysis we infer that the energy gap for the native RC, which will be denoted by $\Delta G_1(N)$, is $\Delta G_1(N)$ = -480 $cm^{-1}$. In view of the spread of the experimental data (Fig. 1) the energetic parameters at room temperature, which provide an adequate fit of these results for the native RC and for the 'good' single-site mutants, are

$$\begin{aligned} \lambda_1 &= 800 \pm 250 \text{ cm}^{-1} \\ \Delta G_1(N) &= -480 \pm 180 \text{ cm}^{-1}. \end{aligned} \tag{4}$$

The (free) energy domain spanned by the native RC and by the 'good' single-site mutants corresponds in range from $\Delta G_1$ = -1100 $cm^{-1}$ to $\Delta G_1$ = 100 $cm^{-1}$. The energetic parameters for the native RC, Eq. (4), are close (within their uncertainty range) to activationless ET, i.e., $-\Delta G_1(N) \simeq \lambda_1$. It is of considerable interest to extend the free energy range for lower $\Delta G_1$ values, which correspond to the strongly exoergic inverted region ($-\Delta G_1 > \lambda_1$). This analysis will clearly bring up the role of quantum effects on the Franck-Condon factor F, which originates from the coupling with the intramolecular vibrational modes of the prosthetic groups [44]. This can be accomplished by the analysis of the $k_{ET}$ vs $\Delta G_1$ relation for the Ni-B RC [40].

## 3. Energetic Parameters for the Ni-B RC

The incorporation of the Ni substituted prosthetic groups into the photosynthetic RC of *R.sphaeroides* R26, replacing the two accessory bacteriochlorophylls $B_A$ and $B_B$, results in a drastic change in the energy of the ion pair state $P^+(Ni\text{-}B)^-$ relative to that of $P^+B^-$ in the native RC. The redox potential of the Ni substituted chlorophyll $(NiCh\ell)^-/(NiCh\ell)$ in solution is lower by 0.29 eV than the corresponding value for the redox potential of $Ch\ell^-/Ch\ell$ [40]. From the thermally activated participation of $P^+(Ni\text{-}B)^-$ in the recombination dynamics of the radical pair $P^+H_A^-$ an energy gap between the two states of less than 500 $cm^{-1}$ can be estimated [45] and we shall tentatively take $\Delta G_1$ = -1500 $cm^{-1}$ for the (free) energy gap for (Ni-B) chemically substituted RC. Obviously, the primary ET in this chemically engineered RC corresponds to the inverted region. In spite of this drastic reduction of $\Delta G_1$, which corresponds to the lowest value of the gap achieved either by mutagenesis and/or chemical substitution, the primary ET lifetime is very close to that of the native RC. The decay of $^1P^*$ in the (Ni-B) substituted RC at T = 295 K is characterized by a biexponential decay with the short component $\tau_{ET}$ = 2.5 ps (amplitude 69%) and a longer component $\tau_L$ = 7 ps (amplitude 31%) [45]. The short decay component for the primary ET in the (Ni-B) substituted RC is very close (within 30%, which corresponds to experimental uncertainty) to $\tau_{ET}$ for the native RC (Fig. 1). The very weak energy gap dependence of $\tau_{ET}$ in the range $\Delta G_1 \simeq -400\ cm^{-1}$ to $\Delta G_1 \simeq -1500\ cm^{-1}$ (Fig. 1), can be well accounted for by the ET theory, i.e., Eqs. (2) and (3) with the invariant value of $\lambda_1$, Eq. (4). In the inverted region, i.e., $-\Delta G_1 \geq \lambda_1$, the contribution of the quantum modes is major, resulting in a marked deviation from the classical result (which corresponds to $S_c = 0$), which is attributed to the intramolecular vibrational excitation of the prosthetic groups induced by ET [44]. As is evident from Fig. 1, the energy gap dependence of $\tau_{ET}$ for the Ni-B substituted RC is reasonably well accounted for by the ET theory, with $\omega_c = 1000\ cm^{-1}$.

## 4. Concluding Remarks

The energetics of the primary ET in the native photosynthetic RC, some of its single-site 'good' mutants and the Ni-B chemically engineered RC, can be accounted for in terms of the quantum free energy relationship, Eqs. (2) and (3). The major results of the analysis, Eq. (4), resulted in the (mutant and chemically substituted invariant) value of the medium reorganization energy $\lambda_1 = 800\pm250\ cm^{-1}$ and in the (free) energy gap $\Delta G_1(N) = -480\pm180\ cm^{-1}$ for the native RC. From these results we conclude that:

1) The reorganization energy $\lambda_1$ for the primary ET incorporates all the "low frequency" vibrational modes of the protein and of the prosthetic groups (for which the frequencies are $\hbar\omega < k_BT$ at T = 295 K) which involve the protein modes and the intramolecular modes of the dimer. The value of $\lambda_1$ obtained herein is considerably higher than the surprisingly low value ($\lambda_1 \leq 250\ cm^{-1}$ at T = 295 K), which was previously inferred from the analysis of single site mutants [25,35-38]. Our value of $\lambda_1$ is larger than the spectroscopic electron-phonon coupling for the electronic excitation $P \rightarrow {}^1P^*$, which gives the spectroscopic reorganization energy $\lambda_S = 250\ cm^{-1}$ [46,47]. Indeed, $\lambda_S$ is expected to provide the lower limit for $\lambda_1$. Finally, we note that $\lambda_1$ is lower than the reorganization energy $\lambda_T > 1300\ cm^{-1}$ inferred from magnetic field effects [48] on the recombination of the ion pair $P^+BH^- \rightarrow {}^3P^*$.
2) The values of $\lambda_1$ and $\Delta G_1(N)$, Eq. (4), imply that ET in the native RC is close (within the uncertainty range of the energetic

parameters) to activationless ET, i.e., $-\Delta G_1 \simeq \lambda_1$. This physical situation corresponds to (nearly) optimal ET (i.e., nearly shortest $\tau_{ET}$) in the native RC.

3) The (free) energy gap relation, which incorporates vibrational quantum effects, predicts a very fast ET for the Ni-B chemically substituted RC. $\tau_{ET}$ for this system, which is characterized by the lowest currently available value of $\Delta G_1$ ($\simeq$ -1500 $cm^{-1}$), is close to the ET lifetime for the native RC. This prediction, pertaining to the inverted region, is borne out by the experimental results.
4) The low value of $\Delta G_1(N)$ for the native RC is in good agreement with the lower limit $\Delta G_1 > -600$ $cm^{-1}$ previously inferred from the analysis of magnetic data for the $P^+BH^-$ ion pair [34] and the value $\Delta G_1 = -450$ $cm^{-1}$ derived from time-resolved data for the pheophytin → H chemically modified RC [27].
5) The low value of $\Delta G_1(N) = -480 \pm 180$ $cm^{-1}$ obtained herein for the native RC is considerably lower in its absolute value than the energy gap $\Delta G = -2000$ $cm^{-1}$ between $^1P^*$ and $P^+BH^-$ [49,50]. This result implies that the first ET step in the native RC at room temperature is not a direct superexchange mediated ET to H.
6) The dominance of the sequential mechanism for primary ET prevails for the room temperature inverted region, i.e., the Ni-B chemically modified RC and the single-site mutants presented in Fig. 1. For larger values of $\Delta G_1$ ($\geq$ -300 $cm^{-1}$), which correspond to double and to triple mutants [39] at room temperature and at low temperatures [39] (which were not analyzed herein), as well as for some single site mutants at low temperatures, the contribution of the superexchange mechanism is substantial [30]. Even for the native RC at low temperature, when heterogeneity effects are incorporated, a contribution (~ 10%) of the superexchange is expected to prevail [30] and to contribute to the long time tail of the decay of $^1P^*$ [30].

There are some pitfalls in our analysis. From the experimental point of view, we have utilized in Fig. 1 experimental data for different samples from different laboratories, increasing experimental uncertainty. In our analysis we have invoked the invariance of $\lambda_1$ for different mutants and for the chemically engineered Ni-B RC. Finally, we have provided only a heuristic account of heterogeneity effects. This important issue is addressed elsewhere [30,51].

## References


1. J. Jortner and M.E. Michel-Beyerle, in: Antennas and Reaction Centers of Photosynthetic Bacteria, ed. M.E. Michel-Beyerle (Springer, Berlin, 1985) p. 345-354.

2. The Photosynthetic Bacterial Reaction Center. Structure and Dynamics, eds. J. Breton and A. Vermeglio (Plenum NATO ASI Series, New York, 1988).
3. Reaction Centers of Photosynthetic Bacteria, ed. M.E. Michel-Beyerle (Springer Verlag, Berlin, 1990).
4. The Photosynthetic Bacterial Reaction Center II, NATO-ASI series A, J. Breton and A. Verméglio (Life Sciences, Plenum Press, New York, 1992) vol. 237.
5. The Photosynthetic Reaction Center, J. Deisenhofer and J.R. Norris (Academic Press, San Diego, 1993).
6. N.W. Woodbury, M. Becker, D. Middenforf and W.W. Parson, Biochemistry 24 (1985) 7516.
7. S.F. Fischer, I. Nussbaum and P.O.J. Scherer, in: Antennas and Reaction Centers of Photosynthetic Bacteria, ed. M.E. Michel-Beyerle (Springer, Berlin, 1985) p. 256.
8. A. Ogrodnik, N. Remy-Richter, M.E. Michel-Beyerle and R. Feick, Chem. Phys. Lett. 135 (1987) 576.
9. J.R. Norris, D.E. Budil, D.M. Tiede, J. Tang, S.V. Kolaczkowski, C.H. Chang and M. Schiffer, in: Progress in Photosynthetic Research, ed. J. Biggins (Martinus Nijhoff, Dordrecht, 1987) vol. I, p. 1.4.363.
10. M.E. Michel-Beyerle, M. Plato, J. Deisenhofer, H. Michel, M. Bixon and J. Jortner, Biochim. Biophys. Acta 932 (1988) 52.
11. M. Bixon, J. Jortner, M. Plato and M.E. Michel-Beyerle, in: The photosynthetic bacterial reaction center. Structure and Dynamics, eds. J. Breton and A. Vermeglio (Plenum NATO ASI Series, New York, 1988) p. 399.
12. M. Plato, K. Möbius, M.E. Michel-Beyerle, M. Bixon and J. Jortner, J. Am. Chem. Soc. 110 (1988) 7279.
13. M. Bixon, M.E. Michel-Beyerle and J. Jortner, Isr. J. Chem. 28 (1988) 155.
14. M. Bixon, J. Jortner, M.E. Michel-Beyerle and A. Ogrodnik, Biochim. Biophys. Acta 977 (1989) 273.
15. R.A. Friesner and Y. Won, Biochim. Biophys. Acta 977 (1989) 99.
16. R.A. Marcus, Isr. J. Chem. 28 (1988) 205.
17. R. Haberkorn, M.E. Michel-Beyerle and R.A. Marcus, Proc. Natl. Acad. Sci. USA 70 (1979) 4185.
18. R.A. Marcus, Chem. Phys. Lett. 133 (1987) 471.
19. S.V. Chekalin, Ya.A. Matveetz, A.Ya. Shkuropatov, V.A. Shuvalov and A.P. Yartzev, FEBS Lett. 216 (1987) 245.
20. R.A. Marcus, Chem. Phys. Lett. 146 (1977) 13.
21. S. Creighton, J.-K. Hwang, A. Warshel, W.W. Parson and J. Norris, Biochem. 27 (1988) 774.
22. W. Holzapfel, U. Finkele, W. Kaiser, D. Oesterhelt, H. Scheer, H.U. Stilz and W. Zinth, Chem. Phys. Lett. 160 (1989) 1.

23. W. Holzapfel, U. Finkele, W. Kaiser, D. Oesterhelt, H. Scheer, H.U. Stilz and W. Zinth, Proc. Nat. Acad. Sci. USA 87 (1990) 5168.
24. C. Lauterwasser, U. Finkele, H. Scheer and W. Zinth, Chem. Phys. Lett. 183 (1991) 471.
25. P. Hamm, K.A. Gray, D. Oesterhelt, R. Feick, H. Scheer and W. Zinth, Biochim. Biophys. Acta 1142 (1993) 99.
26. T. Arlt, S. Schmidt, W. Kaiser, C. Lauterwasser, M. Meyer, H. Scheer and W. Zinth, Proc. Natl. Acad. Sci. USA 90 (1993) 11757.
27. P.M. Schmidt, T. Arlt, P. Hamm, H. Huber, T. Nägele, J. Wachtveitel, M. Meyer, H. Scheer and W. Zinth, Chem. Phys. Lett. 223 (1994) 116
28. M. Bixon, J. Jortner and M.E. Michel-Beyerle, Biochim. Biophys. Acta 1056 (1991) 301.
29. M. Bixon, J. Jortner and M.E. Michel-Beyerle, reference 4 (1992) 291.
30. M. Bixon, J. Jortner and M.E. Michel-Beyerle, Chem. Phys. 197 (1995) 389.
31. W.W. Parson, Z.T. Chu and A. Warshel, Biochem. Biophys. Acta 1017 (1990) 251.
32. M. Mercht, J.N. Gehlen, D. Chandler and M. Newton, J. Am. Chem. Soc. (1993) 4178.
33. A. Warshel, Z.T. Chu and W.W. Parson, Photochem. and Photobiology A82 (1994) 123.
34. M. Volk, A. Ogrodnik and M.E. Michel-Beyerle, in: Anoxygenic Photosynthetic Bacteria, eds. R.E. Blankenship, M.T. Madigan and C.E. Bauer, pp. 595 (Kluwer Acad., Dordrecht, 1995).
35. Y. Jia, T.J. DiMagno, C.-K. Chan, Z. Wang, M. Du, D.K. Hanson, M. Schiffer, J.R. Norris, G.R. Fleming and M.S. Popov, J. Phys. Chem. 97 (1993) 13180.
36. J.C. Williams, R.G. Alden, H.A. Murchison, J.M. Peloquin, N.W. Woodbury and J.P. Allen, Biochemistry 31 (1992) 11029.
37. H.A. Murchison, R.G. Alden, J.P. Allen, J.M. Peloquin, A.K.W. Taguchi, N.W. Woodbury and J.C. Williams, Biochemistry 32 (1993) 3498.
38. V. Nagarajan, W.W. Parson, D. Davis and C.C. Schenck, Biochemistry 32 (1993) 12324.
39. N.W. Woodbury, S. Lin, X. Lin, J.M. Peloquin, A.K.W. Taguchi, J.C. Williams and J.P. Allen, Chem. Phys. 197 (1995) 405.
40. G. Hartwich, Ph.D. Thesis, TU München (1994).
41. G. Hartwich, M. Friese, H. Scheer, A. Ogrodnik, and M.E. Michel-Beyerle, Chem. Phys. (in press) 1995.
42. M. Du, S.J. Rosenthal, X. Xie, T.J. DiMagno, M. Schmidt, D.K. Hanson, M. Schiffer, J.R. Norris and G.R. Fleming, Proc.

Natl. Acad. Sci. USA 89 (1992) 8517.
43. M.G. Müller, K. Griebenow and A.R. Holzwarth, Chem. Phys. Lett. 199 (1992) 465.
44. M. Bixon and J. Jortner, J. Phys. Chem. 95 (1991) 1942.
45. T. Häberle, H. Lossau, M. Friese, G. Hatwich, A. Ogrodnik, H. Scheer and M.E. Michel-Beyerle (this volume).
46. R. Raja, S. Reddy, P.A. Lyle and G.J. Small, Photosynth. Res. 31 (1992) 167.
47. P.A. Lyle, S. Kolaczkowski and G.J. Small, J. Phys. Chem. 97 (1993) 6924.
48. M. Bixon, J. Jortner and M.E. Michel-Beyerle, Z. Phys. Chem. 180 (1993) 193.
49. A. Ogrodnik, M. Volk. R. Letterer, R. Feick and M.E. Michel-Beyerle, Biochim. Biophys. Acta 936 (1988) 361.
50. R.A. Goldstein, L. Takiff and S.G. Boxer, Biochim. Biophys. Acta 934 (1988) 253.
51. M. Bixon, J. Jortner and M.E. Michel-Beyerle (this volume).

# EFFECTS OF STATIC HETEROGENEITY ON THE PARALLEL-SEQUENTIAL-SUPEREXCHANGE MECHANISM FOR THE PRIMARY CHARGE SEPARATION IN BACTERIAL PHOTOSYNTHESIS

M. Bixon,[1] Joshua Jortner,[1] and M.E. Michel-Beyerle[2]
[1]School of Chemistry, Tel Aviv University, Tel Aviv 69978, Israel.
[2]Institut für Physikalische und Theoretische Chemie, Technische Universität München, Lichtenbergstr. 4, D-85748 Garching, Germany.

Key Words: Primary charge separation, Ni-Reaction center, RC mutants, Heterogeneity.

## 1. Introduction

Until the early 1990's, the experimental psec and subpsec data for the kinetics of the primary charge separation process in the photosynthetic RC were interpreted in terms of a single lifetime for each of the electronically excited singlet states of the bacteriochlorophyll dimer ($^1P^*$) and the ion pairs [1-4]. This approach seemed to indicate that heterogeneity effects on the kinetics are minor. This physical situation pointed towards the possibility that the dynamics in membrane proteins is drastically different from that in globular proteins [5-7], where low-temperature chemical dynamics, e.g., the recombination of CO to the heme site of myglobin, is dominated by a nonexponential heterogeneous kinetics [5-7]. Of course, heterogeneous kinetics, involving static inhomogeneity effects, indicates the existence of distinct microenvironments (for the relevant prosthetic groups involved in the reaction), which do not interconvert on the relevant time scale of the dynamic process. The existence of such static distribution of microenvironments, which is ubiquitous in globular proteins, results in a static distribution of the energetics, nuclear Franck-Condon factors or/and the electronic couplings which determine the rates. The apparent absence of inhomogeneous kinetics [1-4] for ET in the photsynthetic RC was surprising. With the refinements of the experimental techniques, evidence for inhomogeneous kinetics in the bacterial RC started emerging, being originally inferred from the dependence of the kinetics on the probing wavelength [8,9]. Recently, long-time tails were observed in the decay of $^1P^*$ as interrogated by stimulated emission and in fluorescence upconversion [10-13]. The decay of $^1P^*$ in the wild type reaction center could be fit by a biexponential with $\tau_1$ = 2.6 ps and $\tau_3$ = 12 ps with relative amplitudes of $A_3/A_1 \sim 0.15$ [10,11]. A similar behavior is exhibited for chemically modified RCs, e.g., for the $13^2$-OH-Ni-bacteriochlorophyll (Ni-B) → bacteriochlorophyll (B) substituted RC [14-16], where $\tau_1$ = 2.5 psec, $\tau_3$ =

7 psec and $A_3/A_1 \sim 0.4$ at T = 285 K. The biexponential pattern prevails also for a variety of local mutants with $A_3/A_1$ being considerably larger [10,11,17-19]. Structural heterogeneity may result in a continuous (or bimodal) distribution of the energy gaps and/or the reorganization energies and/or the electronic couplings. Experimental information on the spread of the energy of the $P^+BH^-$ ion pair state relative to $^1P^*$ has become accessible from the temperature dependence of delayed fluorescence due to recombination of the ion pairs and the magnetic field effect therein [20]. This spread was characterized by a Gaussian distribution with a width parameter $\sigma = 400\ cm^{-1}$ (i.e., the FWHM being $2(2\ell n2)^{1/2}\sigma$) [20]. Concurrently, theoretical expressions were provided [11,21] for the ET rates in an inhomogeneously broadened (dispersive) system.

An outstanding problem in the understanding of the primary charge separation from $^1P^*$ to H across the A branch of the bacterial photosynthetic RC pertains to its mechanism [22-45]. We have advanced the parallel sequential-superexchange mechanism [43-45] for the primary ET, which provides a unified scheme that includes both the unistep superexchange mechanism and the two-step sequential mechanism as limiting cases. Although substantial experimental evidence for the dominance of the sequential mechanism in the native (N) RC at room temperature (T = 280-300 K) is available [10,14,38-42], the superexchange route is of some significance at low temperature native RC [43-45], is important for most single-site mutants at low temperature [45] and may be dominating for mutants with extreme values of the energy gap [45]. The merging between the parallel sequential-superexchange routes is of considerable interest, providing information and experimental guidance for:

(1) The energy gap dependence of the (averaged) initial decay times of $^1P^*$.
(2) The nature of the temporal nonexponential decay pattern of $^1P^*$.
(3) The energy gap dependence of the quantum yield for charge separation.
(4) The relative contribution of the parallel mechanisms (i.e., sequential and superexchange) to the initial decay times, to the pattern of the temporal decay of $^1P^*$ and to the quantum yield for a broad range of energy gaps.

The systems we explore correspond to:

(i) the native (N) RC ($\Delta G_1 \simeq -500\ cm^{-1}$) [42,45-47];
(ii) the Ni-B chemically substituted RC ($\Delta G_1 = -1500\ cm^{-1}$ [14-16]) corresponding to the lowest value of $\Delta G_1$;
(iii) some other chemically engineered RCs [42];
(iv) the single-site 'good' [45,47] mutants [10,11,17,18], for which geometrical changes are minimized and the perturbation corresponds to $^1P^*/P^+$, while the perturbations of the prosthetic

groups of B and H are minor ($\Delta G_1 \simeq -1100\ cm^{-1}$ to $200\ cm^{-1}$);

(v) triple mutants [19] ($\Delta G_1 \simeq 1000\ cm^{-1}$), which are characterized by the largest value of $\Delta G_1$.

The $\Delta G_1$ scale spans the range of $2500\ cm^{-1}$ (8 kcal/mol).

## 2. The parallel sequential-superexchange mechanism

The kinetic scheme for this mechanism [43-45] (Fig. 1) consists of:

(i) the sequential route

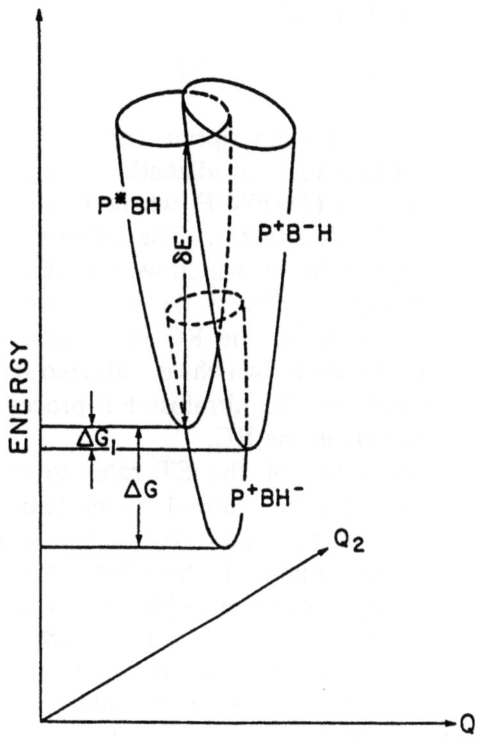


Fig. 1. Schematic potential energy surfaces for the primary charge separation in the RC.

$$^1P^*BH \underset{k_{-1}}{\overset{k_1}{\rightleftharpoons}} P^+B^-H \xrightarrow{k_2} P^+BH^- \qquad (1)$$

(ii) the superexchange route

$$^1P^*BH \underset{k_-}{\overset{k}{\rightleftharpoons}} P^+BH^- \qquad (2)$$

(iii) quinone reduction

$$P^+BH^-Q \xrightarrow{k_Q} P^+BHQ^- \tag{3}$$

(iv) internal conversion of the dimer

$$^1P^*BH \xrightarrow{k_d} PBH \tag{4}$$

(v) recombination of ion pairs

$$P^+B^-H \xrightarrow{k_R} PBH \tag{5}$$

$$P^+BH^- \xrightarrow{k_S} PBH \tag{6}$$

$$P^+BH^- \xrightarrow{k_T} {}^3P^*BH \ . \tag{7}$$

The theoretical description for the ET rates will be given in terms of the multiphonon nonadiabatic theory, which expresses the ET rate in the form $k = (2\pi/\hbar)V^2F$, where V is the electronic coupling and F is the Franck-Condon factor. The validity conditions for the applicability of this simple scheme, which we recently discussed [43-45], are:

1) The applicability of the nonadiabatic limit, which for characteristic parameters for the RC imply that [44] $V < 70\ cm^{-1}$ at $T = 300$ K. This relation, which is inferred from an a posteriori analysis, is obeyed for the ultrafast ET process and for all the subsequent ET reactions in the RC.
2) Insensitivity of the ET rates to medium dynamics. This state of affairs can be realized under two circumstances. (i) Fast medium relaxation dynamics. (ii) Activationless ET. Case (ii) is relevant for the description of the ultrafast primary ET processes. We have recently demonstrated [48] that activationless ET cannot be described in terms of the diffusion towards a point sink at the intersection of the potential nuclear surface at the minimum of the initial nuclear surface (for the medium modes). Rather, for activationless ET the microscopic ET rates are insensitive (within a numerical factor of ~ 2) to the distribution of the initial vibrational states [48]. Accordingly, the (activationless) primary ET processes in the RC are not controlled by medium dynamics and can be reasonably well described in terms of the conventional ET rate from an equilibrated vibrational manifold.

The four microscopic primary ET rate constants in Eqs. (1) and (2) are:

The superexchange rate:

$$k = \left(\frac{2\pi}{\hbar}\right) (V_{PB} V_{BH}/\delta E)^2 F(\lambda,\omega_m, S_c, \hbar\omega_c, \Delta G, T) \tag{8}$$

The first sequential rate:

$$k_1 = \left(\frac{2\pi}{\hbar}\right) V_{PB}^2 \; F(\lambda_1,\hbar\omega_m, S_c, \hbar\omega_c, \Delta G_1, T) \tag{9}$$

The reverse rate:

$$k_{-1} = k_1 \exp(-\Delta G_1/k_B T) \tag{10}$$

The second sequential rate:

$$k_2 = \left(\frac{2\pi}{\hbar}\right) V_{BH}^2 \; F(\lambda_2,\hbar\omega_m, S_c, \hbar\omega_c, \Delta G_2, T) \; . \tag{11}$$

The electronic couplings, the energey gaps and the nuclear coupling parameters which determine the rates are defined in Table 1. The Franck-Condon factors are

$$F(\lambda,\hbar\omega_m, S_c, \hbar\omega_c, \Delta G, T) = \sum_{n=0}^{\infty} F_m(\Delta G - n\hbar\omega_c) F_c(n) \tag{12}$$

with the medium contributions

$$F_m(\Delta E) = (\hbar\omega_m)^{-1} \left(\frac{\bar{v}+1}{\bar{v}}\right)^{p/2} \exp[-S_m(2\bar{v}+1)] I_p(2S_m[\bar{v}(\bar{v}+1)]^{1/2}), \tag{13}$$

where $\bar{v} = [\exp(\hbar\omega_m/k_B T)-1]^{-1}$ is the thermal population of the mode, $p = \Delta G/\hbar\omega_m$, $S_m = \lambda/\hbar\omega_m$ and $I_p(\cdot)$ is the modified Bessel function of order p. The intramolecular high-frequency contribution is

$$F_c(n) = \exp(-S_c) \, S_c^n/n! \; . \tag{14}$$

**Table 1.** Input data for kinetic modelling. Data assembled in reference 45 and references are quoted therein.

| Rate | Energy gap | Medium frequencies | Medium reorganization | Intramolecular mode | Electronic coupling |
|---|---|---|---|---|---|
| $k_1$ | $\Delta G_1$ | $\omega_m$=95cm$^{-1}$ | $\lambda_1$=800 | $\omega_c$=1500 $S_c$=0.5 | $V_{PB}$=20cm$^{-1}$ |
| $k_2$ | $\Delta G_2=\Delta G-\Delta G_1$ | $\omega_m$ | $\lambda_2$=800 | $\omega_c S_c$ | $V_{BH}$=40cm$^{-1}$ |
| $k_{-1}$ | $-\Delta G_1$ | $\omega_m$ | $\lambda_2$ | $\omega_c, S_c$ | $V_{BH}$ |
| $k$ | $\Delta G$ | $\omega_m$ | $\lambda$=1600cm$^{-1}$ | $\omega_c, S_c$ | $V_{PH}=\frac{V_{PB}V_{BH}}{\lvert\Delta G_1+\lambda_1\rvert}$ |
| $k_-$ | $-\Delta G$ | $\omega_m$ | $\lambda$ | $\omega_c, S_c$ | $V_{PH}$ |
| $k_Q$ | (200ps)$^{-1}$ | | | | |
| $k_d$ | (300ps)$^{-1}$ | | | | |
| $k_R$ | (1ns)$^{-1}$ | | | | |
| $k_s$ | (20ns)$^{-1}$ | | | | |
| $k_T$ | (2ns)$^{-1}$ | | | | |

The energetic, electronic and nuclear parameters (Tables 1 and 2), which determine the primary rates (8)-(10), are taken from our recent work [45]. The (free) energy gaps for the chemically engineered and mutant RCs studied herein are summarized in Table 2. The rates for the surprisingly fast dimer internal conversion [49], the recombination of the ion pairs [50] and the quinone reduction rate [1-3] are taken from the experimental data and are summarized in Table 1. The solution of the kinetic equations (1)-(7) resulted in three lifetimes $\tau_j$(j=1-3) and an amplitudes matrix $A_j^{(k)}$ (j, k = 1-3), with the temporal populations

$$[P^*](t) = \sum_{j=1}^{3} A_j^{(1)} \exp(-t/\tau_j) \ , \tag{15a}$$

$$[P^+B^-H](t) = \sum_{j=1}^{3} A_j^{(2)} \exp(-t/\tau_j) \ , \tag{15b}$$

$$[P^+BH^-](t) = \sum_{j=1}^{3} A_j^{(3)} \exp(-t/\tau_j) \quad . \tag{15c}$$

**Table 2.** Energetic data for chemically engineered RC and for some of its mutants.

| RC | References | $\Delta G_1(cm^{-1})$ | $\Delta G_2(cm^{-1})$ | $\Delta G(cm^{-1})$ |
|---|---|---|---|---|
| NATIVE | 42,45-47 | - 500 | -1500 | -2000 |
| NiB→B | 14-16 | -1500 | - 500 | -2000 |
| H → B | | (-2000) | (0) | (-2000) |
| Pheo→H | 42 | - 450 | - 180 | - 630 |
| B → H | | (- 500) | (0) | (- 500) |
| NiB→B Pheo→H | | (-1500) | ( 1000) | (- 500) |
| 'good' single site mutants | 10,11, 17,18 | -900 to 300 | -1500 | -2400 to -1200 |
| triple H bonded mutant | 19 | (+1000) model | (-1500) | (- 500) |

Calculations were performed for room temperature T = 300 K, using the electronic and energetic parameters previously advanced (Tables 1 and 2). In addition, model calculations were performed for a low temperature system at T = 20 K using the same parameters. The low temperature approximation does not account for the anisotropic thermal contraction, which may modify electronic, energetic, nuclear and static heterogeneity parameters (see section 3). Nevertheless, such low temperature simulations will be of use for the elucidation of the gross features of the ET kinetics.

## 3. Kinetic heterogeneity in the parallel mechanism

The parallel-sequential-superexchange scheme is now extended by the incorporation of the energy distribution of the ion pair states $P^+BH^-$

and $P^+B^-H$. The energy distribution functions $p(\Delta G_j')$ for the energy gaps $\{\Delta G_j'\}$ around a mean value $\Delta G_j$, i.e., $\Delta G_1$, $\Delta G_2$ and $p(\Delta G')$ for $\{\Delta G\}$ around the mean value $\Delta G$ are taken to be Gaussian

$$p(\Delta G_j') = (2\pi\sigma^2)^{-1/2} \exp[-(\Delta G_j'-\Delta G_j)^2/2\sigma_j^2] \quad , \tag{16}$$

where $\sigma_1$ and $\sigma_2$ denote the distribution widths parameters around $\Delta G_1$ and $\Delta G_2$, respectively, while $\sigma$ is the distribution width parameter around $\Delta G$. Magnetic data for the recombination of the $P^+BH^-$ have established that static heterogeneity of the energetics of the ion pair can be described in terms of Eq. (16) with $\sigma = 400\ cm^{-1}$. For the treatment of the N RC and of the mutants we have taken the energy levels of the ion pairs $P^+B^-H$ and $P^+BH^-$ to be correlated, i.e., $\Delta G_2' = -1500\ cm^{-1}$. The static heterogeneity parameters were taken to be temperature independent.

The kinetic analysis was conducted for the inhomogeneously broadened system. We calculated the lifetimes' vector $\tau(\Delta G_1') = \{\tau_j(\Delta G_1')\}$ and the amplitude matrix $\underset{\approx}{A}(\Delta G_1') = \{A_j^{(k)}(\Delta G_1')\}$ for each value of $\Delta G_1'$ within a discretized Gaussian distribution ($\sigma = 400\ cm^{-1}$) of 800 values.

The static heterogeneity results in a distribution of the lifetimes, which are characterized by the probability density of the distribution of the decay times of $^1P^*$

$$q(\tau) = \sum_j \int d(\Delta G_1')\delta(\tau-\tau_j(\Delta G_1'))\ A_j^{(1)}(\Delta G_1')p(\Delta G_1') \quad , \tag{17}$$

The time evolution of $^1P^*BH$ in the heterogeneous system is given by

$$P(t) = \int p(\Delta G_1')\ [P^*](\Delta G_1';t)d(\Delta G_1') \quad . \tag{18}$$

where $[P^*]\ (\Delta G_1';t)$ is calculated according to Eq. (15a) for the given values of $\Delta G_1'$. Making use of Eq. (17) the time evolution, Eq. (18), is given by

$$P(t) = \int d\tau\ q(\tau) \exp(-t/\tau) \quad . \tag{19}$$

To make contact with experimental reality we have calculated the quantum yield Y for charge separation

$$Y = \int k_Q[P^+BH^-](t)dt = k_Q \sum_{j=1}^{3} A_j^{(3)} \tau_j \ . \tag{20}$$

To explore the central mechanistic issue of the interplay between the sequential and superexchange charge separation routes, we have calculated the branching ratio $F_{sup}$ for superexchange, which is given by

$$F_{sup} = (1/Y) \sum_j (kA_j^{(1)} - k_- A_j^{(3)})\tau_j \ . \tag{21}$$

where the rates k and $k_-$ are defined by Eq. (2).

## 4. Kinetic modelling

### 4.1 The native RC

The static heterogeneity results, of course, in a distribution of lifetimes. Figure 2 portrays the probability density $q(\tau)$ vs $\tau$, Eq. (17), for the native RC at T = 300 K and at T = 20 K. We note that $q(\tau)$ peaks at $\simeq \tau_1$, i.e., $\tau_1$ = 3.5 ps at room temperature and $\tau_1$ = 1.5 ps at T = 20 K, exhibiting the weak non-Arrhenius temperature dependence of the experimental nearly activationless ET. Longer time tails are exhibited with ~ 20% of the probability density for $\tau_1 \geq$ 6 ps at room temperature, while for T = 20 K about 40% of the probability density is manifested for $\tau_1 \geq$ 6 ps. Thus the effect of the distribution of lifetimes is considerably enhanced at low temperatures. This is apparent from the time dependence of the $^1P^*$ population P(t), which exhibits (Fig. 3) marked nonexponentiality, with dramatic long time tails being manifested at low temperatures. To explore the relative contributions of the sequential and the superexchange mechanisms, we calculated from Eq. (21) the branching ratios for superexchange. At room temperature (T = 300 K) $F_{sup}$ = 0.02, with this minor contribution of superexchange originating from heterogeneity effects as for the homogeneous ($\sigma$ = 0) model system $F_{sup}$ = 0. We infer that for the native RC at room temperature the sequential mechanism dominates. At low temperatures, T = 20 K, we find $F_{seq}$ = 0.1, whereupon the sequential mechanism still dominates but the superexchange makes a nonnegligible contribution. The 10% low-temperature contribution of

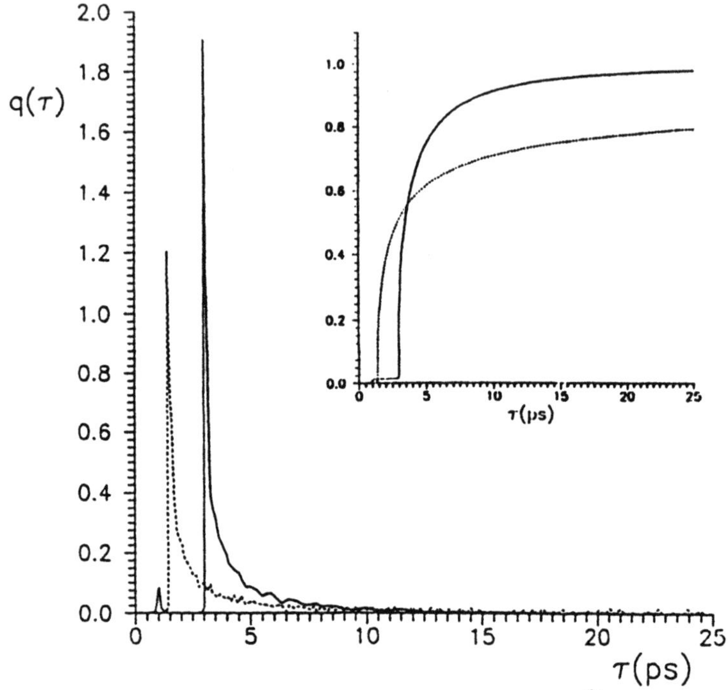


Fig. 2. The probability density $q(\tau)$, Eq. (17), of the $^1P^*$ relaxation time distribution for the native RC ($\Delta G$ = -2000 $cm^{-1}$, $\Delta G_1$ = -500 $cm^{-1}$). The insert shows the accumulated value of $q(\tau)$. (1) Solid line T = 300 K. (2) Dashed line T = 20 K.

superexchange is intimately related to the appearance of the marked long-time tails in P(t) (Fig. 3), which originate from the superexchange channel. This conclusion is in accord with low temperature (T = 80 K) electric field effects on the polarization of the long time fluorescence (time scale 30-500 ps) from the native RC [54], which reveals the formation of the $P^+BH^-$ ion pair via superexchange.

Of considerable interest is the long-time fluorescence (up to t ~ 1 ns) from native and from quinone depleted RCs [20,51-54]. Holzwarth et al [55] attempted to explain these phenomena in terms of a model based on the production of $P^+BH^-$ via superexchange followed by its conformational relaxation. This model seems to be inconsistent with (i) the insensitivity of this "middle" fluorescence component to the quionone content [52] and (ii) the absence of nonphotochemical hole burning [56]. Therefore, this long-time direct fluorescence is attributed to static heterogeneity effects, which are limited by the internal conversion of $^1P^*$ ($k_d$ = (0.3 ns)$^{-1}$). Thus on the time scale up to ~ 1 ns direct dimer fluorescence competes with slow superexchange ET in the heterogeneous system. This physical picture implies that the long-

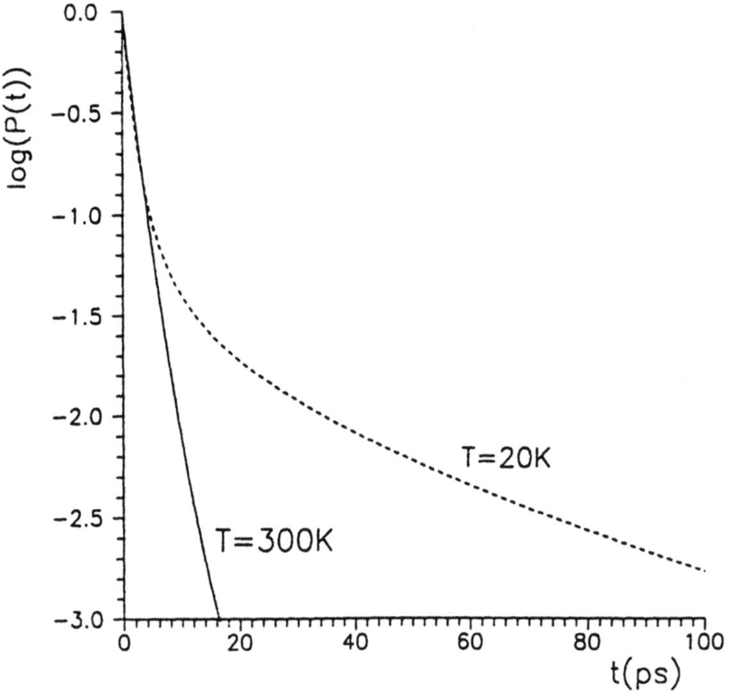


Fig. 3. Decay curves on a logarithmic scale (natural logarithm) of the $^1P^*$ population for the native RC ($\Delta G_1$ = -500 cm$^{-1}$, $\Delta G$ = -2000 cm$^{-1}$) at T = 300 K, in the heterogeneous model: Solid line T = 300 K; dashed line T = 20 K.

time direct fluorescence (up to ~ 1 ns) should be similar for the native RC and in the Q depleted RC, as implied by our simulations (Fig. 4).

Another important experimental observable pertains to the quantum yield for primary charge separation, Eq. (20). Our kinetic modelling for the native RC at T = 300 K results in Y = 0.95, with the reduction of Y from unity being due to competition of ET with the internal conversion of the dimer ($k_d$) and the recombination of $P^+BH^-$ ($k_R$). The pioneering work of Wraight and Clayton [57] gave Y = 1.02 ± 0.04, while recent experimental work [58, 59] resulted in Y = 0.97 ± 0.02 at T = 300 K, in accord with our result. The quantum yield at low temperature (T = 20 K) is predicted to be reduced to Y = 0.90, manifesting more effective competition between internal conversion of $^1P^*$ and superexchange mediated long-time ET in the heterogeneous system.

We conclude that static heterogeneity effects at low temperature in the native RC are responsible for: (1) Superexchange (10%) contribution to the primary process. (2) Long-time (up to 1 ns) tails in the decay of $^1P^*$ manifesting superexchange ET. (3) Long-time direct

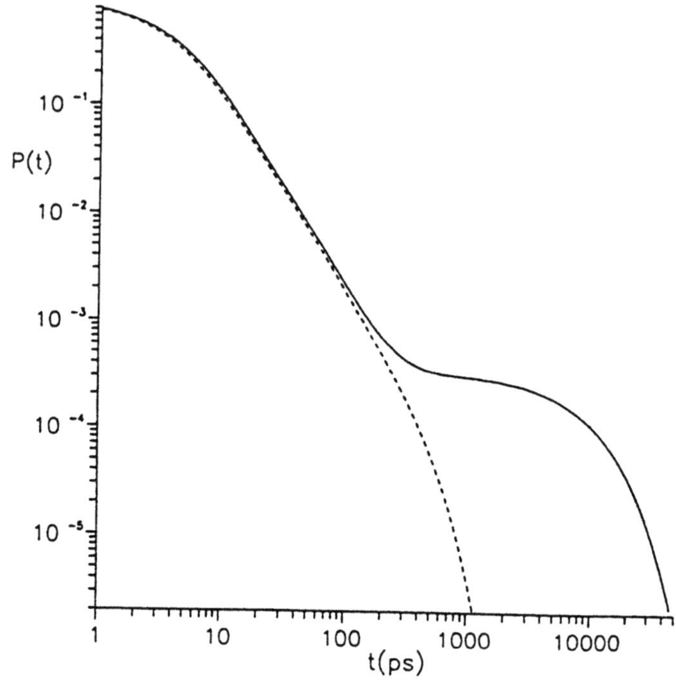


Fig. 4a. T = 300 K. Dashed line: native, and solid line: Q depleted.

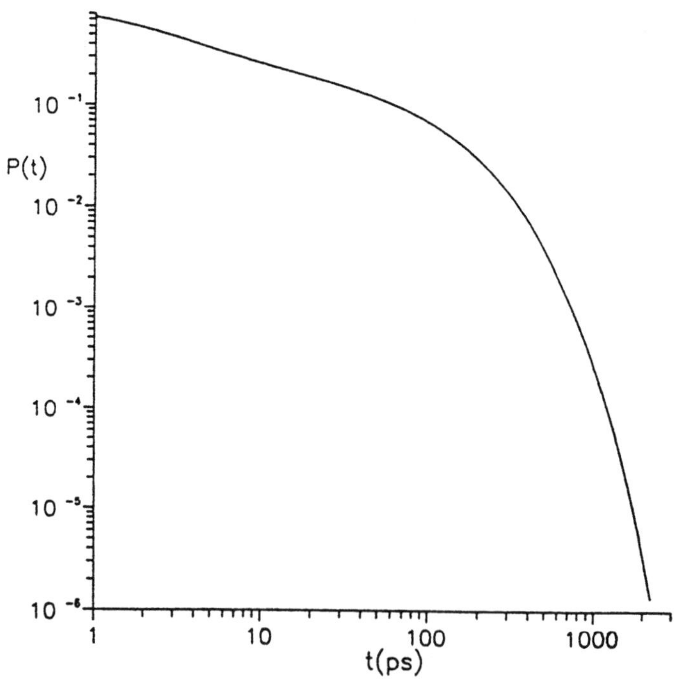


Fig. 4b. T = 20 K. The curves for native and for Q depleted RCs coincide.

Fig. 4. Kinetic modelling of long-time fluorescence decay in native and native-Q depleted RCs.

fluorescence due to competition between superexchange ET and internal conversion. (4) Reduction of the quantum yield for charge separation to $Y \simeq 0.90$.

## 4.2 The Ni-B RC

The replacement of the accessory bacteriochlorophylls $B_A$ and $B_B$ by the Ni-B in the photosynthetic RC of *R.sphaeroides R26* results in a drastic lowering of the energy of the $P^+(\text{Ni-B})^-$ ion pair state (relative to $P^+B^-$) by about 1000 $cm^{-1}$ [14]. This lowering is equivalent to $\Delta G_1 = -1500\ cm^{-1}$ for the Ni-B RC (Table 2). This chemically engineered RC provides the lowest value of $\Delta G_1$. In this kinetic modelling we have accounted for quantum vibrational effects choosing a high frequency mode of $\omega_c = 1000\ cm^{-1}$ [47]. As before, we took $\sigma = 400\ cm^{-1}$. The recombination rates have been determined to be $k_T = 2\ (ns)^{-1}$ and $k_S = 20\ (ns)^{-1}$ [14]. The primary ET in this chemically engineered RC corresponds to the inverted region, where ET induced vibrational excitations of the quantum modes result in a very weak $\Delta G_1$ dependence of the ET rate (Fig. 5). We predict that the primary ET rates of Ni-B ($\Delta G_1 = -1500\ cm^{-1}$) and of the native ($\Delta G_1 = -500\ cm^{-1}$) reaction centers (both with $\lambda_1 = 800\ cm^{-1}$) are close (Fig. 5). This prediction is borne out by the room temperature data for the Ni-B RC at T = 295 K which is characterized by a short decay component of $\tau_1 = 2.5$ ps (amplitude 69%) [15, 16], being very close to $\tau_1$ for the native RC [39-42].

The effects of heterogeneity on the first step of ET are minor [14] as $|\Delta G_1| \gg \sigma$. We expect that the heterogeneity effects are less marked for the Ni-B RC than for the native RC. This conclusion rests on the analysis of $q(\tau)$, Eq. (17) (Fig. 6) and P(t), Eq. (18) (Fig. 7), which reveal less pronounced long ($\tau \geq 6$ ps) temporal component than for the native RC (section (3.1)). In view of the minor heterogeneity effect ($|\Delta G_1| \gg \sigma_1$) we expect that the low temperature long-time (t = 50-1000 ps) fluorescence is absent in the Ni-B RC (Fig. 7), in accord with experimental data. This result provides strong support for our physical picture for the long-time fluorescence tails in the native RC (section 3.1).

The new modified $H \rightarrow B$ RC, where the accessory bacteriochlorophyll(s) are substituted by bacteriopheophytin(s) (Table 2), is characterized by energetic parameters which are close to those of the Ni-B RC. Accordingly, the kinetics of ET in the $H \rightarrow B$ RC is expected to be similar to those predicted for the Ni-B RC.

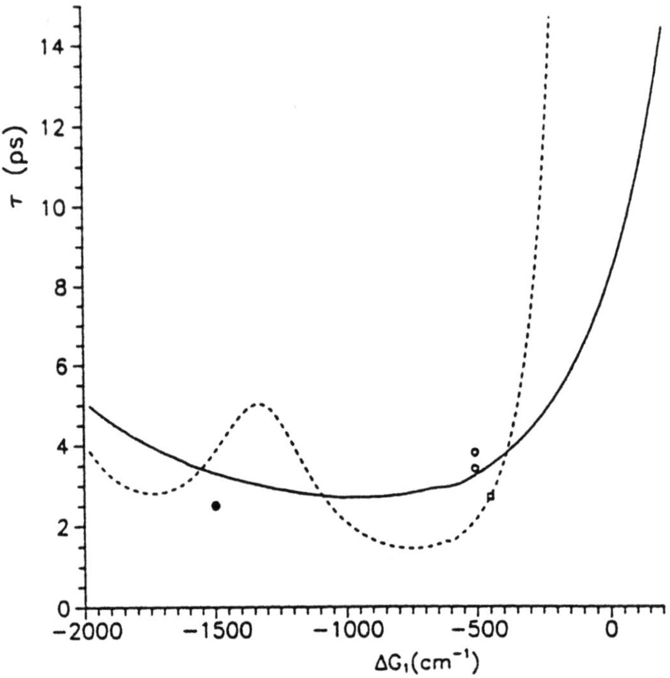


Fig. 5. The energy gap ($\Delta G_1$) dependence of $\tau_1$ in the absence of heterogeneity. Parameters from Tables 1 and 2 and $\omega_c$ = 1000 $cm^{-1}$. Solid line T = 300 K and dashed line T = 20 K. Experimental data for the short-time decay component of $^1P^*$ at T = 300 K are: Native RC (O - references 17, 18 and [] - reference 11). Ni-B substituted RC (• - reference 15, 16).

The marked lowering of $\Delta G_1$ in the chemically engineered RC is of intrinsic interest in the context of the possibility of the induction of the primary charge separation across the B branch of the RC. The symmetry breaking, which induced ET across the A branch in the native RC, originates (at least) from the cumulative effects of reduced electronic coupling across the B branch and a high (unknown) value of the energy of the $P^+B_B^-$ ion pair state. Chemical substitution of $B_B$ by Ni-B or by H will reduce the energy of the corresponding primary ion pair state, i.e., $P^+(Ni\text{-}B)_B^- H_B$ or $P^+H_B^- H_B$ across the B branch by ~ -1000 $cm^{-1}$. It remains to be seen whether this energy reduction is sufficient to break the symmetry breaking for the primary ET.

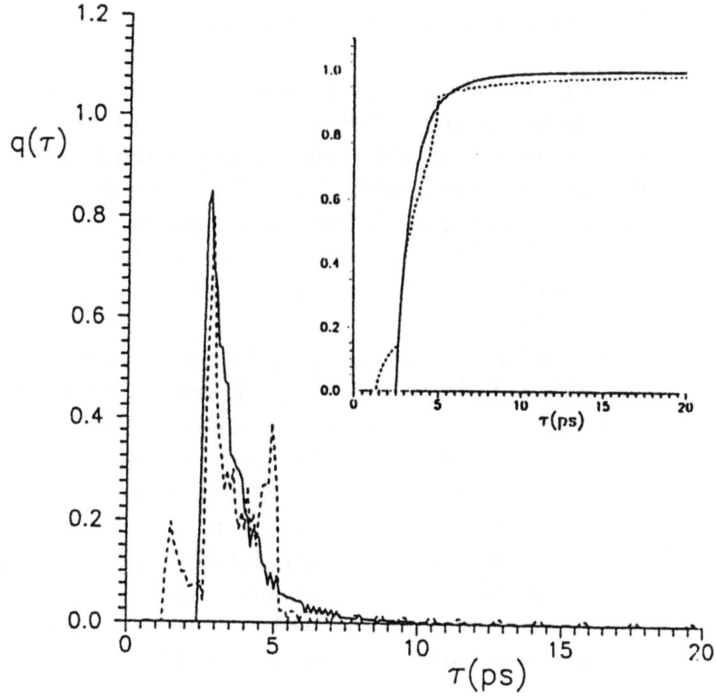


Fig. 6. The probability density q(τ) of the $^1P^*$ relaxation time distribution for the Ni-B substituted RC ($\Delta G_1$=-1500 $cm^{-1}$ and $\Delta G$ =-2000 $cm^{-1}$). The insert shows the accumulated value of q(τ). (1) Solid line T = 300 K. (2) Dashed line T = 20 K.

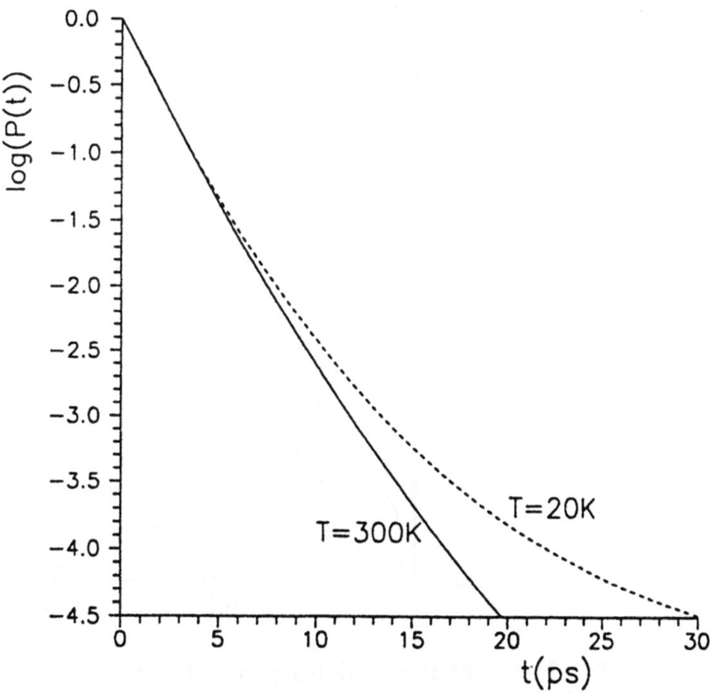


Fig. 7. Decay curve on a (natural) logarithmic scale of the $^1P^*$ population for the Ni-B substituted RC with heterogeneity.

### 4.3 Single-site and triple mutants

We shall consider now primary ET in 'good' [45,47] single-site mutants of *R.capsulatus* and *R.sphaeroides* [10,11,17,18] and the exploration of the primary ET in the triple hydrogen bond [LH(131) + LH)M160) + FH(197)] of *R.sphaeroides* [19]. The relevant energetics is presented in Table 2. Regarding heterogeneity effects, the energy levels of the $P^+B^-H$ and $P^+BH^-$ ion pair states were taken to be correlated with the same distribution ($\sigma_1$ = 400 $cm^{-1}$) and with the energy levels $\Delta G' = \Delta G_1' - 1500$ $cm^{-1}$.

The probability density $q(\tau)$, Eq. (17), for a series of model mutants with $\Delta G_1$ = 0-1500 $cm^{-1}$ (Fig. 8), reveals broad distributions. When the temporal decay of $^1P^*$ in the homogeneous system is dominated by a single exponential, $q(\tau)$ in the heterogeneous system peaks around $\sim \tau_1$. For $\Delta G_1$ = -500 $cm^{-1}$ (native RC) to 700 $cm^{-1}$ the distribution $q(\tau)$ is skewed towards higher values of $\tau$ (Fig. 8). For $\Delta G_1$ = 1000 $cm^{-1}$, when near degeneracy of the kinetic matrix prevails, a very broad distribution of $q(\tau)$ is exhibited, spanning the entire domain of $\tau$. Finally, for the large $\Delta G_1$ = 1500 $cm^{-1}$ the distribution of the decay times peaks around the upper limit of $k_D^{-1}$, being skewed towards a low value of $\tau$.

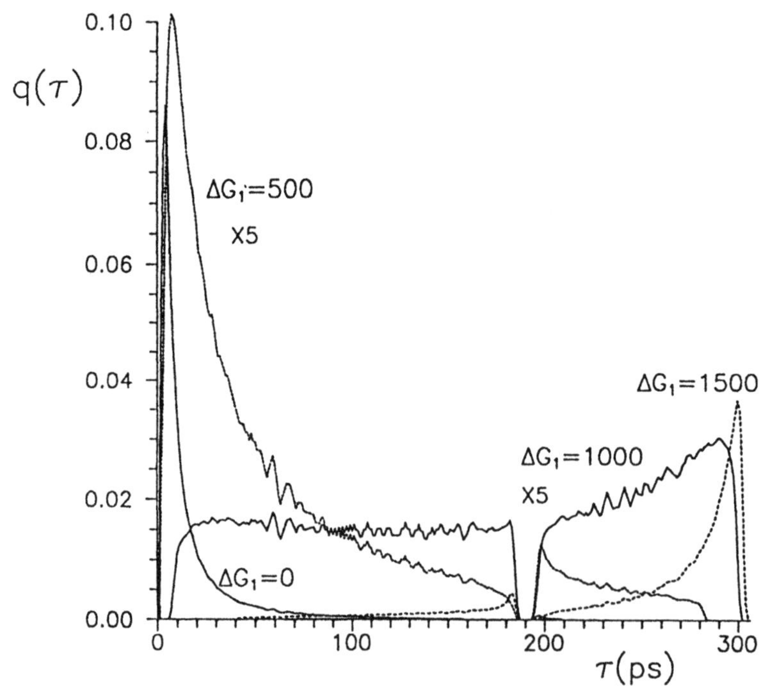


Fig. 8. The probability density $q(\tau)$ for the distribution of relaxation times of $^1P^*$ at T = 300 K for four values of $\Delta G_1$ ($\Delta G_1$ = 0 $cm^{-1}$, 500 $cm^{-1}$, 1000 $cm^{-1}$, 1500 $cm^{-1}$).

We have specified the (initial) dispersive decay of P(t), Eq. (19) (Fig. 9), in the heterogeneous system in terms of the lifetime $\tau(e^{-1})$ for which P(t) reaches the value of $e^{-1}$. In figures 10 and 11 we display the energy gap $\Delta G_1$ dependence of the quantum yield for T = 300 K and for T = 20 K together with the characteristic lifetimes $\tau(e^{-1})$ for the decay of $^1P^*$ in the heterogeneous system. The values of $\tau(e^{-1})$ at T = 300 K span the range from 3.5 ps for $\Delta G_1$ = -1000 $cm^{-1}$ (inverted region through $\tau(e^{-1})$ = 4 ps at $\Delta G_1$ = -500 $cm^{-1}$ (native RC)) and then gradually increase towards $\tau(e^{-1})$ = 200 ps for $\Delta G_1$ = 2000 $cm^{-1}$ (normal region). At low values of $\Delta G_1$ the ET dynamics is dominated by the short sequential lifetimes $k_1^{-1}$ and $k_2^{-1}$. On the other hand, for large values of $\Delta G_1$ (> 1000 $cm^{-1}$) the ET dynamics is dominated by the $^1P^*$ internal conversion lifetime $k_d^{-1}$ (Fig. 8).

We now turn to quantum yield data. The weak dependence of Y on $\Delta G_1$ for $\Delta G_1$ < 0 indicates that for a large number of single-site mutants the primary charge separation is efficient at room temperature. Only for mutants with $\Delta G_1 \geq 0$ one expects a marked reduction of Y with increasing $\Delta G_1$ in the range where the lifetime $\tau(e^{-1})$ exceeds 20 ps or so (Fig. 10), i.e., where internal conversion competes with charge separation. The effects of heterogeneity on Y at room temperature are small for $\Delta G_1 < 0$ becoming somewhat more marked (20%) at $\Delta G_1$ = 0-500 $cm^{-1}$, while in the extreme case of $\Delta G_1$ = 1500 $cm^{-1}$ the heterogeneity enhances Y by a numerical factor of 2 (Fig. 10).

The energy gap dependence of the quantum yield at low temperatures T = 20 K (Fig. 11) is more pronounced, exhibiting a nearly linear decrease from Y = 0.90 for $\Delta G_1$ = -500 $cm^{-1}$ (mimicking the native RC) to Y = 0.20 for $\Delta G_1$ = 500 $cm^{-1}$. At $\Delta G_1$ = 1000 $cm^{-1}$ the yield further reduces to Y = 0.1, while photosynthetic charge separation at low temperatures is terminated for large $\Delta G_1 \geq$ 1500 $cm^{-1}$. The heterogeneity effects on Y at T = 20 K are not pronounced for $\Delta G_1$ < 500 $cm^{-1}$, however, for large values of $\Delta G_1$ > 500 $cm^{-1}$ the distribution of the energy gaps considerably enhances the quantum yield for charge separation, while the decay lifetimes (which are dominated by $k_d^{-1}$) are only weakly affected by heterogeneity. Finally, we turn to the mechanistic issues. The energy gap dependence of $F_{sup}$ (Fig. 12) is classified according to the following ranges. (I) Dominance of the sequential mechanism, $F_{sup} \leq 0.1$. (II) Superposition of superexchange and sequential mechanisms, $0.1 < F_{sup} \leq 0.8$. (III) Dominance of the superexchange, $F_{sup} > 0.8$. At room temperature range (I) prevails for $\Delta G_1$ < 500 $cm^{-1}$ and only at large $\Delta G_1$ = 500-1500 $cm^{-1}$ range (II) is realized. At T = 300 K, heterogeneity effects are minor in range (I), while in range (II) heterogeneity effects enhance the contribution of the sequential channel (Fig. 12). At T = 20 K range (II) is realized in the

Fig. 9a.
a) $\Delta G_1$=-500cm$^{-1}$;
b) $\Delta G_1$=-200cm$^{-1}$;
c) $\Delta G_1$= 0;
d) $\Delta G_1$=+300cm$^{-1}$.

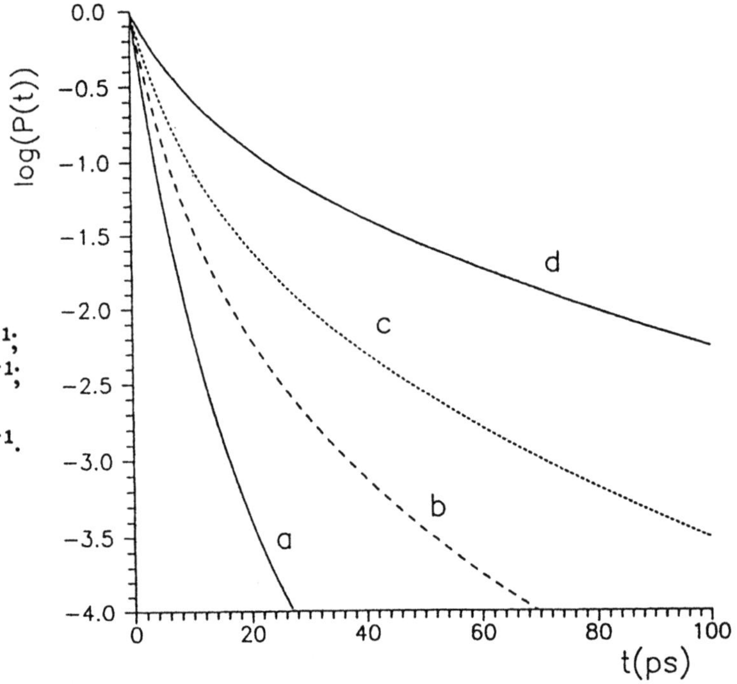


Fig. 9b.
a) $\Delta G_1$=+ 500cm$^{-1}$;
b) $\Delta G_1$=+ 700cm$^{-1}$;
c) $\Delta G_1$=+1000cm$^{-1}$;
d) $\Delta G_1$=+1500cm$^{-1}$.

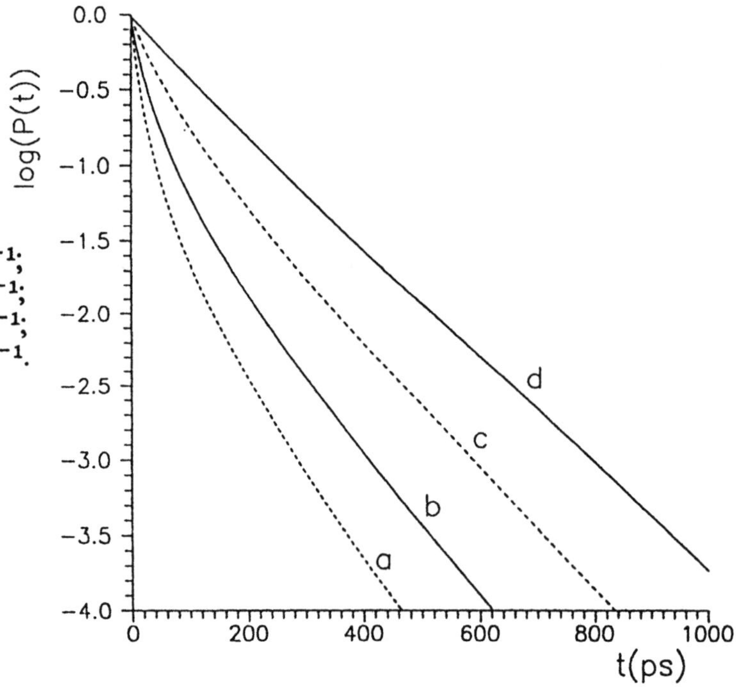


Fig. 9. Decay curves on a logarithmic scale (natural logarithm) of the $^1P^*$ population at T = 300 K, in the heterogeneous model.

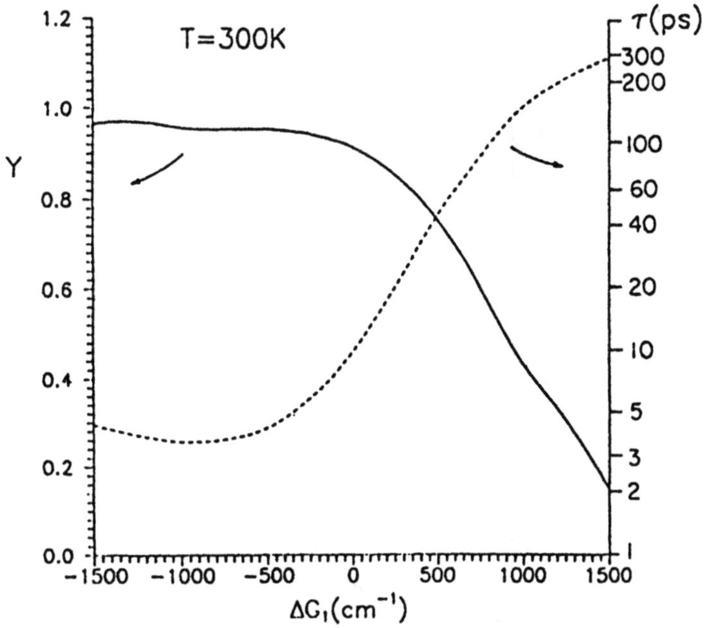


Fig. 10. The quantum yield (solid line) and the relaxation time ($\tau(e^{-1})$) of $^1P^*$ (dashed line) as function of the free energy gap $\Delta G_1$ for the heterogeneous model at T = 300 K.

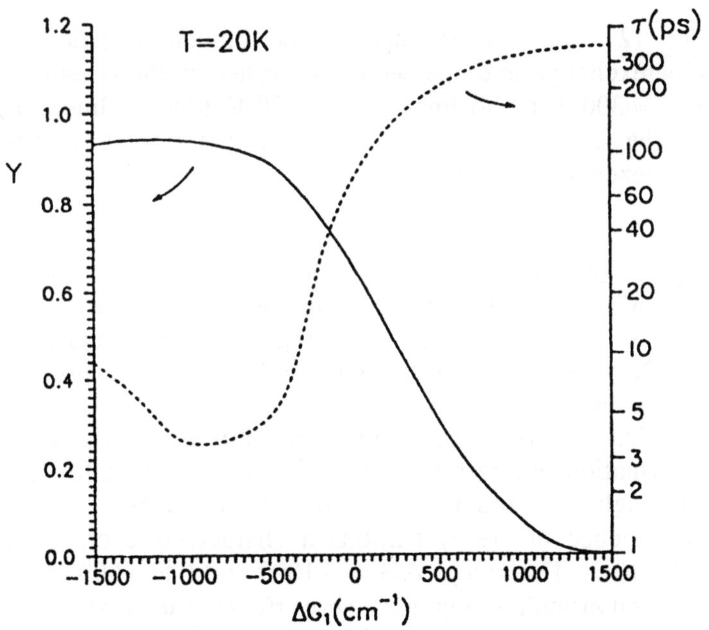


Fig. 11. The quantum yield (solid line) and the relaxation ($\tau(e^{-1})$) time of $^1P^*$ (dashed line) as function of the free energy gap $\Delta G_1$ for the heterogeneous model at T = 20 K.

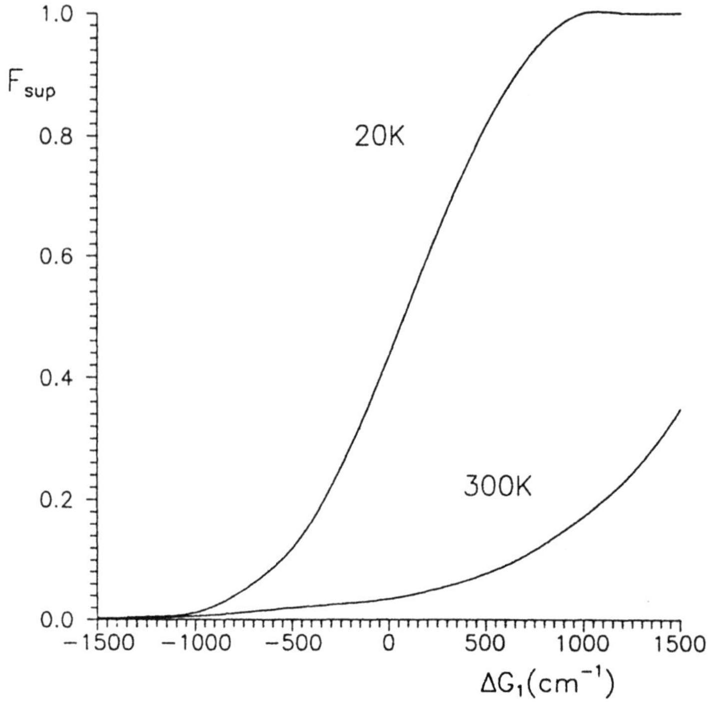


Fig. 12. The dependence of the branching ratio $F_{sup}$ between the superexchange and the sequential rates on the (mean) energy gap $\Delta G_1$ at T = 300 K (solid line) and T = 20 K (dashed line). $F_{sup}$ is expressed as the fraction of the quantum yield that originates from the superexchange channel.

region $-500\ cm^{-1} < \Delta G_1 < 400\ cm^{-1}$. For $\Delta G_1 > 400\ cm^{-1}$ range (III) is exhibited at T = 20 K. Finally, at very large $\Delta G_1 \geq 1000\ cm^{-1}$ $F_{sup} \simeq 1.0$ and the superexchange channel exclusively dominates at T = 20 K, however, in this energy domain the quantum yield for charge separation is low.

What are the implications for mutagenesis? The most extensive information emerges from the kinetic modelling for the native RC and its single-site mutants at room temperature. The calculated $\Delta G_1$ dependence of $\tau(e^{-1})$ exhibits a characteristic energy gap dependence (Fig. 10). The range $\Delta G_1 = -1000\ cm^{-1}$ to $-500\ cm^{-1}$ corresponds to the acitvationless region where $\tau(e^{-1})$ varies slowly, while a strong increase of $\tau(e^{-1})$ with increasing $\Delta G_1$ is manifested for $\Delta G_1 = -500\ cm^{-1}$ to $300\ cm^{-1}$. Regarding the efficiency of the charge separation, the quantum yield in the relevant range $\Delta G_1 = -900$ to $300\ cm^{-1}$ is high, Y = 0.95-0.90, with Y only slightly decreasing with increasing $\Delta G_1$. From the mechanistic point of view the charge separation for the native RC and its single site-mutants is dominated by the sequential route [range (I)].

Some heuristic information emerges on the charge separation in the native RC and its single-site mutants at low temperature (20 K). For mutants $\tau(e^{-1})$ increases with increasing $\Delta G_1$ (Fig. 11) exhibiting a low-temperature energy gap dependence. In the range $\Delta G_1 = -300\ cm^{-1}$ to $300\ cm^{-1}$, $\tau(e^{-1})$ at 20 K is higher than the corresponding value at 300 K, so that these mutants exhibit an ordinary activated temperature dependence for the primary charge separation. The yield of the charge separation at 20 K reduces from Y = 0.9 for the native RC to Y = 0.4 for $\Delta G_1 = 300\ cm^{-1}$. The role of superexchange is more pronounced at low temperatures, corresponding to the coexistence of the superexchange of sequential channels [range (II)]. For the single-site mutants, where $F_{sup} \simeq 0.1\text{-}0.6$ at T = 20 K, the effects of superexchange are marked.

The information which emerges from the kinetic modelling on the primary ET in the triple hydrogen bond mutant [19] is of considerable interest. The experimental values of the quantum yields for the charge separation in the triple mutant [19], i.e., Y = 0.4 at T = 300 K and Y = 0.10-0.15 at T = 20 K fitted well the kinetic modelling for $\Delta G_1 = 1000\ cm^{-1}$. For this large value of $\Delta G_1$ the temporal decay of $^1P^*$ does not exhibit marked deviations from exponentiality (Fig. 9b) and the lifetimes $\tau(e^{-1})$ are long. The experimental lifetimes [19] are $\tau = 66$ ps at T = 300 K and $\tau = 290$ ps at 20 K. These results are in reasonable agreement with the lifetimes emerging from our kinetic simulations for $\Delta G_1 = 1000\ cm^{-1}$, $\tau(e^{-1}) = 160$ ps for $\Delta G_1 = 1000\ cm^{-1}$ (while $\tau(e^{-1}) = 83$ ps for $\Delta G_1 = 700\ cm^{-1}$) at T = 300 K and $\tau(e^{-1}) = 270$ ps for $\Delta G_1 = 1000\ cm^{-1}$ at 20K. In the triple mutant the effects of superexchange are marked ($F_{sup} \simeq 0.2$) at room temperature and dominant ($F_{seq} \simeq 1.0$) at low temperature.

## References


1. The Photosynthetic Bacterial Reaction Center. Structure and dynamics, eds. J. Breton and A. Vermeglio (Plenum NATO ASI Series, New York, 1988).
2. Reaction Centers of Photosynthetic Bacteria, ed. M.E. Michel-Beyerle (Springer Verlag, Berlin, 1990).
3. The Photosynthetic Bacterial Reaction Center II, eds. J. Breton and A. Verméglio NATO-ASI series A (Life Sciences, Plenum Press, New York, 1992) vol. 237.

4. The Photosynthetic Reaction Center, eds. J. Deisenhofer and J.R. Norris (Academic Press, San Diego, 1993).
5. H. Frauenfelder, G.U. Nienhaus and J.B. Johnson, Ber. Bunsenges. Phys. Chem. 95 (1991) 272.
6. H. Frauenfelder and P.G. Wolynes, Phys. Roday 47 (1994) 58.
7. Spin Glasses and Biology (World Scientific Press, Singapore, 1992).
8. C. Kirmaier, D. Holten and W.W. Parson, Biochim. Biophys. Acta 810 (1985) 49.
9. C. Kirmaier and D. Holten, Proc. Natl. Acad. Sci. USA 87 (1990) 355, 2.
10. P. Hamm, K.A. Gray, D. Oesterhelt, R. Feick, H. Scheer and W. Zinth, Biochim. Biophys. Acta 1142 (1993) 99.
11. Y. Jia, T.J. DiMagno, C.-K. Chan, Z. Wang, M. Du. D.K. Hanson, M. Schiffer, J.R. Norris, G.R. Fleming and M.S. Popov, J. Phys. Chem. 97 (1993) 13180.
12. M. Du, S.J. Rosenthal, X. Xie, T.J. DiMagno, M. Schmidt, D.K. Hanson, M. Schiffer, J.R. Norris and G.R. Fleming, Proc. Natl. Acad. Sci. USA 89 (1992) 8517.
13. M.G. Müller, K. Griebenow and A.R. Holzwarth, Chem. Phys. Lett. 199 (1992) 465.
14. T. Häberle, H. Lossau, M. Friese, G. Hartwich, A. Ogrodnik, H. Scheer and M.E. Michel-Beyerle (this volume).
15. G. Hartwich, Ph.D. Thesis, TU München (1994).
16. G. Hartwich, M. Friese, H. Scheer, A. Ogrodnick and M.E. Michel-Beyerle, Chem. Phys. 197 (1995) 423.
17. J.C. Williams, R.G. Alden, H.A. Murchison, J.M. Peloquin, N.W. Woodbury and J.P. Allen, Biochemistry 31 (1992) 11029.
18. H.A. Murchison, R.G. Alden, J.P. Allen, J.M. Peloquin, A.K.W. Taguchi, N.W. Woodbury and J.C. Williams, Biochemistry 32 (1993) 3498.
19. N.W. Woodbury, S. Lin, X. Lin, J.M. Peloquin, A.K.W. Taguchi, J.C. Williams and J.P. Allen, Chem. Phys. 197 (1995) 506.
20. A. Ogrodnik, W. Keupp, M. Volk, G. Aumeier and M.E. Michel-Beyerle, J. Phys. Chem. 98 (1994) 3432.
21. G.J. Small, J.M. Hayes and R. Silbey, J. Phys. Chem. 96 (1992) 7499.
22. N.W. Woodbury, M. Becker, D. Middendorf and W.W. Parson, Biochemistry 24 (1985) 7516.
23. S.F. Fischer, I. Nussbaum and P.O.J. Scherer, in: Antennas and Reaction Centers of Photosynthetic Bacteria, ed. M.E. Michel-Beyerle (Springer, Berlin, 1985) p. 256.
24. A. Ogrodnik, N. Remy-Richter, M.E. Michel-Beyerle and R. Feick, Chem. Phys. Lett. 135 (1987) 576.
25. J.R. Norris, D.E. Budil, D.M. Tiede, J. Tang, S.V. Kolaczkowski, C.H. Chang and M. Schiffer, in: Progress in Photosynthetic

Research, ed. J. Biggins (Martinus Nijhoff, Dordrecht, 1987) vol. I, p. 1.4.363.
26. M.E. Michel-Beyerle, M. Plato, J. Deisenhofer, H. Michel, M. Bixon and J. Jortner, Biochim. Biophys. Acta 932 (1988) 52.
27. M. Bixon, J. Jortner, M. Plato and M.E. Michel-Beyerle, in: The Photosynthetic Bacterial Reaction Center. Structure and Dynamics, eds. J. Breton and A. Vermeglio (Plenum NATO ASI Series, New York, 1988) p. 399.
28. M. Plato, K. Möbius, M.E. Michel-Beyerle, M. Bixon and J. Jortner, J. Am. Chem. Soc. 110 (1988) 7279.
29. M. Bixon, M.E. Michel-Beyerle and J. Jortner, Isr. J. Chem. 28 (1988) 155.
30. M. Bixon, J. Jortner, M.E. Michel-Beyerle and A. Ogrodnik, Biochim. Biophys. Acta 977 (1989) 273.
31. R.A. Friesner and Y. Won, Biochim. Biophys. Acta 977 (1989) 99.
32. R.A. Marcus, Isr. J. Chem. 28 (1988) 205.
33. R. Haberkorn, M.E. Michel-Beyerle and R.A. Marcus, Proc. Natl. Acad. Sci. USA 70 (1979) 4185.
34. R.A. Marcus, Chem. Phys. Lett. 133 (1987) 471.
35. S.V. Chekalin, Ya.A. Matveetz, A.Ya. Shkuropatov, V.A. Shuvalov and A.P. Yartzev, FEBS Lett. 216 (1987) 245.
36. R.A. Marcus, Chem. Phys. Lett. 146 (1977) 13.
37. S. Creighton, J.-K. Hwang, A. Warshel, W.W. Parson and J. Norris, Biochem. 27 (1988) 774.
38. W. Holzapfel, U. Finkele, W. Kaiser, D. Oesterhelt, H. Scheer, H.U. Stilz and W. Zinth, Chem. Phys. Lett. 160 (1989) 1.
39. W. Holzapfel, U. Finkele, W. Kaiser, D. Oesterhelt, H. Scheer, H.U. Stilz and W. Zinth, Proc. Nat. Acad. Sci. USA 87 (1990) 5168.
40. C. Lauterwasser, U. Finkele, H. Scheer and W. Zinth, Chem. Phys. Lett. 183 (1991) 471.
41. T. Arlt, S. Schmidt, W. Kaiser, C. Lauterwasser, M. Meyer, H. Scheer and W. Zinth, Proc. Natl. Acad. Sci. USA 90 (1993) 11757.
42. P.M. Schmidt, T. Arlt, P. Hamm, H. Huber, T. Nägele, J. Wachtveitel, M. Meyer, H. Scheer and W. Zinth, Chem. Phys. Lett. 223 (1994) 116
43. M. Bixon, J. Jortner and M.E. Michel-Beyerle, Biochim. Biophys. Acta 1056 (1991) 301.
44. M. Bixon, J. Jortner and M.E. Michel-Beyerle, reference 3 (1992) 291.
45. M. Bixon, J. Jortner and M.E. Michel Beyerle, Chem. Phys. 197 (1995) 389.
46. A. Ogrodnik, M. Friese, P. Gast, A.T. Hoff and M.E. Michel-Beyerle, Biophys. J. (1966) in press.
47. M. Bixon, J. Jortner and M.E. Michel-Beyerle (this volume).

48. M. Bixon and J. Jortner, Chem. Phys. 176 (1993) 467.
49. U. Eberl, M. Gilbert, W. Keupp, T. Langenbacher, J. Siege, I. Sinning, A. Ogrodnik, S.J. Robles, J. Breton, D.C. Youvan and M.E. Michel-Beyerle, reference 15, p. 253.
50. M. Bixon, J. Jortner and M.E. Michel-Beyerle, Z. Phys. Chem. 180 (1993) 193.
51. A. Ogrodnik, Mol. Cryst. Liq. Cryst. 230 (1993) 35.
52. G. Hartwich, H. Lossau, A. Ogrodnik and M.E. Michel-Beyerle (this volume).
53. J.M. Peloquin, Jo Ann C. Williams, X. Lin, R.G. Alden, A.K.W. Taguchi, J.P. Allen and N.W. Woodbury, Biochemistry 33 (1994) 8089.
54. U. Eberl, A. Ogrodnik and M.E. Michel-Beyerle, Z. Naturf. 45a (1995) 763.
55. M.G. Müller, K. Griebenow and A.R. Holzwarth, Chem. Phys. Lett. 199 (1992) 465.
56. N. Raja, S. Reddy, S.V. Kolaczkowski and G.J. Small, J. Phys. Chem. 97 (1993) 6934.
57. C.A. Wraight and R.K. Clayton, Biochim. Biophys. Acta 333 (1973) 246.
58. M. Volk, G. Scheidel, A. Ogrodnik, R. Feick and M.E. Michel-Beyerle, Biochem. Biophys. Acta 1058 (1991) 217.
59. P. Müller, A. Ogrodnik and M.E. Michel-Beyerle (to be published).

# Structural and Functional Properties of the State $P^{+\cdot}Q^{-\cdot}$ from Transient EPR Spectroscopy

A. van der Est and D. Stehlik

Fachbereich Physik, Freie Universität Berlin,Arnimallee 14, 14195 Berlin, Germany

**Abstract.** The structure of the charge separated state $P^{+\cdot}Q^{-\cdot}$ reaction centres (RC's) of *Rhodobacter sphaeroides* R-26 in which, the non-heme iron has been replaced by zinc (Zn-bRC's) is studied using transient EPR spectroscopy. A qualitative description of the spin polarized transient EPR spectra of $P^{+\cdot}Q^{-\cdot}$ is presented and used to draw conclusions about the relative orientations of $P_{865}^{+\cdot}$ and $Q_A^{-\cdot}$. Experimental and calculated spectra are presented for fully deuterated and protonated Zn-bRC's at three microwave frequencies: X-band (9 GHz), K-band (24 GHz) and W-band (95 GHz). The simulated spectra are calculated on the basis of independently determined experimental parameters. From the good agreement between the experimental spectra and those calculated using the ground state x-ray structure, it is concluded that no significant light induced structural changes occur. K-band spectra at 300 K and 50 K are compared and it is suggested that the narrowing of the spectrum at room temperature is a result of changes in the hyperfine couplings due to motion of the chromophores as well as a change in the charge distribution on $P_{865}^{+\cdot}$.

**Keywords.** Transient EPR, structure, quinone orientation, spin polarization, *Rb. sphaeroides*, photosystem I

# 1 Introduction

Photosynthetic reaction centres (RC's) have been studied extensively using EPR spectroscopy. This is not only because the EPR spectra of the chromophore radical ions can be used to determine their magnetic, electronic and structural properties but also because the electron transfer induced by a laser pulse produces effects closely related to those induced by the microwave pulses in a pulsed EPR experiment. In time resolved experiments, the first charge separated state which is long lived enough to be detected by EPR methods is $P^{+\cdot}Q^{-\cdot}$. This radical pair is created in a spin state which is far from equilibrium immediately following the laser pulse and its transient EPR signal can be followed as it relaxes. These EPR transients are dependent on the geometry of the radical pair and can be analysed to obtain structural information about the charge separated state [1, 2]. The spin dynamics of this system are also of fundamental interest to magnetic resonance spectroscopists because features such as quantum beats [3–6], nuclear coherences [6,7], orientation dependent transient nutations [8,9] and out-of-phase spin echos [10–12] occur which are not easily produced in other systems. Although only two weakly coupled electron spins are involved, a complete description of the EPR signals is a rather complicated time dependent density matrix calculation. However, if the spin polarized transient EPR spectra obtained by integrating the signals in a given time window are considered and the lifetime of the precursor states is $<\sim 500$ ps, then the time dependence of the problem does not need to be taken explicitly into account. This simplifies the calculation considerably and makes the dependence of the data on the structural parameters more obvious.

In this contribution, we will present a non-mathematical description of the spin polarized EPR spectra of $P^{+\cdot}Q^{-\cdot}$ and will show that important structural features of the pair can be deduced by simple inspection. Our goal, here, is to make the analysis of the data understandable to as wide an audience as possible. Following this qualitative description of the form of the spectra, we will compare the experimental results for *Rhodobacter (Rb.) sphaeroides* with calculated spectra. Because the magnetic and structural properties of the chromophores in the reaction centres of this species are known in considerable detail, it is possible to calculate the spectra with essentially no adjustable parameters. This not only allows us to demonstrate the validity of our approach but also permits more subtle questions to be addressed such as whether light inducted structural changes occur. The most complete set of experimental and calculated spectra of $P_{865}^{+\cdot}Q_A^{-\cdot}$ available will be presented. This includes fully deuterated and protonated Zn-substituted RC's (D-Zn-bRC's and H-Zn-bRC's respectively) at X-band (9 GHz), K-band (24 GHz) and W-band (95 GHz). We will also compare room temperature and low temperature K-band spectra of H-Zn-bRC's. These results indicate that no significant light induced reorientation of $Q_A^{-\cdot}$ occurs. However, the spectra show that the contribution from $P_{865}^{+\cdot}$ changes considerably with temperature. Some possible origins of this temperature dependence will be discussed.

# 2 Spin Polarized EPR Spectra of $P^{+\cdot}Q^{-\cdot}$

The spin polarized EPR spectrum of the state $P^{+\cdot}Q^{-\cdot}$ has been described in detail in several publications [1,13–19]. This state is referred to as a correlated coupled radical pair (CCRP) and consists of two weakly coupled spins. The determination of the energy levels of such a system is treated in most introductions to magnetic resonance and is usually the first matrix eigenvalue problem encountered by physical chemistry or physics students. One obtains the four eigenstates given in Fig. 1 and there are four allowed EPR transitions as shown. This leads to the familiar doublet of doublets spectrum shown in the top part of Fig. 2A. This spectrum applies at thermal equilibrium with a

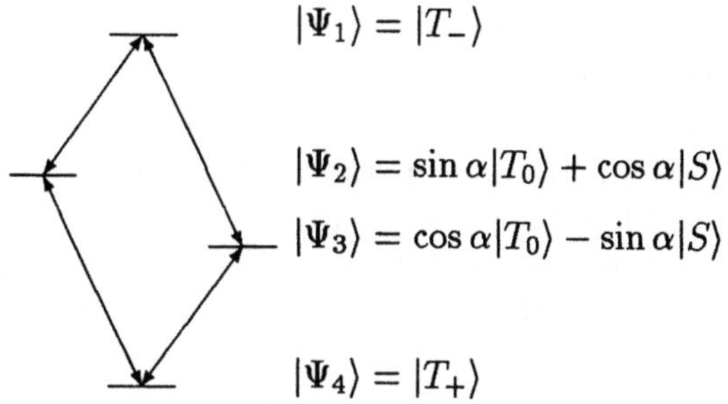


**Fig. 1** Energy level diagram for a weakly coupled two spin system

Boltzmann population distribution. However, the state $P^{+\cdot}Q^{-\cdot}$ is generated from the excited singlet state, $P^*$, on a picosecond timescale. Because of this, the eigenstates in Fig. 1 are populated according to their singlet character so that only $|\Psi_2\rangle$ and $|\Psi_3\rangle$ are occupied. The spectrum which results from this population distribution is shown in the lower part of Fig. 2A. As can be seen, the intensity of the lines increases in comparison to the thermal equilibrium spectrum and two of the transitions become emissive. Because the singlet state precursor has no net spin polarization the intensity of the absorptive and emissive contributions are equal. This is the spectrum one would observe for a single orientation of the RC's relative to the magnetic field. However,

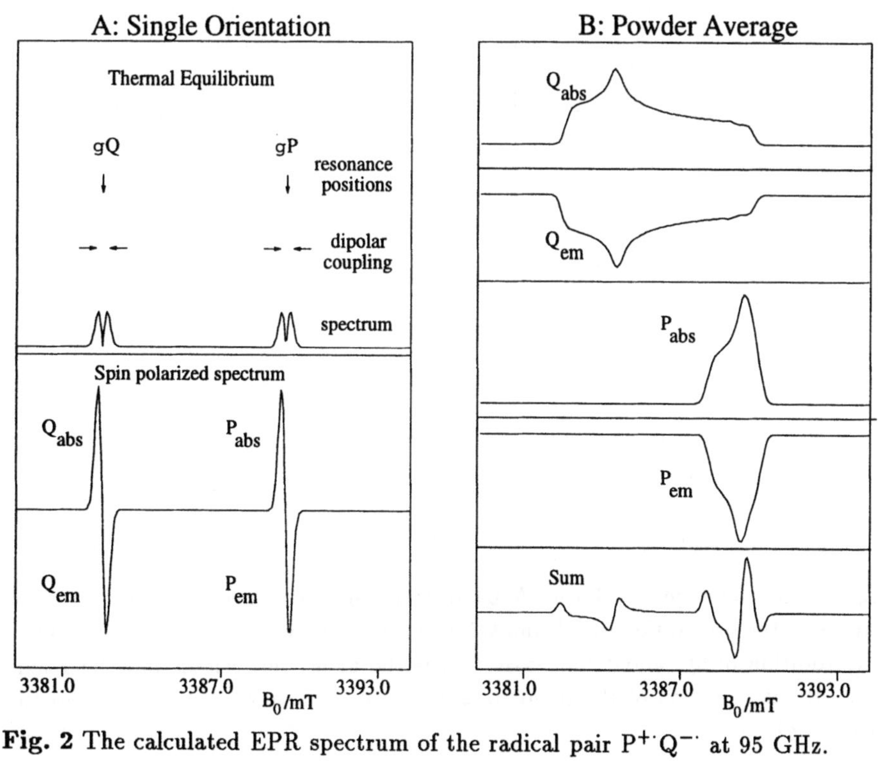


**Fig. 2** The calculated EPR spectrum of the radical pair $P^{+\cdot}Q^{-\cdot}$ at 95 GHz.

in liquid or frozen solutions the spectrum is a sum over the random distribution of orientations present in the sample. In Fig. 2B this sum has been carried out individually for each of the four lines in Fig. 2A. The form of the four spectra is common in solid state NMR and is referred to as a powder pattern. The width of the powder pattern is determined by the anisotropy of the corresponding $g$-tensor and it is clear from the figure that this anisotropy is much larger for $Q^{-\cdot}$ than for $P^{+\cdot}$. The sum of the four powder patterns shown at the bottom of Fig. 2B is very sensitive to small differences between the oppositely polarized contributions which almost cancel one another. For this reason the observed spectra are very sensitive to the values of the magnetic and structural parameters which describe them. The most important of these is the orientation of the two $g$-tensors, $g$P and $g$Q, relative to the dipolar vector, $z_d$, which joins $P^{+\cdot}$ and $Q^{-\cdot}$. In order to demonstrate this dependence, two series of calculated spectra are shown in Fig. 3. On the left side of the figure the orientation of $Q^{-\cdot}$ has been varied and on the right the same is done for $P^{+\cdot}$. The relative orientation of $z_d$ and the $g$-tensor of

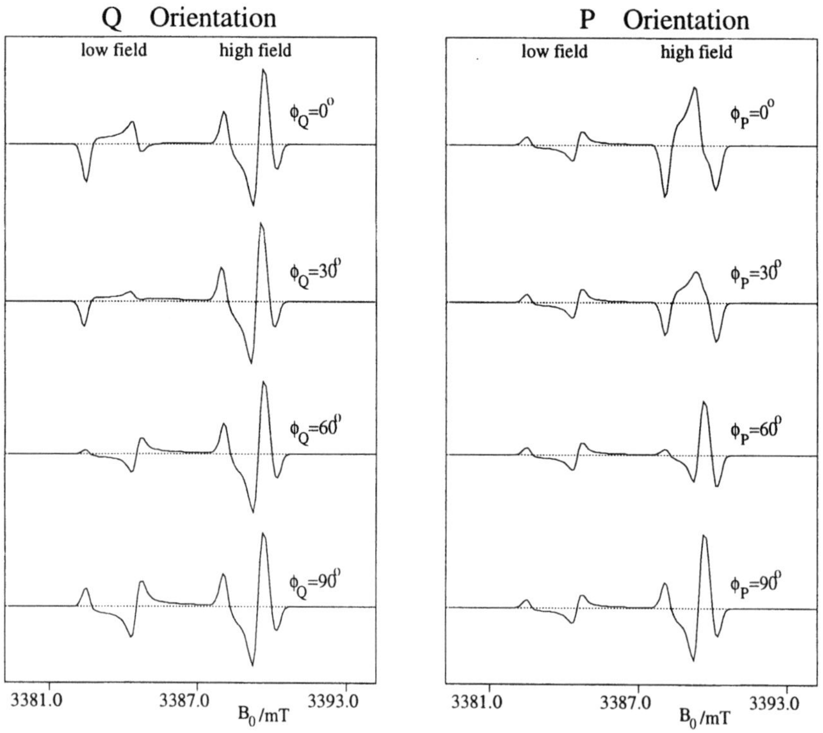


**Fig. 3** The dependence of the W-band transient EPR spectrum of $P^{+\cdot}Q^{-\cdot}$ on changes in the orientation of $P^{+\cdot}$ and $Q^{-\cdot}$ relative to the vector joining $P^{+\cdot}$ and $Q^{-\cdot}$. Left: rotation of $Q^{-\cdot}$ about its $z$-axis. $\phi_Q$ is the angle between $z_d$ and the $x$-axis of $g$Q. Right: rotation of $P^{+\cdot}$ about its $z$-axis. $\phi_P$ is the angle between $z_d$ and the $x$-axis of $g$P.

either $Q^{-\cdot}$ or $P^{+\cdot}$ can be described by a polar angle $\theta$ and an azimuthal angle $\phi$. In both cases the $z$-axis of the $g$-tensor being varied has been chosen to be perpendicular to $z_d$ i.e. $\theta = 90°$. The angle $\phi$ then describes the orientation of the $x$- and $y$-axes relative to $z_d$. Spectra ranging from $x \,||z_d$, to $y \,||z_d$, are shown. All other parameters have been held constant at the values obtained from independent experiments as tabulated in [20, 21] (see also Fig. 4). From Fig. 3 it is apparent that changes in the orientation of the $x$- and $y$-axes of $Q^{-\cdot}$ affect only the low field part of the spectra whereas changes in the orientation of $P^{+\cdot}$ affect only the high field region. In both cases, the respective parts of the spectrum start on the left with an emissive contribution when the corresponding $x$-axis is parallel to $z_d$. In contrast, an absorptive contribution is observed when the $x$-axis is perpendicular to $z_d$. Thus, from the sign of the spin polarization one can make an estimate of the orientation fo the $x$-axes of the two $g$-tensors (radical ions) relative to $z_d$ (the axis joining them). The spectra in Fig. 3 have been calculated for a microwave frequency of 95 GHz (W-band) because the spectral components are more clearly resolved at higher frequencies and the orientation dependence is thus more easily demonstrated. The same orientation dependence is observed at lower frequencies [21, 22], although the contributions to the spectra become overlapped so that the interpretation is not quite as straight forward.

## 3 Results for $P_{865}^{+\cdot}Q_A^{-\cdot}$ in *Rb. sphaeroides* R-26

The most extensively characterized photosynthetic reaction centre is that of *Rb. sphaeroides* R-26 for which three independent X-ray structure determinations have been performed [23–25]. The native reaction centre contains a paramagnetic Fe atom which is coupled to $Q_A^{-\cdot}$. Thus, the state $P_{865}^{+\cdot}Q_A^{-\cdot}$ cannot be observed without modifying the RC. Spin polarized EPR spectra of $P_{865}^{+\cdot}Q_A^{-\cdot}$ have been reported for iron depleted RC's [26]. However, removal of the iron slows the electron transfer rate to $Q_A^{-\cdot}$ [27]which complicates the analysis of the spectra considerably [28]. In RC's in which the Fe has been replaced with diamagnetic Zn, the electron transfer rate is the same as in native RC's [27, 29] so that both the interference of the Fe atom and the problem of the long lived intermediate state are avoided. The magnetic properties of $P_{865}^{+\cdot}$ [30] and $Q_A^{-\cdot}$ [31] have been studied using high field EPR. Thus, essentially all of the parameters required to simulate the spin polarized EPR spectra of $P_{865}^{+\cdot}Q_A^{-\cdot}$ are available from independent experiments. In the case of $P_{865}^{+\cdot}$, cw-EPR studies on single crystals [30] yield a fourfold ambiguity in the assignment of the $g$-tensor axes to the molecular frame. Two of these possibilities could be ruled out by simulation of the spin polarized K-band spectra of $P_{865}^{+\cdot}Q_A^{-\cdot}$ in D-Zn-bRC's [20]. Subsequent W-band transient EPR experiments on the same sample [32] yielded an unambiguous choice for the orientation of the $g$-tensor axes.

In Fig. 4, experimental and calculated spectra of $P_{865}^{+\cdot}Q_A^{-\cdot}$ are shown at

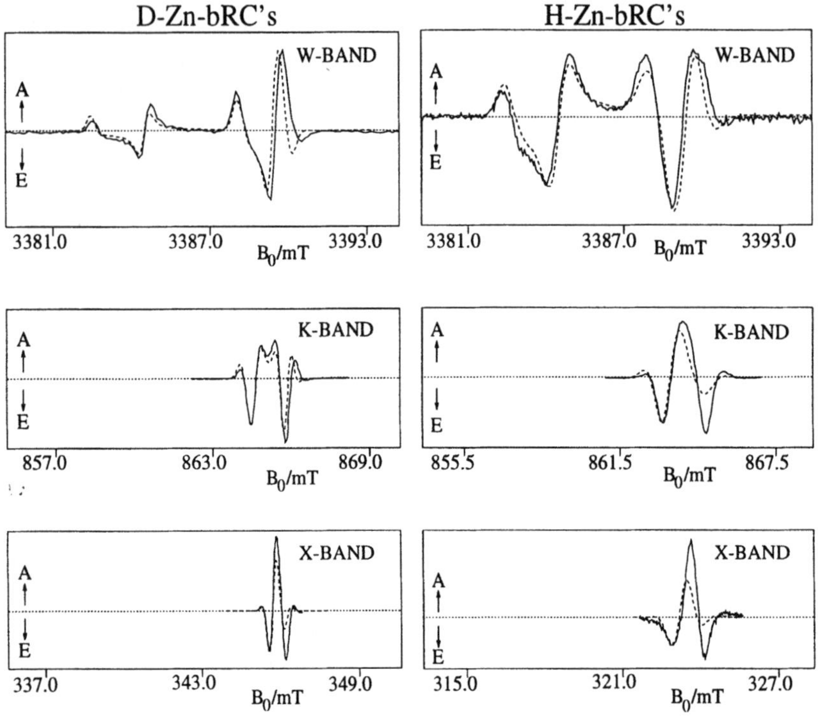


**Fig. 4** Comparison of the spin polarized EPR spectra of $P^{+\cdot}_{865}Q^{-\cdot}_{A}$ in deuterated and protonated bRC's from *Rb. sphaeroides* R-26 at three microwave frequencies. Experimental spectra (solid curves): W-band: field swept two pulse echo spectra (see [32]), K-band and X-band: transient EPR spectra taken 1-4 $\mu$s after the laser pulse see ([20] and [22]). Calculated spectra (dashed curves). Simulation parameters (see [21,32]): $gP^{+\cdot}_{865}$ 2.0033, 2.0025, 2.0021, $gQ^{-\cdot}_{A}$ 2.0066, 2.0054, 2.0022; D=-0.124 mT; J=0.0 mT; $\Delta B_Q$ = 0.30 mT (D-Zn-bRC's), 0.7 mT (H-Zn-bRC's); $\Delta B_P$ = 0.33 mT (D-Zn-bRC's), 1.0 mT (H-Zn-bRC's); Euler angles: $\alpha = 23.0°, \beta = 119.3°, \gamma = 18.3°, \theta = 71.6°, \phi = 68.5°$

X-, K- and W-band (9, 24 and 95 GHz, respectively) for H-Zn-bRC's and D-Zn-bRC's. The figure summarizes the results of several previous publications [20, 22, 32].

By comparison with the left side of Fig. 3 one sees immediately that the quinone $x$-axis makes an angle of roughly 60° with $z_d$ whereas the $g$-tensor $x$-axis of $P^{+\cdot}_{865}$ is roughly perpendicular to $z_d$. This prediction is in good agreement with the values of 69° and 76° respectively obtained from the most recent X-ray structure [25].

The calculated spectra in Fig. 4 are based on the ground state x-ray structure whereas the experimental spectra correspond to the charge separated state. Thus, the good agreement between the calculated spectra (dashed curves) and the experimental results (solid curves), particularly at W-band, shows that no appreciable light induced structural changes occur. At present, such a comparison of the structure of the two states is limited by the accuracy of the ground state structure. Closer inspection of Fig. 4 shows that the agreement is not as good at X-band particularly with H-Zn-bRC's. There are several possible explainations for this behaviour such the influence of a small J coupling [33]. However, the dependence on the microwave frequency and whether the sample is deuterated or not suggests that the hyperfine couplings (hfc's) to the nuclear spins are involved. Because the hfc's do not depend on the strength of the external they dominate the X-band spectra whereas at W-band the $g$-anisotropy plays a more important role. Thus, the poor agreement at X-band likely occurs because the hfc's have not been taken explicitly into account but rather taken as a Gaussian lineshape. The values of the linewidths, $\Delta B_P$ and $\Delta B_Q$, used in the simulations are given in the caption to Fig. 4. For the deuterated samples they are roughly the same whereas for the H-Zn-bRC's $\Delta B_P$ must be taken to be considerably larger than $\Delta B_Q$. This is probably due to the fact that the anisotropy in the hfc's for $P^{+\cdot}$ and $Q^{-\cdot}$ are different. In addition, strong nuclear coherence oscillations are observed in these samples at X-band [6] but not at K-band. This difference in the behaviour of the transients is expected because these coherences are particularly strong when the local hyperfine field exactly or nearly cancels the external field as is known to occur at X-band for the nitrogen hfc in $P_{865}^{+\cdot}$ [34]. In the case of exact cancelation, the form of the radical pair spectrum is expected to change considerably.

Fig. 4 demonstrates the importance of measuring the EPR spectra at as many microwave frequencies as possible. Using only the data at X-band, it is not clear which of the parameters need to be adjusted in order to obtain better agreement. Similarly, it is not obvious from the W-band spectra that an adjustment of the parameters is required. An additional indication that the anistropic hfc's play an important role in determining the spectra comes from the temperature dependence of the K-band results. In Fig. 5 the spectra of H-Zn-bRC's are shown at 50 K and at 300 K. As can be seen, there is a noticeable narrowing of the high field part of the spectrum at room temperature. From ENDOR experiments on $P^{+\cdot}$ it is known that the hfc's are considerably anisotropic [35] but that in RC solutions at room temperature the anisotropy is averaged by the molecular motions [36, 37]. The narrowing in Fig. 5 is certainly at least partly caused by this averaging. However, it is important to point out that the motions responsible for the averaging of the hfc's are different from those which average the $g$-anisotropy and the dipolar coupling. As discussed in [22] if the whole RC tumbles at a rate rapid enough to average the anisotropic hfc's then no spin polarized EPR spectrum should be observable. Thus, taken together the ENDOR and transient EPR results suggest that the averaging of the hfc's is due to local fluctuations of the chromophores and protein surroundings. Another effect which has been observed in ENDOR and ESEEM experiments on $P_{865}^{+\cdot}$ is that upon freezing RC's without cryoprotectants the charge on $P_{865}^{+\cdot}$ becomes localized on the L side of the dimer [34, 37]. This leads to larger hfc's in frozen solution in

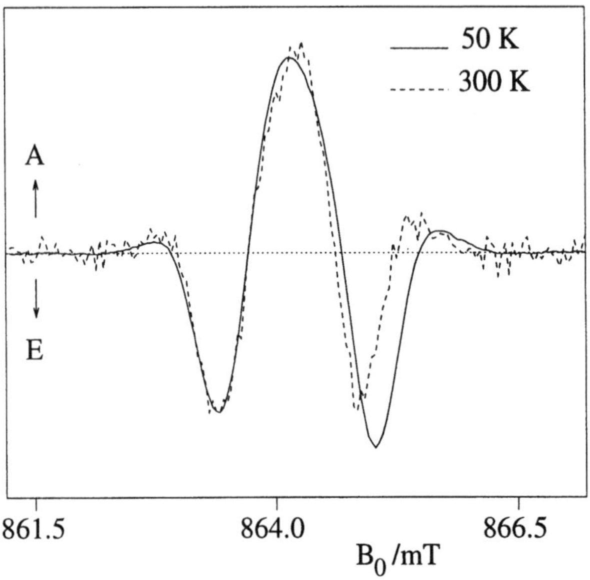


**Fig. 5** Comparison of the K-band transient EPR spectra of $P^{+\cdot}_{865}Q^{-\cdot}_{A}$ in H-Zn-bRc's at room temperature and low temperature. Both spectra correspond to a time gate 1-4 $\mu$s after the laser pulse. Microwave frequency: 24.2256 GHz, microwave power: 6.3 mW (300 K) and 0.32 mW (50 K)

agreement with the broader transient EPR spectrum observed at 50 K. In addition, the $z$ component of the $g$-tensor of $Q^{-\cdot}_{A}$ also contributes to the high field part of the $P^{+\cdot}_{865}Q^{-\cdot}_{A}$ spectrum. In recent Q-band experiments on $Q^{-\cdot}_{A}$ in single crystals [38] this component shifts to lower values (higher field) upon freezing. The differences in Fig. 5 must also be related to this effect. However, the magnitude of the $g$-factor shift is too small to account entirely for the narrowing in Fig. 5.

# 4 Comparison of bRC's and PS I

The spectra in Fig. 4 demonstrate that the general form of the polarization pattern can be used to draw conclusions about the relative orientation of $P^{+\cdot}$ and $Q^{-\cdot}$. This is of particular importance for the analysis of the spectra of $P^{+\cdot}_{700}A^{-\cdot}_{1}$ in photosystem I (PS I) for which no precise x-ray data are available yet. A comparison of the spectra of $P^{+\cdot}_{865}Q^{-\cdot}_{A}$ and $P^{+\cdot}_{700}A^{-\cdot}_{1}$ shows that the low field regions are considerably different [21, 22] which indicates that the orientation of the dipolar vector relative to the quinone accetors is different in bRC's and PS I. From analyses of the transient EPR data it has been concluded [5, 18, 22] that the angle between the quinone $x$-axis and the dipolar vector, $z_d$, is between 0° and 30° in PS I. The high field region of the spectra which is dominated by $P^{+\cdot}$ is very similar for the two RC's. Thus, one can conclude that the position of $Q^{-\cdot}$ is relative to $P^{+\cdot}$ is similar in both cases but that the orientation of the two quinones in their respective binding sites is different. The reason for this appears to be a marked difference in the binding of $A_1$ and $Q_A$. Recent quinone exchange experiments [21, 39] produced the remarkable result that different quinones have different orientations in PS I but all have essentially the same orientation in Zn-bRC's. Moreover, the quinone $g$-factors are different in the two RC's such that the anisotropy of the $g$-tensors is considerably larger in PS I. These results can be accounted for [21] by assuming that the quinone binding in PS I is due to $\pi-\pi^*$ interactions and not due to H-bonds as is clearly the case in bRC's.

# 5 Summary

The transient EPR spectra of $P^{+\cdot}Q^{-\cdot}$ are sensitive to the orientation of the two chromophores and we have shown that the geometry of the radical pair can be estimated by inspection of the spectra. The W-band spectra are particularly well suited for this because the individual $g$-tensor components are well separated and the hyperfine couplings do not play as important a role as at lower frequencies. However, in order to obtain reliable values for the structural and magnetic parameters by simulating the transient EPR spectra, it is important that data at several microwave frequencies be compared.

# 6 Acknowledgements

All of the results presented here were obtained from samples provided by W. Lubitz (T.U. Berlin) and prepared with the help of E. Abresch (U.C. San Diego). The W-band spectra were measured in collaboration with T. Prisner and K. Möbius (F.U. Berlin) and the analysis of the spectra was done with the help of R. Bittl (T.U. Berlin). This work was supported by grants from the Deutsche Forschungsgemeinschaft (Sfb 312 and Sfb 337).

# References

1 Hore, P. J. (1989) in Advanced EPR, Applications in Biology and Biochemistry (Hoff, A.J., ed.), Ch. 12, pp. 405–440, Elsevier, Amsterdam.

2 Snyder, S. W. and Thurnauer, M. C. (1993) in The Photosynthetic Reaction Center (Norris, J. R. and Deisenhofer, J., eds.), Vol. II Ch. 11 Academic Press, New York.

3 Salikhov, K. M., Bock, C. H. and Stehlik, D. (1990) Appl. Magn. Reson. 1,195–211.

4 Bittl, R. and Kothe, G. (1991) Chem. Phys. Letters 177,547–553.

5 Kothe, G., Weber, S., Bittl, R., Ohmes, E., Thurnauer, M. C. and Norris, J. R. (1991) Chem. Phys. Lett. 186,474–480.

6 Bittl, R., van der Est, A., Kamlowski, A., Lubitz, W. and Stehlik, D. (1994) Chem. Phys. Lett. 226,349–358.

7 Weber, S. (1994) PhD thesis, Universität Stuttgart.

8 Gierer, M., van der Est, A. and Stehlik, D. (1991) Chem. Phys. Lett. 186,238–247.

9 Timmel, C.R. and Hore, P.J. (1994) Chem. Phys. Lett. 226,144–150.

10 Thurnauer, M.C. and Norris, J.R. (1980) Chem. Phys. Lett. 76, 577–579.

11 Tang, J., Thurnauer, M.C. and Norris, J.R. (1994) Chem. Phys. Lett. 219,283–291.

12 Moënne-Loccoz, P., Heathcote, P., Maclachlan, D., Berry, M., Davis, I.H. and Evans, M.C.W. (1994) Biochemistry 33,10037–10042.

13 Thurnauer, M. C. and Norris, J. R. (1980) Chem. Phys. Lett. 76,557–561.

14 Thurnauer, M. C. and Meisel, D. (1983) J. Am. Chem. Soc. 105,3729–3731.

15 Hore, P. J., Hunter, D. A., McKie, C. D. and Hoff, A. J. (1987) Chem. Phys. Lett. 137,495–500.

16 Buckley, C. D., Hunter, D. A., Hore, P. J. and McLauchlan, K. A. (1987) Chem. Phys. Lett. 135,307–312.

17 Closs, G. L., Forbes, M. D. E. and Norris, J. R. (1987) J. Phys. Chem. 91,3592–3599.

18 Stehlik, D., Bock, C. H. and Petersen, J. (1989) J. Phys. Chem. 93,1612–1619.

19 Norris, J. R., Morris, A. L., Thurnauer, M. C. and Tang, J. (1990) J. Chem. Phys. 92,4239–4249.

20 van der Est, A.J., Bittl, R., Abresch, E.C., Lubitz, W. and Stehlik, D. (1993) Chem. Phys. Lett. 212,561–568.

21 van der Est, A., Sieckmann, I., Lubitz, W. and Stehlik, D. (1995) Chem. Phys. 194,349–359.

22 Füchsle, G., Bittl, R., van der Est, A., Lubitz, W. and Stehlik, D. (1993) Biochim. Biophys. Acta 1142,23–35.

23 Chang, C. H., El-Kabbani, O., Tiede, D., Norris, J. and Schiffer, M. (1991) Biochemistry 30,5352–5360 For earlier work see references cited therein. The coordinates are registered at the Brookhaven Protein Data Bank, preliminary entry P2RCR.

24 Komiya, H., Yeates, T. O., Rees, D. C., Allen, J. P. and Feher, G. (1988) Proc. Natl. Acad. Sci. USA 85,9012–9017 This is part 6 in a series of papers. References for parts 1–5 are given therein. The coordinates are registered at the Brookhaven Protein Data Bank, preliminary entries P1RCR and P4RCR.

25 Ermler, U., Fritzsch, G., Buchanan, S. and Michel, H. (1992) in Research in Photosynthesis (Murata, N., ed.), Vol. I p. 341, Kluwer Academic Publishers, Dordrecht.

26 Feezel, L. L., Gast, P., Smith, U. H. and Thurnauer, M. C. (1989) Biochim. Biophys. Acta 974,149–155.

27 Kirmaier, C., Holton, D., Debus, R. J., Feher, F. and Okamura, M. Y. (1986) Proc. Natl. Acad. Sci. USA 83,6407–6411.

28 Morris, A.L., Snyder, S.W., Zhang, Y., Thurnauer, M.C., Dutton, P.L., Robertson, D. E. and Dutton, M.R. (1995) J. Phys. Chem. 99,3854–3866.

29 Debus, R. J., Feher, G. and Okamura, M. Y. (1986) Biochemistry 25,2276–2287.

30 Klette, R., Törring, J., Plato, M., Bönigk, B., Lubitz, W. and Möbius, K. (1993) J. Phys. Chem. 97,2015–2020.

31 Burghaus, O., Lubitz, W., Plato, M., MacMillan, F., Rohrer, M. and Möbius, K. (1993) J. Phys. Chem. 97,7639–7647.

32 Prisner, T.F., van der Est, A., Bittl, R., Lubtiz, W., Stehlik, D. and Möbius, K. (1995) Chem. Phys. 194,361–370.

33 van den Brink, J. S., Hulsebosch, R. J., Gast, P., Hore, P. J. and Hoff, A. J. (1994) Biochemistry 33,13668–13677.

34 Käß, H., Rautter, J., Bönigk, B., Höffer, P. and Lubitz, W. (1995) J. Phys. Chem. 99,436–448.

35 Lendzian, F., Huber, M., Isaacson, R.A., Endeward, B., Plato, M., Bönigk, B., Möbius, K., Lubitz, W. and Feher, G. (1993) Biochim. Biophys. Acta 1183,139–160.

36 Lendzian, F., Lubitz, W., Scheer, H., Bubenzer, C. and Möbius, K. (1981) J. Am. Chem. Soc. 103,4635–4637.

37 Rautter, J.R., Lendzian, F., Lubtiz, W., Wang, S. and Allen, J.P. (1994) Biochemistry 33,12077–12084.

38 Isaacson, R.I., Abresch, E.C., Feher, G., Lendzian, F. and Lubitz, W. (1995) Biophys. J. (in press).

39 Sieckmann, I., K.Brettel, , Bock, C., van der Est, A. and Stehlik, D. (1992) Biochemistry 32,4842–4847.

# Light-induced changes in transient EPR spectra of $P_{865}^{+\cdot}Q_A^{-\cdot}$

Robert Bittl*, Stephan G. Zech and Wolfgang Lubitz
Max-Volmer-Institut für Biophysikalische und Physikalische Chemie
Technische Universität Berlin, Straße des 17. Juni 135, 10623 Berlin, Germany

**Abstract.** Transient EPR spectra of the spin-polarized padical pair state $P_{865}^{+\cdot}Q_A^{-\cdot}$ in fully deuterated Zn-substituted reaction centers of *Rhodobacter sphaeroides* R-26 have been recorded for samples frozen in the dark and samples frozen under illumination. Small changes in the spectra were found. These changes show no correlation with changes in the recombination kinetics of $P_{865}^{+\cdot}Q_A^{-\cdot}$ observed by EPR.

**Keywords.** Radical pair geometry, radical pair recombination, light-induced changes, time-resolved EPR

## 1 Introduction

Reaction centers (RCs) of *Rhodobacter sphaeroides* R-26 show a remarkably prolonged lifetime of the radical pair (RP) state $P_{865}^{+\cdot}Q_A^{-\cdot}$ when cooled to cryogenic temperatures under illumination [1]. This change of the lifetime, compared with RCs frozen in the dark, has been attributed to a light-induced structural change in the RCs. A direct investigation of a conformational difference between the RC with $P_{865}Q_A$ in the ground state and the charge-separated state $P_{865}^{+\cdot}Q_A^{-\cdot}$ by x-ray crystallography has not yet been possible due to lack of resolution. In the study of Kleinfeld *et al.* [1] the changed recombination kinetics have been discussed in terms of a shift of the two chromophores resulting in a larger distance between them. A shift of the order of 1 Å was found to be sufficient to model the observed kinetic changes. Crystallographic data suggest that a $Q_A$ shift of this size is possible in the binding pocket [7]. However, spectroscopic investigations by FTIR [2–4], time-resolved FTIR [5] and time-resolved EPR [6] have not shown large changes of the two chromophores involved.

The transient EPR spectroscopy of the highly spin-polarized RP state $P_{865}^{+\cdot}Q_A^{-\cdot}$, as used by van den Brink *et al.* [6] is extremely sensitive to small changes in the relative orientation of the chromophores [8] and almost insensitive to small changes in the distance which conserve the relative orientation [9]. The transient EPR spectra were found to be virtually identical for RCs

*To whom correspondence should be addressed.

frozen in the dark and RCs frozen under illumination [6]. Therefore, it was concluded that no major reorientation of $Q_A$ is responsible for the kinetic changes. For the protonated samples used in ref. [6] the signal/noise ratio of the spectra was poor compared with spectra obtained for fully deuterated samples [10] and did not allow resolution of small changes. The use of fully deuterated RCs results in a considerably increased intensity and spectral resolution of transient EPR experiments (see e.g. ref. [8]). Here we report on experiments similar to those of van den Brink *et al.*, however using fully deuterated Zn-substituted RCs of *Rhodobacter sphaeroides* R-26 containing the H-subunit. The H-subunit was missing in the samples of ref. [6].

## 2 Materials and Methods

For transient EPR of the RP state $P_{865}^{+\cdot}Q_A^{-\cdot}$ the removal or substitution of the non-heme high-spin $Fe^{2+}$ coupled to $Q_A^{-\cdot}$ is of great importance. If the non-heme $Fe^{2+}$ is present only the part ascribed to $P_{865}^{+\cdot}$ of the radical pair spectrum can be observed. The fully deuterated Zn-subsituted RCs of *Rhodobacter sphaeroides* R-26 used here are identical to those in serveral previous studies [10–12] and their preparation has been described in detail [11]. Sample volumes and capillaries were identical to the ones used in [10]. To prevent electron transfer past $Q_A^{-\cdot}$ at room temperature an about 200-fold excess of *o*-phenantroline was added to the samples. The samples were either rapidly frozen in the dark or rapidly frozen under continuous illumination. The continous illumination of the sample while freezing was performed by a focussed 100 W tungsten halogen lamp filtered to allow transmission of light in the range of 830–900 nm. The samples were illuminated for a few seconds at room temperature. A transparent dewar filled with $LN_2$ was then raised over the sample while the illumination continued.

The time-resolved EPR experiments were performed on a BRUKER X-band spectrometer ESP 380E with a ER 4118 X-MD-5W1 dielectric ring probehead and a helium cryostat (Oxford CF935). Recombination kinetics of $P_{865}^{+\cdot}Q_A^{-\cdot}$ were measured using field modulation and lock-in amplification. The spin-polarized RP signals have been observed in the direct detection mode using the cw microwave arm of the bridge. In both operating modes of the spectrometer the kinetic traces were recorded using a LeCroy 9450A 350 MHz digital oscilloscope, equipped with the WP01 waveform package. The control of the whole experiment by the spectrometer data system was faciliated by a home-written acquisition module. Complete 2-D data sets (time/magnetic field) were acquired and processed on a personal computer.

For the light excitation in the EPR experiments a Q-switched Spectra Physics GCR 130 Nd-YAG laser (approx. 10 ns pulse width at 532 nm) was used. The typical pulse energy at the sample was smaller than 5 mJ. Due to the extremely slow recombination kinetics of $P_{865}^{+\cdot}Q_A^{-\cdot}$ after freezing under illumination the laser repetition rate was set to 1.3 Hz (for RCs frozen in the

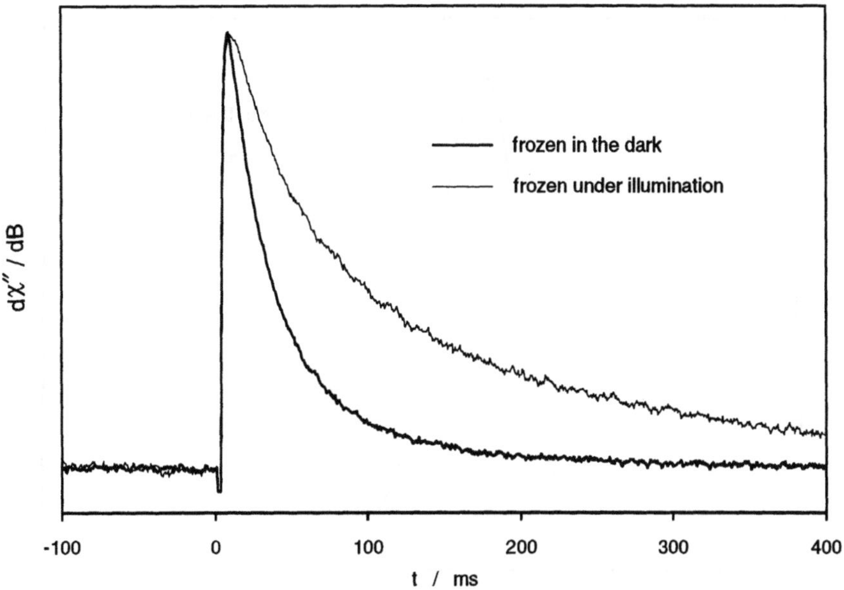


Fig. 1: Recombination kinetics of $P_{865}^{+\cdot}Q_A^{-\cdot}$ for samples frozen in the dark and samples frozen under illumination. The kinetic traces have been recorded by field modulation EPR. Field modulation frequency 100 kHz, modulation amplitude 0.1 mT, microwave power 224 $\mu$W. 4096 traces have been averaged at 1.3 Hz laser repetition rate, $T = 77$ K.

dark and RCs frozen under illumination). Measurements reported here have been performed at 77 K and at 100 K.

# 3 Results and Discussion

The differences in the recombination kinetics of $P_{865}^{+\cdot}Q_A^{-\cdot}$ for RCs frozen in the dark vs. frozen in the light as observed by Kleinfeld *et al.* [1] could be reproduced here and are shown in Fig. 1. The times $\tau$ after which the signals have fallen to $e^{-1}$ of their initial values are $\tau_d = 27$ ms and $\tau_l = 101$ ms for the RCs frozen in the dark and frozen under illumination, respectively. These values are in good agreement with the values reported in ref. [1] ($\tau_d = 25$ ms and $\tau_l = 120$ ms, respectively) for protonated and $Fe^{2+}$ containing RCs. The decay for RCs frozen under illumination is clearly non-exponential. A biexponential fit yields the time constants $\tau_{d1} = 25$ ms (97%), and $\tau_{d2} = 90$ ms and $\tau_{l1} = 46$ ms (49%), $\tau_{l2} = 204$ ms for RCs frozen in the dark and RCs frozen in the light, respectively. Similar results of recombination kinetics measured by time-resolved EPR have been reported by van den Brink *et al.* [6].

At a temperature of 100 K we obtained recombination kinetics identical to those at 77 K. During the measurement time of a few hours, no change of

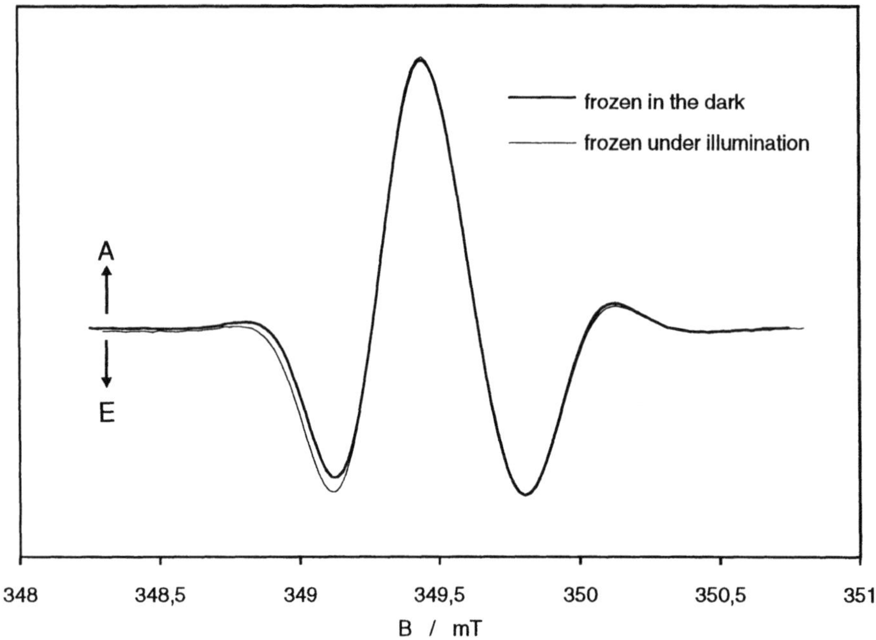


Fig. 2: X-band transient EPR spectra of $P_{865}^{+\cdot}Q_A^{-\cdot}$ for samples frozen in the dark and samples frozen under illumination. The spectra have been extracted from 2-D (time/magnetic field) data sets at a microwave power of 224 $\mu$W and microwave frequency of 9.8 GHz in the time window 0.5–6 $\mu$s after the laser flash. 64 transients have been averaged at each field point at 1.3 Hz laser repetition rate, $T = 100$ K.

the recombination kinetics for both types of samples was observed at 100 K. The change of the recombination kinetics for the samples frozen under illumination above $\sim$ 90 K observed in ref. [1] has to be extremely slow on the time scale of hours at 100 K.

Having established the change in the recombination kinetics for our samples changes of the spin-polarized RP spectra of $P_{865}^{+\cdot}Q_A^{-\cdot}$ can now be investigated. Figure 2 shows the transient spectra for samples frozen in the dark and frozen under illumination.

The spectrum of the sample frozen in the dark is identical to the one recorded previously for a sample from the same batch on a different spectrometer [10]. A clear, but small deviation of the spectrum for the samples frozen under illumination from the spectrum for the sample frozen in the dark is visible. Both types of spectra are extremely well-reproducible, i.e. spectra extracted from different 2-D data sets are indistinguishable. The spectra of the dark frozen and the light-frozen RCs differ only in the low-field part. This part of the spectra can be attributed to $Q_A^{-\cdot}$. Therefore it is tempting to assign the observed difference in the spectra to a structural change of $Q_A$ in the RCs. In Fig. 2 the two spectra have been normalized to equal intensity. This is necessary for a comparison of the spectra. In agreement with ref. [6], the observed signal intensities are smaller for the samples frozen under illumination.

In ref. [6] this effect was discussed in terms of small changes in the relative orientation of $P_{865}^{+\cdot}$ and $Q_A^{-\cdot}$ which do not affect the spectral shape but the signal intensity. Another possible geometry change which would only decrease the signal intensity is an increase of the distance between $P_{865}^{+\cdot}$ and $Q_A^{-\cdot}$ without a change of the relative orientation of the cofactors. However, a shift of $Q_A^{-\cdot}$ in its binding pocket along the axis connecting $P_{865}^{+\cdot}$ and $Q_A^{-\cdot}$ without a reorientation is unlikely. In addition, it is important to note that concomitant with the reduction of the intensity of the spin-polarized signals the signal intensity of kinetic traces observed by field modulation decreased in our experiments. A similar effect seems to have occured also in the experiments of van den Brink *et al.*, where the signal/noise ratio in the kinetic traces is better for the samples frozen in the dark. The intensity of the kinetic traces, however, is to a good approximation independent of the relative orientation and distance of $P_{865}^{+\cdot}$ and $Q_A^{-\cdot}$. Therefore, the intensity decrease of the spin-polarized spectra after freezing under illumination is probably not caused by a change of the RP geometry but by the inhibition of cyclic electron transfer under these conditions in part of the sample.

This observation and the small spectral changes restrict the narrow range of allowed reorientations compatible with the spectra, as discussed in ref. [6], even further. Therefore, we agree with the conclusion of ref. [6] that it is extremely unlikely that the change in the recombination kinetics is caused by such a small reorientation of the quinone.

Experimental evidence for this conclusion is here provided by the observation of a different relaxation behaviour of the kinetic and spectral changes. After warming the sample frozen under illumination to temperatures above 150 K for approximately one hour and then recooling to 100 K the recombination kinetics of $P_{865}^{+\cdot}Q_A^{-\cdot}$ is found to be drastically accelerated as compared with the kinetics observed immediately after freezing. Thus, the change of the recombination kinetics for the samples frozen under illumination above $\sim$ 90 K observed in ref. [1] has to be fast on the time scale of hours at temperatures above 150 K. The kinetic change, however, was not found to be accompanied by a change of the spin-polarized $P_{865}^{+\cdot}Q_A^{-\cdot}$ spectrum. The transient spectrum was found to be identical to the one recorded prior to warming and recooling the sample frozen under illumination. The spectrum of dark frozen samples reappeares only after thawing the sample frozen under illumination and freezing again in the dark. After such a cycle the recombination kinetics and the spectrum are identical to those of dark frozen samples which had never been frozen under illumination. Therefore, we conclude that the change of the recombination kinetics induced by freezing under illumination has a different origin than the very small possible reorientation of $Q_A^{-\cdot}$ compatible with the observed spectral changes for $P_{865}^{+\cdot}Q_A^{-\cdot}$.

At present we have not yet tried to simulate the spectral change in terms of the correlated coupled radical pair (CCRP) model (see e.g. [13]) to extract changes of the radical pair geometry. Simulation of X-band spectra is difficult, while spectra obtained in higher frequency bands, i.e. K-band (24 GHz) [11] and W-band (95 GHz) [12] can be simulated satisfactorily on the basis of independently known parameters. This is probably due to effects of the hyperfine interaction which is not explicitely treated in the simulations. These effects are less important at higher frequencies. Similar experiments using time-resolved EPR at Q-band (34 GHz) are, therefore, in progress in our laboratory. In additon we are performing pulsed EPR experiments in X-band which are, in contrast to the direct detection method, very sensitive to the coupling between the spins of the radical pair [14, 15] and, thus, very sensitive to the distance between $P_{865}^{+\cdot}$ and $Q_A^{-\cdot}$.

The observation of uncorrelated kinetic and spectral changes leads to the conclusion that the different kinetics are caused by light-induced changes of the protein/pigment complex other than a reorientation of the cofactors.

## 4 Acknowledgemets

We are grateful to E.C. Abresch (UC San Diego) for his help in preparing the fully deuterated Zn-substituted RCs of *Rhodobacter sphaeroides* R-26. This work was supported by Deutsche Forschungsgemeinschaft (Sfb 312, TP A4), NATO (CRG910468) and Fonds der Chemischen Industrie (to WL).

## References

[1] Kleinfeld, D., Okamura, M.Y., and Feher, G. (1984) Biochemistry 23, 5780–5786.

[2] Bagley, K.A., Abresch, E.C., Okamura, M.Y., Feher, G., Bauscher, M., and Mäntele, W. (1990) in (Baltscheffsky, M., ed.), Proceedings of the VIII International Congress on Photosynthesis, Kluwer, Dordrecht.

[3] Breton, J., Thibodeau, D. L., Berthomieu, C., Mäntele, W., Verméglio, A., and Nabedryk, E. (1991) FEBS Lett. 278, 257–260.

[4] Buchanan, S., Michel, H., and Gerwert, K. (1992) Biochemistry 31, 1314–1322.

[5] Thibodeau, D. L., Nabedryk, E., Hienerwadel, R., Lenz, F., Mäntele, W., and Breton, J. (1990) Biochim. Biophys. Acta 1020, 253–259.

[6] van den Brink, J.S., Hulsebosch, R.J., Gast, P., Hore, P.J., and Hoff, A.J. (1994) Biochemistry 33, 13668–13677.

[7] Allen, J.P., Feher, G., Yeates, T.O., Komiya, H., and Rees, D.C. (1988) Proc. Nat. Acad. of Sci., USA 85, 8487–8491.

[8] Füchsle, G., Bittl, R., van der Est, A., Lubitz, W., and Stehlik, D. (1993) Biochim. Biophys. Acta 1142, 23–35.

[9] Stehlik, D., Bock, C.H., and Petersen, J. (1989) J. Phys. Chem. 93, 1612–1619.

[10] Bittl, R., van der Est, A., Kamlowski, A., Lubitz, W., and Stehlik, D. (1994) Chem. Phys. Lett. 226, 349–358.

[11] van der Est, A., Bittl, R., Abresch, E.C., Lubitz, W., and Stehlik, D. (1993) Chem. Phys. Lett. 212, 561–568.

[12] Prisner, T.F., van der Est, A., Bittl, R., Lubitz, W., Stehlik, D., and Möbius, K. (1995) Chem. Phys. 194, 361–370.

[13] Hore, P.J. (1989) in (Hoff, A.J., ed.), Advanced EPR in Biology and Biochemistry, Ch. 12, pp. 405–440, Elsevier, Amsterdam.

[14] Salikhov, K.M., Kandrashkin, Yu.E., and Salikhov, A.K. (1992) Appl. Magn. Res. 3, 199–216.

[15] Tang, J., Thurnauer, M.C., and Norris, J.R. (1994) Chem. Phys. Lett. 219, 283–290.

# QUANTUM BEATS AS PROBES OF THE PRIMARY EVENTS IN BACTERIAL PHOTOSYNTHESIS

Stefan Weber[1], Ernst Ohmes[2], Marion C. Thurnauer[3], James R. Norris[1,3] and Gerd Kothe[2]

[1] Department of Chemistry, The University of Chicago, Chicago IL 60637, USA

[2] Department of Physical Chemistry, University of Freiburg, Albertstr. 21, D-79104 Freiburg, Germany

[3] Chemistry Division, Argonne National Laboratory, Argonne IL 60439, USA

**Abstract.** Light-induced radical pairs in fully deuterated iron-containing bacterial reaction centers of *Rhodobacter sphaeroides* have been studied by transient EPR following pulsed laser excitation. *Quantum beat oscillations* are observed in the transverse magnetization at low temperatures. Due to its spin-correlated generation, the secondary radical pair $P^+ \left[Q_A\,Fe^{2+}\right]^-$ is expected to start out in a coherent superposition of eigenstates, as observed experimentally. Thorough investigation of light-induced *nuclear coherences*, detected in the transverse electron magnetization, can provide detailed information on the electronic structure of the primary donor, which is essential for a better understanding of the primary events of bacterial photosynthesis.

**Keywords.** Deuterated reaction centers, iron-quinone complex, nuclear modulations, primary donor, quantum beats, spin-correlated radical pairs, transient EPR

## 1.Introduction

In the reaction center of purple bacteria *Rhodobacter sphaeroides* the primary energy conversion steps are [1, 2]

$$P\,H\,Q_A\,Fe^{2+} \overset{h\nu}{\to} {}^1P^*\,H\,Q_A\,Fe^{2+} \overset{2.8ps}{\to} P^+\,H^-\,Q_A\,Fe^{2+} \overset{200ps}{\to}$$

$$P^+\,H\,\left[Q_A\,Fe^{2+}\right]^- \qquad (T < 50\,K) \qquad (0.1)$$

where $P$ is a special pair of bacteriochlorophylls, $H$ is bacteriopheophytin, and $Q_A$ is ubiquinone. In its reduced state $Q_A$ interacts with high-spin $Fe^{2+}$ forming the $\left[Q_A\,Fe^{2+}\right]^-$ complex [3, 4]. As a result, light-induced polarized signals from native bacterial reaction centers (BRCs) are not observed except at cryogenic temperatures [5-7].

Apparently, the primary processes of bacterial photosynthesis proceed via radical pairs as intermediates, which are generated with correlated spins [5-23]. If the initial configuration of the radical pair (e.g. a singlet) is not an eigenstate of the corresponding spin Hamiltonian, the radical pair starts out in a coherent superposition of two of the four spin states [24-27], which can manifest itself as *quantum beats* in an EPR experiment with adequate time resolution [28-33]. In the following we report observation of *quantum beats* for the light-induced radical pairs of fully deuterated (99.7 %) iron-containing BRCs of *Rb. sphaeroides* R 26 and suggest how they may be used to obtain new information about the primary events of bacterial photosynthesis.

## 2. Materials and Methods

Deuterated reaction centers were isolated from whole cells of *Rb. sphaeroides* R 26 which were grown in $D_2O$ (99.7 %) on deuterated substrates [34]. Reaction centers were isolated according to the procedures described by Wraight [35]. Samples for the EPR experiments were prepared by filling the reaction center solutions (deuterated tricine buffer) into a quartz tube (2 mm inner diameter) located in the symmetry axis of the microwave resonator. The temperature of the samples was controlled using a helium flow cryostat (Oxford CF-935) and was stable to ± 0.1 K.

The EPR experiment performed has been described previously [36]. A modified X-band spectrometer (Bruker ER-200 D) was used, equipped with a fast microwave preamplifier (36 dB, 1.8 dB noise figure) and a broad band video amplifier (band width 200 Hz - 200 MHz). The sample was irradiated in a home-built split ring resonator with 2 ns pulses of a Nd:YAG pumped dye laser (Spectra Physics, 580 nm, 10 mJ/pulse) at a repetition rate of 10 Hz. The split ring resonator exhibits a high filling factor at low Q (unloaded $Q \approx 500$) and provides an easy means of sample irradiation.

The time-dependent EPR signal was digitized in a transient recorder (LeCroy 9450 digital oscilloscope) at a rate of 2.5 ns/12 bit sample. The time resolution of the experimental setup is estimated to be in the 10 ns range. Typically, 512 transients were accumulated at off-resonance conditions und subtracted from those on resonance to get rid of the $B_o$ independent background signal induced by the laser pulse.

## 3. Results and Discussion

In our EPR experiments, the sample is irradiated with a short laser pulse and the time evolution of the transverse magnetization is detected in the presence of a weak microwave magnetic field. Generally, a complete data set consists of transient EPR signals taken at equidistant magnetic field points covering the total spectral width. This implies a two-dimensional variation of the signal intensity with respect to both the magnetic field and time axis. Such a complete data set is shown in Figure 1 for native iron-containing BRCs from fully deuterated purple bacteria *Rb. sphaeroides*. The plot refers to a constant microwave field of $B_1 = 0.08$ mT and T = 10 K. Note that a positive signal indicates absorptive (a) and a negative emissive (e) spin polarization.

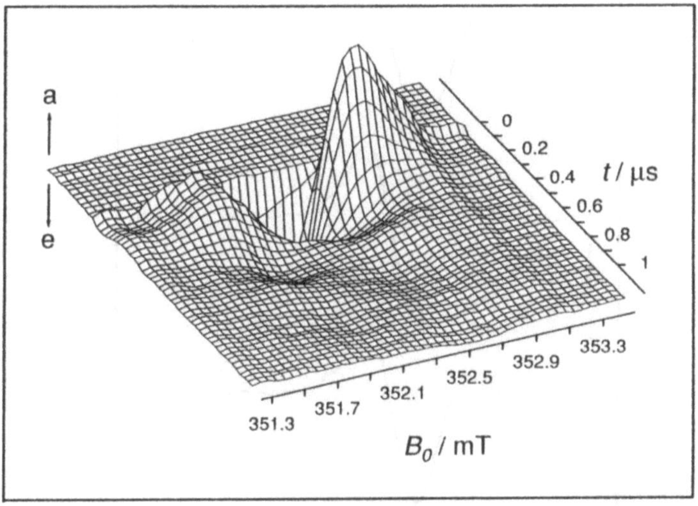


**Fig. 1.** Complete data set of the transient EPR signal of the light-induced radical pair $P^+\ [Q_A\,Fe^{2+}]^-$ in 99.7 % deuterated bacterial reaction centers of *Rhodobacter sphaeroides* R 26. Positive and negative signals indicate absorptive (a) and emissive (e) polarization, respectively. Microwave frequency $\omega/2\pi = 9.8820$ GHz, microwave field, $B_1 = 0.08$ mT, temperature T = 10 K.

Transient spectra can be extracted from this plot at any fixed time after the laser pulse as slices parallel to the magnetic field axis. Likewise, the time evolution of the transverse magnetization may be obtained for any given field as a slice along the time axis. Typical lineshapes, observed 40, 60, 80 and 100 ns after the laser pulse, are shown in Figure 2. Each spectrum is integrated over a narrow time window of 10 ns. Apparently, the early spectrum is much broader than the later ones. A detailed analysis reveals that lifetime broadening actually dominates the early spectrum [30]. The distinct a/e/a pattern, observed for the later spectra, corresponds to previous observations [5-7]. Therefore, the transient lineshapes can be assigned to the secondary radical pair $P^+\ [Q_A\,Fe^{2+}]^-$ .

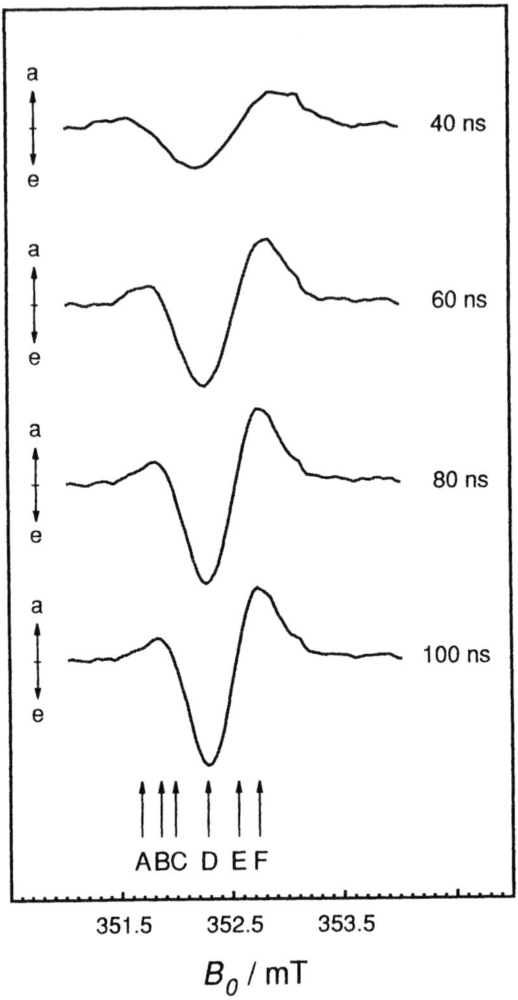


**Fig. 2.** Transient EPR lineshapes of the light-induced radical pair $P^+$ $[Q_A\,Fe^{2+}]^-$ in 99.7 % deuterated bacterial reaction centers of *Rhodobacter sphaeroides* R 26 at various times after the laser pulse. Time window = 10 ns. Positive and negative signals indicate absorptive (a) and emissive (e) polarization, respectively. Microwave frequency $\omega/2\pi$ = 9.8820 GHz, microwave field, $B_1$ = 0.08 mT, temperature T = 10 K.

Figure 3 depicts the time evolution of the transverse magnetization, measured with a time resolution of 10 ns. The transients refer to a constant microwave field of $B_1$ = 0.08 mT, T = 10 K, and six selected field positions (A-F, Fig. 2). Apparently, all transients exhibit oscillatory behavior at early times after the laser pulse. Note, however, that the frequency of these oscillations varies across the spectrum, indicating an anisotropic nature of the transients. Generally, oscillation frequencies of 20-30 MHz can be extracted from the time profiles. In addition, slower oscillations with a frequency of 2 to 3 MHz can be seen.

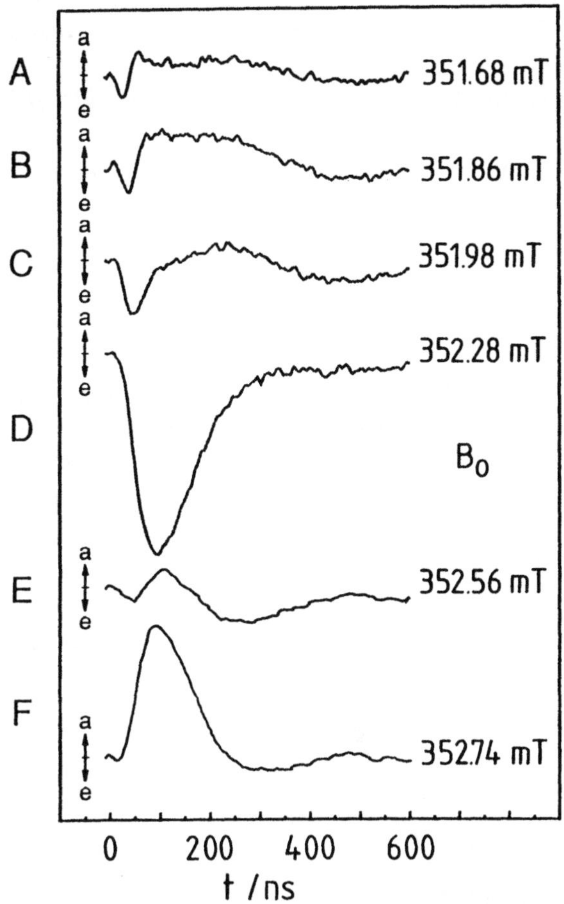


**Fig. 3.** Time evolution of the transverse magnetization of the light-induced radical pair $P^+ \ [Q_A\, Fe^{2+}]^-$ in 99.7 % deuterated bacterial reaction centers of *Rhodobacter sphaeroides* R 26 at six different static magnetic fields $B_o$ (see positions A - F in Fig. 2). Positive and negative signals indicate absorptive (a) and emissive (e) polarization, respectively. Microwave frequency $\omega/2\pi = 9.8820$ GHz, microwave field, $B_1 = 0.08$ mT, temperature T = 10 K.

Basically, the fast oscillations represent *quantum beats*, associated with the spin-correlated generation of $P^+ \ [Q_A\, Fe^{2+}]^-$. Because of the short lifetime of the precursor spin pair, $P^+ \ [Q_A\, Fe^{2+}]^-$ is generated in a virtually pure singlet state [5, 22]. Figure 4 depicts a schematic of the corresponding level diagram. One sees that such a singlet radical pair is formed with spin-correlated populations of two of the four spin levels [8-11] and *zero quantum coherence* (ZQC) between these states [24-27, 30]. In the absence of a microwave field, the eigenstate populations are constant in time, while the *zero quantum coherence* oscillates at a frequency, $\nu_{ZQ}$, given by [30]:

$$\nu_{ZQ} = (1/h)\{[2J + \frac{2}{3} D^{zz}(\Omega)]^2 + [(g_1^{zz}(\Omega) - g_2^{zz}(\Omega))\beta B_o + \sum_k A_{1k}^{zz}(\Omega) M_{1k}^i - \sum_l A_{2l}^{zz}(\Omega) M_{2l}^j]^2\}^{1/2} \quad (0.2)$$

$$M_{1k}^i = I_{1k}, I_{1k} - 1, ...., -I_{1k} \quad (0.3)$$

$$M_{2l}^j = I_{2l}, I_{2l} - 1, ...., -I_{2l} \quad (0.4)$$

Here $J$, $D^{zz}(\Omega), g_i^{zz}(\Omega), \beta, B_o, A_{ik}^{zz}(\Omega)$ and $I_{ik}$ are the electron exchange interaction, the $zz$ component of the dipolar coupling tensor, the $zz$ component of the g-tensor of radical $i$, the Bohr magneton, the static magnetic field oriented along the laboratory $z$-axis, the $zz$ compound of the hyperfine tensor of nucleus $k$ in radical $i$, and the corresponding nuclear spin quantum number, respectively.

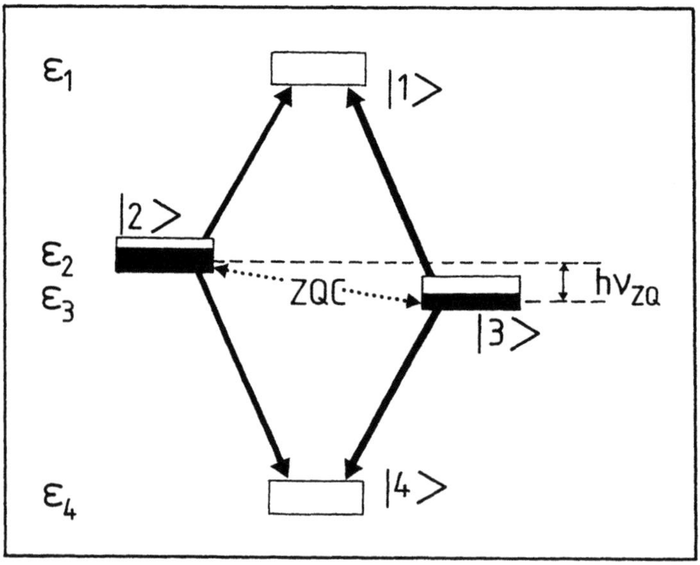


**Fig. 4.** Energy level diagram for a spin-correlated radical pair, created in the singlet state. Full arrows: Single quantum coherences. Dotted arrow: Zero quantum coherence. $\varepsilon_i$ = energy of eigenstate | i>.

Application of the continuous microwave field has two effects. First, it converts the longitudinal magnetization associated with the population difference between neighboring states into transverse magnetization [27, 37]. This gives isotropic *Torrey oscillations* with a frequency of

$$\nu_{TO} = \frac{1}{2}(g_1 + g_2)\,\beta\, B_1\,/\,h. \tag{0.5}$$

Second, it converts the *zero quantum coherence* into observable single quantum coherence [24, 27], whose frequency varies with the orientation of the radical pair [29-31, 33]. Because of hyperfine interactions, rapid averaging of these oscillations occurs even in the case of narrow-band microwave excitation. Thus, manifestations of *zero quantum coherence* are likely to be observed only immediately after the laser pulse.

Preliminary model calculations for $P^+ \left[Q_A\, Fe^{2+}\right]^-$, based on the known X-ray structure of the reaction center complex [38-40] and on the published g-tensor parameters for $P^+$ [21, 41] and the lowest doublets in $\left[Q_A\, Fe^{2+}\right]^-$ [4, 42], are in qualitative agreement with the experimental time profiles. We therefore conclude, that *quantum beat oscillations* have been detected in native iron-containing BRCs. This result confirms the previous assignment of the transient spectra (see Fig. 2) to the $P^+$ part of $P^+ \left[Q_A\, Fe^{2+}\right]^-$ [5-7]. Moreover, it provides a means for direct measurement of a number of important structural and magnetic parameters.

As noted above, the *zero quantum frequency*, $\nu_{ZQ}$, depends on the orientation $\Omega$ of the radical pair in the laboratory frame (see eqs. 0.2-0.4). The weak $B_1$ field, commonly employed in transient EPR, allows for only a small range of orientations to meet the resonance condition. Consequently, we expect the *quantum beat oscillations* to vary significantly with $B_o$ across the powder spectrum. The pronounced variation of these *oscillations* (see Fig. 3) has been used to evaluate the geometry of the secondary radical pair in plant photosystem I [30] even without having high resolution X-ray structure. It is worthwile to note that single crystals are not required for this accurate structural technique.

Although the basic features of the magnetic interactions in $[Q_A\; Fe^{2+}]^-$ are now resolved [3, 4], the g-tensor orientations in the lowest doublets are not known in any great detail [42]. Analysis of the *quantum beat oscillations* as a function of the external field could provide this information. Studies along these lines are in progress. From the results we expect to assess the effect of high-spin $Fe^{2+}$ on the spin dynamics of the secondary radical pairs in bacterial reaction centers.

At closer inspection an additional coherence phenomenon can be observed, recently reported for Zn-substituted BRCs [32] and plant photosystem I [43]. Figure 5 (upper curve) depicts the time evolution of the transverse magnetization, measured at a field position intermediate between E and F (see Fig. 2). The transient refers to $B_1 = 0.07$ mT and T $= 10$ K. Oscillations with frequencies of about 2 MHz can be seen. Interestingly, however, these slow oscillations cannot all be assigned to Torrey precessions (see eq. 0.5). Rather, there are additional modulations of the transverse magnetization. This is clearly seen in the power spectrum (lower curve), obtained by Fourier transformation of the time profile. Only the peak marked with an arrow indicates Torrey oscillations. The remaining oscillations, indicative of a powder sample, exhibit frequencies between 1 and 3 MHz.

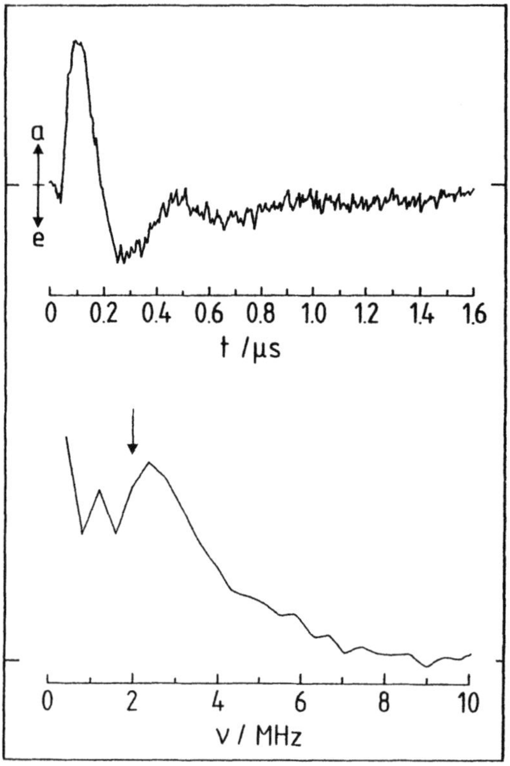


**Fig. 5.** Time evolution of the transverse magnetization of the light-induced radical pair $P^+ \ [Q_A\, Fe^{2+}]^-$ in 99.7 % deuterated bacterial reaction centers of *Rhodobacter sphaeroides* R 26. The transient was taken at a field position intermediate between E and F (see Fig. 2). Microwave frequency $\omega/2\pi =$ 9.8820 GHz, microwave field, $B_1 = 0.07$ mT, temperature T $= 10$ K. Upper curve: Time profile. Lower curve: Power spectrum. Torrey oscillations are indicated by an arrow.

In order to assign these oscillations, a comprehensive study has been carried out in case of plant photosystem I [43]. The results clearly show that light-induced *nuclear coherences* are observed in the transverse magnetization of the secondary radical pair. Since the initial nuclear spin configurations are not eigenstates of the radical pair Hamiltonian, the light pulse induces a coherent time evolution of the nuclear spin system [44]. This *nuclear coherence* is then transferred to observable electron coherence by means of the continuous microwave magnetic field. In this case, the frequencies of the oscillations (see Fig. 5) are expected to correspond to ENDOR frequencies of $^{14}N$ nuclei in the primary donor [43].

At present, the electronic structure of $P^+$ is not fully understood. In particular, the asymmetry of the spin density distribution in the special pair [45], considerably increasing at lower temperatures [46], is still the subject of discussions. Additional studies of the hyperfine interactions in $P^+$, employing light-induced *nuclear quantum beats* might resolve the problem. Studies along these lines, involving deuterated $^{15}N$ enriched BRC preparations, are currently in progress. From the results we expect detailed information on the electronic structure of $P^+$, which is essential for a better understanding of the primary events of bacterial photosynthesis.

## 4. Conclusion

Using high time resolution continuous wave EPR we have been able to detect *quantum beat oscillations* in the transverse magnetization of the secondary radical pair $P^+ \left[Q_A Fe^{2+}\right]^-$, generated by pulsed laser excitation of native iron-containing BRCs. Thorough investigation of this electron *zero quantum coherence* allows a more detailed characterization of the magnetic interactions in the $\left[Q_A Fe^{2+}\right]^-$ complex. At closer inspection, light-induced *nuclear modulations* are observed in the transverse magnetization of $P^+ \left[Q_A Fe^{2+}\right]^-$. From an analysis of these *nuclear coherences* as a function of the static and microwave magnetic field we expect valuable information on the electronic structure of the primary donor.

**Acknowledgment**
Financial support by the U.S. Department of Energy, Office of Basic Energy Sciences, Division of Chemical Sciences (Contract W-31-109-ENG-38) and by the Deutsche Forschungsgemeinschaft is gratefully aknowledged. J.R.N. and S.W. greatly acknowledge support from the Humboldt Foundation.

## References

1 Kirmaier, C., Holton, D. and Parson, W. W. (1985) Biochim. Biophys. Acta 810, 33-48.
2 Deisenhofer J. and Norris, J. R. (Eds.) (1993) The Photosynthetic Reaction Center, Academic Press, New York.
3 Butler, W. F., Johnston, D. C., Shore, H. B., Fredkin, D. R., Okamura, M. Y. and Feher, G. (1980) Biophys. J. 32, 967-992.

4 Butler, W. F., Calvo, R., Fredkin, D. R., Isaacson, R. A., Okamura, M. Y. and Feher, G. (1984) Biophys. J. 45, 947-973.
5 Snyder, S. W., Morris, A. L., Bondeson, S. R., Norris, J. R. and Thurnauer, M.C. (1993) J. Am. Chem. Soc. 115, 3774-3775.
6 Proskuryakov, I. I., Shkuropatow, A. Ya., Sarrazian, N. A. and Shuvalov, V. A. (1991) Dok. Akad. Nauk SSSR (Russian) 320, 1006-1008.
7 Proskuryakov, I. I., Klenina, I. B., Shkuropatow, A. Ya., Shkuropatova, V. A. and Shuvalov, V. A. (1993) Biochim. Biophys. Acta 1142, 207-210.
8 Thurnauer, M. C. and Norris J. R. (1980) Chem. Phys. Lett. 76, 557-561.
9 Closs, G. L., Forbes, M. D. E. and Norris J. R. (1987) J. Phys. Chem. 91, 3592-3599.
10 Buckley, C. D., Hunter, D. A., Hore, P. J. and MacLauchlan, K. H. (1987) Chem. Phys. Lett. 135, 307-312.
11 Hore, P. J., Hunter, D. A., McKie, C. D. and Hoff, A. J. (1987) Chem. Phys. Lett. 137, 495-500.
12 Hore, P. J. (1989) Advanced EPR, Applications in Biology and Biochemistry (Hoff, A. J., ed.), pp. 405-440, Elsevier, Amsterdam.
13 Feezel, L. L., Gast, P., Smith, U. H. and Thurnauer, M. C. (1989) Biochim. Biophys. Acta 974, 149-155.
14 Morris, A. L., Norris, J. R. and Thurnauer M. C. (1990) Reaction Centers of Photosynthetic Bacteria (Michel-Beyerle, M.-E., ed.), pp. 423-435, Springer Series in Biophysics, Springer, Berlin.
15 Norris, J. R., Morris, A. L., Thurnauer, M. C. and Tang, J. (1990) J. Chem. Phys. 92, 4239-4249.
16 Snyder, S. and Thurnauer, M. C. (1993) The Photosynthetic Reaction Center (Deisenhofer, J. and Norris, J. R., eds.) pp. 285-330, Academic Press, New York.
17 Füchsle, G., Bittl, R., van der Est, A., Lubitz, W. and Stehlik, D. (1993) Biochim. Biophys. Acta 1142, 23-35.
18 Van der Est, A., Bittl, R., Abresch, E. C., Lubitz, W. and Stehlik, D. (1993) Chem. Phys. Lett. 212, 561-568.
19 Van der Brink, J. S., Hulsebosch, R. J., Gast, P., Hore, P. J. and Hoff, A. J. (1994) Biochemistry 33, 13668-13677.
20 Tang, J., Thurnauer, M. C. and Norris, J. R. (1994) Chem. Phys. Lett. 219, 283-290.
21 Prisner, T. F., van der Est, A., Bittl, R., Lubitz, W., Stehlik, D. and Möbius, K. (1995) Chem. Phys. 194, 361-370.
22 Morris, A. L., Snyder, S. W., Zhang, Y., Tang, J., Thurnauer, M. C., Dutton, P. L., Robertson, D. E. and Gunner, M. R. (1995) J. Phys. Chem. 99, 3854-3866.
23 Dzuba, S. A., Gast, P. and Hoff, A. J. (1995) Chem. Phys. Lett. 236, 595-602.
24 Salikhov, K. M., Bock, C. H. and Stehlik, D. (1990) Appl. Magn. Reson. 1, 195-211.
25 Bittl, R. and Kothe, G. (1991) Chem. Phys. Lett. 177, 547-553.

26 Wang, Z., Tang, J. and Norris, J. R. (1992) J. Magn. Reson. 97, 322-334.
27 Zwanenburg, G. and Hore, P. J. (1993) Chem. Phys. Lett. 203, 65-74.
28 Kothe, G., Weber, S., Bittl, R., Norris, J. R., Snyder, S. S., Tang, J., Thurnauer, M. C., Morris, A. L., Rustandi, R. R. and Wang, Z. (1991) Spin Chemistry (I'Haya, Y. J., ed.), pp. 420-434, The Oji International Conference on Spin Chemistry, Tokyo.
29 Kothe, G., Weber, S., Bittl, R., Ohmes, E., Thurnauer, M. C. and Norris, J. R. (1991) Chem. Phys. Lett. 186, 474-480.
30 Kothe, G., Weber, S., Ohmes, E., Thurnauer, M. C. and Norris, J. R. (1994) J. Phys. Chem. 98, 2706-2712.
31 Kothe, G., Weber, S., Ohmes, E., Thurnauer, M. C. and Norris, J. R. (1994) J. Am. Chem. Soc. 116, 7729-7734.
32 Bittl, R., van der Est, A., Kamlowski, A., Lubitz, W. and Stehlik, D. (1994) Chem. Phys. Lett. 226, 349-358.
33 Laukenmann, K., Weber, S., Kothe, G., Oesterle, C., Angerhofer, A., Wasielewski, M. R., Svec, W. A. and Norris, J. R. (1995) J. Phys. Chem. 99, 4324-4329.
34 Crespi, H. L. (1982) Methods in Enzymology (Packer, L., ed.) pp. 3-5, Vol. 88, Academic Press, New York.
35 Wraight, C. A. (1979) Biochim. Biophys. Acta 548, 309-327.
36 Münzenmaier, A., Rösch, N., Weber, S., Feller, C., Ohmes, E. and Kothe, G. (1992) J. Phys. Chem. 96, 10645-10653.
37 Gierer, M., van der Est, A. and Stehlik, D. (1991) Chem. Phys. Lett. 186, 238-247.
38 Deisenhofer, J., Epp, O., Miki, K., Huber, R. and Michel, H. (1984) J. Mol. Biol. 180, 385-398.
39 Chang, C. H., Tiede, D. M., Tang, J., Smith, U., Norris, J. R. and Schiffer, M. (1986) FEBS Lett. 205, 82-86.
40 Allen, J. P., Feher, G., Yeates, T. V., Rees, C. D., Deisenhofer, J., Michel, H. and Huber, R. (1986) Proc. Natl. Acad. Sci. U.S.A. 83, 8589-8593.
41 Klette, R., Törring, J. T., Plato, M., Möbius, K., Bönigk, B. and Lubitz, W. (1993) J. Phys. Chem. 97, 2015-2020.
42 Evelo, R. G., Nan, H. M. and Hoff, A. J. (1988) FEBS lett. 239, 351-357.
43 Weber, S., Ohmes, E., Thurnauer, M.C., Norris, J. R. and Kothe, G. (1995) Proc. Natl. Acad. Sci. U.S.A., in press.
44 Salikhov, K. M. (1993) Chem. Phys. Lett. 201, 261-264.
45 Lendzian, F., Huber, M., Isaacson, R. A., Endeward, B., Plato, M., Bönigk, B., Möbius, K., Lubitz, W. and Feher, G. (1993) Biochim. Biophys. Acta 1183, 139-160.
46 Käss, H., Rautter, H., Bönigk, B., Höfer, P. and Lubitz, W. (1995) J. Phys. Chem. 99, 436-448.

# Asymmetry of the Binding Sites of $Q_A^{-\cdot}$ and $Q_B^{-\cdot}$ in Reaction Centers of *Rb. sphaeroides* Probed by Q-Band EPR with $^{13}C$-Labeled Quinones

R.A. Isaacson[1], E.C. Abresch[1], F. Lendzian[2], C. Boullais[3], M.L. Paddock[1], C. Mioskowski[3], W. Lubitz[2] and G. Feher[1]

1 Department of Physics, UC San Diego, La Jolla, CA 92093, USA

2 Max-Volmer-Institut für Biophysikalische und Physikalische Chemie, Technische Universität Berlin, 10623 Berlin, Germany

3 Département de Biologie Cellulaire et Moléculaire, CEA-Saclay, 91191 Gif-sur-Yvette, France

**Abstract.** Ubiquinone-3 labeled with $^{13}C$ at either position 1 or 4 was incorporated into Zn-containing reaction centers of *Rb. sphaeroides* in the mutant His→Cys (M266). Q-band EPR of the primary and secondary quinone anion radicals, $Q_A^{-\cdot}$ and $Q_B^{-\cdot}$, revealed differences in the $^{13}C$-hyperfine couplings of the carbons of the two carbonyl groups. The asymmetry of the spin density distribution was particularly pronounced for $Q_A^{-\cdot}$ and was explained by a strong hydrogen bond from the protein to one of the carbonyl oxygens in accord with the findings of van den Brink et al (FEBS Lett, 1994, 353, 273-276). The binding of $Q_B^{-\cdot}$ is only slightly asymmetric and more similar to that found for the ubiquinone-3 anion radical in an alcoholic solvent matrix. The hydrogen bond donors to $Q_A^{-\cdot}$ and $Q_B^{-\cdot}$ of the amino acid framework are identified and consequences for the function of the quinones in the electron transfer process are briefly discussed.

**Keywords**: EPR, quinone anion radicals, $^{13}C$ hyperfine couplings, bacterial photosynthesis, hydrogen bonds, quinone binding sites

# 1 Introduction

In bacterial reaction centers (bRCs) of *Rhodobacter (Rb.) sphaeroides* the primary and secondary quinones, $Q_A$ and $Q_B$, are both ubiquinone-10 ($UQ_{10}$) molecules. These quinones act sequentially in the electron transfer process in the bRC. The two quinones have different redox potentials. Further, $Q_A$ accepts only one electron, whereas $Q_B$ can accept two electrons and two protons [1-3]. Since $Q_A$ and $Q_B$ are chemically identical molecules, their different properties in the electron transfer (ET) chain result from specific interactions with the protein environment. These interactions are believed to fine-tune the electronic structures of $Q_A$ and $Q_B$ for optimum function in the bRC.

Details of the electronic structure of the two quinones are obtainable from the hyperfine coupling constants (hfc's) determined by EPR and, in particular, ENDOR spectroscopy of the radical anions, $Q_A^{-\cdot}$ and $Q_B^{-\cdot}$, formed in the ET process [4,5]. For such measurements the high spin $Fe^{2+}$ coupled to the quinones had to be replaced by a diamagnetic metal ion, e.g. $Zn^{2+}$ [6]. The metal exchange did not alter the functional properties, i.e. the ET rates of the bRC [7].

Our earlier $^1H$ ENDOR investigations of $Q_A^{-\cdot}$ and $Q_B^{-\cdot}$ in frozen ZnbRCs of *Rb. sphaeroides* clearly showed that the electronic spin density distribution of $Q_A^{-\cdot}$ is considerably different from that of $Q_B^{-\cdot}$ [4,5]. This was attributed to stronger and more asymmetric hydrogen bonds between the protein and $Q_A^{-\cdot}$ as compared with $Q_B^{-\cdot}$. This interpretation was corroborated by the determination of $^{17}O$ hf-tensor components in these molecules labeled with $^{17}O$ at *both* carbonyl oxygens. Due to the lack of a label at a specific position it was not possible to *directly* identify the carbonyl oxygen (O1 or O4, see Fig. 1) to which the strong(er) H-bond is formed.

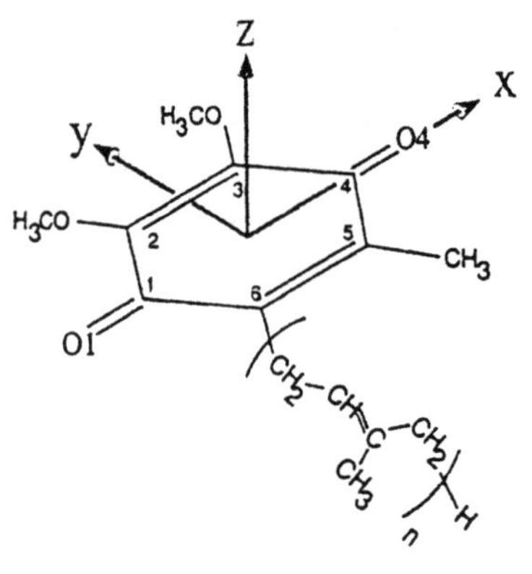


*Fig. 1: Molecular structure and numbering scheme for ubiquinone-n ($UQ_n$), where n=3 or 10 as discussed in the text. The carbonyl oxygens (O1, O4) are labeled according to the carbon atom to which they are attached. The g-tensor axes and $^{13}C$ hf tensor axes (pos. 1 and 4) are assumed to be collinear with the molecular axes (x,y,z) (see text).*

A very sensitive structural probe for changes of the environment of a carbonyl group is the $^{13}C$ hfc [8]. Consequently, ubiquinones specifically labeled with $^{13}C$ at positions 1 or 4 (see Figure 1) are ideally suited to sense the shift of spin density in the respective CO group caused by H-bonds of different strengths.

Recently, specifically labeled $^{13}$C-ubiquinones were synthesized by two groups and incorporated into native bRCs of *Rb. sphaeroides* [9,10]. Both groups performed Fourier transform infrared (FTIR) spectroscopy on these bRCs and identified a strong H-bond to the CO group of $Q_A$ in position 4 (Fig. 1) [9,10]. Van den Brink et al. [11] investigated $Q_A^{-\cdot}$ in ZnbRCs in which the native $UQ_{10}$ was replaced with $UQ_{10}$ specifically $^{13}$C-labeled at various ring positions by Q-band (35 GHz) EPR. The analysis of the spectra showed that the $^{13}$C hfc is larger for the CO group at position 4 than for position 1, again indicating a stronger hydrogen bond to the oxygen at position 4. This oxygen of the ubiquinone points towards His (M219) (Fig. 2) that was postulated as a H-bond donor. This proposal was corroborated by the same group [11] who detected a $^{14}$N hyperfine and quadrupole coupling to a nitrogen of the protein matrix in the electron spin echo envelope modulation (ESEEM) spectra of $Q_A^{-\cdot}$ [12]. The $^{14}$N quadrupole parameters were very similar to those of a histidine $N^{\delta(+)}$-H group.

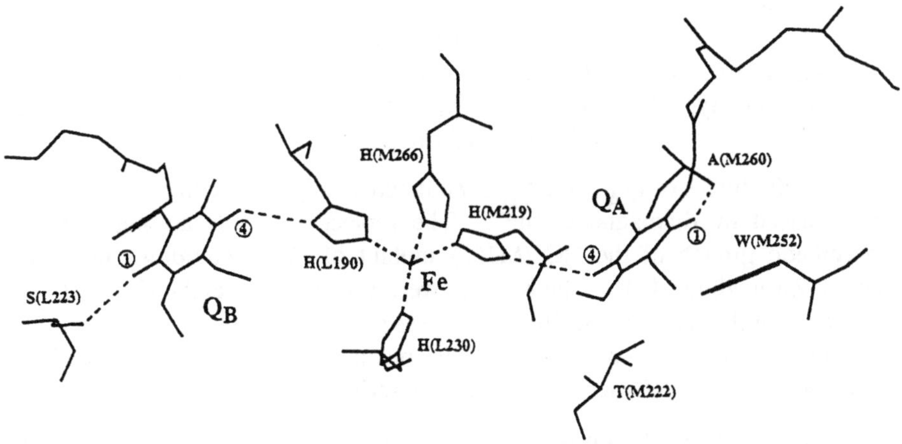


*Fig. 2: Structure of the $Q_A$, $Q_B$ and Fe site in the bRC of Rb. sphaeroides [32]. The $Fe^{2+}$ is ligated to four histidines and the bidentate ligand of Glu (M234) (omitted for clarity); the isoprenoid chains of the quinones are truncated. The two carbonyl oxygens of both quinones are labelled 1 and 4 (see Fig 1) and the proposed H-bonds and ligation to the iron are indicated by dashed lines. In the static ground state X-ray structure of ref. [32] $Q_B$ is H-bonded to His (L190) and Ser (L233), and $Q_A$ to His (M219) and the backbone NH of Ala (M260). Spectroscopic evidence from EPR (this work and [11]) and FTIR [9,10] suggests, at most, a very weak hydrogen bond at O1 of $Q_A^{-\cdot}$. In the mutant used in this study [HC (M266)] His at M266 is replaced by Cys.*

Most of the experiments described above were performed on a ZnbRC preparation that did not bind $Q_B$ [11]. This is probably due to the loss of the H protein subunit in the majority of the bRCs (≥95%) which frequently occurs during the chemical replacement of $Fe^{2+}$ by $Zn^{2+}$ [6]. We have therefore developed a procedure to biosynthetically incorporate $Zn^{2+}$ to a high degree into intact bRC. This procedure uses a mutant, His → Cys (M266), see Fig. 2, which allows the iron site to incorporate other divalent metal ions [13]. It was shown that the ET properties of ZnbRCs of mutant HC (M266) are essentially the same as those of wild type [Paddock, unpublished]. In such bRCs unlabeled $UQ_3$ or $UQ_3$ $^{13}C$-labeled in position 1 or 4 was incorporated in both the $Q_A$ and $Q_B$ sites. Independent FTIR experiments on native bRCs have shown that $UQ_3$ is bound in the same way as $UQ_{10}$ [10].

In this paper we present Q-band EPR data of the radical anions of $^{13}C$-labeled $UQ_3$ in the $Q_A$ and $Q_B$ sites of the ZnbRCs in mutant HC (M266) in organic solvent. Our results corroborate and extend the work of van den Brink et al. [11].

## 2 Materials and Methods

$UQ_3$, selectively $^{13}C$-labeled at carbon positions 1 and 4, was synthesized as described in [10]. The isotopic enrichment (>99%) was checked by NMR and mass spectrometry. The structure of $UQ_n$ is shown in Fig. 1 where n = 10 in the native bRC and n = 3 in the $^{13}C$-labeled samples .

For EPR studies of $Q_A^{-\cdot}$ and $Q_B^{-\cdot}$, the magnetically interacting $Fe^{2+}$ in bRCs was replaced by diamagnetic $Zn^{2+}$. The replacement was accomplished by biosynthetic growth of the HC (M266) mutant of *Rb. sphaeroides* in a low-Fe/high-Zn medium. [14]. The bacteria were grown semi-aerobically [15] with the exception that no iron and 10 ppm Zn was added to the media resulting in 0.4 ± 0.1 ppm Fe and 10 ppm Zn in the final media. Cells were harvested after 5 days of growth and bRCs were prepared as described in [15].

The native $Q_A$ ($UQ_{10}$) was removed from HC (M266) ZnbRCs by binding the ZnbRCs to a 1-2 ml DEAE Toyopearl (Toyosoda) column and washing with approximately 250 ml of $UQ_0$ (Sigma) in 10 mM Tris-Cl pH 8, 0.1% LDAO, 0.1 mM $ZnCl_2$ (TLZ) for 18 hours at 23 °C. The column was then washed with 200 ml TLZ to remove $UQ_0$. The bRCs were eluted with 1 M NaCl in TLZ and dialyzed overnight against 10 mM Tris-Cl pH 8, 0.025% LDAO, 0.1 mM $ZnCl_2$.

Quinone was removed from the $Q_B$ site in HC (M266) ZnbRCs by the following method: ZnbRCs were diluted to A800 =1 and 5 μM stigmatellin (Fluka) was added. The ZnbRCs were immediately put onto a 5 ml DEAE Toyopearl column and washed with 1 L TLZ over 2-4 hours. The ZnbRCs were then eluted with 1 M NaCl in TLZ and dialyzed overnight against 10 mM Tris-Cl pH 8, 0.025% LDAO, 0.1 mM $ZnCl_2$.

The $Q_A$ and $Q_B$ depleted ZnbRCs were exchanged into $D_2O$ buffer by diluting 10-fold with 10 mM Tris-Cl pH 8, 0.1 mM $ZnCl_2$ in $D_2O$ and concentrating in a Centricon-30 cell (Amicon). This procedure was repeated to bring the $D_2O$ concentration to about 99%.

The $Q_A^{-\cdot}$ samples were made from 20 μl of 300 μM HC (M266) ZnbRCs ($Q_A$ depleted) in $D_2O$. 3mM cytochrome *c* (Sigma, horse heart Type VI) and 1 mM of the appropriate $^{13}C$ $UQ_3$ was added to the ZnbRCs. The $Q_A$ and $Q_B$ sites were checked by optical spectroscopy and found to have full $Q_A$ and $Q_B$ occupancy after adding the quinone. The samples were illuminated for 5 s with white light from a Zeiss 500 W projector and frozen in liquid nitrogen while still being illuminated.

The $Q_B^{-\cdot}$ samples were made from 20 μl of 300 μM HC (M266) ZnbRCs ($Q_B$ depleted) in $D_2O$. 3 mM cytochrome *c* and ≈6 mM of the appropriate $^{13}C$ quinone was added to the ZnbRCs. The samples were given one saturating laser pulse (0.3 J in 0.5 μs, Phase-R model DL 2100C, Lumen-X Inc. New Durham, NH) and immediately frozen in liquid nitrogen.

The home-built Q-band (35 GHz) superheterodyne EPR spectrometer, using a cylindrical $TE_{011}$ cavity and an immersion dewar system for temperature control, has been described previously [16]. A model QLN-3635-AA low noise microwave preamplifier (Quinstar Technology, Torrance, CA, USA) has been added to improve the signal-to-noise of the spectrometer by a factor of five. A phosphorous-doped silicon (P-Si) powder sample was used as a g-marker [17]. Its g-value was determined from a calibration against a Li/Li F g-standard [18] to be 1.99891 ± 0.00003 at T = 80K. The P-Si marker was permanently mounted on the top surface of the cavity and usually measured together with the sample.

The Q-band EPR powder spectra were analyzed by using a simulation and fit program based on the work of Rieger [19] that utilizes a modified Levenberg-Marquardt nonlinear least squares method [20]. This program includes second order effects and can handle an arbitrary number of nuclei with non-collinear g- and hf-tensors.

## 3 Results

The spectra of $Q_A^{-\cdot}$ (Fig. 3) and $Q_B^{-\cdot}$ (Fig. 4) were compared with those obtained from the $UQ_3$ radical anion in an organic solvent that is capable of H-bond formation. Figure 5 shows a comparison of the Q-band EPR spectra of $^{12}C$-, $^{13}C1$- and $^{13}C4$-$UQ_3^{-\cdot}$ in frozen deuterated isopropanol. The positions corresponding to the g-tensor principal values are indicated; the additional narrow line at high-field is the P-Si g-marker.

For the fit of the powder EPR spectrum of the unlabeled $^{12}C$-$UQ_3^{-\cdot}$ an orthorhombic g-tensor was assumed [16,21]. Hyperfine interactions resulting from the protons of $UQ_3^{-\cdot}$ and the deuterons of the hydrogen bonds formed to

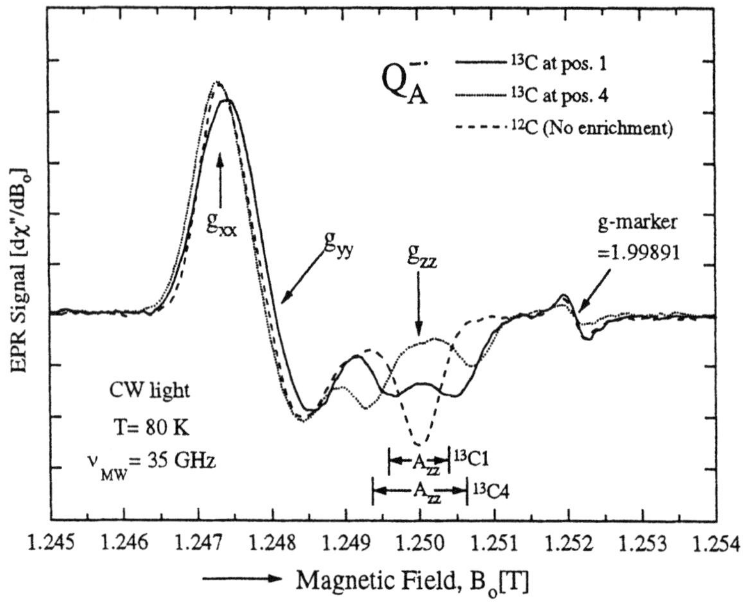


***Fig. 3:*** *Q-band (35 GHz) EPR spectra of $Q_A^{-\cdot}$ in ZnbRCs of Rb. sphaeroides mutant HC (M266) in $D_2O$ buffer, reconstituted with $UQ_3$ $^{13}C$ enriched at position 1, position 4 and with unlabeled $UQ_3$ ($^{12}C$). Experimental conditions: MW power = 0.3 μW; field modulation: amplitude = 0.2 mT, frequency = 400 Hz; scan time = 30 sec, 16 scans each; T= 80 K.*

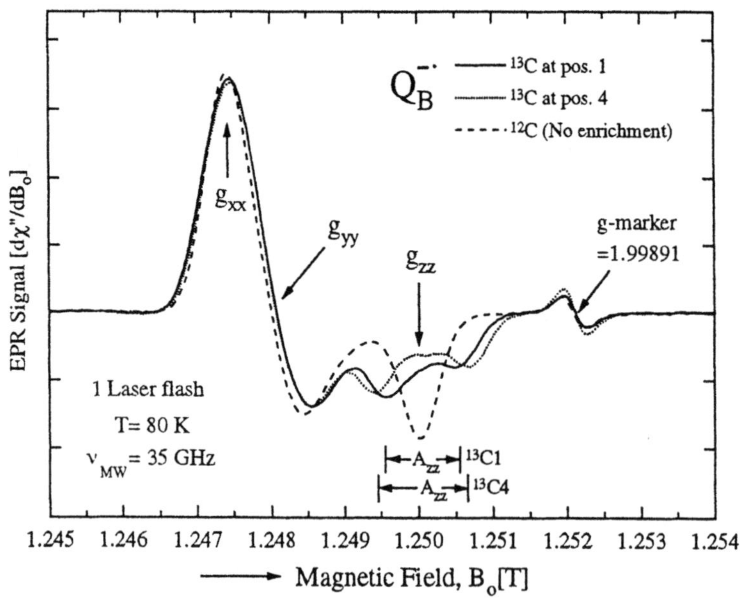


***Fig. 4:*** *Q-band (35 GHz) EPR spectra of $Q_B^{-\cdot}$ in ZnbRCs of Rb. sphaeroides mutant HC (M266) in $D_2O$ buffer, reconstituted with $UQ_3$ $^{13}C$ enriched at position 1, position 4 and with unlabeled $UQ_3$ ($^{12}C$). Experimental conditions: see Figure 3.*

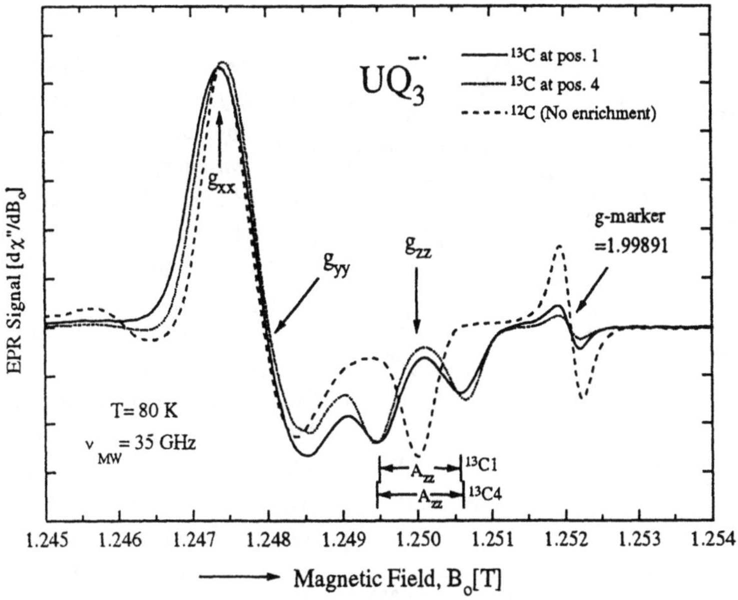


*Fig. 5: Q-band (35 GHz) EPR spectra of* $UQ_3^{-\cdot}$ $^{13}C$ *enriched at positions 1, 4, and of unlabeled* $UQ_3^{-\cdot}$ *(*$^{12}C$*) in isopropanol-*$d_8$*. Experimental conditions: see Figure 3.*

isopropanol-$d_8$ were included in the simulations. By this approach excellent agreement was obtained between the fits and the experimental spectra for $^{12}C$-$UQ_3^{-\cdot}$ in isopropanol-$d_8$ as well as for $Q_A^{-\cdot}$ and $Q_B^{-\cdot}$ reconstituted with $^{12}C$-$UQ_3$ (for an example, see Fig. 6). The principal values of the g-tensor were in good agreement with those obtained recently by Q-band EPR for fully deuterated $UQ_{10}^{-\cdot}$ [16] (see Table, last column).

The spectra of $^{13}C1$- and $^{13}C4$-$UQ_3^{-\cdot}$ were fitted by retaining the g- and hf-tensors of $^{12}C$-$UQ_3^{-\cdot}$, and adding the $^{13}C$-hf tensor, $A_{xx}$, $A_{yy}$, and $A_{zz}$ of $^{13}C$ as a free parameter. From symmetry considerations, this tensor was assumed to be collinear with the g-tensor axes. Calculations of the magnitude and orientation of the g-tensor have shown that these axes indeed coincide with the molecular axes within a few degrees even in the case of asymmetric hydrogen bonding to the CO groups [16].

From the results in Figures 3-5 it is clear that a large splitting due to the $^{13}C$ coupling occurs only for the $g_{zz}$ component of $^{13}C1$- and $^{13}C4$-$UQ_3^{-\cdot}$. In the $g_{xx}$ to $g_{yy}$ range only a broadening is observed. Therefore, $|A_{xx}|$ and $|A_{yy}|$ are much smaller than $|A_{zz}|$. The principal values of the $^{13}C$-hf tensor obtained from the fits are given in the Table. The signs of the principal values $A_{ii}$ (i = x, y, z) are discussed below.

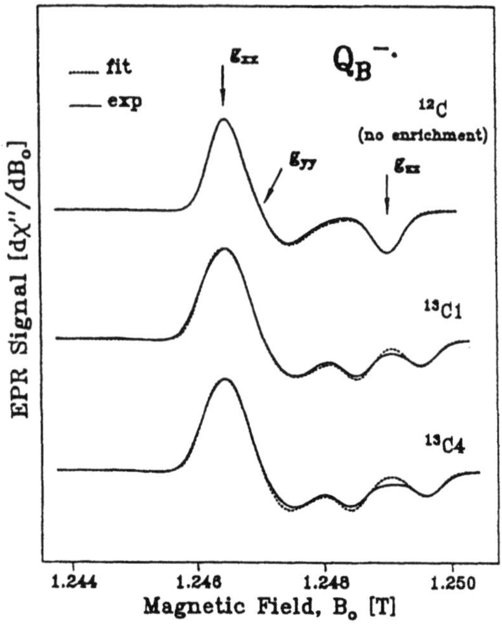


*Fig. 6: Experimental spectra (solid line) compared with fitted spectra (dashed) for $Q_B^{-\cdot}$ reconstituted with $UQ_3$. Top: unenriched ($^{12}C$), center: $^{13}C1$- enriched, and bottom: $^{13}C4$-enriched $UQ_3$. Agreements of similar quality between fits and experimental spectra were obtained for $UQ_3^{-\cdot}$ in isopropanol-$d_8$ and $Q_A^{-\cdot}$ (not shown). The g-tensor and $^{13}C$ hf tensor values are given in the Table. A basic Gaussian linewidth due to unresolved $^1H$ hf interactions (0.38 mT) and 2 additional D hf tensors ($A_{xx}$ = -0.11 mT, $A_{yy}$ = 0.21 mT, $A_{zz}$ = -0.10 mT) were obtained from the fit of the $^{12}C$-spectrum and were kept constant for the fits of the $^{13}C$-spectra (see text).*

A comparison of the Q-band EPR spectra of $Q_A^{-\cdot}$ in ZnbRCs, mutant HC (M266), reconstituted with $^{12}C$-, $^{13}C1$- and $^{13}C4$-$UQ_3$ is shown in Figure 3. $^{13}C1$-$Q_A^{-\cdot}$ clearly exhibits a much smaller $^{13}C$ splitting of the $g_{zz}$ component than $^{13}C4$-$Q_A^{-\cdot}$. This is in contrast with the spectrum of $UQ_3^{-\cdot}$ in isopropanol, where both positions show almost the same splitting (see Figure 5). The spectra were analyzed by using the same strategy as described for $UQ_3^{-\cdot}$ in isopropanol (data not shown),. The principal values of the g- and $^{13}C$-hf tensor are presented in the Table, together with the data of $Q_A^{-\cdot}$ reconstituted with $^{13}C$-labeled $UQ_{10}^{-\cdot}$ by van den Brink et al. [11].

The Q-band EPR spectra of $^{12}C$-, $^{13}C1$- and $^{13}C4$-$Q_B^{-\cdot}$ in ZnbRCs mutant HC (M266) are shown in Fig. 4. They are similar to the respective spectra of $UQ_3^{-\cdot}$ in isopropanol (see Figure 5). The $^{13}C$ hf splitting of the $g_{zz}$ component differs only slightly for $^{13}C1$- and $^{13}C4$-$Q_B^{-\cdot}$. Both, the $g_{xx}$ and $g_{yy}$ components of the $^{13}C1$- and $^{13}C4$-$Q_B^{-\cdot}$ spectrum are broadened, compared with $^{12}C$-$Q_B^{-\cdot}$. This was also found for $UQ_3^{-\cdot}$ in isopropanol. The spectra have been analyzed by fitting as described for $UQ_3^{-\cdot}$. The experimental and simulated spectra of $Q_B^{-\cdot}$ are compared in Fig. 6, demonstrating the quality of the fits. The g- and $^{13}C$ hf-tensor principal values are given in the Table.

**Table 1.** Principal values of the g -and $^{13}C$-hf tensors of $Q_A^{\cdot-}$ and $Q_B^{\cdot-}$ in ZnRCs mutant HC (M266) containing $UQ_3$ specifically $^{13}C$-labeled at positions 1 and 4 and of $^{13}C$-$UQ_3^{\cdot-}$ in isopropanol-$d_8$.

| | | $^{13}C$-$UQ_3^{\cdot-}$ [a] | | | $^{13}C$-$UQ_{10}^{\cdot-}$ [b] |
|---|---|---|---|---|---|
| | | isopropanol-$d_8$ | $Q_B^{\cdot-}$ | $Q_A^{\cdot-}$ | $Q_A^{\cdot-}$ |
| | $g_{xx}$ [c] | 2.00627(5) | 2.00624(5) | 2.00645(5) | 2.00649(5) |
| | $g_{yy}$ | 2.00531(5) | 2.00526(5) | 2.00532(5) | 2.00532(5) |
| | $g_{zz}$ | 2.00213(5) | 2.00213(5) | 2.00215(5) | 2.00210(5) |
| $^{13}C$-O position 1 | $A_{xx}(1)$ [d] | - 0.43 | - 0.39 | - 0.45 | 0.55(5) |
| | $A_{yy}(1)$ | - 0.37 | - 0.47 | - 0.52 | 0.65(5) |
| | $A_{zz}(1)$ | + 1.09(2) | + 0.99(2) | + 0.81(2) | 0.80(3) |
| | 1/3Tr**A**(1) [e] | + 0.10 | + 0.04 | - 0.05 | nd[h] |
| | ρ (C1)[f] | + 0.129 | + 0.123 | + 0.113 | nd |
| | ρ (O1)[g] | nd | + 0.178 | + 0.208 | nd |
| $^{13}C$-O position 4 | $A_{xx}(4)$ [d] | - 0.40 | - 0.36 | - 0.33 | < 0.25 |
| | $A_{yy}(4)$ | - 0.35 | - 0.37 | - 0.35 | < 0.25 |
| | $A_{zz}(4)$ | + 1.15(2) | + 1.15(2) | + 1.25(2) | 1.27(3) |
| | 1/3 Tr**A**(4)[e] | + 0.13 | + 0.14 | + 0.19 | nd |
| | ρ (C4)[f] | + 0.133 | + 0.131 | + 0.138 | nd |
| | ρ (O4)[g] | nd | + 0.178 | + 0.148 | nd |

a this work

b g-values from Q-band EPR on ZnRCs of *Rb. sphaeroides* reconstituted with fully deuterated $^{12}C$-$UQ_{10}$ [17]. $^{13}C$-hf tensor principal values $A_{ii}$ from van den Brink et al. [11]. The authors gave no signs for the $^{13}C$-hf tensor elements.

c g-values from fits of the respective $^{12}C$ spectra, see Figure 6, caption. Numbers in brackets are the errors in the last digit.

d $^{13}C$-hf tensor principal values [mT] from fits of the $^{13}C$-EPR spectra, leaving all other fit parameters obtained from the respective $^{12}C$-spectra constant (see text). Errors: ±0.02 mT for $A_{zz}$, the estimated errors for $A_{xx}$ and $A_{yy}$ are larger by a factor of 2 to 3.

e isotropic part of the respective $^{13}C$ hf-tensors

f Carbon π-spin density estimated from the traceless part A' of the hf tensor assuming axial symmetry and using the "uniaxial hf constant" $b_0$ ($^{13}C$) = 3.83 mT given in [22]. A' for $\rho_c(\pi) = 1$ is given by $A'_{11} = 2b_0$ and $A'_{22} = A'_{33} = -b_0$ [22]. Approximately 5% larger ρ(C)-values are estimated when effects of spin densities at the oxygens are additionally considered.

g Oxygen π-spin density estimated from the traceless $^{17}O$-hf tensor using the $^{17}O$-hf tensor components $A_{zz}$ (85 MHz for O1 and O4 in $Q_B$ and 75 MHz for O4 and 95 MHz for O1 in $Q_A$) and an isotropic value of -25 MHz from ref. [5] and a "uniaxial hf constant" $b_0$ ($^{17}O$) = - 6.01 mT from ref. [22].

h nd = not determined

# 4 Discussion

In this work we are interested in changes of the electronic structure of $Q_A^{-\cdot}$ and $Q_B^{-\cdot}$ in the bRC as a consequence of the protein surrounding. In principle, such changes should be reflected in all parameters of the spin Hamiltonian, i.e. the g- and hyperfine-tensor values.

The g-tensor components obtained from the spectral fits (see Table) show that the values for $Q_B^{-\cdot}$ are close to those of $UQ_3^{-\cdot}$ in isopropanol glass; whereas, the g-tensor of $Q_A^{-\cdot}$ shows deviations from that of $UQ_3^{-\cdot}$ *in vitro*. This g-shift of $Q_A^{-\cdot}$, in particular the increase of $g_{xx}$, was observed earlier in ZnbRCs containing deuterated $UQ_{10}$ and has been explained by a special environment of $Q_A^{-\cdot}$ altering its electronic structure [16]. Increased $g_{xx}$ values have been observed for quinone anion radicals in aprotic solvents in which no H-bonds are formed [21]. The detected shift for $Q_A^{-\cdot}$ (see Table) may therefore be explained by a strongly asymmetric binding situation which essentially leaves one of the carbonyl oxygens non or only weakly hydrogen bonded.

Effects of the environment on the g-tensor of an organic radical are generally fairly small and unspecific since the g-value represents only an integral property of the electronic wave function. Effects on the hf tensor values of the individual nuclei are expected to be more pronounced, in particular those of the CO groups. In the following the $^{13}C$ hfc's will therefore be analyzed in detail.

In our spectra only the $^{13}C$-hf tensor components along $g_{zz}$ could clearly be resolved, the $A_{xx}$ and $A_{yy}$ values obtained from the simulations have larger errors since they contribute only to a broadening of the respective superimposed $g_{xx}$ and $g_{yy}$ components in the powder EPR spectra. All the $^{13}C$ hf tensors show approximately axial symmetry (see Table).

The π-spin density at the carbon atoms can be obtained from the anisotropic components of the hf tensor $A'_{ii} = A_{ii} - 1/3\ \mathrm{TrA}$ ($i = x, y, z$) by assuming a uniaxial tensor and using the hf constant for $\rho_C\ (p_z) = 1$ given in [22], see Table. For the calculation of the $A'_{ii}$ values a knowledge of $1/3\ \mathrm{TrA} = a_{iso}$ is necessary, which in turn requires a determination of the signs of all $A_{ii}$ components in the Table. The isotropic $^{13}C$ hf values in solution were shown to be relatively small in various ubiquinone radical anions in alcohols and ethers [23]. The recent work of Nimz [24] showed that the sign of the isotropic $^{13}C$ hfc in the CO group of $UQ_0^{-\cdot}$ is small and positive in isopropanol and negative in aprotic solvents[1]. The small magnitude of the isotropic $^{13}C$ hfc requires that the principal values $A_{xx}$ and $A_{yy}$ in the Table are both negative, if $A_{zz} > 0$, since $\mathrm{Tr}A'_{ii} = 0$. A positive $A_{zz}$ value can safely be assumed for the CO carbon in benzoquinone anion radicals having a positive π-spin density at this carbon atom. This

---

[1] $A_{iso}$ is +0.05 mT in isopropanol and -0.14 mT in 1,2-dimethoxyethane/2-methyltetrahydrofurane for $^{13}C1$-labeled $UQ_0^{-\cdot}$.

assignment allowed us to calculate $a_{iso}$ and the $A'_{ii}$ values and to estimate the π-spin densities both at the C1 and C4 positions in all radicals studied. They are given in the Table together with some values for the oxygen π-spin densities[2] that were obtained earlier from independent experiments on $^{17}O$-$UQ_{10}$ anion radicals in the $Q_A$ and $Q_B$ sites of *Rb. sphaeroides* [5].

The analysis shows that the spin density in $UQ_3^{-\cdot}$ in isopropanol is only slightly asymmetric with respect to the two CO groups. This can be attributed to the asymmetric substitution pattern of the molecule and slightly different interactions with the protic alcoholic matrix. In the protein a stronger asymmetry is generally expected depending on the specific bonding situation.

For $Q_A^{-\cdot}$ a rather large difference between the carbon and oxygen spin densities at positions 1 and 4 is observed. The values obtained for $UQ_3$ containing bRCs, obtained in this work, are in very good agreement with those reported by van den Brink et al. [11] for $UQ_{10}$ in the $Q_A$ site. The strongly asymmetric spin density distribution in $Q_A^{-\cdot}$ with respect to the quinone anion radical *in vitro,* can only be explained by an asymmetry of the binding pocket, which is most probably caused by a strongly asymmetric H-bonding.

The shift of spin density in semiquinone anion radicals can be understood from a simple valence bond model [8,25]. The two schemes describing resonant structures associated with the asymmetric hydrogen bonds at the two carbonyl oxgens are shown below:

I

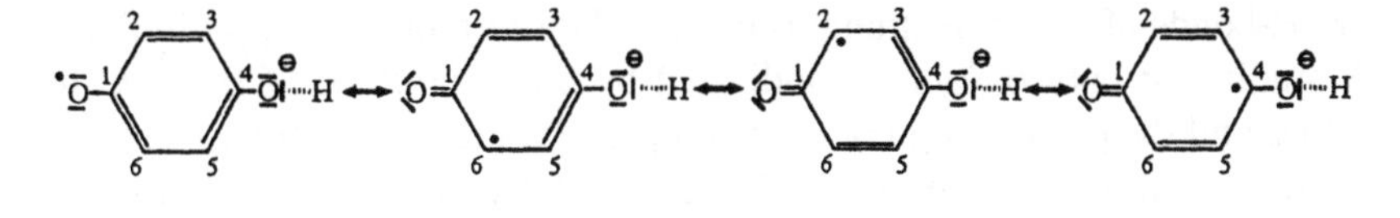

II

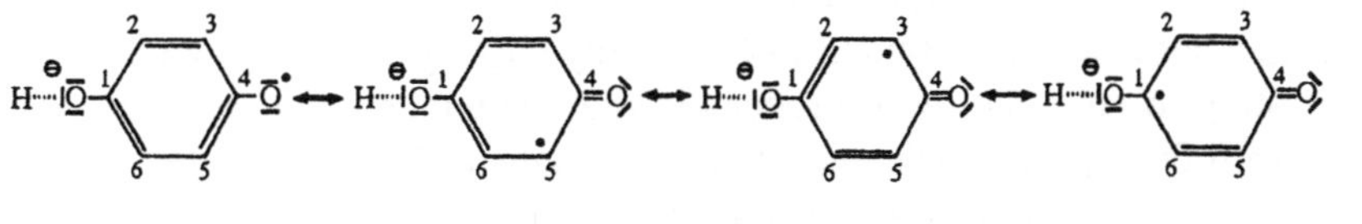

Scheme I represents a strong hydrogen bond to O4. This is equivalent to the approach of a partial positive charge, which increases the negative charge at the oxygen. This results in an increase in spin density at carbon 2,4,6 and O1. Scheme II represents a strong hydrogen bond to O1 resulting in an increase in

2 An assignment to O1 and O4 has only been possible by a concomitant determination of the $^{13}C$ hfc's [11].

spin density at carbon 1,3,5 and O4. Experimentally we observed a larger hf splitting with C4 than with C1 (see Table), showing that scheme I applies, i.e. the strong hydrogen bond is to the carbonyl oxygen O4.

The results of the EPR studies of $Q_A^{-\cdot}$ are in agreement with recent FTIR studies of $Q_A$ reconstituted with selectively $^{13}$C-labeled ubiquinones, in which a strong H-bond was reported for O4 [9,10]. No indication for hydrogen bonding was found for O1. This is consistent with binding affinity studies indicating that quinones in the $Q_A$ binding site are attached to the protein by only one of the CO groups [26,27]. In our earlier ENDOR studies of $Q_A^{-\cdot}$ in ZnbRCs we postulated two H-bonds of different strengths to $Q_A^{-\cdot}$ [5]. We are currently reinvestigating this problem by studying $Q_A^{-\cdot}$ in ZnbRC single crystals [28]. If there are indeed two H-bonds, it is possible that they are both formed to the same oxygen (O4), leaving O1 unattached to the protein. It has been shown in X-ray crystallographic studies of bRCs of *Rb. sphaeroides* that the binding pocket of $Q_A$ is quite large, allowing for some variability in quinone position [29-32]. From the X-ray studies H-bonds have been postulated to the backbone of Ala (M260) [29-32], to His (M219) [31,32] and also to Thr (M222) [29,30], see Fig. 2. In the light of the above-mentioned EPR [5,11,12] and FTIR data [9,10], His (M219) is the most likely candidate for a hydrogen bond with the Thr (M222) perhaps forming a second bond.

Inspection of the situation for $Q_B^{-\cdot}$ reveals only a slight bonding asymmetry as judged from the $^{13}$C spin densities. This is in agreement with the finding that two H-bonds of different strengths exist to the CO groups of $Q_B^{-\cdot}$ in the bRC [5]. A binding of both oxygens is indicated by the magnitude of the $^{13}$C hf tensor values and the positive sign of the isotropic $^{13}$C hfc's (see Table). These findings are in accord with the recent FTIR work by Brudler et al. [33] and Breton et al. [34] on $Q_B$ reconstituted with $^{13}$C1- and $^{13}$C4-labeled ubiquinone. Their data indicate weak symmetrical H-bonding to $Q_B$ and $Q_B^{-\cdot}$ [33,34]. Likely candidates for the two H-bonds are the amino acids Ser (L223) and His (L190), see Fig. 2. Our data show that the H-bond to O4 is only slightly stronger than to O1. It is interesting to note that the hydrogens in these H-bonds show different exchange rates with $D_2O$ [35]. Such studies are important with respect to the protonation of reduced $Q_B$ and nearby amino acid residues [2,3].

## 5 Conclusions

The analysis of the $^{13}$C hf tensors of $Q_A^{-\cdot}$ and $Q_B^{-\cdot}$ in ZnbRCs mutant HC (M266) reconstituted with selectively labeled $UQ_3$ at positions 1 and 4 have revealed the symmetry of the binding sites for both ubiquinones in the bRC. The symmetry of the electronic structure of both quinones is largely influenced by formation of H-bonds between the CO groups of the quinones and specific protein residues. For $Q_A^{-\cdot}$ it was shown that the binding is very asymmetric caused by formation of one (or two) strong H-bonds to only one CO group. The postulated very tight H-bridge to His (M219) may facilitate the electron transfer from $Q_A^{-\cdot}$ to $Q_B$ by electron delocalization onto the histidine. The asymmetry should also have a major impact on the electronic structure of this species changing the redox potential to the required value.

In contrast to $Q_A^{-\cdot}$, the secondary quinone $Q_B^{-\cdot}$ is much more symmetric and the two different H-bonds are weaker than the strong H-bond in $Q_A^{-\cdot}$. Consequently, $Q_B$ functions more like a quinone in protic solution and shows the usual two-electron-reduction and subsequent protonation to form the hydroquinone in the bRC.

The spectroscopic results presented in this work are still incomplete. To obtain more detailed information it is necessary to measure the full $^{13}C$ hf tensors for $Q_A^{-\cdot}$ and $Q_B^{-\cdot}$ in ZnbRC single crystals reconstituted with the respective labeled quinones, using both EPR and ENDOR techniques. An alternative approach would be to determine the hf tensor values from powder spectra by high field EPR [36].

The results demonstrate that magnetic resonance, together with other spectroscopic techniques, yields detailed information about specific cofactor-protein interactions (e.g. hydrogen bonding) in the different charged states of the bRC. This information is difficult, if not impossible, to obtain from x-ray diffraction data. Furthermore, by studying the electronic structure of the cofactors, the functional details of the electron transfer process can be revealed. In future studies of ZnbRCs containing $^{13}C$-labeled quinones the $^{13}C$ hfc will also be used as a very sensitive local probe to detect possible light-induced structural changes.

**Acknowledgements**

The authors are grateful to J.C. Williams and J.P. Allen for supplying the HC(M266) mutant, to R. Bittl and R. Fiege (TU Berlin) for developing the simulation and fit program and to M. Plato (FU Berlin) for providing a program for calculating dipolar hyperfine couplings. Helpful discussions concerning quinone binding in reaction centers with M.Y. Okamura (UC San Diego) and J. Breton (CEA Saclay) are gratefully acknowledged. This work was supported by DFG (Sfb 312, TP A4), NATO (CRG910468), NIH (GM13191) and Fonds der Chemischen Industrie.

## References


[1] Okamura, M.Y., Debus, R.J., Kleinfield, D. and Feher, G. (1982) in Function of Quinones in Energy Conserving Systems (Trumpower, B.L., ed.), Academic Press, New York

[2] Okamura, M.Y. and Feher, G. (1992) Annu. Rev. Biochem. 61, 861-896

[3] Shinkarev, V.P. and Wraight, C.A. (1993) in The Photosynthetic Reaction Center (Deisenhofer, J. and Norris, J.R., eds.), Vol. 1, pp. 194-255, Academic Press, San Diego

[4] Lubitz, W., Abresch, E.C., Debus, R.J., Isaacson, R.A., Okamura, M.Y. and Feher, G. (1985) Biochim. Biophys. Acta 808, 464-469

[5] Feher, G., Isaacson, R.A., Okamura, M.Y. and D.C. Rees (1989) in Antennas and Reaction Centers of Photosynthetic Bacteria (Michel-Beyerle, M.E., ed.), pp. 174-189, Springer Verlag, Berlin

[6] Debus, R.J., Feher, G. and Okamura, M.Y. (1986) Biochemistry 25:2276-2287

[7] Kirmaier, C., Holten, D., Debus, R.J., Feher, G. and Okamura, M.Y. (1986) Proc. Natl. Acad. Sci USA 83, 6407-6411

[8] Lubitz, W., Plato, M., Möbius, K. and Biehl, R. (1979) J. Phys. Chem. 83, 3402-3413

[9] Brudler, R., de Groot, H.J.M., van Liemt, W.B.S., Steggerda, W.F., Esmeijer, R., Gast, P., Hoff, A.J., Lugtenburg, J. and Gerwert, K. (1994) EMBO J. 13, 5523-5530

[10] Breton, J., Boullais, C., Burie, J.-R., Nabedryk, E. and Mioskowski, C. (1994) Biochemistry 33, 14378-14386

[11] van den Brink, J.S., Spoyalov, A.P., Gast, P., van Liemt, B.S.W., Rasp, J., Lugtenburg, J. and Hoff, A.J. (1994) FEBS Lett. 353, 273-276

[12] Bosch, M.K., Gast, P., Hoff, A.J.. Spoyalov, A.P. and Tsvetkov, Yu.D. (1995) Chem. Phys. Lett. 239, 306-312

[13] Williams, J.C., Paddock, M.L., Feher, G. and Allen, J.P. (1991) Biophys. J. 59, 142a

[14] K. Ferris (1987), PhD Thesis, Princeton University pp 139-143

[15] Paddock, M.L., Rongey, S.H., Feher, G. and Okamura, M.Y. Okamura (1989) Proc. Natl. Acad. Sci. USA 86, 6602-6606

[16] Isaacson, R.A., Lendzian, F., Abresch, E.C., Lubitz, W. and Feher, G. (1995) Biophys. J. 69, 311-322

[17] A. Stesmans and G. De Vos (1986) Phys. Rev. B. 34:6499-6502

[18] A. Stesmans and van Gorp, G. (1989) Rev. Sci. Instr. 60, 2949-2952

[19] Rieger, P.H. (1982) J. Magn. Reson. 50, 485-489

[20] Press, W.H., Flannery, B.P., Teukolsky, S.A. and Vetterling, W.T. (1988) Numerical Recipes in C: The Art of Scientific Computing, Cambridge University Press, New York

[21] Burghaus, O., Plato, M., Rohrer, M., Möbius, K., MacMillan, F. and Lubitz W. (1993) J. Phys. Chem. 97, 7639-7647

[22] Wertz, J.E. and Bolton, J.R. (1972) Electron Spin Resonance, McGraw-Hill, New York
[23] Samoilova, R.I., van Liemt, W., Steggerda, W.F., Lugtenburg, J., Hoff, A.J., Spoyalov, A.P., Tyryshkin, A.M., Gritzaqn, N.P. and Tsvetkov, Y.D. (1994) J. Chem. Soc. Perkin. Trans. 2, 609-614
[24] Nimz, O. (1995) Diploma Thesis, Technische Universität Berlin
[25] MacMillan, F., Lendzian, F., Renger, G. and Lubitz, W. (1995) Biochemistry 34, 8144-8156
[26] Warncke, K. and Dutton, P.L. (1993) Proc. Natl. Acad. Sci. USA 90, 2920-2924
[27] Warncke, K., Gunner, M.R., Braun, B.S., Gu, L., Yu, C.-A., Bruce, J.M. and Dutton, P.L. (1994) Biochemistry 33, 7830-7841
[28] Isaacson, R.A., Abresch, E.C., Feher, G., Lendzian, F. and Lubitz W. (1994) Biophys. J. 66, A229
[29] Allen, J.P., Feher, G., Yeates, T.O., Komiya, H. and Rees D.C. (1988) Proc. Natl. Acad. Sci. USA 85, 8487-8491
[30] El-Kabbani, O., Chang, C.-H., Tiede, D., Norris, J. and Schiffer, M. (1991) Biochemistry 30, 5361-5369
[31] U. Ermler, Fritzsch, G., Buchanan, S.K. and Michel, H. (1994) Structure 2, 925-936
[32] Chirino, A.J., Lous, E.J., Huber, M., Allen, J.P., Schenck, C.C., Paddock, M.L., Feher, G. and Rees, D.C: (1994) Biochemistry 33, 4584-4593
[33] Brudler, R., de Groot, H.J.M., van Liemt, W.B.S., Steggerda, W.F., Esmeijer, R., Gast, P., Hoff, A.J., Lugtenburg, J. and Gerwert, K. (1995) FEBS Lett. 370, 88-92
[34] Breton, J., Boullais, C., Berger, G., Mioskowski, C. and Nabedryk, E. (1995) Biochemistry, in press
[35] Paddock, M.L, Abresch, E., Isaacson, R.A:, Feher, G. and Okamura, M.Y. (1995) Biophys. J. 68, A246
[36] Möbius, K. (1993) in Biological Magnetic Resonance (Berliner, L.J., Reuben, J., eds.), Vol. 13, pp. 253-274, Plenum Press, N.Y.

# Alternate Reaction Center Acceptors: Spectral Signatures and Redox Properties of Pheoporphyrin, Zinc and Copper Bacteriochlorin Anion Radicals

M. W. Renner,[1] Y. Zhang,[1] D. Noy,[2] A. Scherz,[2] K. M. Smith[3] and J. Fajer[1]

[1] Department of Applied Science, Brookhaven National Laboratory, Upton, New York 11973, USA
[2] Biochemistry Department, Weizmann Institute of Science, 76100 Rehovot, Israel
[3] Department of Chemistry, University of California, Davis, California 95616, USA

**Abstract**. Optical and EPR spectra and redox data are presented for pheoporphyrin $a_5$, zinc and copper bacteriochlorophylls and their anion radicals. The compounds are considered as candidates for replacement of the bacteriopheophytins or accessory bacteriochlorophylls in isolated photosynthetic reaction centers where they would offer specific redox or spectral advantages for studies of the primary charge separation.

**Key words**. Anion radicals, EPR, optical spectra, redox potentials, pheoporphyrin $a_5$, copper bacteriochlorophyll, zinc bacteriochlorophyll

## 1. Pheoporphyrin $a_5$

Marine algae contain, besides the ubiquitous chlorophyll *a* (Chl), up to three slightly different Chl *c* chromophores in their light harvesting complexes.[1] Unlike Chl *a* (or *b*), the Chls *c* are porphyrins rather than chlorins but retain the exocyclic ring V found in all photosynthetic Chl derivatives. The chromophores are thus classified as pheoporphyrins.

As part of an investigation of the physical and chemical properties of pheoporphyrins, we have determined[2] the molecular structure of a pheoporphyrin[3] that serves as a simple model of Chl *c*. Besides the obvious change of ring IV to a pyrrole ring that establishes the chromophore as a porphyrin rather than a chlorin, the molecule retains many of the structural features due to the exocyclic ring V found in chlorophylls and bacteriochlorophylls.[4]

A direct consequence of the porphinoid electronic structure is a blue-shift of the low energy absorption bands of the Chls *c* and the model, and the loss of the high oscillator strengths of these Qy transitions that enhance the light harvesting properties of (bacterio)chlorophylls.

The pheoporphyrin model of Chl *c* (or the authentic chromophores themselves) provide the interesting possibility of replacing the native electron acceptors in bacterial reaction centers (RCs). Zinth, Scheer and coworkers[5,6] have shown recently that the bacteriopheophytin *a* (BPheo) acceptor in *Rb. sphaeroides* can be replaced by pheophytin *a* (Pheo), a *chlorin* derivative. This exchange offers two advantages: 1) Pheo is harder to reduce than BPheo by 0.13 V[7,8] and thus raises the energy level of the new acceptor closer to that of the bacteriochlorophyll (BChl) "primary" acceptor in the electron transfer sequence: $(BChl)_2 \rightarrow BChl \rightarrow Pheo$. 2) Because the Qy band of Pheo is blue-shifted relative to that of BPheo,[7,8] the exchange of Pheo for BPheo unmasks the absorption bands of the BChl anion, and Zinth et al.[5,6] conclude that *BChl is a real electron acceptor* in the primary charge separation of bacterial RCs as opposed to simply participating in a super exchange mechanism.

Molecular orbital calculations[9] predict that the reduction potentials of analogous porphyrins, chlorins and bacteriochlorins should be the same. Indeed, the reduction potential of the free base pheoporphyrin falls within 40 mV of that of Pheo *a*. (The differences in reduction potentials of BPheo *a* and Pheo *a* are not due to the fact that one is a bacteriochlorin and the other a chlorin. The difference results from the presence of an acetyl group on ring I of the BPheo *a*. A modified BPheo with a vinyl group should have the same reduction potential as Pheo *a* and, conversely, a pheophytin with an acetyl group should have the same reduction potential as BPheo *a*. A significant effect on rates of electron transfer due to the introduction of a modified 3-vinyl BChl in RCs has already been reported by Finkele et al.[10] In further agreement with these trends, $3^1$-hydroxy BChl *a* is harder to reduce than BChl *a* by 0.13 V.[11,12])

*In vitro* reduction potentials obtained in dimethylformamide by cyclic voltammetry are: BPheo *a*: -0.51V, Pheo *a*: -0.64V and pheoporphyrin: -0.60V (vs NHE). Substitution of the pheoporphyrin should therefore have the same effect as Pheo on the energetics of electron transfer in an exchanged RC and, in addition, would have the additional benefit of the further blue-shifted absorption bands of the new acceptor as shown in the optical spectra of the pheoporphyrin and its anion radical (Figure 1). Note the significant blue-shifts of the low energy bands and their low oscillator strengths that would no longer interfere with the BChl absorption bands in a reconstituted RC. (Equivalent data for the reductions of BPheo and Pheo *a* can be found in references 7 and 8).

It is perhaps not too surprising that Pheo *a* can replace BPheo *a* in RCs. A comparison of the crystallographic data for methyl bacteriopheophorbide *a* and methyl pheophorbide *a* show their molecular dimensions to be similar, except for the orientations of the substituents.[4] The pheoporphyrin considered here should also fit the same BPheo pocket in the RC: an overlap of the BPheo a[4] and the pheoporphyrin[2] skeletons shows the two structures to be comparable. The BPheo protein pocket should therefore readily accommodate a pheoporphyrin substitution.

## 2. Zinc and Copper Bacteriochlorophylls *a*

Exchange of the accessory BChls in RCs by other metallobacteriochlorophylls offers additional spectral probes and signatures for the controversial role of these chromophores as electron acceptors in the primary charge separation in bacterial RCs.

Zn derivatives are appealing as exchange candidates, particularly since a Zn $13^2$-hydroxy BChl *a* (Zn-HOB) has already been successfully substituted for BChl *a* in *R. sphaeroides* RCs using the methods developed by Struck and Scheer.[13,14] Zn bacteriochlorins offer several advantages: a) Their redox chemistry closely parallels that of Mg derivatives as do the electronic profiles of the anion and cation $\pi$ radicals.[15,16] b) Unlike isolated BChls that form penta- and hexacoordinated complexes in solution, Zn porphyrins, chlorins and bacteriochlorins preferentially form pentacoordinated species with nitrogenous axial ligands[4], and thus closely mimic the pentacoordinated Mg of BChls *a* and *b* in wild-type RCs.[17,18] c) An extensive body of high precision structural data already exists for isolated pentacoordinated Zn porphyrins and chlorins and it provides a benchmark for the use of Zn BChls as EXAFS probes of the metal environment in reconstituted RCs.[13]

In contrast to the diamagnetic BChl and Zn BChl, Cu BChl is paramagnetic and, as shown further on, it becomes diamagnetic upon reduction to the π anion radical because of antiferromagnetic coupling of the Cu and radical spins. Cu porphyrins bind axial ligands and the shifts of the Qx absorption band of Cu BChl in tetrahydrofuran (THF) relative to $CH_2Cl_2$ suggest that the molecule is also ligated. Cu BChl is therefore likely to replace an accessory BChl in RCs where it would act as an EPR reporter of its redox state. Furthermore, as shown in Table 1, Cu BChl is considerably easier to reduce than BChl and, in fact, has practically the same reduction potential as BPheo in THF.[12] Reaction centers reconstituted with Cu BChl in the accessory BChl sites should therefore exhibit rates of primary charge separation similar to those recently found for RCs reconstituted with $13^2$-hydroxy Ni BChl.[19] The reduction potential of the latter is reported as ~300 mV more positive than that of BChl,[19,20] i.e. approximately equal to or actually *lower* than that of BPheo[12,21]

Redox potentials for the modified BChls discussed above are listed in Table 1 and compared there with those of BChl and BPheo *a*. The trend towards less negative reduction potentials, i.e. BChl>Zn BChl>Cu BChl ≈BPheo is that expected for chlorins and porphyrins,[15,21,22] and generally correlates with the electronegativity[22] of the metals. Also included in Table 1 are the redox data for $3^1$-OH BChl *a*[11] which illustrate the effects of replacing the electron-withdrawing 3-acetyl group that facilitates reduction with a hydroxy substituent thereby making the molecule *harder* to reduce (and easier to oxidize).

Optical data that reflect the reduction of Zn BChl in the presence of imidazole are presented in Figure 2 which also displays the difference spectrum between the anion radical and the parent compound, i.e. the spectral changes that would be expected upon reduction of Zn BChl ligated by a histidine in a reconstituted RC. Also shown in Figure 2 are the difference spectra obtained in the *absence* of imidazole: note the shift of the Qx band with the ligand that is typical of pentacoordinated Zn porphyrin derivatives.

The EPR spectrum of the Zn BChl π anion radical in THF at 298K is shown in Figure 3. Addition of imidazole does not alter the EPR spectrum setting an upper limit of ~0.5G for a hyperfine coupling constant to the axial nitrogen, based on computer simulations. The spectrum of the Zn anion radical is remarkably similar to that previously found for BPheo and, not surprisingly, Monte Carlo simulations predict the major hyperfine coupling constants of the nitrogens, methyl and meso protons to be comparable in the two compounds, and in BChl as well.[21,23]

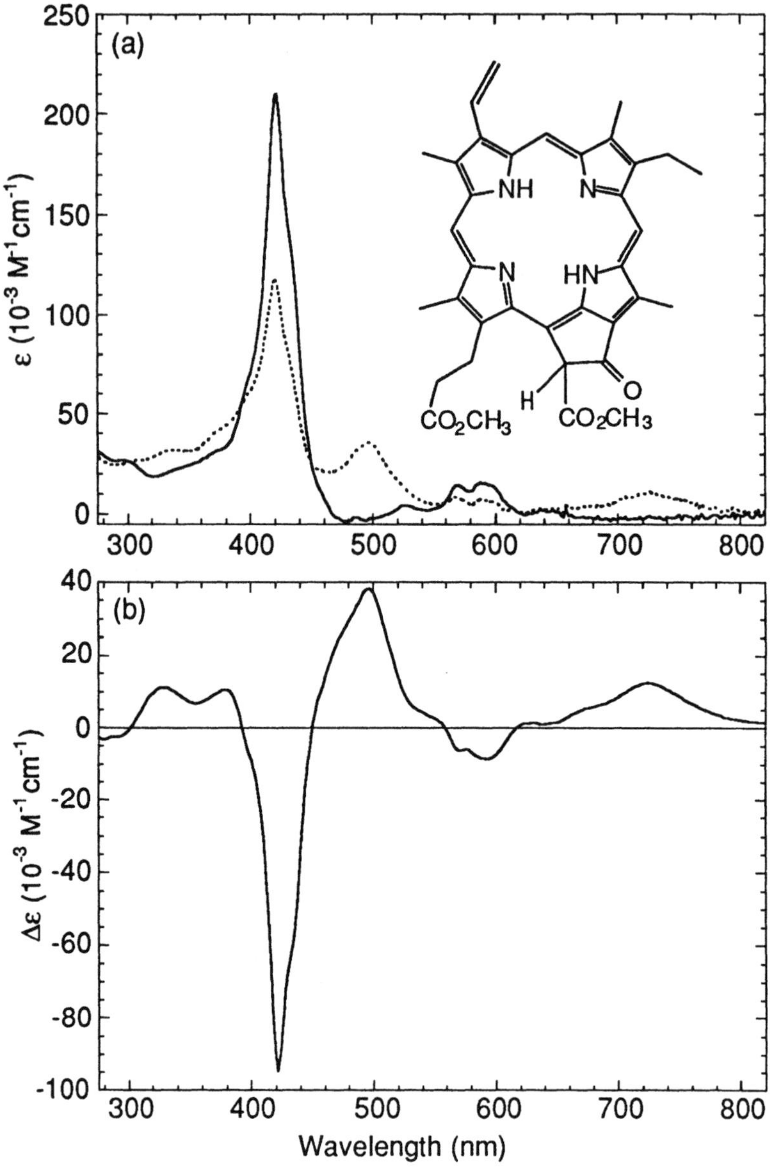


**Figure 1.** a) Optical spectra in dimethylformamide containing 0.1M $Bu_4NClO_4$: Pheoporphyrin $a_5$ ( —— ), anion radical ( ---- ). b) Difference spectrum: anion minus parent.

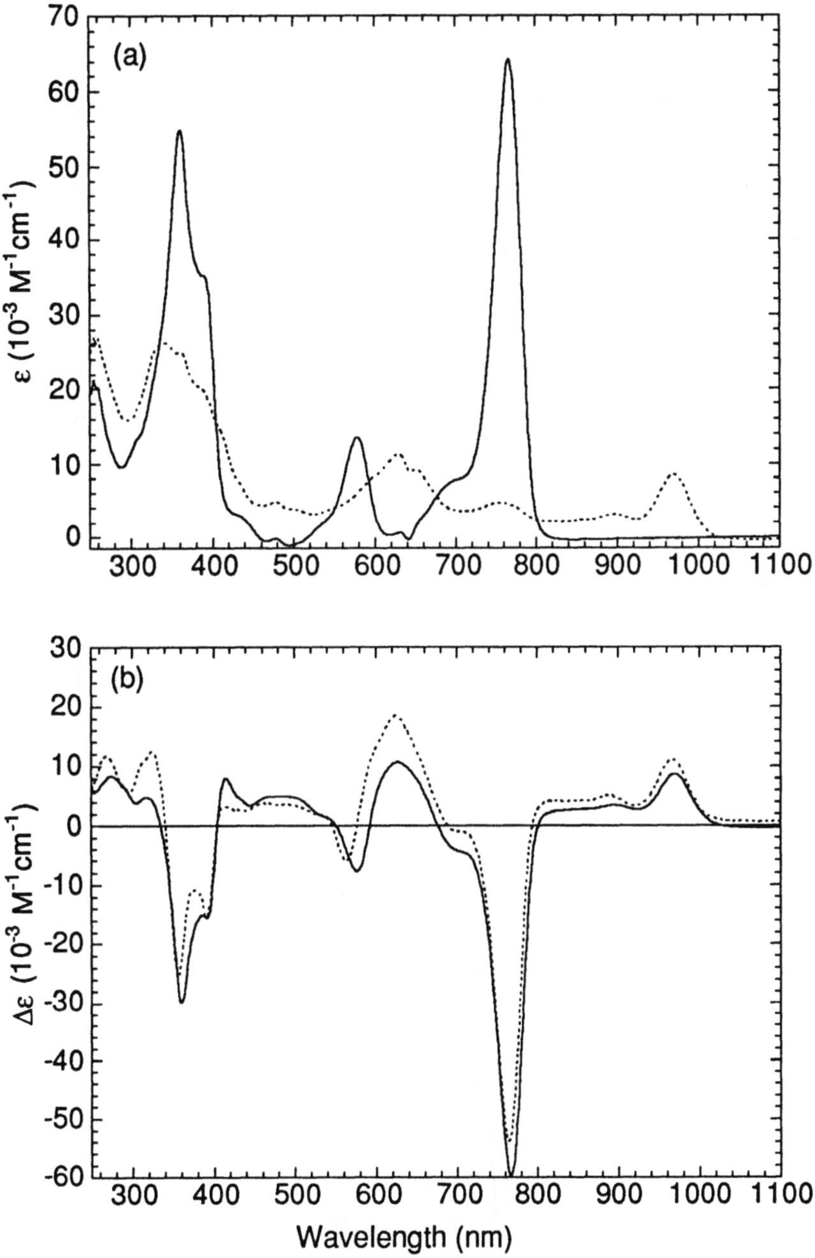


**Figure 2.** Optical spectra in THF containing 0.1M $Bu_4ClO_4$. a) Zn BChl + Imidazole ( —— ), anion radical ( ---- ). b) Anion minus parent difference spectra in the presence of Im ( —— ) and the absence of Im ( ---- ).

**Table 1**. Reduction and Oxidation Potentials of Modified BChls (vs NHE)[a]

| Compound | Solvent | $E^1_{red}$ | $E^2_{red}$ | $E^1_{ox}$ | $E^2_{ox}$ |
|---|---|---|---|---|---|
| BChl *a*[12] | THF | -0.83 | -1.17 | 0.76 | 1.07 |
| Zn BChl | THF | -0.67 | -1.07 | 0.84 | 1.19 |
| Zn BChl | THF/Im | -0.73 | -- | 0.77 | -- |
| Cu BChl | THF/Im | -0.57 | -0.93 | 0.88 | -- |
| Cu BChl | THF | -0.57 | -0.95 | 0.91 | 1.29 |
| BPheo[12] | THF | -0.60 | -0.93 | 1.12 | -- |
| $3^1$-OH BChl[11] | THF | -0.96 | -- | 0.64 | 0.97 |

[a] $E_{1/2}$ values were measured by cyclic voltammetry in tetrahydrofuran containing 0.1M tetrabutyl ammonium perchlorate. THF/Im solutions contain 1% imidazole. $E^1$ and $E^2$ are the first and second reduction or oxidation steps, respectively.

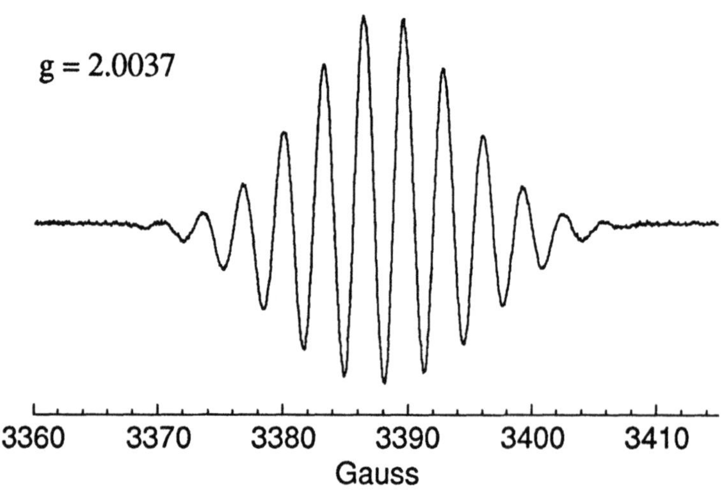


**Figure 3.** Second derivative EPR spectrum of the Zn BChl anion radical in THF at room temperature. The counterion is $Na^+$.

Besides the differences in redox potentials between BChl and Zn BChl that may alter rates of electron transfer in modified RCs, reconstitution with Zn chromophores allows the use of synchrotron radiation (EXAFS and XANES) to uniquely focus on the Zn and probe the immediate environment of the putative electron acceptor in its resting state and, possibly, in its (trapped) reduced state.[13,24]

Crystallographic studies of synthetic Zn porphyrins offer some insights into what to expect for Zn BChl in a RC. In Zn(II) -5,10,15,20-tetraphenylbacteriochlorin (ZnTPBC) which is pentacoordinated with pyridine as axial ligand,[4] the Zn-N distances range between 2.043 and 2.122(4)Å for an average of 2.08(4)Å, with a distance of 2.164(5)Å to the axial pyridine. For Zn BChl in toluene at 110°, our EXAFS results yield an environment of 4.0 nitrogens with average Zn-N distances of 2.05Å (with $\Delta\ \sigma^2 = 0.004Å^2$ and $\chi^2=0.009$). Upon addition of the strong nitrogenous base piperidine, the EXAFS results can be fitted to 4.0 nitrogens with Zn-N=2.05Å ($\Delta\ \sigma^2=0.007$) and 1.0 nitrogen with Zn-N=2.11Å ($\Delta\ \sigma^2 = -0.006$), in reasonable agreement with the crystallographic Zn-N distances in ZnTPBC·pyridine (Recall that EXAFS distances are typically reliable to ±0.02Å[24])

Chen et al. recently reported an EXAFS study of Zn-$13^2$-hydroxy bacteriochlorophyll.[13] The Zn-N distances for the isolated chromophore (as a solid) were the same as found here for Zn BChl: 2.05(2)Å. In RCs reconstituted with the Zn HOBChl, the Chen et al.[13] EXAFS results yielded 5(±1) nitrogens with Zn-N=2.01(4)Å. A pedestrian reason for the shorter distances in the RC is simply a larger error inherent to the more dilute RCs. Other possibilities include a shorter Zn-N distance to the histidine that would shorten the observed average Zn-N since metal-imidazole distances in synthetic porphyrins tend to be ~0.1Å shorter than metal-pyridine distances.[4] Another interesting possibility for the shorter Zn-N distances is that the chromophore is more distorted in the RC. In nonplanar ruffled porphyrins,[26,27] conformations not unlike those observed in bacterial antenna and RC complexes,[17,18,25] metal-nitrogen distances tend to shorten by as much as 0.05Å. A combination of these two trends would seem to offer a reasonable explanation for the Zn-$13^2$-hydroxy BChl results in the modified RC. Chen et al.[13] also reconstituted RCs with Ni(II) HOBChl and found the Ni-N distances to *increase* upon insertion into the RC. Based on the changes in XANES, they also concluded that the Ni(II) becomes pentacoordinated in the RC. Our own crystallographic results for axially coordinated nonplanar Ni(II) porphyrins[26,27] provide a ready explanation for the effect observed by Chen et al.: the longer Ni-N distances are due principally to a *change in spin state* of the Ni(II) (from low spin to high spin) and also reflect a Ni-N distance of ~2.1Å to the axial histidine.[28]

Optical spectra obtained on electrochemical reduction of Cu BChl in THF are shown in Figure 4. The optical changes are similar to those observed on reduction of ZnBChl and are characteristic of bacteriochlorin reductions[7,12,13,20,21] Unlike the Zn BChl whose Qx band in THF shifts on addition of imidazole, the Cu derivative does not. However, its Qx absorption band is already shifted relative to that in $CH_2Cl_2$ which indicates that Cu BChl is already ligated in THF and that imidazole does not alter its coordination number.

Cu(II) porphyrin derivatives are paramagnetic because of the $d^9$ configuration of Cu(II). The unpaired electron resides in the $d_{x^2-y^2}$ orbital which leads to characteristic

Cu(II) porphyrin and chlorin EPR spectra: four peaks due to the nuclear spin of 3/2 of the Cu, and superhyperfine interactions with the four nitrogens of the porphyrin. The EPR spectrum of CuBChl in a 2-methyl THF glass at 110°K is shown in Figure 5. The general features of the spectrum resemble those recently reported for Cu(II) chlorophylls.[29] However, simulations of the spectrum yield smaller Cu parameters: $A_\perp$=19G, $A_\parallel$=163G with $g_\perp$=2.045 and $g_\parallel$=2.212. Interestingly, comparable Cu parameters and g-values: $A_\perp$=19.0G, $A_\parallel$=183G, $g_\perp$=2.050 and $g_\parallel$=2.216 have been observed in Cu(II) cytochrome *c*[30] in which the iron of the heme enzyme has been replaced by Cu but the metal coordination is retained, i.e. the Cu is *hexacoordinated.* We noted above that the Qx absorption band of CuBChl in THF indicates that the Cu is complexed by THF. The similarity between the Cu BChl and Cu cytochrome *c* EPR parameters thus reinforce this conclusion.

Chemical reduction of Cu BChl with sodium amalgan causes its EPR spectrum to diminish as illustrated in Figure 5. Since the optical spectrum of the reduced product (Figure 4) is clearly that of a bacteriochlorin π anion, the product is *not* a Cu(I) species and must therefore be a Cu(II) π anion radical. No EPR signal is detected at half-field that would indicate either a triplet or a dimeric species in which the porphyrin spins are coupled intermolecularly. The lack of an EPR signal in the Cu BChl anion is thus ascribable to either fast relaxation due to the neighboring unpaired electrons on the Cu and the BChl or to *antiferromagnetic* coupling between them. An equivalent antiferromagnetic coupling that yields diamagnetic species has been observed in Cu(II) porphyrin π cation radicals.[31]

Further reduction of the Cu BChl anion should yield a Cu(II) bacteriochlorin π *dianion* which should again be paramagnetic and exhibit a new Cu EPR spectrum because the bacteriochlorin π dianion is expected to be diamagnetic. Indeed, a new Cu signal begins to appear as reduction of the parent Cu BChl proceeds (Figure 5).

The disappearance of the very characteristic EPR signal of Cu BChl on reduction to its π anion radical suggests that the compound would act as a very specific EPR reporter of its redox state in a reconstituted RC. As noted above, the tendency of the compound to bind axial ligands suggests that it would substitute at the accessory BChl site(s). (Note also that Cu porphyrins have been successfully substituted into cytochrome *c*[30]). If so, the "resting" RC should exhibit a characteristic Cu bacteriochlorin EPR spectrum, particularly at low temperatures. Recall also that low temperature techniques have allowed the primary acceptors to be trapped and examined by EPR.[32] Therefore, if the electron ejected from the special pair in the primary charge separation is localized on the Cu BChl at low temperatures, its EPR signal should disappear (or diminish by 50% if both accessory BChls have been replaced).

**Acknowledgements**

We thank K.M. Barkigia, L.R. Furenlid and D. Melamed for their assistance and the authors of reference 20 for communicating their results prior to publication. The work at Brookhaven National Laboratory was supported by the Division of Chemical Sciences, U.S. Department of Energy, under Contract DE-AC02-76CH00016.

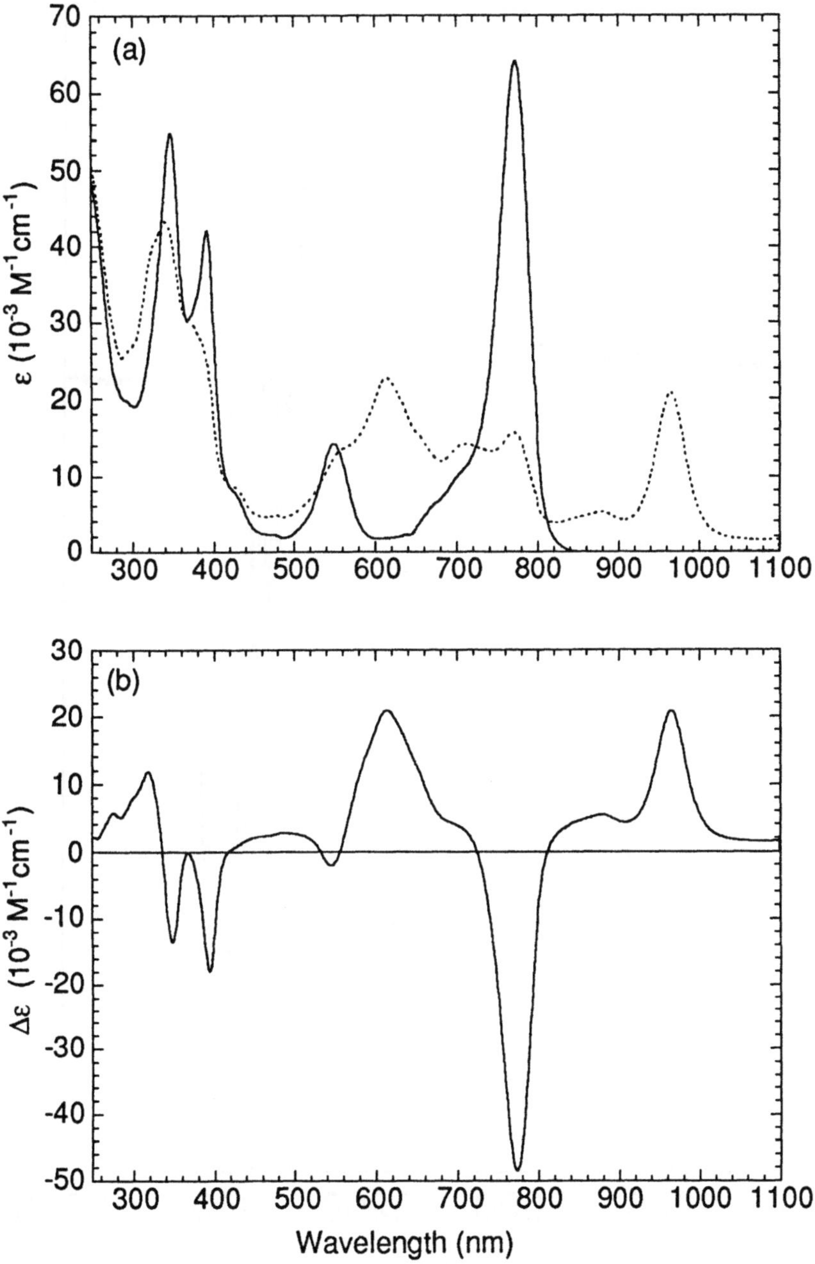


**Figure 4.** Optical spectra in THF containing 0.1M $Bu_4NClO_4$. a) Cu BChl ( —— ), Cu BChl anion ( ---- ). b) Difference spectrum: anion minus parent.

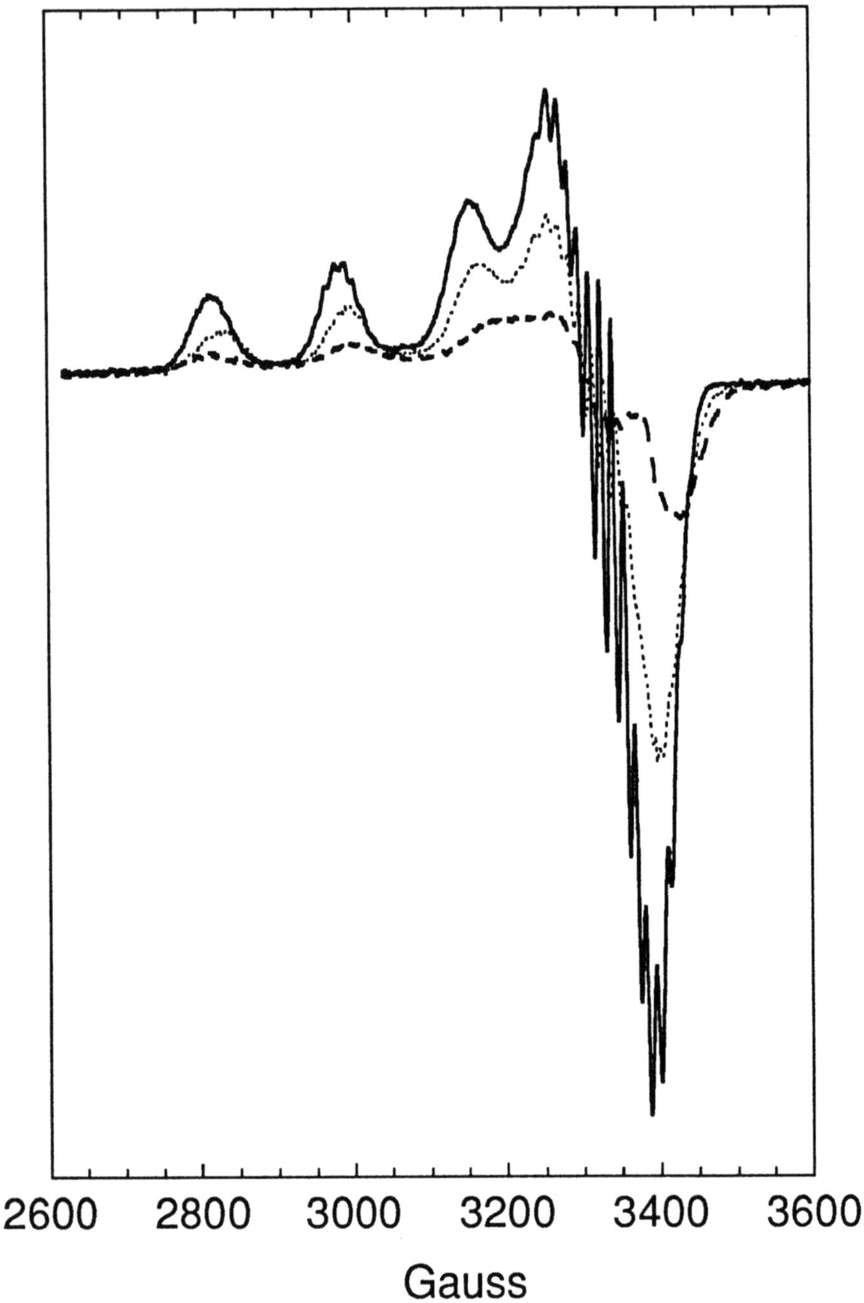


**Figure 5.** EPR spectra in 2-methyl THF at 110 °K. Cu BChl ( —— ), intermediate stage in the reduction of the parent compound to the monoanion ( ..... ), onset of the dianion features at ~25 % of the initial Cu BChl concentration ( ---- ). Counterions are $Na^+$.

**References**


1. Owens, T.G. and Wold, E.R. (1986) *Plant Physiol.*, 80, 732.
2. Barkigia, K.M.; Melamed, D., Smith, K.M. and Fajer, J. Unpublished results.
3. Kenner, G.W., McCombie, S.W. and Smith, K.M. (1974) *J. Chem. Soc., Perkins Trans. 1*, 527.
4. Barkigia, K.M. and Fajer, J. (1993) In *The Photosynthetic Reaction Center*, J. Deisenhofer and J.R. Norris, Eds., Academic Press, San Diego, Vol. II, p. 513.
5. Schmidt, S., Arlt, T., Hamm, P., Huber, H., Nagele, T., Wachtveitl, J., Meyer, M., Scheer, H. and Zinth, W. (1994) *Chem. Phys. Letters*, 223, 116.
6. See also: Shkuropatov, A.Y. and Shuvalov, V.A. (1993) *FEBS Letters*, 322, 168.
7. Fajer, J., Brune, D.C., Davis, M.S., Forman, A. and Spaulding, L.D. (1975) *Proc. Natl. Acad. Sci. USA*, 72, 4956.
8. Fujita, I., Davis, M.S. and Fajer, J. (1978) *J. Am. Chem. Soc.*, 100, 6280.
9. Fajer, J. (1991) *Chemistry & Industry*, 869.
10. Finkele, U., Lauterwasser, C., Struck, A., Scheer, H. and Zinth, W. (1992) *Proc. Nat'l. Acad. Sci. USA* 89, 9514.
11. Zhang, Y., Moore, L., Boxer, S.G. and Fajer, J. Unpublished results.
12. Cotton, T.M. and Van Duyne, R.P. (1979). *J. Am. Chem. Soc.* 101, 7605.
13. Chen, L.X., Wang, Z., Hartwich, G., Katheder, I., Scheer, H., Scherz, A., Montano, P.A. and Norris, J.R. (1995) *Chem. Phys. Letters* 234, 437.
14. Struck, A. and Scheer, H. (1990) *FEBS Letters* 261, 385.
15. Fajer, J., Borg, D.C., Forman, A., Dolphin, D. and Felton, R.H. (1973) *J. Am. Chem. Soc.* 95, 2739.
16. Fajer, J. and Davis, M.S. (1979) In *The Porphyrins*, Dolphin, D. Ed., Academic Press, New York, Vol. IV, p. 197.
17. Deisenhofer, J., Epp, O., Sinning, I. and Michel, H. (1995) *J. Mol. Biol.* 246, 429.
18. Ermler, U., Fritzch, G., Buchanan, S.K. and Michel, H. (1994) *Strucure* 2, 925.
19. Frieze, M., Hartwich, G., Ogrodnik, A., Scheer, H. and Michel- Beyerle, M.E. (1995) *Biophys. J.* 68, A367.
20. Geskes, C., Hartwich, G., Scheer, H., Mantele, W. and Heinze, J. Private communication.
21. Fajer, J., Brune, D.C., Davis, M.S., Forman, A. and Spaulding, L.D. (1975) *Proc. Nat'l. Acad. Sci. USA* 72, 4956.
22. Watanabe, T. and Kobayaski, M. (1991) in *Chlorophylls*, Scheer, H., Ed., CRC Press, Boca Raton, p. 287.
23. Lubitz, W., Lendzian, F. and Möbius, K. (1981) *Chem. Phys. Letters* 81, 235.
24. Fajer, J., Hanson, L.K., Zerner, M.C. and Thompson, M.A. (1992) In *The photosynthetic bacterial reaction center II*, Breton J. and Vermiglio, A. Eds., Plenum Press, New York, p. 33.
25. Tronrud, D.E., Schmid, M.F. and Matthews, B.W. (1986) *J. Mol. Biol.* 188, 443.
26. Barkigia, K.M., Renner, M.W., Furenlid, L.R., Medforth, C.J., Smith, K.M., Fajer, J. (1993) *J. Am. Chem. Soc.* 115, 3627.

27. Renner, M.W., Furenlid, L.R., Barkigia, K.M., Forman, A., Shim, H.K., Simpson, D.J., Smith, K.M. and Fajer, J. (1991). *J. Am. Chem. Soc.* 113, 6891.
28. Renner, M.W., Melamed, D., Barkigia, K.M. and Fajer, J. Unpublished results for high spin Ni(II) porphyrins with imidazole axial ligands.
29. Nonomura, Y., Yoshioka, N. and Inoue, H. (1994) *Inorg. Chim. Acta* 224, 181.
30. Findlay, M.C., Dickinson, L.C. and Chien, J.C.W. (1977). *J. Am. Chem. Soc.* 99, 5168.
31. Renner, M.W., Barkigia, K.M., Zhang, Y., Medforth, C.J., Smith, K.M. and Fajer, J. (1994) *J. Am. Chem. Soc.* 116, 8582.
32. See, for example, Davis, M.S., Forman, A., Hanson, L.K., Thornber, J.P. and Fajer, J. (1979) *J. Phys. Chem.* 83, 3325.

# PROTEIN–QUINONE INTERACTIONS IN PHOTOSYNTHETIC BACTERIAL REACTION CENTERS INVESTIGATED BY LIGHT–INDUCED FTIR DIFFERENCE SPECTROSCOPY

Jacques Breton[1], Eliane Nabedryk[1], Charles Mioskowski[2], & Claude Boullais[2]
[1]SBE, [2]SMM, DBCM, CEA–Saclay, 91191 Gif–sur–Yvette, France

**Abstract.** FTIR spectroscopy has been used to assess the bonding interactions of the quinone carbonyls of $Q_A$ (asymmetrical binding) and $Q_B$ (more symmetrical). These interactions are compared to those proposed in the X–ray structures.

**Keywords.** Selective isotope labeling, quinone photoreduction, reaction center

## 1. INTRODUCTION

In the reaction center (RC) of photosynthetic purple bacteria, two quinone molecules ($Q_A$ and $Q_B$) play an essential role in coupling the electron and proton transfer reactions leading to the conversion of light energy into chemical energy (for a review, see Ref. 1). Following absorption of a photon, electron transfer occurs between a dimer of bacteriochlorophyll molecules and a bacteriopheophytin in ≈3 ps. This is followed by electron transfer to $Q_A$ in ≈200 ps and then to $Q_B$ in 20–100 $\mu$s. $Q_A$ acts as a one–electron acceptor only, while $Q_B$ plays the role of a two–electron gate.

The crystal structure of the RC of the two species of photosynthetic bacteria that have been investigated the most, namely *Rhodobacter* (*Rb.*) *sphaeroides* and *Rhodopseudomonas* (*Rp.*) *viridis*, has been solved to different levels of atomic resolution in several laboratories [2–10]. For these two highly homologous proteins the derived structural models reveal that $Q_A$ and $Q_B$, which in *Rp. viridis* are menaquinone–9 and ubiquinone–9 ($Q_9$), respectively, while in *Rb. sphaeroides* they are both $Q_{10}$, differ significantly in the polarity of their protein binding sites. Notably, the lining of the $Q_A$ site is mostly apolar, while $Q_B$ is surrounded by several protonatable residues. The structures suggest that the bonding interactions between the protein and the quinones are important both for the geometrical organization of the electron transfer pathway and for fine tuning the energy levels of these cofactors. The bonding interactions of $Q_A$ with the protein are relatively well defined (within the ± 0.25–0.5 Å precision on the position of the non–hydrogen atoms in the X–ray data) although the details vary among the various structures, notably with respect to the identity of the proton–donating residues involved in hydrogen bonding of the two carbonyls and in the relative strength of the two hydrogen bonds. In contrast, the description of the binding site of $Q_B$ differs wildly for the various structural models. In a recent study the X–ray structures of the different RCs were overlaid to show that from one structure to the other, the position of the center of $Q_B$ varies almost continuously along a ≈5–Å path [5]. To make this situation even worse, the two extreme locations of $Q_B$ are found in the two highest resolution structures presently available, one being for *Rb. sphaeroides* [9] and the other for *Rp. viridis* [5]. Although the binding of $Q_B$ could be different in the RC of the two species, these discrepancies between the various structural models are more likely due, at least in part, to the fact that a large fraction of $Q_B$ is frequently lost under the conditions required for crystallization and X-ray study of the RC protein. Pending the availability of higher resolution X-ray structures and of a better parameterization of the cofactors, the determination of the details and relative strength of the bonding interactions of each carbonyl of $Q_A$ and $Q_B$ with the protein rely on structural spectroscopy methods.

Among the spectroscopic techniques that can selectively probe the bonding interactions of the quinones with the protein, light-induced FTIR difference spectroscopy appears well suited to investigate both the neutral and reduced forms of the quinones [11-32]. Analyzing the light-induced $Q_A^-/Q_A$ FTIR difference spectra of $Q_A$-depleted *Rb. sphaeroides* RCs reconstituted with isotopically labeled ubiquinones, the C=O and C=C vibrational modes of $Q_A$ were revealed for the first time and an asymmetry in the bonding interactions of the two carbonyls of $Q_A$ could be determined [19]. FTIR investigation of RCs reconstituted with the chainless symmetrical 2,3-dimethoxy-5,6-dimethyl-1,4-benzoquinone in the $Q_A$ site has further shown that the asymmetry is not caused by the difference in the substituents at the 5- and 6-positions of the ubiquinone (for numbering of quinone atoms, see Ref. 19) but rather by the different proteic environment of the two carbonyls [20]. The unambiguous assignment of the frequency of each of the two carbonyl modes of $Q_A$ in *Rb. sphaeroides* RCs was recently achieved by site-specific isotope labeling of one or the other of the carbonyls [21, 24].

In the present study, the issues of the degree of symmetry of the carbonyl vibrations of the neutral state of $Q_A$ [21] and $Q_B$ in their respective binding site in *Rb. sphaeroides* as well as of a possible difference in the bonding interactions of the carbonyls of $Q_B$ in *Rb. sphaeroides* and in *Rp. viridis* RCs [22] are being addressed by analyzing the light-induced FTIR $Q^-/Q$ difference spectra of RCs reconstituted with site-specific $^{13}C$-labeled ubiquinone.

## 2. RESULTS

### 2.1 *$Q_A^-/Q_A$ FTIR Difference Spectra in Rb. sphaeroides*:

The $Q_A^-/Q_A$ spectrum of $Q_A$-depleted *Rb. sphaeroides* RCs reconstituted with unlabeled $Q_3$ (Figure 1a) is essentially indistinguishable from that of native RCs and of RCs reconstituted with $Q_6$ or $Q_8$ [15, 18, 19]. In these spectra, the bands of the neutral $Q_A$ state appear as negative signals while the positive bands belong to the $Q_A^-$ state. When the RCs are reconstituted with selectively $^{13}C$-labeled $Q_3$, the position of some of the bands in the $Q_A^-/Q_A$ spectra is significantly different from that found with unlabeled $Q_3$ and depends whether the label is at the 1- (Figure 1b) or the 4-position (Figure 1c). A negative band at 1628 $cm^{-1}$ in the $Q_A^-/Q_A$ spectrum for unlabeled $Q_3$ (Figure 1a) disappears upon site-specific $^{13}C$ labeling (Figure 1b,c). New negative bands are seen at 1619 $cm^{-1}$ upon $^{13}C_1$ labeling (Figure 1b) and at 1611 and 1578 $cm^{-1}$ upon $^{13}C_4$ labeling (Figure 1c). While the 1601-$cm^{-1}$ band is unaffected by the $^{13}C_1$ labeling, it disappears upon $^{13}C_4$ labeling. Conversely, an increase of the positive band at 1658 $cm^{-1}$ is observed for $^{13}C_1$ labeling but not for $^{13}C_4$ labeling.

Isotope-sensitive vibrations from the quinone itself in the $Q_A^-/Q_A$ spectra can be separated from those of the protein by calculating the double-difference spectrum

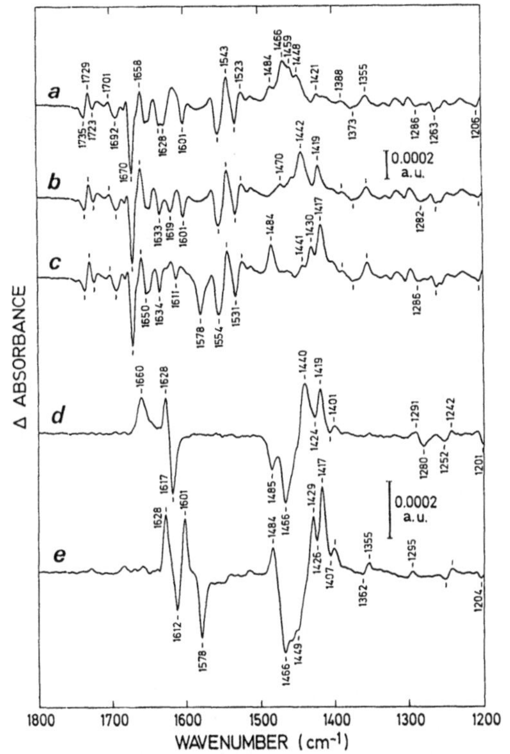


**Figure 1**: Light-induced $Q_A^-/Q_A$ FTIR difference spectra at 5°C of $Q_A$-depleted *Rb. sphaeroides* RCs reconstituted with (a) unlabeled Q3, (b) $^{13}C_1$-labeled Q3, and (c) $^{13}C_4$-labeled Q3. Double-difference spectra ($^{13}C$-labeled-minus-unlabeled) obtained for $^{13}C_1$-labeled Q3 (d) and $^{13}C_4$-labeled Q3 (e). 4-$cm^{-1}$ resolution. The frequency of the IR bands is given with an accuracy of $\pm$ 1 $cm^{-1}$.

between a pair of $Q_A^-/Q_A$ spectra recorded with RCs reconstituted with isotopically labeled and unlabeled quinones [19]. Such double-difference spectra are shown for $^{13}C_1$ (Figure 1d) and $^{13}C_4$ labeling (Figure 1e). In these double-difference (isotopically labeled-minus-unlabeled) spectra, the IR bands of the neutral unlabeled $Q_A$ appear with a positive sign while the downshifted bands of the labeled quinone exhibit a negative sign. A reverse situation is found for the semiquinone bands. In addition, only those vibrations of the quinone *in vivo* that are affected by the labeling will contribute. The decrease of intensity of the vibrational modes upon isotope labeling as well as the overlap of the positive and negative bands can lead to an apparent cancellation of some of the bands, as, e.g., for the negative band corresponding to the downshifted 1660-$cm^{-1}$ band in Figure 1d.

The C=O and C=C vibrations of the neutral unlabeled $Q_A$ lead to the three positive bands at 1660, 1628, and 1601 $cm^{-1}$ in the double-difference spectra of the non-specifically labeled quinones [19]. In the case of the selectively labeled Q3 only two out of these three positive bands are seen (Figure 1d,e). Upon $^{13}C_4$ labeling the 1660-$cm^{-1}$ band is missing (Figure 1e) showing that this $Q_A$ vibration is unaffected by the labeling at the 4-position. The two positive bands at 1628 and 1601 $cm^{-1}$ are downshifted to 1612 and 1578 $cm^{-1}$, respectively. Upon $^{13}C_1$ labeling, it is the band at 1601 $cm^{-1}$ that disappears in the double-difference spectrum (Figure 1d) demonstrating that this vibration is unaffected by labeling at the 1-position. The positive bands at 1660 and 1628 $cm^{-1}$ give rise to a single asymmetrical negative band of large amplitude at 1617 $cm^{-1}$.

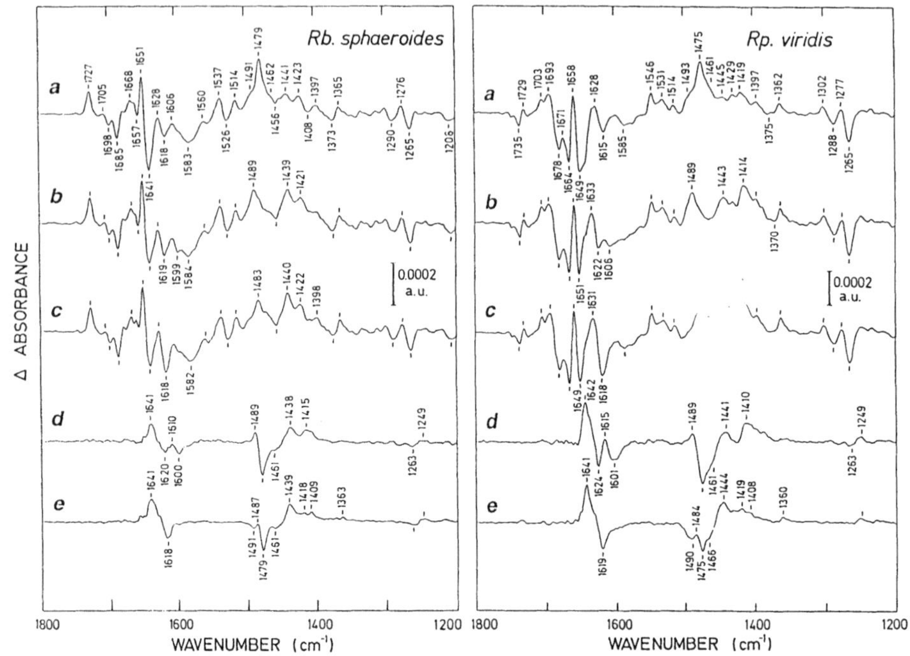


**Figures 2** (left, *Rb. sphaeroides*) **and 3** (right, *Rp. viridis*): Light-induced $Q_B^-/Q_B$ FTIR difference spectra at 15°C of $Q_B$-depleted RCs reconstituted with (a) unlabeled $Q_3$, (b) $^{13}C_1$-labeled $Q_3$, and (c) $^{13}C_4$-labeled $Q_3$. Double-difference spectra (isotopically labeled-minus-unlabeled) obtained for $^{13}C_1$-labeled $Q_3$ (d) and $^{13}C_4$-labeled $Q_3$ (e).

## 2.2 $Q_B^-/Q_B$ *FTIR Difference Spectra in Rb. sphaeroides and Rp. viridis*:

For both *Rb. sphaeroides* (Figure 2a) and *Rp. viridis* (Figure 3a) the $Q_B^-/Q_B$ spectra of $Q_B$-depleted RCs reconstituted with unlabeled $Q_3$ are essentially indistinguishable from those recorded with $Q_6$, $Q_8$, or $Q_{10}$ [16, 22]. When the RCs are reconstituted with selectively $^{13}C$-labeled $Q_3$, pronounced effects are seen, notably on the 1641-$cm^{-1}$ band that decreases in amplitude for both $^{13}C_1$ and $^{13}C_4$ labeling. A single negative band at 1618 $cm^{-1}$ is seen for $^{13}C_4$ labeling (Figures 2c and 3c) while two negative bands at 1619–1622 and at 1599–1606 $cm^{-1}$ are observed for $^{13}C_1$ labeling (Figures 2b and 3b) demonstrating some asymmetry in the bonding interactions of the neutral $Q_B$ with the protein.

These effects are best revealed in the double-difference spectra which show an identical frequency of the unlabeled $C_1=O$ and $C_4=O$ modes (1641 $cm^{-1}$) of $Q_B$ for both *Rb. sphaeroides* (Figure 2d,e) and *Rp. viridis* (Figure 3d,e). On the other hand, an inequivalence of the quinone modes for labeling at the two positions is detected in the 1630- to 1580-$cm^{-1}$ frequency range of the C=C vibrations where the downshift to 1600 $cm^{-1}$ of a mode absorbing at 1610–1615 $cm^{-1}$ is observed only for labeling at $C_1$. This inequivalence is clearly seen when the effect of selective labeling on the C=C vibrations of the isolated quinones are compared to the corresponding effects in the RCs. The pattern of two negative bands at 1620–1621 and 1600 $cm^{-1}$ with a positive band in between at 1610 $cm^{-1}$ seen in the difference spectra of the isolated $Q_3$ [22] is recognized for labeling at the 1-position *in vivo* (with negative bands at

1620 and 1600 $cm^{-1}$ in *Rb. sphaeroides* or at 1624 and 1601 $cm^{-1}$ in *Rp. viridis* and a positive one at 1610 $cm^{-1}$ in *Rb. sphaeroides* or 1615 $cm^{-1}$ in *Rp. viridis*) but is missing for labeling at the 4-position. In this case a single negative band at 1618–1619 $cm^{-1}$, assigned to the $^{13}C_4$=O mode, is observed (Figures 2e and 3e).

## 3. DISCUSSION

### 3.1 *Vibrations of Neutral $Q_A$ in Rb. sphaeroides*:

In a previous FTIR study of the $Q_A^-/Q_A$ spectra of *Rb. sphaeroides* RCs reconstituted with $Q_6$ labeled with $^{18}O$ on both carbonyls or with fully $^{13}C$-labeled $Q_8$, three bands at 1660, 1628, and 1601 $cm^{-1}$ were recognized as the C=O and C=C modes of the neutral unlabeled $Q_A$ [19]. Reconstitution with $Q_3$ selectively labeled with $^{13}C$ at the 1- or the 4-position reveals the contribution of the same bands and thus definitely support the $Q_A$ origin of these bands. The 1660-$cm^{-1}$ band has been assigned previously to one of the C=O modes of $Q_A$ [19]. The present data show that the 1660-$cm^{-1}$ band is unaffected by the labeling at the 4-position and downshifts by 37 ± 5 $cm^{-1}$ upon selective $^{13}C$ labeling at the 1-position, as expected for a pure C=O mode. These observations provide compelling evidence for an assignment of the 1660-$cm^{-1}$ band to the $C_1$=O carbonyl of $Q_A$, i.e., to the carbonyl proximal to the isoprenoid chain.

In the previous investigation using non-specifically labeled quinones, it was recognized that the two other $Q_A$ bands at 1628 and 1601 $cm^{-1}$ exhibit such a highly mixed C=O and C=C character that it was not possible to propose an assignment on the basis of the isotope effects alone. The 1628-$cm^{-1}$ band of $Q_3$ in the $Q_A$ site of *Rb. sphaeroides* RCs shifts to 1617 $cm^{-1}$ upon $^{13}C_1$ labeling and to 1612 $cm^{-1}$ upon $^{13}C_4$ labeling. In contrast, the 1601-$cm^{-1}$ band is unaffected by the labeling of $C_1$ and is downshifted by 23 $cm^{-1}$ upon labeling of $C_4$. This behavior of the 1601-$cm^{-1}$ band strengthens more the C=O character than the C=C character of this band. If the 1601-$cm^{-1}$ band were a C=C mode coupled to the C=O modes, a downshift upon labeling at both positions would be expected. Conversely, the shift of the 1628-$cm^{-1}$ band by 11 and 16 $cm^{-1}$ upon $^{13}C_1$ and $^{13}C_4$ labeling, respectively, appears more compatible with its assignment to a C=C mode coupled to C=O modes than to a C=O mode. The simplest interpretation of the selective isotope labeling data is thus to assign the 1628-$cm^{-1}$ band to the C=C mode and the 1601-$cm^{-1}$ band to the $C_4$=O mode of $Q_A$, keeping in mind that they both have a strongly mixed C=O and C=C character. The assignment of the 1601-$cm^{-1}$ band to the $C_4$=O mode implies that this mode is not coupled to the 1660-$cm^{-1}$ $C_1$=O mode as the labeling of one carbonyl has no effect on the frequency of the other. Although essentially the same uncoupled behavior of the C=O modes has also been demonstrated for the ubiquinones *in vitro* [20, 21], an opposite view is propounded by Brudler et al., [24].

### 3.2 *Conformation of $Q_A$*:

The clearest result emerging from the study of the $Q_A^-/Q_A$ spectra of *Rb. sphaeroides* RCs reconstituted with $Q_3$ selectively $^{13}C$-labeled at the 1- or the 4-position is obtained for the $C_1$=O vibration. This mode is responsible for the whole 1660-$cm^{-1}$ band and is not coupled to the $C_4$=O mode. It is close in frequency to the carbonyls of $Q_3$ *in vitro* and is thus assigned to an essentially free carbonyl group.

While the $C_1$ carbonyl of $Q_A$ appears little affected by the protein, a very different situation is encountered for the C4 carbonyl. The $Q_A^-/Q_A$ FTIR difference spectrum of selectively $^{13}C_4$-labeled Q3 shows a very large perturbation of the C=O and C=C modes associated with labeling at this position. The C4=O vibration is well localized at 1601 $cm^{-1}$ with a strong admixture of C=C character and the C=C mode is found at 1628 $cm^{-1}$ with a strong C=O character. All these observations point to an unusual environment of the C4 carbonyl *in vivo*. Although quinones *in vitro* are rather insensitive to the hydrogen bonding properties of the solvent, large downshifts (20 to 180 $cm^{-1}$) of their carbonyl frequency have been reported under special conditions leading to the formation of strong hydrogen bonds such as complexation of quinone with dihydroquinone [33, 34], intramolecular hydrogen bonding [35, 36], or interaction with Lewis acids [37]. Compared to the frequency found *in vitro*, the ≈50- to 60-$cm^{-1}$ downshift of the C4=O mode of ubiquinone observed upon binding to the $Q_A$ proteic site is thus not unprecedented. This shift would correspond to a very strong hydrogen bond of the order of −6 to −8 kcal·$mol^{-1}$ [19–21, 38]. Such a strong hydrogen bonding is often achieved in conjugated chelation systems when resonance involving delocalized charges comes into play to enhance the strength of the hydrogen bond [36]. The assignment of the 1601-$cm^{-1}$ band to a strongly hydrogen bonded carbonyl of $Q_A$ agrees with the binding affinity studies of Gunner & coll. who have inferred the presence of only one strong hydrogen bond between the carbonyls of various quinones and the $Q_A$ site with a binding free energy of −3 to −7 kcal·$mol^{-1}$ [39–41].

3.3 *Vibrations of Neutral $Q_B$ in Rb. sphaeroides and Rp. viridis*:

The light-induced FTIR $Q_B^-/Q_B$ difference spectra of $Q_B$-depleted RCs of *Rb. sphaeroides* and *Rp. viridis* reconstituted with uniformly $^{13}C$-labeled Q8 reveal several bands which are not affected by the isotope substitutions [22]. They can be assigned to modes corresponding to stable structural rearrangements of the protein which accompany the photoreduction of $Q_B$. As previously noticed for $Q_A$ photoreduction [19; see also Figure 1], the amplitude of the protein bands in the $Q_B^-/Q_B$ spectra is roughly comparable to that of the quinone modes, suggesting similar overall perturbation of the protein and of the quinone. The large differential signals arising from the protein, notably in the 1650- to 1700-$cm^{-1}$ frequency range, represent a serious difficulty when generating and analyzing the double-difference spectra. A specific problem is encountered around 1650–1670 $cm^{-1}$ which represents the most variable region of the $Q_B^-/Q_B$ difference spectra. The FTIR signals in this frequency range have been observed to vary slightly from sample to sample even when the very same treatment is applied to a given batch of RCs. They notably appear sensitive to "aging" effects of the sample*. In the present study, this problem has been overcome by averaging a large number of individual spectra taken (i) with different

----------------

* In a poster at the Feldafing III Meeting, Brudler et al. reported light-induced $Q_B^-/Q_B$ FTIR difference spectra of *Rb. sphaeroides* RCs reconstituted with selectively $^{13}C$-labeled $Q_{10}$. For the labeling at the 1- or the 4-position, the double-difference spectra were generally very similar to the ones that we have presented at the same meeting (see also Figure 2d,e) except in the region from 1670 to 1640 $cm^{-1}$ where two bands of significant amplitude at ≈1665 and 1652 $cm^{-1}$ were seen instead of the single band at 1641 $cm^{-1}$ in our spectra. In our opinion, this discrepancy most probably originates from an incomplete cancellation of the variable protein bands in their spectra.

samples and (ii) for each sample, at various delay times after preparation and over a 2–3 days period. On the other hand and as previously noticed for the $Q_A^-/Q_A$ spectra of *Rb. sphaeroides* RCs reconstituted with different ubiquinones [19, 21], varying between 3 and 10 the number of isoprene units of the chain appears not to affect the $Q_B^-/Q_B$ spectra. The same conclusion has been recently extended to $Q_1$ in the $Q_B$ site of *Rp. viridis* (J. B., unpublished).

The single positive band of large amplitude observed at 1641 $cm^{-1}$ in the double-difference spectra obtained upon reconstitution of the RCs of both species with non-specifically labeled quinones [22] falls in the typical frequency range of the C=O vibrations of quinones in solution although it is downshifted by 10–20 $cm^{-1}$ compared to that of isolated ubiquinone. This band shifts to 1621 or ≈1607 $cm^{-1}$ upon $^{18}O$ labeling of both carbonyls or uniform $^{13}C$ labeling, respectively. The isotopic shift of 20 $cm^{-1}$ found upon $^{18}O$ labeling is smaller than that (31 ± 3 $cm^{-1}$) observed for $Q_6$ in solution and the ≈34-$cm^{-1}$ shift upon uniform $^{13}C$ labeling is also smaller than that (42 ± 1 $cm^{-1}$) observed *in vitro* [19]. The corresponding calculated shifts for a pure C=O stretching mode are of 40 and 37 $cm^{-1}$, respectively. The downshifts of the ubiquinone vibrational modes in the $Q_B$ site are thus consistent with an assignment of the 1641-$cm^{-1}$ band to a C=O mode significantly coupled to the C=C mode as previously found for the C=O modes of $Q_A$ [19, 21]. The double-difference spectra obtained with the selectively labeled $Q_3$ (Figures 2d,e and 3d,e) also show that for both isotopomers a positive band of comparable amplitude at ≈1641 $cm^{-1}$ is downshifted by 22 ± 1 $cm^{-1}$ for *Rb. sphaeroides* and by 20 ± 2 $cm^{-1}$ for *Rp. viridis*. This demonstrates that both carbonyls of $Q_B$ do indeed contribute similarly to the 1641-$cm^{-1}$ band. In this case the isotopic shifts observed *in vivo* are also significantly smaller than the ≈32-$cm^{-1}$ shift observed for the isolated $Q_3$ molecules.

The presence of a small derivative signal positive at 1609 ± 1 $cm^{-1}$ and negative at 1604 $cm^{-1}$ in the double-difference spectra obtained upon $^{18}O$ labeling of both carbonyls of $Q_B$ as well as the appearance of a negative band at 1560 $cm^{-1}$ upon uniform $^{13}C$ labeling [22] provide clear signatures for a C=C mode in the $Q_B$ state. A differential signal positive at 1610–1615 $cm^{-1}$ and negative at ≈1600 $cm^{-1}$ is observed upon $^{13}C_1$ labeling (Figures 2d and 3d) but is absent upon $^{13}C_4$ labeling (Figures 2e and 3e). A strikingly similar signal is also observed at 1610/1600 $cm^{-1}$ in the two difference spectra calculated for the equivalent isotope effect on the isolated quinones [22]. In this case the differential signal can be unambiguously ascribed to the downshift of the C=C mode upon labeling. The observation that this differential signal is present *in vitro* for both labeling positions of the quinone while it is only observed for labeling at $C_1$ *in vivo* indicates a very peculiar asymmetry of the two carbonyls in the $Q_B$ site. It appears that the $Q_B$ C=C mode absorbing at ≈1617 $cm^{-1}$ involves motion of the $C_1$-atom but not of the $C_4$-atom, in contrast to the 1611-$cm^{-1}$ C=C band of ubiquinone in solution for which the motion of both atoms is involved. The identical extent of perturbation of the frequency of the two carbonyls of $Q_B$ upon binding to the RC is thus contrasted by a strong inequivalence of their coupling to the C=C mode.

On the basis of a comparison of the $Q_B^-/Q_B$ difference spectra obtained with RCs of both *Rb. sphaeroides* and *Rp. viridis* reconstituted either with $Q_{10}$ or with $Q_0$, it has been proposed that the $Q_B$ carbonyl vibrations contribute to the band at

1640 cm$^{-1}$ while the C=C vibrations absorb at 1616–1618 cm$^{-1}$ [16]. Furthermore, the relative amplitude of bands at 1636 and 1618 cm$^{-1}$ in the $P^+Q_B^-/PQ_B$ spectrum of *Rb. sphaeroides* RCs has been observed to vary depending upon the isotope composition of the $Q_{10}$ molecule used to reconstitute the $Q_B$ site [11]. In addition, a prominent positive band at ≈1640 cm$^{-1}$ also appears in the $Q_A^-Q_B/Q_AQ_B^-$ double-difference spectra obtained by a time-resolved (rapid-scan) FTIR technique for both *Rb. sphaeroides* [30, 31] and *Rp. viridis* [32]. In contrast, none of the $Q_B$ or $Q_B^-$ bands are seen in a previously reported $Q_A^-Q_B/Q_AQ_B^-$ double-difference spectrum calculated from $P^+Q^-/PQ$ spectra of *Rp. viridis* [26], indicating that these authors most probably failed to generate significant amounts of the $P^+Q_B^-/PQ_B$ state.

3.4 *Interaction of $Q_B$ with the Protein*:
The present study demonstrates that the $C_1$ and $C_4$ carbonyls of $Q_B$ contribute equally to the band at 1641 cm$^{-1}$. Taken together with the finding that the two C=O groups of ubiquinones both in solution [20] and in the $Q_A$ site of *Rb. sphaeroides* [21] behave spectroscopically like isolated vibrators, this observation shows that the overall strength of the interaction that each carbonyl of $Q_B$ is engaging with the protein at the binding site is equivalent for the $C_1$=O and the $C_4$=O groups. Furthermore, the observation of an identical frequency for the C=O vibrations of $Q_B$ in the RCs of *Rb. sphaeroides* and of *Rp. viridis* strongly suggests identical interactions in both RCs. The C=O groups of ubiquinone *in vitro* absorb at 1663 and 1650 cm$^{-1}$ depending whether the methoxy group proximal to each carbonyl is in the quinone ring plane or out of this plane, respectively [21, 22, 37]. Thus the carbonyl band at 1641 cm$^{-1}$ in the RCs is downshifted by either 9 or 22 cm$^{-1}$ depending on the actual conformation of the methoxy groups in the binding site. Assuming that no other factor affects the comparison between the *in vivo* and *in vitro* data, these shifts would be indicative of weak to moderate hydrogen bonding interactions of the carbonyls of $Q_B$ with the protein.

The most unexpected result emerging from the present study concerns the very different behavior of the C=C vibrations of $Q_B$ upon selective $^{13}C$ labeling of each carbonyl. The comparison of the isotopic shifts observed *in vivo* and *in vitro* [22] shows that the protein at the $Q_B$ site imparts a specific configuration to the quinone so that the $C_4$ atom is specifically perturbed compared to the isolated molecule. In contrast, the $C_1$ atom of $Q_B$ behaves as it does for $Q_3$ *in vitro*. The reasons for this clear differential effect of the protein on $C_1$ and $C_4$ in the $Q_B$ site while the vibrational frequency of the two C=O groups is the same are still obscure. It is however striking that in the $Q_A$ site of *Rb. sphaeroides* the $C_4$ atom of the ubiquinone is also dramatically perturbed, while the $C_1$ atom behaves essentially as in the isolated molecule. While the strong hydrogen bond at the $C_4$ carbonyl of $Q_A$ has been taken as responsible for the large effect on the C=C vibrations coupled to the $C_4$=O mode [21], this cannot be the reason for the differential effect seen on the C=C band of $Q_B$. In order to explain this effect of the protein on the differential coupling of the C=C mode of $Q_B$ to vibrations involving displacements of the $C_1$ and $C_4$ atoms, another possibility is to invoke the interactions of the peripheral substituents with the protein at the binding site. In this respect it is worthy of note that the response of the 1611-cm$^{-1}$ C=C band of the isolated ubiquinone to isotope labeling

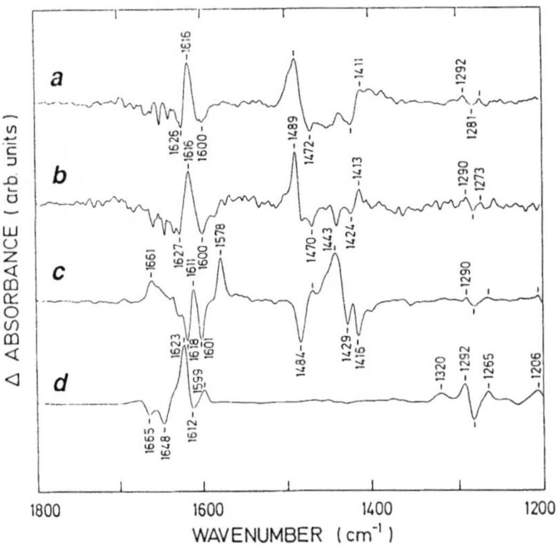


**Figure 4:** Double-difference spectra obtained for $^{13}C_1$-labeled and $^{13}C_4$-labeled Q3 ($^{13}C_1$-minus-$^{13}C_4$) for (a) $Q_B^-/Q_B$ spectra of *Rp. viridis*, (b) $Q_B^-/Q_B$ spectra of *Rb. sphaeroides*, and (c) $Q_A^-/Q_A$ spectra of *Rb. sphaeroides*. (d) Difference between the absorption spectrum of isolated $^{13}C_4$-labeled and $^{13}C_1$-labeled Q3.

is quite complex indicating that the mode responsible for this band involves the motion of many of the quinone atoms [22]. In addition, this band *in vitro* is unaffected by $^{13}C$ labeling of the methyl at the 5'-position [42] while it downshifts by 2 and 4 cm$^{-1}$ upon $^{18}O$ labeling of both methoxy groups and uniform $^{2}H$ labeling, respectively (unpublished results). This last observation is specially worthy of interest as it shows that the 1611-cm$^{-1}$ C=C mode can be affected by changing the mass of atoms which are not part of the quinone ring. It can thus be surmised that a specific anchoring of the 5'-methyl and/or of the methoxy group at C3 into a tight binding niche could perturb the C=C mode coupled to the C4=O mode much more than that coupled to the C1=O mode provided the chain and the other methoxy group are less constrained. Some evidence for such asymmetrical anchoring of the substituents at C3 and/or at C5 has been derived from FTIR measurements [22].

The close similarity in the pattern and the relative amplitudes of the positive and negative bands observed in the double-difference spectra for the two species (Figures 2d,e and 3d,e) points to a close correspondence of the $Q_B$ and $Q_B^-$ vibrations and thus a close equivalence of the geometry of the ubiquinone at the $Q_B$ site in the RCs of *Rb. sphaeroides* and *Rp. viridis*. This result becomes even sharper when the double-difference spectra are calculated between the $Q_B^-/Q_B$ spectra obtained for the selective $^{13}C$ labeling at the 1- or the 4-position (e.g., spectrum 2b-minus-spectrum 2c). Although in this case the interpretation of the double-difference spectrum is even more complex than for the usual double-difference spectra, this spectrum represents a very specific fingerprint of the effect of moving a single neutron from C1 to C4 on the vibrations of $Q_B$ and $Q_B^-$. In brief, the contribution of the unlabeled and labeled carbonyls of each of the specifically labeled quinones will cancel as they are approximately equivalent in amplitude and frequency and only the differential effects of the selective labeling on the C=C vibrations will show up in such a spectrum. When calculated for the RCs of both species, these spectra (Figure 4a,b) display a remarkable similarity with a positive band at 1616 cm$^{-1}$ flanked by two smaller negative bands at ≈1626 and 1600 cm$^{-1}$. In the anion absorption region, the similarity is also striking with a large differential signal at ≈1489/1471 cm$^{-1}$ and a smaller one at ≈1412/1424 $cm^{-1}$. Such a striking similarity for the differential effect of selective labeling on the semiquinone C⋯C vibrations was not obvious from the direct comparison of the two corresponding sets of double-difference spectra (Figures 2d,e and 3d,e). This appears to be due essentially to a slight difference in the width and the peak frequency of the $Q_B^-$ C⋯O mode of the unlabeled Q3 in the RCs of the two species. As these C⋯O bands essentially cancel each other in the double-difference spectrum for the differential effect of the selective labeling, only the contribution from the C⋯C modes is revealed. The close identity of the fingerprints of the differential effect of the selective labeling on the semiquinone C⋯C modes for the two species provides compelling evidence for a very similar mode of binding of $Q_B^-$ in *Rb. sphaeroides* and *Rp. viridis* and is contrasted by the very different fingerprints observed for the corresponding spectra for $Q_A$ in *Rb. sphaeroides* (Figure 4c) and for the isolated Q3 in solution (Figure 4d). The main remaining difference between the vibrational properties of the $Q_B^-$ semiquinone in the RCs of both species is thus the larger width and the 4-$cm^{-1}$ downshift of the C⋯O mode in *Rp. viridis* (1475 $cm^{-1}$) compared to *Rb. sphaeroides* (1479 $cm^{-1}$). This can be related to the difference in the response of the protein to $Q_B^-$ formation which is clearly detected above 1650 $cm^{-1}$ in the $Q_B^-/Q_B$ FTIR difference spectra (Figures 1a and 2a). Both of these effects are likely to be due to some of the few residues lining the $Q_B$ pocket which differ between the two organisms, notably, Val L194 in *Rb. sphaeroides* (Ile in *Rp. viridis*), Asp L213 (Asn), Phe L215 (Tyr), and Thr L226 (Ala) although conserved residues in a different conformation and/or environment could also contribute.

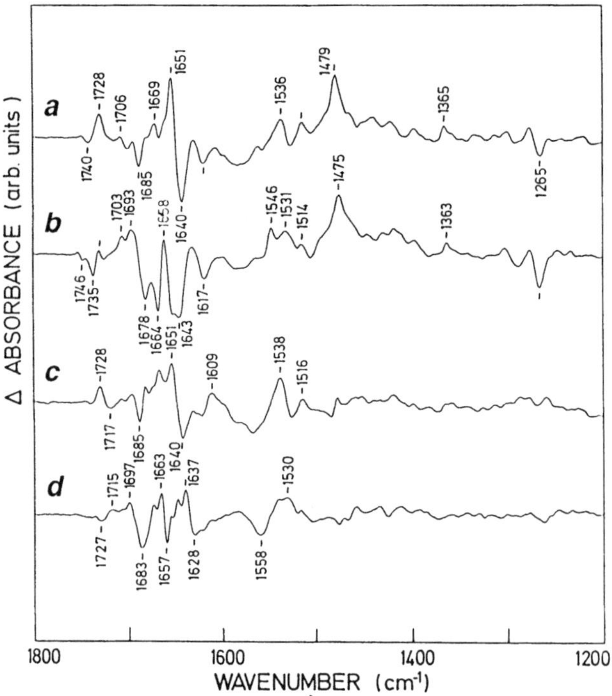


**Figure 5:** Light-induced $Q_B^-/Q_B$ FTIR difference spectra in $^1H_2O$ at 15°C of RCs from (a) *Rb. sphaeroides* and (b) *Rp. viridis*. Double-difference spectra ($^1H_2O$-minus-$^2H_2O$) for (c) *Rb. sphaeroides* and (d) *Rp. viridis*.

3.5 *Protein Response upon $Q_B$ Reduction*:

Further information on the protein bands that are affected upon $Q_B^-$ formation can be obtained by analyzing the effect of $^1H/^2H$ exchange on the FTIR difference spectra. Such an analysis has been recently carried out for the $Q_A^-/Q_A$ spectra of *Rb. sphaeroides* and *Rp. viridis* [23] as well as for the $Q_B^-/Q_B$ spectra of *Rb. sphaeroides* Wt and of several proton transfer mutants at Glu L212, Asp L213, and Asp L210 [43, 44]. In the latter study, the comparison of the $Q_B^-/Q_B$ FTIR spectra in $^1H_2O$ and $^2H_2O$ for the various mutants has notably led to the assignment of the positive 1728-$cm^{-1}$ band in the $^1H_2O$ $Q_B^-/Q_B$ spectrum of *Rb. sphaeroides* Wt (Figure 5a) to an increased protonation ($\approx$0.4 $H^+/Q_B^-$ at pH 7) of Glu L212 upon $Q_B^-$ formation. This COOH band decreases in amplitude and shifts to 1717 $cm^{-1}$ in $^2H_2O$, leading to the 1728/1717-$cm^{-1}$ differential signal in the $^1H_2O$-minus-$^2H_2O$ double-difference spectrum (Figure 5c). Although Glu L212 is also present in *Rp. viridis*, it is worthy of note that this residue appears to protonate only in *Rb. sphaeroides*. The small 1727/1715-$cm^{-1}$ differential signal in the $^1H_2O$-minus-$^2H_2O$ double-difference spectra for *Rp. viridis* (Figure 5d) would indicate instead an environmental shift or even a slight deprotonation of Glu L212 upon $Q_B^-$ formation, assuming an identical frequency for the COOH group in both species.

In the other spectral regions, the double-difference spectra upon $^1H/^2H$ exchange also differ significantly for the RCs of the two species. The sharp contrast between the quasi-identical conformations of both $Q_B$ and $Q_B^-$ in *Rp. viridis* and *Rb. sphaeroides* and their very different protein response to $Q_B$ reduction is striking. Hence, a same pattern of bonding interactions of $Q_B$ and $Q_B^-$ with the RC protein of the two bacteria can nevertheless lead to large differences in the effect of $Q_B$ reduction on the peptide groups and/or the amino acid side chains of the few residues that are not homologous in the two species. In this sense, it seems interesting to investigate the $Q_B^-/Q_B$ FTIR spectra of *Rb. sphaeroides* proton transfer mutants that have been made to partly mimic the $Q_B$ site of *Rp. viridis* RC.

3.6 *Comparison of the FTIR Data with the X-ray Structures*:

One of the interesting issue addressed by the present work is that of the nature, strength, geometry, and role of the bonding interactions between the protein and the cofactors as they can be proposed from the analysis of the X-ray data or as they are derived from light-induced FTIR spectroscopy. The X-ray structures of the RCs provide information on the direction of a putative hydrogen bond and on the distance between the hydrogen bond donor and acceptor with a mean coordinate error on each of the non-hydrogen atoms which is 0.25–0.3 Å for the best available structures [5]. On the other hand, the frequency of the IR vibrations provides precise information on the perturbation of a given chemical group by the environment. Although, in the long term, these two pieces of information will hopefully converge to give a detailed description of the parameters involved in the binding of the cofactors to the protein, this is not yet always the case. The reasons for these discrepancies are not really understood, although a large part of them probably originates from the parameterization of the electron density of the cofactors serving as initial input to analyze the measured electron density map of the crystal as well as from the intrinsic blur in the atomic coordinates. On the other hand, the IR results obtained with the selectively labeled quinones which, *per se*, do provide the information on the very large difference in the perturbation of the $C_1$ and $C_4$ carbonyls at the $Q_A$ site of *Rb. sphaeroides*, are totally silent on which residue is likely to provide such interactions. Thus, the information arising from both techniques has to be combined.

The $C_1$=O group corresponds to that proximal to the isoprenoid chain which, according to the description of the X-ray structures at the $Q_A$ site of *Rb. sphaeroides*, is the carbonyl group in strong hydrogen bonding interaction with the peptide NH of Ala M260 (donor–acceptor distance 2.5–2.8 Å). The $C_4$=O group is proposed to interact more weakly with either Thr M222 [6–8, 10] or His M219 [9]. In the highest resolution X-ray structure available for *Rb. sphaeroides* [9], the $C_4$ carbonyl is positioned in hydrogen bonding interaction (distance 3.2 Å) with the proton (at Nδ1) of the His M219 residue which is also ligand (at N∈2) to the non-heme $Fe^{2+}$ atom. A comparable positioning of the carbonyls of the menaquinone is also proposed for the $Q_A$ site of *Rp. viridis* with a distance of 3.1 Å for both hydrogen bonds [4, 5]. The interaction of the histidine-$Fe^{2+}$ with the quinone ring could thus favor electronic resonance within the imidazole ring and it has been suggested that the dramatic downshift of the frequency of the $C_4$ carbonyl of $Q_A$ may reflect the formation of a specific hydrogen bond at the Nδ1 atom of the side chain of His M219 stabilized by resonance interaction to the $Fe^{2+}$ atom via the imidazole ring system [21]. On the other hand, the FTIR data have indicated a $C_1$=O group essentially free from interaction with the protein and do not support at all the notion of a strong hydrogen bond to the $C_1$ carbonyl of ubiquinone in the $Q_A$ site of *Rb. sphaeroides* based on the analysis of the X-ray results [6–10].

In the best available X-ray structure of the $Q_B$ site [5], the carbonyl oxygen at $C_1$ is proposed to accept two hydrogen bonds from the peptide nitrogens of Ile L224 (at a distance of 3.1 Å) and Gly L225 (distance 3.0 Å) while the $C_4$ carbonyl accepts a hydrogen bond from the Nδ1 atom of His L190 (distance 2.7 Å), which is also a ligand (at N∈2) to the non-heme $Fe^{2+}$. This pattern of interactions of His L190 with both the non-heme $Fe^{2+}$ and the $C_4$=O group of $Q_B$ in *Rp. viridis* is thus very close to that proposed for His M219 with the non-heme $Fe^{2+}$ and the $C_4$=O group of $Q_A$ in *Rb. sphaeroides* [9] and in *Rp. viridis* [5]. Owing to the pseudo-twofold symmetry of the RCs, the overall orientation of the hydrogen bond partners is also conserved at the $Q_A$ and $Q_B$ sites. The finding that the frequency of the $C_4$=O group is much less perturbed for $Q_B$ than for $Q_A$ shows that electronic coupling of the $C_4$=O of $Q_A$ with the non-heme $Fe^{2+}$ atom through the imidazole ring of His M219 cannot be the main reason for the strong downshift of the $C_4$=O of $Q_A$.

Taken together with the evidence for identical bonding interactions of $Q_B$ in the RC of both species derived from the present FTIR data, the highest resolution X-ray results for $Q_A$ [9] and $Q_B$ [5] suggest that the main difference in the bonding interaction between the $C_4$ carbonyl of $Q_A$ and $Q_B$ in *Rb. sphaeroides* resides in the shorter hydrogen bond length proposed for $Q_B$ (2.7 Å) than for $Q_A$ (3.2 Å). Thus, the best X-ray structures for the quinone binding sites in RCs give no clues to the large differences in the vibrational frequency of the $C_4$ carbonyl of $Q_A$ and $Q_B$ detected by light-induced FTIR difference spectroscopy. It is particularly striking that the $C_4$=O mode of $Q_A$ experiences upon binding a much larger downshift (50 $cm^{-1}$) than that of $Q_B$ (9 $cm^{-1}$) while, considering the present X-ray structures, one would have made the opposite prediction [22].

Compared to the large variations in the proposed bonding interactions of $Q_B$

derived from the interpretation of the X-ray data [2–10], the remarkable equivalence of the various sets of double-difference spectra presented here unambiguously demonstrates the almost identical bonding pattern and conformation of the neutral $Q_B$ in the RCs of the two species. Taking the newest structure of *Rp. viridis* [5] as the most reliable representation presently available of the $Q_B$ pocket in bacterial RC, it can thus be stated that it also provides the description of the native $Q_B$ site in the RCs of *Rb. sphaeroides*. In our opinion, higher resolution structures should further confirm the close identity of the $Q_B$ conformation in the RCs of both species. In addition, a number of the unresolved issues outlined in the present work can be addressed by performing further FTIR experiments with site-directed mutants and with selective labeling of either the cofactors or the protein.

ACKNOWLEDGMENT

The authors are grateful to S. Andrianambinintsoa and D. Dejonghe for preparing the $Q_A$- and $Q_B$-depleted RCs, to G. Berger for purification and labeling of quinones, to P. Tavan and M. Nonella for their help with normal-mode calculations on quinones, to G. Fritsch and H. Michel for providing coordinates of the X-ray structure of *Rb. sphaeroides*, to G. Feher and M. Okamura for sustained interest in this work, and to J.-R. Burie and R. Lancaster for stimulating discussions.

REFERENCES

1 Feher, G., Allen, J. P., Okamura, M. Y., & Rees, D. C. (1989) *Nature* 339, 111–116.
2 Michel, H., Epp, O., & Deisenhofer, J. (1986) *EMBO J.* 5, 2445–2451.
3 Deisenhofer, J. & Michel, H. (1989) *EMBO J.* 8, 2149–2169.
4 Deisenhofer, J., Epp, O., Sinning, I., & Michel, H. (1995) *J. Mol. Biol.* 246, 429–457.
5 Lancaster, C. R. D., Ermler, U., & Michel, H. (1995) in *Anoxygenic Photosynthetic Bacteria* (Blankenship, R. E., Madigan, M. T., & Bauer, C. E., Eds.) Kluwer, Dordrecht, The Netherlands, in press.
6 Allen, J. P., Feher, G., Yeates, T. O., Komiya, H., & Rees, D. C. (1988) *Proc. Natl. Acad. Sci. U.S.A.* 85, 8487–8491.
7 El-Kabbani, O., Chang, C.-H., Tiede, D., Norris, J., & Schiffer, M. (1991) *Biochemistry* 30, 5361–5369.
8 Chirino, A. J., Lous, E. J., Huber, M., Allen, J. P., Schenck, C. C., Paddock, M. L., Feher, G., & Rees, D. (1994) *Biochemistry* 33, 4584–4593.
9 Ermler, U., Fritzsch, G., Buchanan, S. K., & Michel, H. (1994) *Structure* 2, 925–936.
10 Arnoux, B. & Reiss-Husson, F. (1995) *EMBO J.*, in press.
11 Bagley, K., Abresch, E., Okamura, M. Y., Feher, G., Bauscher, M., Mäntele, W., Nabedryk, E., & Breton, J. (1990) in *Current Research in Photosynthesis* (Baltscheffsky, M., Ed.) pp. 77–80, Kluwer, Dordrecht, The Netherlands.
12 Bauscher, M., Leonhard, M. Moss, D. A., & Mäntele, W. (1993) *Biochim. Biophys. Acta* 1183, 59–71.
13 Berthomieu, C., Nabedryk, E., Mäntele, W., & Breton, J. (1990) *FEBS Lett.* 269, 363–367.
14 Berthomieu, C., Nabedryk, E., Breton, J., & Boussac, A. (1992) in *Research in Photosynthesis* (Murata, N., Ed.) Vol. II, pp. 53–56, Kluwer, Dordrecht, The Netherlands.
15 Breton, J., Thibodeau, D. L., Berthomieu, C., Mäntele, W., Verméglio, A., & Nabedryk, E. (1991a) *FEBS Lett.* 278, 257–260.
16 Breton, J., Berthomieu, C., Thibodeau, D. L., & Nabedryk, E. (1991b) *FEBS Lett.* 288, 109–113.

17 Breton, J., Bauscher, M., Berthomieu, C., Thibodeau, D. L., Andrianambinintsoa, S., Dejonghe, D., Mäntele, W., & Nabedryk, E. (1991c) in *Spectroscopy of Biological Molecules* (Hester, R. E., & Girling, R. B., Eds.) pp. 43–46, The Royal Society of Chemistry, Cambridge.
18 Breton, J., Burie, J.-R., Berthomieu, C., Thibodeau, D. L., Andrianambinintsoa, S., Dejonghe, D., Berger, G., & Nabedryk, E. (1992) in *The Photosynthetic Bacterial Reaction Center II: Structure, Spectroscopy, and Dynamics* (Breton, J. & Verméglio, A., Eds.) pp. 155–162, Plenum Press, New York.
19 Breton, J., Burie, J.-R., Berthomieu, C., Berger, G., & Nabedryk, E. (1994a) *Biochemistry* 33, 4953–4965.
20 Breton, J., Burie, J.-R., Boullais, C., Berger, G., & Nabedryk, E. (1994b) *Biochemistry* 33, 12405–12415.
21 Breton, J., Boullais, C., Burie, J.-R., Nabedryk, E., & Mioskowski, C. (1994c) *Biochemistry* 33, 14378–14386.
22 Breton, J., Boullais, C., Mioskowski, C., & Nabedryk, E. (1995a) *Biochemistry* 34, in press.
23 Breton, J. & Nabedryk, E. (1995b) in *Proc. 10th Int. Photosynthesis. Congr.* (Mathis, P., Ed.) Kluwer, Dordrecht, The Netherlands, in press.
24 Brudler, R., de Groot, H. J. M., van Liemt, W. B. S., Steggerda, W. F., Esmeijer, R., Gast, P., Hoff, A. J., Lugtenburg, J., & Gerwert, K. (1994) *EMBO J.* 13, 5523–5530.
25 Buchanan, S., Michel, H., & Gerwert, K. (1990) in *Reaction Centers of Photosynthetic Bacteria* (Michel-Beyerle, M.-E., Ed.) pp. 75–85, Springer, Berlin.
26 Buchanan, S., Michel, H., & Gerwert, K. (1992) *Biochemistry* 31, 1314–1322.
27 Mäntele, W., Leonhard, M., Bauscher, M., Nabedryk, E., Breton, J., & Moss, D. A. (1990) in *Reaction Centers of Photosynthetic Bacteria* (Michel-Beyerle, M.-E., Ed.) pp. 31–44, Springer, Berlin.
28 Nabedryk, E., Bagley, K., Thibodeau, D. L., Bauscher, M., Mäntele, W., & Breton, J. (1990) *FEBS Lett.* 266, 59–62.
29 Nabedryk, E., Berthomieu, C., Verméglio, A., & Breton, J. (1991) *FEBS Lett.* 293, 53–58.
30 Thibodeau, D. L., Breton, J., Berthomieu, C., Bagley, K., Mäntele, W., & Nabedryk, E. (1990a) in *Reaction Centers of Photosynthetic Bacteria* (Michel-Beyerle, M.-E., Ed.) pp. 87–98, Springer, Berlin.
31 Thibodeau, D. L., Nabedryk, E., Hienerwadel, R., Lenz, F., Mäntele, W., & Breton, J. (1990b) *Biochim. Biophys. Acta* 1020, 253–259.
32 Thibodeau, D. L., Nabedryk, E., Hienerwadel, R., Lenz, F., Mäntele, W., & Breton, J. (1992) in *Time-Resolved Vibrational Spectroscopy V* (Takahashi, H., Ed.) pp. 79–82, Springer, Berlin.
33 Slifkin, M. A. & Walmsley, R. H. (1970) *Spectrochim. Acta* 26A, 1237–1242.
34 Kruk, J., Strzalka, K., & Leblanc, R. M. (1993) *Biophys. Chem.* 45, 235–244.
35 Hadzi, D. & Sheppard, N. (1954) *Trans. Faraday Soc.* 50, 911–918.
36 Bloom, H., Briggs, L. H., & Cleverley, B. (1959) *J. Chem. Soc.* 178–185.
37 Burie, J.-R. (1994) Ph. D. Thesis, University of Paris VI.
38 Badger, R. M. & Bauer, S. H. (1937) *J. Chem. Phys.* 5, 839–851.
39 Gunner, M. R., Braun, B. S., Bruce, J. M., & Dutton, P. L. (1985) in *Antennas and Reaction Centers of Photosynthetic Bacteria* (Michel-Beyerle, M.-E., Ed.) pp. 298–305, Springer, Berlin.
40 Warncke, K. & Dutton, P. L. (1993) *Proc. Natl. Acad. Sci. U.S.A.* (1993), 90, 2920–2924.
41 Warncke, K., Gunner, M. R., Braun, B. S., Gu, L., Yu, C.-A., Bruce, J. M., & Dutton, P. L. (1994) *Biochemistry* 33, 7830–7841.
42 van Liemt, W. B. S. (1994) Ph. D. Thesis, University of Leiden.
43 Nabedryk, E., Breton, J., Hienerwadel, R., Fogel, C., Mäntele, W., Paddock, M. L., & Okamura, M. Y. (1995a) *Biochemistry*, submitted.
44 Nabedryk, E., Breton, J., Hienerwadel, R., Fogel, C., Mäntele, W., Paddock, M. L., & Okamura, M. Y. (1995b) in *Proc. 10th Int. Photos. Congr.* (Mathis, P., Ed.) Kluwer, in press.

# FTIR spectroscopic characterization of the $Q_A$ and $Q_B$ binding sites in *Rhodobacter sphaeroides R26* reaction centres

R. Brudler[1], H.J.M. de Groot[2], W.B.S. van Liemt[2], P. Gast[3], A.J. Hoff[3], J. Lugtenburg[2], K. Gerwert[1]

[1]Lehrstuhl für Biophysik, Ruhr-Universität Bochum, Postfach 102148, 44780 Bochum, Germany
[2]Leiden Institute of Chemistry, Gorlaeus Laboratories, Leiden University, P.O. Box 9502, 2300 RA Leiden, The Netherlands
[3]Department of Biophysics, Huygens Laboratory, Leiden University, P.O. Box 9504, 2300 RA Leiden, The Netherlands

**Abstract**
The $Q_A$ and the $Q_B$ binding sites of *Rhodobacter sphaeroides R26* reaction centres have been investigated by FTIR spectroscopy, using site specifically $^{13}C$-labelled ubiquinone-10. The C=O and C=C (neutral state) and the C┄O and C┄C (semiquinone state) stretching vibrations of $Q_A$ and $Q_B$ have been assigned. At the $Q_A$ site asymmetric binding of the 1- and the 4-C=O groups has been detected. The 4-C=O group is strongly bound to the RC, most probably via a hydrogen bond to His M219. The 1-C=O group is only weakly hydrogen bonded. In the charge separated state the asymmetry is largely maintained. At the $Q_B$ site two different fractions of ubiquinone-10 have been observed. One fraction (~25 %) has no specific interactions with the protein, the other (~75 %) shows distinct but less strong hydrogen bonding to the RC than the 4-C=O group of $Q_A$. Both carbonyl groups are symmetrically bound to the RC at the $Q_B$ site. In the charge separated state the carbonyl groups of $Q_B^-$ are nearly equivalent and equally hydrogen bonded to the protein. The different C=O protein interactions contribute to the factors determining different functions of ubiquinone-10 at the $Q_A$ and the $Q_B$ site, respectively.

**Key words:** Bacterial reaction centre, Ubiquinone, isotopic labelling, Fourier transform infrared spectroscopy, electron transfer

## Introduction

A striking feature of reaction centres (RCs) is the presence of several identical cofactors, which show different electrochemical and electron-transfer properties in the protein. For instance, there are four hemes in the cytochrome subunit and four bacteriochlorophylls in the RC core with different functions [1, 2]. Furthermore, at the end of the electron transport chain there are two quinones, $Q_A$ and $Q_B$, which are both ubiquinone-10 ($UQ_{10}$) molecules in *Rhodobacter sphaeroides* and display distinct binding and electrochemical specifities: $Q_A$ is tightly and permanently bound to the RC. It accepts only one electron that is transferred to $Q_B$ in about 200 μs. $Q_B$ is weaker bound to the protein. It successively accepts two electrons and two protons and is released from the RC as quinol.

In order to elucidate the protein-cofactor interactions that determine the different functions of the quinones in their binding sites, FTIR difference spectroscopy has been applied [3]. By use of $UQ_{10}$, selectivley $^{13}C$-labelled at the ring positions 1, 2, 3 and 4, the 1- and 4-C=O and the 2/3-C=C stretching vibrations of $UQ_{10}$ bound at the $Q_A$ site have been specifically assigned for the first time [4] and asymmetric binding of the 1- and 4-C=O group of $Q_A$ has been detected [4, 5]. The isotpe labels have now been incorporated at the $Q_B$ binding site. In this summary we focus on the results of 1- and 4-$^{13}C$-labelled $UQ_{10}$ and their implications for the different functions of $UQ_{10}$ at the $Q_A$ and $Q_B$ binding sites.

## Materials and methods

*Rb. sphaeroides R26* RCs were purified according to standard procedures [6]. $UQ_{10}$, selectively $^{13}C$-labelled at positions 1, 2, 3, 4 and 3' was synthesized as described by van Liemt et al. [7]. Incorporation of native and specifically $^{13}C$-labelled $UQ_{10}$ at the $Q_A$ site is described in [4]. Removal of $Q_B$ and reconstitution of the $Q_B$ site with native and $^{13}C$-labelled $UQ_{10}$ was performed according to [8]. Recording of the FTIR spectra of native and $^{13}C$-labelled $UQ_{10}$ and of $Q_A^- - Q_A$ and $Q_B^- - Q_B$ difference spectra was performed as described in [4]. The actinic light intensity was tenfold decreased for measuring the $Q_B^- - Q_B$ difference spectra. Subtraction of the $Q_A^- - Q_A$ ($Q_B^- - Q_B$) difference spectra with native or $^{13}C$-labelled $UQ_{10}$ at the $Q_A$ ($Q_B$) site was done as described [4]. Normalization of the $Q_A^- - Q_A$ ($Q_B^- - Q_B$) difference spectra was performed between 1800 - 1675 $cm^{-1}$ (1800 - 1700 $cm^{-1}$) where no bandshifts were observed. The spectra of unbound unlabelled and unbound $^{13}C$-labelled $UQ_{10}$ have been normalized on the bands at 1450 and 1436 $cm^{-1}$ that are unaffected by the labelling. Spectral resolution was 4 $cm^{-1}$.

## Results

Fig. 1a shows the difference spectrum between unbound unlabelled $UQ_{10}$ and 1-$^{13}$C-labelled $UQ_{10}$ in the region of the C=O and C=C stretching vibrations. It is identical to the difference spectrum between unbound unlabelled and 4-$^{13}$C-labelled $UQ_{10}$ (data not shown). Isotope labelling induces a shift of the labelled group absorption to lower frequencies and thereby allows the specific assignment of the quinone vibrations. In Fig. 1a positive bands belong to unlabelled $UQ_{10}$, while negative signals represent the shifted bands of 1-$^{13}$C-$UQ_{10}$. The two bands at 1666 and 1648 cm$^{-1}$ are shifted to 1621 cm$^{-1}$ for 1- and 4-$^{13}$C-labelled $UQ_{10}$ and are dominated by the 1- and 4-C=O stretching vibrations. A distinction between the 1- and 4-C=O group is not possible due to mixing of the 1- and 4-C=O vibrations. A band at 1610 cm$^{-1}$ is shifted to 1601 cm$^{-1}$ and is dominated by the 2/3-C=C stretching vibration [4]. As discussed in [4], isotope labelling of the C=O groups shifts also bands of C=C vibrations and vice versa due to coupling of the C=O and C=C vibrations of $UQ_{10}$.

$Q_A^-$ - $Q_A$ and $Q_B^-$ - $Q_B$ FTIR difference spectra with native, 1- and 4-$^{13}$C-labelled $UQ_{10}$ at the $Q_A$ and the $Q_B$ site, respectively, have been recorded. In order to visualize selectively the labelling-induced band shifts, double difference spectra between $Q_A^-$ - $Q_A$ ($Q_B^-$ - $Q_B$) difference spectra with specifically labelled and unlabelled $UQ_{10}$ at the $Q_A$ ($Q_B$) site have been calculated as described in "Materials and methods". In Fig. 1b the double difference spectrum between $Q_A^-$ - $Q_A$ difference spectra with 1-$^{13}$C-labelled and unlabelled $UQ_{10}$ at the $Q_A$ site, respectively, is shown. In the double difference spectra the bands of the C=O and C=C stretching vibrations of $UQ_{10}$ in the neutral ground state (1670 - 1570 cm$^{-1}$) are positive for unlabelled $UQ_{10}$ and negative for the corresponding shifted bands of labelled $UQ_{10}$. Inversely, the bands of the C┅O and C┅C stretching vibrations of the semiquinone (1500 - 1400 cm$^{-1}$) [9] are negative for unlabelled $UQ_{10}$, while the shifted bands of labelled $UQ_{10}$ are positive. In Fig. 1b the bands of neutral, unlabelled $UQ_{10}$ are seen at 1660 and 1628 cm$^{-1}$. The band at 1628 cm$^{-1}$ is shifted to 1618 cm$^{-1}$. The band at 1660 cm$^{-1}$ is most probably shifted to about 1630 cm$^{-1}$ where it is masked due to overlap with the positive band at 1628 cm$^{-1}$. In agreement with this assumption, the integral intensity of the band at 1628 cm$^{-1}$ is less compared with its shifted signal at 1618 cm$^{-1}$. In the region of the semiquinone vibrations negative bands of unlabelled $UQ_{10}$ appear at 1487 and 1466 cm$^{-1}$, positive bands of 1-$^{13}$C-labelled $UQ_{10}$ are seen at 1441 and 1420 cm$^{-1}$. The strong negative band at 1466 cm$^{-1}$ seems to be shifted to 1441 cm$^{-1}$. The negative signal at 1487 cm$^{-1}$ appears shifted to 1470 cm$^{-1}$. The positive band at 1470 cm$^{-1}$ is masked by the strong negative band at 1466 cm$^{-1}$ in Fig. 1b but can be clearly seen in the $Q_A^-$ - $Q_A$ difference spectrum with 1-$^{13}$C-$UQ_{10}$ at the $Q_A$ site [4]. In addition, the $Q_A^-$ - $Q_A$ difference spectrum shows that the band at 1420

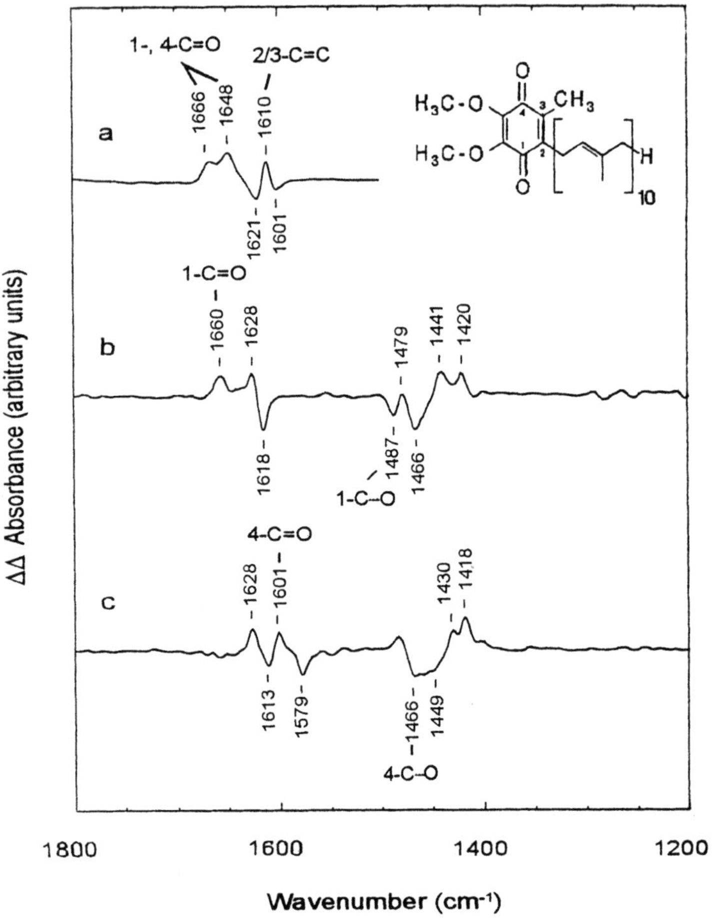


**Fig. 1**: (a) Difference spectrum between unlabelled and 1-$^{13}$C-labelled $UQ_{10}$. (b) Double difference spectrum between $Q_A^{\cdot-}$ - $Q_A$ difference spectra with 1-$^{13}$C-labelled and unlabelled $UQ_{10}$ bound at the $Q_A$ site. (c) Double difference spectrum between $Q_A^{\cdot-}$ - $Q_A$ difference spectra with 4-$^{13}$C-labelled and unlabelled $UQ_{10}$ at the $Q_A$ site. The assignments of the dominant vibrations are marked in the figures. Inset: Structural formula of $UQ_{10}$.

$cm^{-1}$ (Fig. 1a) is most probably shifted from 1447 $cm^{-1}$. In Fig. 1b the band at 1447 $cm^{-1}$ coincides with the positive signal at 1441 $cm^{-1}$.

In Fig. 1c the double difference spectrum between $Q_A^{\cdot-}$ - $Q_A$ difference spectra with 4-$^{13}$C-labelled and unlabelled $UQ_{10}$ at the $Q_A$ site, respectively, is displayed. For the neutral state (1670 - 1570 $cm^{-1}$) two bands of unlabelled $UQ_{10}$ appear at 1628 and 1601 $cm^{-1}$. They are shifted to 1613 and 1579 $cm^{-1}$, respectively. For the semiquinone state (1500 - 1400 $cm^{-1}$) negative bands of unlabelled $UQ_{10}$ are seen at 1466 and 1449 $cm^{-1}$, which seem to be shifted to 1430 and 1418 $cm^{-1}$, respectively.

Despite the strong coupling of the C=O and C=C and of the C⋯O and C⋯C stretching vibrations of $UQ_{10}$ bound at the $Q_A$ site, the dominant vibrations could be assigned and are indicated in Fig. 1. The band at 1660 $cm^{-1}$ appears upon 1-$^{13}$C-labelling (Fig. 1b) but not upon 4-$^{13}$C-labelling (Fig. 1b). The situation is inverse for the band at 1601 $cm^{-1}$ (compare Fig. 1b and 1c). Therefore, the band at 1660 $cm^{-1}$ is assigned to the 1-C=O and the band at 1601 $cm^{-1}$ to the 4-C=O stretching vibration. The band at 1628 $cm^{-1}$ appears upon 1-, 4- and 2-, 3-$^{13}$C-labelling [4] and represents a highly mixed C=O/C=C stretching vibration. In the semiquinone spectral region the band at 1487 $cm^{-1}$ is shifted upon 1-$^{13}$C- ($\Delta\nu$ = -16 $cm^{-1}$) and 3-$^{13}$C-labelling ($\Delta\nu$ = -8 $cm^{-1}$) [4]. It therefore seems to be dominated by the 1-C⋯O vibration. The band at 1466 $cm^{-1}$ is dominated by the 4-C⋯O vibration ($\Delta\nu$ = -36 $cm^{-1}$ upon 4-$^{13}$C-labelling) but has a large contribution from the 1-C⋯O vibration ($\Delta\nu$ = -25 $cm^{-1}$ upon 1-$^{13}$C-labelling) and of the 1/2-C⋯C ring vibration ($\Delta\nu$ = -21 $cm^{-1}$ upon 2-$^{13}$C-labelling). The band at 1447 $cm^{-1}$ is approximately equally shifted upon 1-, 2- and 4-$^{13}$C-labelling and seems to represent a highly mixed C⋯C/C⋯O vibration.

Fig. 2a shows the double difference spectrum between $Q_B^-$ - $Q_B$ difference spectra with 1-$^{13}$C-$UQ_{10}$ or unlabelled $UQ_{10}$ at the $Q_B$ site. The bands of neutral unlabelled $UQ_{10}$ appear at 1664, 1651, 1641 and 1611 $cm^{-1}$. Negative bands of the shifted signals are found at 1620 and 1601 $cm^{-1}$. The bands at 1664 and 1651 $cm^{-1}$ are only small as compared to the band at 1641 $cm^{-1}$ but have been consistently resolved in every measurement. The bands at 1664 and 1651 $cm^{-1}$ and the shifted signal at 1620 $cm^{-1}$ (Fig. 2a) have counterparts in the difference spectrum of unbound $UQ_{10}$ (Fig. 1a). They are therefore assigned to C=O stretching vibrations of $UQ_{10}$ at the $Q_B$ site. However, the largest band in the region of the carbonyl vibrations appears at 1641 $cm^{-1}$ (Fig. 2a). The maximal shift for a C=O vibration upon $^{13}$C-labelling is 36 $cm^{-1}$. It seems that the band at 1641 $cm^{-1}$ is shifted to about 1610 $cm^{-1}$, underneath the band at 1611 $cm^{-1}$ (Fig. 2a). In analogy to unbound $UQ_{10}$ (Fig. 1a), the difference band at 1611/1601 $cm^{-1}$ (Fig. 2a) is assigned to the 2/3-C=C stretching vibration. The decreased intensity of the band at 1611 $cm^{-1}$ (Fig. 2a) as compared to Fig. 1a is in agreement with the assumption that the band at 1641 $cm^{-1}$ is shifted to about 1610 $cm^{-1}$, underneath the band at 1611 $cm^{-1}$. For the semiquinone state the double difference spectrum shows a band at 1479 $cm^{-1}$ of unlabelled $UQ_{10}$ and shifted bands of 1-$^{13}$C-$UQ_{10}$ at 1438 and 1416 $cm^{-1}$.

In Fig. 2b the double difference spectrum between $Q_B^-$ - $Q_B$ difference spectra with 4-$^{13}$C-labelled and unlabelled $UQ_{10}$ at the $Q_B$ site, respectively, is shown. It is similar to the double difference spectrum with 1-$^{13}$C-$UQ_{10}$ at the $Q_B$ site (Fig. 2a). Positive bands appear at 1664, 1651 and 1641 $cm^{-1}$ and are correspondingly

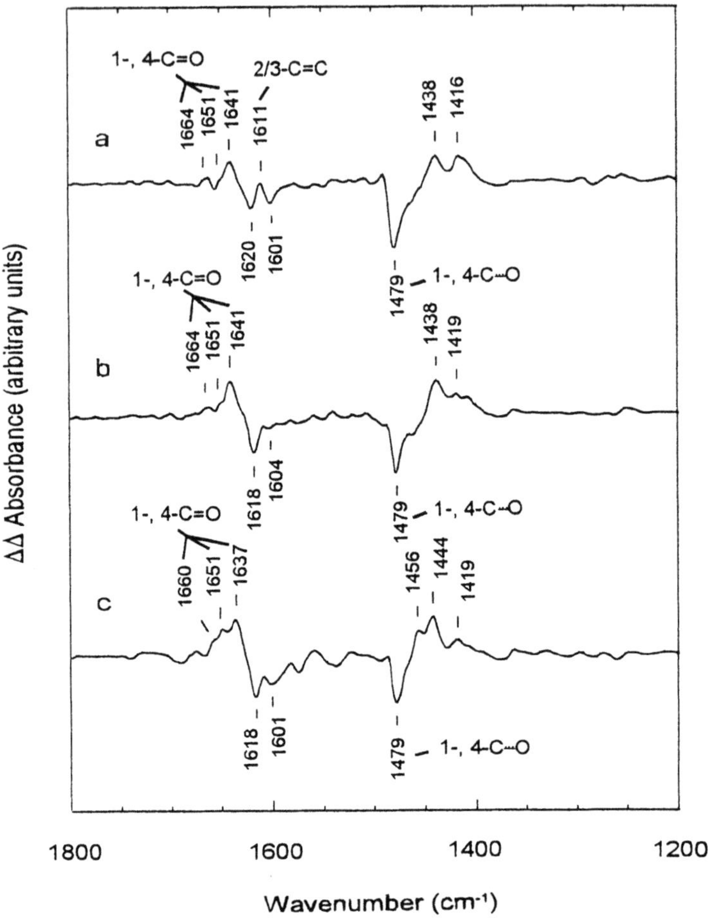


**Fig. 2**: (a) Double difference spectrum between $Q_B^{\cdot-}$ - $Q_B$ difference spectra with 1-$^{13}$C-labelled and unlabelled $UQ_{10}$ bound at the $Q_B$ site. (b) Double difference spectrum between $Q_B^{\cdot-}$ - $Q_B$ difference spectra with 4-$^{13}$C-labelled and unlabelled $UQ_{10}$ at the $Q_B$ site. (c) The same as (b) but measured in $D_2O$ instead of $H_2O$. The assignments of the dominant vibrations are marked in the figures.

assigned to C=O stretching vibrations. As for unbound $UQ_{10}$, and in contrast to $UQ_{10}$ bound at the $Q_A$ site, the 1- and 4-C=O vibrations of of $Q_B$ cannot be distinguished. The shifted signal of the bands at 1664 and 1651 $cm^{-1}$ is seen at 1618 $cm^{-1}$. The difference band at 1610/1601 $cm^{-1}$, that is observed in the difference spectrum between unbound and 1- or 4-$^{13}$C-labelled $UQ_{10}$ (Fig. 1a), is missing in Fig. 2b. A shift of the C=O vibration from 1641 $cm^{-1}$ to around 1610 $cm^{-1}$ and of the C=C vibration from 1610 to 1601 $cm^{-1}$ seem to cancel each other more than in the case of 1-$^{13}$C-labelling (compare Fig. 2a and 2b). In order to clarify this, the samples have been measured in $D_2O$ where the bands shift slightly differently (Fig. 2c). The difference band at 1611/1601 $cm^{-1}$ is better resolved than in $H_2O$ (Fig. 2b). In addition, the bands at 1664, 1651, 1641 and 1618 $cm^{-1}$ in $H_2O$ (Fig. 2b) are seen in $D_2O$ as well at 1660, 1651, 1637 and 1618 $cm^{-1}$, respectively (Fig. 2c). They are also found in the double difference spectrum between the $Q_B^-$ - $Q_B$ difference spectra with 1-$^{13}$C-$UQ_{10}$ and unlabelled $UQ_{10}$ at the $Q_B$ site, measured in $D_2O$ (not shown). This excludes that they are caused by water. In $D_2O$, the absorbtion of the O-H bending vibration of water at 1645 $cm^{-1}$ is replaced by the O-D bending vibration at 1215 $cm^{-1}$. For the semiquinone state the double difference spectra show a band at 1479 $cm^{-1}$ (Fig. 2b, 2c). Shifted bands of 4-$^{13}$C-$UQ_{10}$ are observed at 1438 and 1419 $cm^{-1}$ in $H_2O$ (Fig. 2b) and at 1456, 1444 and 1419 $cm^{-1}$ in $D_2O$ (Fig. 2c). The band at 1479 $cm^{-1}$ is shifted upon 1- and 4-$^{13}$C-labelling (Fig. 2a - c) and is assigned to the 1- and 4-C$\cdots$O stretching vibrations. It is not clear, to which frequency the band at 1479 $cm^{-1}$ is shifted. Interestingly, 2- and 3-$^{13}$C-labelled $UQ_{10}$ show no band shifts in the $Q_B$ site (data not shown). Hence the shifted signals at 1438 or 1416/1419 $cm^{-1}$ are not attributed to coupled $^{13}$C$\cdots$C stretching vibrations.

## Discussion

### *Carbonyl vibrations of neutral $Q_A$ and $Q_B$*

The equivalence of the carbonyl groups 1 and 4 of unbound $UQ_{10}$, which both absorb at 1666 and 1648 $cm^{-1}$ (Fig. 1a) is disturbed upon binding at the $Q_A$ site (Fig. 1b, 1c). The 1-C=O group absorbs at 1660 $cm^{-1}$ (Fig. 1b), at a similar frequency as for unbound $UQ_{10}$. It is therefore only weakly bound to the protein. In contrast, the absorption frequency of the 4-C=O group is drastically downshifted to 1601 $cm^{-1}$ (Fig. 1c), indicating a strong decrease of bond order, which can be attributed to a strong hydrogen bond to His M219 [10]. The strong hydrogen bond binds $Q_A$ tightly to the protein and governs the fast electron transfer to $Q_B$ [4, 5]. The finding of asymmetric binding of the 1- and 4-C=O groups at the $Q_A$ site is in agreement with former measurements of the binding free energies of different quinones and related compounds at the $Q_A$ site. It turned out that the loss of one carbonyl oxygen atom does not significantly influence the binding strength at the $Q_A$ site [11-13].

At the $Q_B$ binding site surprisingly three carbonyl bands are observed at 1664, 1651 and 1641 $cm^{-1}$ (Fig. 2a, 2b). The two bands at 1664 and 1651 $cm^{-1}$ agree in frequencies with the carbonyl vibrations of unbound $UQ_{10}$ (1666 and 1648 $cm^{-1}$; compare Fig. 1a with Fig. 2a and 2b). The band at 1641 $cm^{-1}$ is slightly downshifted as compared to unbound $UQ_{10}$. It is proposed that the three carbonyl bands represent two fractions of $UQ_{10}$ at the $Q_B$ site: The two bands at 1664 and 1651 $cm^{-1}$ are attributed to a fraction with no specific interactions to the protein environment. The slight downshift of the C=O vibration of the second fraction to 1641 $cm^{-1}$ indicates specific hydrogen bonding to the protein. In contrast to the asymmetric binding of the 1- and the 4-C=O group at the $Q_A$ site, both carbonyl groups are symmetrically bound at the $Q_B$ site. Comparison of the integral intensity of the carbonyl bands results in roughly 25 % for the loosely bound fraction and roughly 75 % for the fraction absorbing at 1641 $cm^{-1}$. Both fractions represent reacting $UQ_{10}$ because only bands of functionally active groups appear in the double difference spectra. However, it is not clear, whether both fractions have physiological relevance.

Heterogeneous binding of ubiquinone at the $Q_B$ site has been also detected by X-ray crystallography. Eight possible binding positions have been observed for $UQ_1$ at the $Q_B$ site of *Rhodopseudomonas viridis* RCs [14]. For *Rb. sphaeroides* two different binding positions of $Q_B$ have been reported. Ermler et al. [15] have positioned the $Q_B$ molecule at the more hydrophobic entrance of the $Q_B$ binding pocket, whereas Allen et al. [16] and El-Kabbani et al. [17] have located it deeper in the protein. Consequently, different hydrogen bond donors have been suggested. In [16] and [17] the 1- and the 4-C=O groups are in hydrogen bonding distance to Ser L223 and His L190, respectively. In [15] only the 1-C=O group could form a hydrogen bond to Ile L224. The relatively high temperature factor for the $Q_B$ site [15] might be explained by the differently bound fractions of $UQ_{10}$ seen in the IR experiments. In addition, Giangiacomo and Dutton [18] have shown that the $Q_B$ site is not highly specific for the native $UQ_{10}$ molecule and functions as well with various substituted quinones.

*Semiquinone vibrations of $Q_A^-$ and $Q_B^-$*

The close similarity of the 1- and 4-$^{13}$C-induced shifts of the band at 1479 $cm^{-1}$ (Fig. 2a, 2b) indicates that the carbonyl groups of $Q_B^-$ are nearly equivalent and equally hydrogen bonded to the protein. This observation is in agreement with results of ENDOR measurements, performed on Zn reconstituted RCs [19]. In contrast, at the $Q_A$ site the asymmetry of the carbonyl groups is maintained as well in the semiquinone state [20, 4, 5]. The absorption frequency of the 1- and 4-C$\cdots$O vibrations of $Q_B$ at 1479 $cm^{-1}$ indicates that the hydrogen bonds are weaker than for the 4-C$\cdots$O group of $Q_A$ that absorbs at 1466 $cm^{-1}$.

**Acknowledgements**

We thank A.H.M. de Wit and S.J. Jansen for culturing the cells and isolating the RCs and W.F. Steggerda and R. Esmeijer for $^{13}$C-labelling of $UQ_{10}$. P.G. is a recipient of a Royal Netherlands Academy of Arts and Sciences (KNAW) career fellowship. P.G., A.J.H. and J.L. acknowledge support by the Foundation for Chemical Research (SON), H.J.M.dG. support by the Foundation for Life Sciences (SLW), both financed by the Netherlands Organization of Pure and Applied Research (NWO). K.G. acknowledges the DFG for financial support (Ge-599/7-1).

### References

[1] Feher, G., Allen, J.P., Okamura, M.Y. and Rees, D.C. (1989) Nature 339, 111-116

[2] Deisenhofer, J. and Michel, H. (1989) EMBO J. 8, 2149-2170

[3] Gerwert, K. (1993) Curr. Opin. Struct. Biol. 3, 769-773

[4] Brudler, R., de Groot, H.J.M., van Liemt, W.B.S., Steggerda, W.F., Esmeijer, R., Gast, P., Hoff, A.J., Lugtenburg, J. and Gerwert, K. (1994) EMBO J. 13, 5523-5530

[5] Breton, J., Boullais, C., Burie, J.-R., Nabedryk, E. and Mioskowski, C. (1994) Biochemistry 33, 14378-14386

[6] Feher, G. and Okamura, M.Y. (1978) in The Photosynthetic Bacteria (Clayton, R.K., Sistrom, W.R., eds.), pp. 349-386, Plenum Press, New York

[7] van Liemt, W.B.S., Steggerda, W.F., Esmeijer, R. and Lugtenburg, J. (1994) Recl. Trav. Chim. Pays-Bas 113, 153-161

[8] Okamura, M.Y., Isaacson, R.A. and Feher, G. (1975) Proc. Natl. Acad. Sci. USA 72, 3491-3495

[9] Buchanan, S., Michel, H. and Gerwert, K. (1992) Biochemistry 31, 1314-1322

[10] Bosch, M.K., Gast, P., Hoff, A.J., Spoyalov, A.P. and Tsvetkov, Yu. D. (1995) Chem. Phys. Lett., in the press

[11] Gunner, M.R. (1991) Curr. Top. Bioenerg. 16, 319-367

[12] Warncke, K. and Dutton, L. (1993a) Biocemistry 32, 4769-4779

[13] Warncke, K. and Dutton, L. (1993b) Proc. Natl. Acad. Sci. USA 90, 2920-2924

[14] Michel, H., Epp, O. and Deisenhofer, J. (1986) EMBO J. 5, 2445-2451

[15] Ermler, U., Fritzsch, G., Buchanan, S. and Michel, H. (1994) Structure 2, 925-936

[16] Allen, J.P., Feher, G., Yeates, T.O., Komiya, H. and Rees, D.C. (1988) Proc. Natl. Acad. Sci. USA 85, 8487-8491

[17] El-Kabbani, O., Chang, C.-H., Tiede, D., Norris, J. and Schiffer, M. (1991) Biochemistry 30, 5361-5369

[18] Giangiacomo, K. and Dutton, P.L. (1989) Proc. Natl. Acad. Sci. USA 86, 2658-2662

[19] Feher, G., Isaacson, R.A., Okamura, M.Y. and Lubitz, W. (1985) in Antennas and Reaction Centers of Photosynthetic Bacteria (Michel-Beyerle, M.E., ed.), pp. 174-189, Springer, Berlin
[20] van den Brink, J.S., Spoyalov, A.P., Gast, P., van Liemt, W.B.S., Raap, J., Lugtenburg, J. and Hoff, A.J. (1994) FEBS Lett. 353, 273-276

# Site-Directed Isotope Labelling as a Tool in Spectroscopy of Photosynthetic Preparations. Investigations on Quinone Binding in Bacterial Reaction Centers

A.J. Hoff,[1] T.N. Kropacheva,[2] R.I. Samoilova,[3] N.P Gritzan,[3] J. Raap,[4] J.S. van den Brink,[1] P. Gast[1] and J. Lugtenburg[4]

[1] Department of Biophysics, Leiden University, Leiden, The Netherlands
[2] Department of Chemistry, Udmurt State University, Izhevsk, Russian Federation
[3] Institute of Chemical Kinetics and Combustion, Academy of Sciences, Novosibirsk, Russian Federation
[4] Department of Chemistry, Leiden University, Leiden, The Netherlands

**Abstract.** The strategy of combining site-directed isotope labelling with various spectroscopic techniques for investigating structural details of complex biosystems is illustrated with a study of the binding of selectively $^{13}$C-labelled ubiquinone-10 cofactors to the bacterial photosynthetic reaction center protein. It is shown that the 4-C=O group of $Q_A^-$ of *Rb. sphaeroides* is hydrogen-bonded to the imidazole $N^{\delta(1)}$ of HisM219. For the neutral $Q_A$ moiety, FTIR shows a distinct asymmetry of the 1- and 4-C=O groups, the latter carbonyl having a bond order about 0.1 lower than that of the 1-C=O group. On the basis of solid-state NMR spectroscopy and quantum chemical calculations it is suggested that this bond order difference might be largely due to a protein-induced change in $sp_2$ hybridization of the 4-carbon, and that formation of the semiquinone leads to a considerable strengthening of hydrogen bond(s) to the 4-carbonyl group.

**Keywords.** Isotope labelling, $^{13}$C hyperfine couplings, quinone binding, bacterial photosynthesis, hydrogen bonds

## 1 Introduction

Isotope labelling in spectroscopy has been used for a long time to solve problems related to assignment, resolution enhancement, spectral simplification, etc. In biological spectroscopy commonly organisms are grown on media containing one or a few compounds enriched in a particular isotope, for example $^2$H or $^{15}$N, so that all, or a particular subset, of these nuclides are labelled. When looking at one particular entity in the sea of labelled biomolecules, for example a cofactor of some enzyme, this may not matter too much, but more often than not spectroscopically separating such an entity from its surroundings is not possible, and the desired goal of firm assignment, enhanced resolution, etc. is not or only incompletely achieved.

To overcome these problems, in recent years isotope labelling methods have focused on *selective, site-directed* labelling, combined with *selective bio-*

*incorporation*. For example, in vision research the retinal moiety of rhodopsin has been labelled with *one* $^{13}C$ nucleus at predetermined carbon positions in the isoprenoid chain. Reconstituting the apoprotein one by one with retinals out of an extensive set labelled at different sites (obtained by total synthesis), has made it possible to determine with solid-state NMR, Raman and FTIR spectroscopy the precise configuration of the cofactor in the various stable intermediaries of the photochemical cycle of light perception [1]. This provided a crucial contribution to the unravelling of the molecular mechanism of vision.

We have now adopted the strategy of *site-directed isotope labelling* combined with *bio-incorporation* and *site-directed mutagenesis* to address the relations between protein structure, cofactor binding, protein dynamics, etc., and electron transport in photosynthetic reaction centers. In this note we focus on the binding of the $Q_A$-cofactor to the reaction center protein of the purple bacterium *Rhodobacter sphaeroides*. We report a number of recent experiments, which will highlight the guiding principle of our work: Only by using not a *single*, but *several complementary* spectroscopic techniques is it possible to gain detailed understanding of the aforementioned relationships.

## 2 Site-directed $^{13}C$-labelling of ubiquinone-10

From the start it was decided to use for the reconstitution of labelled ubiquinone (UQ) compounds only UQ-10, even while this complicated considerably the organic synthesis routes. Although several studies indicated that the length of the isoprenoid chain beyond the first unit was not crucial for binding to the RC and for the electron transport properties of reconstituted $Q_A$ [2,3,4,5] this was much less certain for $Q_B$, while for ascertaining the intricate details of binding of *both* quinones it clearly would always be necessary to refer to the natural quinone. The synthesis of UQ-10 singly labelled with $^{13}C$ at the 1-, 2-, 3-, 3', or 4-C position (Fig. 1) is described in detail in [6]. Briefly, first methylsuccinic anhydride labelled in a desired carbon position was synthesized, starting from labelled synthons such as KCN, $CH_3I$, diethyl(2-$^{13}C$)-methylmalonate, etc. Then, using the anhydride as starting compound, dichloromethylquinone was prepared, which was then reacted with cyclopentadiene to form the Diels-Alders adduct, from which the chlorines were replaced by methoxy groups to form the ubiquinone de-

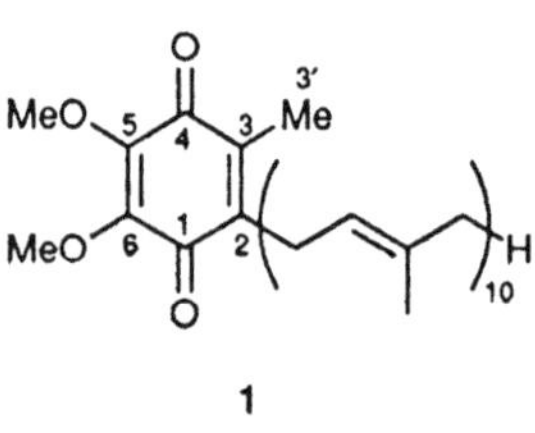

1

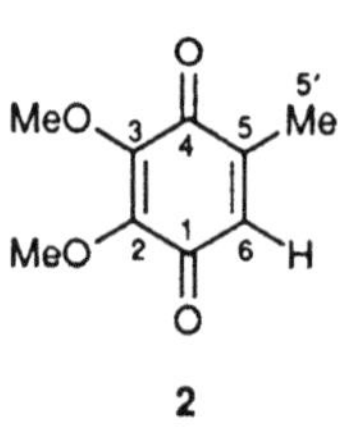

2

*Fig. 1. Ubiquinone-10 (**1**) and ubiquinone-0 (**2**), structure and IUPAC numbering.*

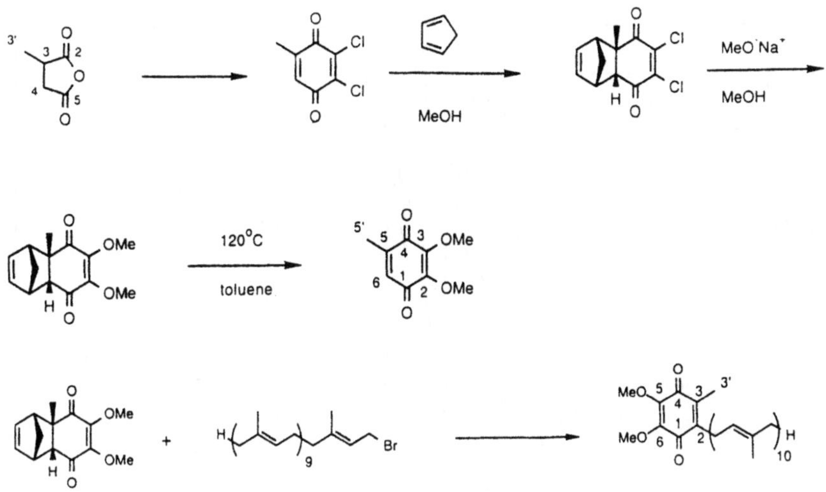


*Fig. 2. Synthesis of ubiquinone-0 and ubiquinone-10, labelled at one of the 1-, 2-, 3-, 3'-, 4-, 5-, 5'- or 6-C positions, starting from methylsuccinic anhydride labelled at the 2-, 3-, 3'-, 4- or 5-C position.*

caprenyladduct. This was then reacted with toluene to form the labelled UQ-0, or with bromide (synthesized from solanesol) to form the labelled UQ-10. The synthesis scheme, which is schematically depicted in Fig. 2, afforded the preparation of UQ-0 or UQ-10 labelled in six carbon positions in the quinone moiety, without any scrambling or dilution of the label.

## 3 EPR of $^{13}$C-labelled UQ-10 *in vitro*

Clearly, it is the RC protein that by its interaction with the quinone cofactors confers to $Q_A$ and $Q_B$ their different, specific properties: $Q_A$ has a more negative redox midpoint potential than $Q_B$, and it is a firmly bound one-electron gate, which is not protonated. This contrasts with $Q_B$, which accepts two electrons, and after protonation (presumably in two steps, one when a semiquinone anion to form the neutral UQH$^{\cdot}$ radical, and one to neutralize UQH$^-$ to UQH$_2$ [7]), leaves the RC.

One way to gain insight in the binding of UQ-10 to the RC is the determination of the EPR parameters of the anion, $Q_A^{\cdot-}$, such as its (anisotropic) *g*- and hyperfine (hf) tensors. It is well known that especially the hf-tensor of the $^{13}$C nucleus is very sensitive to the local environment, and to details of local molecular structure, as bond angles, bond length, etc. [8,9]. Thus, by using a single $^{13}$C-nucleus as a "spy" at the various carbon positions of the UQ ring and the side groups, one should in principle be able to determine quantitatively structural changes of the quinone anion induced by its binding to the protein.

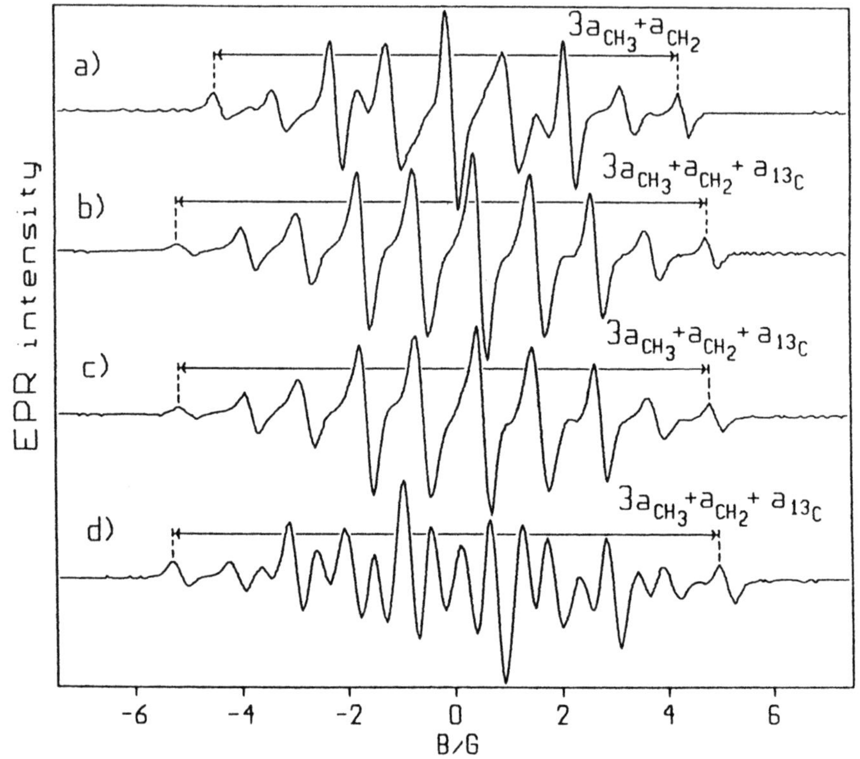


*Fig. 3. EPR spectra of the radical anion of ubiquinone-10 at room temperature in hexamethylphosphoric triamide (HMPT). (a) unlabelled, (b) [1-$^{13}$C], (c) [4-$^{13}$C], (d) [$^{13}$C]methyl. From [11].*

A full determination of the above-mentioned anisotropic tensors, especially the hf-tensor, involves a huge amount of work. Although this remains the ultimate goal, a simpler though less complete way to assess the influence of binding to the RC protein is to compare the EPR properties of $^{13}$C-UQ-10$^-$ bound in the $Q_A$-binding site with those of $^{13}$C-UQ-10$^-$ *in vitro*. As a first step we have determined the *isotropic* $^1$H- and $^{13}$C-hf coupling constants for the various $^{13}$C-labelled UQ-10 semiquinone anions in organic solvents and compared them with those of similarly labelled UQ-0 semiquinones [10,11]. As an example, Fig. 3 gives the EPR spectra of UQ-10$^-$ in HMPT, unlabelled and labelled in the 1-, 4-, and methyl C-position.

Fig. 4 shows the EPR spectrum of unlabelled UQ-0$^-$ and of [1-$^{13}$C]UQ-0$^-$ in HMPT and in EtOH. It is immediately seen that the presence of a polar solvent has a large effect on the [1-$^{13}$C]-hf-coupling constant. This effect, which was earlier observed for natural-abundance $^{13}$C-splittings [12,13], has been attributed to hydrogen bonding of solvent hydroxyl groups to the two carbonyls of the (unprotonated) ubisemiquinone.

We have investigated the effect of solvent polarity on the carbonyl $^{13}C$-splitting of UQ-10$^-$ and UQ-0$^-$ in more detail by monitoring the isotropic hf-constant of [4-$^{13}C$] carbonyl in mixtures of DME and isopropylalcohol (IPA) [14,15]. Fig. 5 shows that for both quinones the effect is considerable. In fact, for UQ-10 the (negative) hf-constant becomes

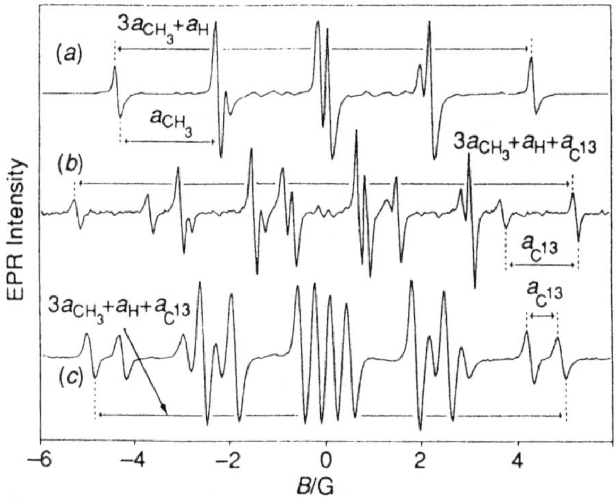


*Fig. 4. EPR spectra of the radical anion of ubiquinone-0 at room temperature. (a) unlabelled in EtOH, (b) [1-$^{13}$C] in HMPT, (c) [1-$^{13}$C] in EtOH. From [10].*

less negative with increasing IPA concentration, is zero at about 30% IPA and becomes increasingly *positive* when more IPA is added. (The *sign* of the hf-constant was determined from the *alternating linewidth effect* [16,17].) The measured curves of Fig. 5 could be simulated very well when it was assumed that the ubisemiquinones are present in non-, mono- and disolvated complexes, with the measured hf-constant a statistical average of the hf-constants of the complexes.

It is possible to calculate the spin density on the carbonyl carbon from its isotropic hf-splitting $a_{CO}$, using the treatment of Karplus and Fraenkel [8]. The hf-coupling is a function of the spin density in the 1*s*-orbital and of the polarization of the $sp_2$ hybrid bonds due to spin density in the $p_z$ orbitals of the carbonyl carbon and of the neighbouring ring carbons and oxygen atom. With the experimental proportionality constants (Q-factors) given in [8] and [13] one obtains for the hf-coupling

$$a_{CO} = 33.8\ \rho^{\pi}_{CO} - 27.8\ \rho^{\pi}_{C} - 27.1\ \rho^{\pi}_{O}\ \mathrm{G}, \qquad (1)$$

where $\rho^{\pi}_{C}$ is the spin density on the two neighbouring ring carbons (which, since they are small anyway, are taken equal). The spin densities on the carbonyl carbon and the carbonyl oxygen, when changed by hydrogen bonding, are to good approximation linearly related [9,18]:

$$\rho^{\pi}_{CO} = -1.115\ \rho^{\pi}_{O} + 0.3406\ . \qquad (2)$$

Substituting (2) in (1), we obtain

$$a_{CO} = 58.1\ \rho^{\pi}_{CO} - 27.8\ \rho^{\pi}_{C} - 8.28\ \text{G}. \quad (3)$$

For UQ-10 Eq. 3 can be further simplified using the experimental value of the hf-coupling of the methyl group, which is practically independent of hydrogen bonding to the carbonyls: $a_{CH3}$(UQ-10) = 2.11 G, which with the Q-factor for a rotating methyl group (0.5 x 300 MHz, i.e 53.6 G) gives $\rho^{\pi}_{C} = 0.04$, and

$$a_{CO} = 58.1\ \rho^{\pi}_{CO} - 9.39\ \text{G}. \quad (4)$$

The experimental values of $a_{CO}$(UQ-10) are −0.93 G and +0.63 G for DME (non protic) and 80% IPA, respectively, which with Eq. 4 gives $\rho^{\pi}_{CO}$(DME) = 0.146 and $\rho^{\pi}_{CO}$(IPA) = 0.172. Thus, the hydrogen-bond induced change in $\rho^{\pi}_{CO}$ is small, but sufficient to change the sign of the hf-coupling, leading to a striking difference in alternating linewidth effect of the solution EPR spectra.

## 4 MO calculations of hydrogen bond geometry and spin densities.

The influence of a polar medium on the hf-couplings was assessed theoretically with MO calculations for complexes of UQ-0 with one or two hydrogen-bonded methanol molecules [11]. First, the geometry and the energy of the complexes were optimized with the PM3 (Unrestricted Hartree-Fock) method. Then, with the

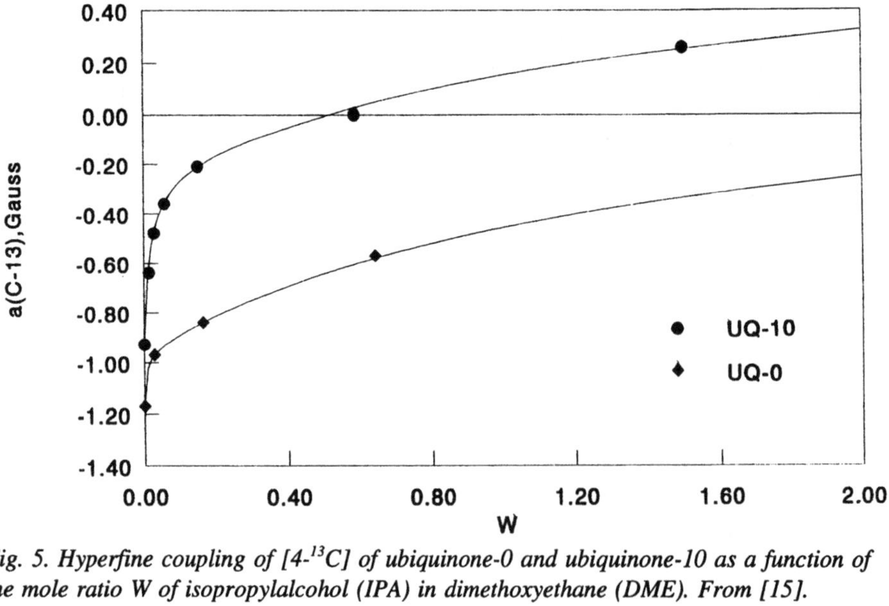


*Fig. 5. Hyperfine coupling of [4-$^{13}$C] of ubiquinone-0 and ubiquinone-10 as a function of the mole ratio W of isopropylalcohol (IPA) in dimethoxyethane (DME). From [15].*

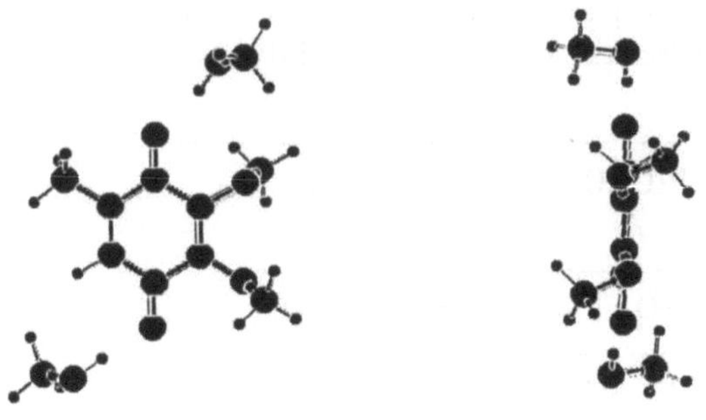


*Fig. 6. Geometry of the ubiquinone-0 anion complexed with two methanol molecules as calculated by the PM3 method. (1) front view, (2) side view. From [11].*

optimal geometry, the spin densities and corresponding hf-couplings were calculated with an INDO (UHF) method. Some pertinent results are summarized in Tables 1,2. The enthalpy of H-bond formation is fairly high, and in the same range as found for anion-molecule complexes, for example the heat of formation of the hydrogen bond between the carboxyl group of acetate and the amide of an alanine model compound is 31.4 kcal/mol [19]. It is seen that complexation indeed has a strong calculated influence on the hf-couplings, agreeing with the measured values for high methanol concentration. For the one-MeOH complex, the hf-coupling with the H-bonded carbonyl carbon considerably decreases (i.e. becomes less negative), while that of the opposing carbonyl substantially increases (i.e. becomes more negative). When two H-bonds are formed, the opposing effects lead to a decrease in the hf-coupling with *both* carbonyl carbons. Experimentally, we have seen that the [4-$^{13}$C]-splitting becomes *positive* for high concentrations of EtOH, at which the UQ will certainly be mostly disolvated. As indicated above, this effect is due to the influence of the spin density on the oxygen and, to a lesser extent, on the neighbouring ring carbons. The geometry of the optimized structure

**Table 1.** Experimental and calculated (INDO) values of the $^{13}$C hyperfine coupling (G) for the radical anion of ubiquinone-0 and its complex with two methanol molecules.

| Carbon number | Experimental | | Calculated | |
|---|---|---|---|---|
| | HMPT | MeOH | Anion | Complex |
| 1 | −1.5 | −0.66 | −1.82 | −1.51 |
| 4 | −1.25 | −0.8 | −1.64 | −1.41 |
| -Me | 1,75 | 1,70 | −0.29 | −0.31 |

**Table 2.** Enthalpy of complex formation between the radical anion of ubiquinone-0 and two methanol molecules (ΔH), and calculated (PM3) H-bond distances.

| ΔH kcal/mol | R(O...O) Å | | R(O...H) Å | |
|---|---|---|---|---|
| | $C_1$=O | $C_4$=O | $C_1$=O | $C_4$=O |
| −18.6 | 2.73 | 2.74 | 1.76 | 1.79 |

of mono- and disolvated UQ-0 is depicted in Fig. 6. Note that for UQ-10 the addition of the isoprenoid chain will change the probability of complexation of the two carbonyls; the general conclusions, however, appear to remain valid.

## 5 EPR of $^{13}$C-labelled UQ-10 *in vivo*

The above EPR experiments were all carried out on ubisemiquinones in solution, and consequently yielded the isotropic hf-couplings. For the ubiquinones in the $Q_{A,B}$ binding sites of the RC protein in liquid or solid solution, EPR experiments yield the powder lineshape, as the rotational correlation time of the protein is much too slow to average out the anisotropic parts of the *g*- and hf-tensors. Individual hf-lines are smeared out, and hf-coupling constants can only be estimated by spectral simulation, involving a large parameter space spanned by the principal elements of all tensors, and their mutual orientations. Although it is in principle possible to limit the dimension of this space, for example by using known orientations and anisotropies of the hf-tensors, and the principal elements of the *g*-tensor derived from a high-frequency EPR powder spectrum of the (preferably perdeuterated, non-$^{13}$C-labelled) model compound *in vitro*, the resulting parameter space is still very large, and the hf-tensors derived from the simulation have at best a large error margin, and may not properly reflect possible protein-induced distortions (which after all is the information of interest).

Fortunately, the above bleak picture is brightened by the advent of very high frequency EPR (95, 135 and 240 GHz). At such high frequencies, the field range spanned by the *g*-tensor is so large that at the low- and high-field edges practically only RCs are observed that have the *x*- or *z*-axis of their *g*-tensor oriented along the magnetic field. In favourable cases one may then directly measure the hf-splittings along these directions, either directly from the EPR spectrum or with ENDOR. As a first step along this line of approach, we recently recorded the Q-band (35 GHz) EPR spectrum of $Q_A^-$ in $Zn^{++}$-substituted RCs of *Rb. sphaeroides* R26 [20]. (It will be recalled that the presence of the non-heme iron severely broadens the EPR line of $Q_A^-$; it is therefore necessary to remove the iron and

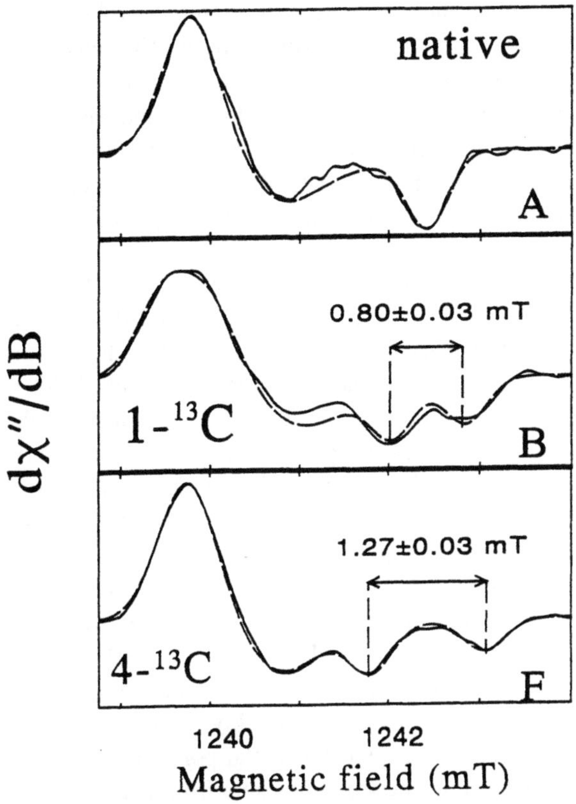


*Fig. 7. Q-band EPR spectra of the radical anion of ubiquinone-10 in RCs of Rb. sphaeroides R26 at 100 K. A, unlabelled; B, [1-$^{13}$C]UQ-10; F, [4-$^{13}$C]UQ-10. Adapted from [20].*

reconstitute the RCs with a diamagnetic divalent ion as $Zn^{++}$.) The spectra of 1- and 4-$^{13}$C-labelled $Q_A^-$, reproduced in Fig. 7, clearly show a splitting of the peak at $g_{zz}$, due to the $z$-component of the $^{13}$C-hf-tensor (the $z$-axis of which is to good approximation parallel to the $z$-axis of the quinone $g$-tensor). The two splittings are quite different (0.80±0.03 and 1.27±0.03 mT, respectively). This is in marked contrast with the two splittings for UQ-10$^-$ in frozen solution, which are virtually identical (1.13±0.04 and 1.10±0.04, respectively).

The marked asymmetry of the $a_{zz}$ component of the $^{13}$C=O hyperfine tensors for $Q_A^-$ must be induced by interactions with the protein, and it is tempting to attribute the higher value for [4-$^{13}$C] to a hydrogen bond. Note that we are here considering the *anisotropic* hf-coupling, not the isotropic coupling, and that therefore the considerations on the effect of hydrogen bonds on the hf-coupling mentioned above do not directly apply. In fact, one may to first approximation attribute the carbonyl $^{13}$C-$a_{zz}$ component to the sum of the isotropic coupling and the axial component of a uniaxial dipolar coupling between the unpaired electron spin in the $p_z$ orbital of the $^{13}$C nucleus and its nuclear spin, neglecting the coupling with the $p_z$-spin density on the neighbouring ring carbons, which is low (*vide supra*). One may then directly apply the calculated value for the axial coupling for $\rho(p_z) = 1$ ($a_\parallel$ = 76.6 G [21]), once the value for the isotropic coupling is known. The latter can be evaluated from the sum of the principal components of the hf-tensor, following the reasoning by Isaacson et al. [22] (this volume) that the large $a_{zz}$ component is certainly positive (the π-spin density on the carbonyl carbons in semiquinones is known to be positive), and since the trace of the dipolar part of the hf-tensor must be zero, the $a_{xx}$ and $a_{yy}$ components must be both negative. For [1-$^{13}$C]UQ-10 as $Q_A$ this gives $a_{iso} = (8.0 - 5.5 - 6.5)/3 = -1.3$ G, so that $\rho^{\pi}_{CO} = 9.3/76.6 = 0.121$; for [4-$^{13}$C]UQ-10 we obtain similarly a lower limit of $a_{iso} = +2.6$ G and a corresponding upper limit for $\rho^{\pi}_{CO} = 10.1/76.6 = 0.132$.

The spin density for 1-$^{13}$C=O of UQ-10 as $Q_A$ is somewhat higher than that found by Isaacson et al. [22] for RCs reconstituted with [1-$^{13}$C]UQ-3 (0.113), the difference seeming to be outside the margin of error and possibly due to the difference in side chain. The upper limit for the spin density on the 4-$^{13}$C is likewise a little higher than the value quoted in [22] (0.138), but here the difference is within the margin of error. Both spin densities are significantly smaller than the spin densities obtained above for UQ-10 in DME or in IPA. Whether this is real, and pointing to a somewhat different spin density distribution for UQ-10$^-$ *in vivo* compared to liquid solution, possibly because of a protein-induced distortion, or an artefact caused by inaccuracies in the numerical values of the constants in Eq. 3 is at present difficult to decide. The calculated values of $a_{iso}$ can be compared with the measured values for UQ-10 *in vitro*, –0.93 G for DME and +0.63 G for DME-80% IPA [14,15]. For 1-$^{13}$C the calculated $a_{iso}$ values are negative: –1.3 G (–0.5 G), pointing to a non- (or weakly) hydrogen-bonded carbonyl; those for 4-$^{13}$C are positive, ≥2.6 G (+0.19 G), and pointing to a strong (or medium-strong) hydrogen-bond (values between brackets from [22]).

A similar exercise as above can be carried out for $Q_B^-$, be it that here it is necessary to use a mutant of *Rb. sphaeroides* that is able to bio-incorporate $Zn^{++}$ instead of $Fe^{++}$ [22], as the reconstitution of isolated RCs leads to loss of the RC H-subunit and concomitant loss of $Q_B$ binding. The result is that the anisotropic hf-components for the 1- and 4-$^{13}$C in RCs reconstituted with UQ-3 are quite similar (see Isaacson et al. [22], this volume), $a_{iso}$ being slightly positive, indicating weak hydrogen bonds to both carbonyls.

## 6 Electron spin echo envelope modulation (ESEEM) spectroscopy of $Q_A^-$

Candidates for the strong hydrogen bond inferred above for 4-C=O of $Q_A^-$ are furnished by the X-ray structure. Both HisM219 and ThrM222 are within hydrogen-bonding distance, the bond being to the imidazole $N^{\delta(1)}$ or to the hydroxyl, respectively [23,24,25]. The method of choice to discriminate between the two amino acids is ESEEM spectroscopy. When the unpaired electron couples to a $^{14}$N, under certain conditions (isotropic hf-coupling approximately equal to the nuclear Zeeman interaction), the Fourier Transform spectrum contains three peaks stemming from the quadrupole coupling of the $^{14}$N ($I$=1) nucleus. From the three frequencies the quadrupole coupling and anisotropy parameters *eqQ* and $\eta$ can be determined accurately. These two parameters are sensitive to the very local environment of the $^{14}N$ nucleus, providing a fingerprint of the group structure containing the nitrogen. In a recent ESEEM study we have obtained unequivocal evidence that $Q_A^-$ is bound to an imidazole $N^{\delta(1)}$ [26]. On basis of the crystal structure we assign this nitrogen to HisM219.

## 7 Fourier-Transform Infra-Red spectroscopy of $Q_A$ and $Q_B$

One of the most important spectroscopic techniques to assess hydrogen bonding to the cofactors in bacterial and plant RCs is Fourier-transform infra-red spectroscopy, FTIR [27]. The IR absorption line of the carbonyl stretch, for example, is exquisitely sensitive to changes in bond order, and therefore to hydrogen bonding [28]. Obviously, to fish the relevant carbonyl out of the sea of carbonyl resonances of the entire protein requires some form of difference spectroscopy. This is accomplished by subtracting spectra of neutral RCs and of RCs in which $Q_A$ or $Q_B$ is singly reduced. Because the semiquinone has little absorbance in the 1700-1500 $cm^{-1}$ region where the carbonyls absorb, the difference spectrum reveals directly bands belonging to the carbonyl groups of the quinone. Further differentiation is achieved by taking the difference of such a spectrum for unlabelled RCs and for RCs reconstituted with UQ-10 selectively $^{13}C$-labelled at one of several carbon positions: the double-difference spectrum. The first such spectra were obtained by Brudler et al. [29] for UQ-10 labelled at the 1-, 2-, 3- and 4-C position, and later independently by Breton et al. [30], the latter work employing the non-native UQ-3 labelled at the 1- and 4-C positions only. A first result of Brudler et al. [29] is that the C=O and C=C stretches are severely mixed. Nevertheless, it is possible to discriminate two resonances that are dominated by the 1-C=O and the 4-C=O stretches, respectively: 1660 and 1601 $cm^{-1}$, which shift to 1630 and 1579 $cm^{-1}$ upon $^{13}C$-labelling, respectively [29] (see also Brudler et al., this volume). These frequencies have to be compared with those of UQ-10 *in vitro*: 1650/1666 $cm^{-1}$ [29], which shift to 1620 $cm^{-1}$ upon 1- or 4-$^{13}C$-labelling.

The shift of ~60 $cm^{-1}$ for the 4-C=O stretch indicates a change in bond order of ~0.1 [28], which could be induced by a strong hydrogen bond. On the other hand, a change in bond order may also be occasioned by a distortion of the molecule, for example a twist of the ring structure that changes the $sp_2$ hybridization. From FTIR data alone, it cannot be decided which effect predominates. The EPR results, especially the $^{14}N$-quadrupole parameters obtained with ESEEM spectroscopy (see above), prove that for the *semi*quinone, there is indeed a hydrogen bond to the imidazole $N^{\delta(1)}$ of HisM219, but additional contributions of molecular distortion cannot be excluded. Another spectroscopy that potentially can discriminate between distortion and hydrogen bonding is NMR: a distortion does not necessarily lead to a change in charge on the carbonyl; a hydrogen bond, however, induces additional electronegativity on the carbonyl oxygen, which should be reflected in a shift of the $^{13}C$ isotropic chemical shift for the hydrogen-bonded carbonyl. Note that also NMR spectroscopy yields results for the *neutral* quinone. This work will be discussed below.

Analogous FTIR work as above has been carried out for $^{13}C$-labelled $Q_B$. Now, the two carbonyls show resonances at 1665/1651 $cm^{-1}$, which shift to 1620 (1-$^{13}C$) and 1618 (4-$^{13}C$) $cm^{-1}$ upon $^{13}C$-labelling, respectively [31] (see also Brudler et al. and Breton et al., this volume). Thus, the bond orders of the two carbonyls in $Q_B$ are indistinguishable from those of UQ-10 *in vitro* in a non-protic solvent, indicating no, or very weak hydrogen bonding for both of them. This is probably related to the much weaker binding of $Q_B$ compared to $Q_A$. Interestingly, the EPR results on $Q_B^-$ (see above) seem to indicate weak hydrogen bonds. The difference with the FTIR results is probably related to the fact that the energy of H-bond formation with an anion is generally much larger than for neutral molecules [11,19], viz. about 15 kcal/mol for UQ-10$^-$ [11] compared to 3-5 kcal/mol for the average H-bond with neutral molecules. Consequently, the hydrogen bond with an anion is much stronger than for the neutral molecule, and the semiquinone will "pull" the hydrogen donating groups of the protein polypeptides closer. For the 4-C=O group of $Q_A$ this effect might not be important, as it is already strongly bound in the neutral quinone.

# 8 Solid-state CP-MAS NMR of $^{13}C$-labelled $Q_A$

Another spectroscopic method to assess hydrogen bonding is Magic-Angle Spinning Cross-Polarization solid-state NMR of site-selected $^{13}C$-labelled compounds. In principle, the trace and the anisotropy of the carbonyl $^{13}C$-chemical shift tensor are sensitive to the presence of a hydrogen bond [32,33]. We have recently measured the $^{13}C$ MAS-NMR spectrum of 1- and 4-$^{13}C$ labelled UQ-10 *in vitro* and in the $Q_A$-binding site of RCs of *Rb. sphaeroides* R26 using difference spectroscopy [34]. The isotropic chemical shifts of 1- and 4-$^{13}C$=O were almost identical (183.8 and 183.1 ppm, respectively; measuring accuracy ±0.2 ppm), and differed little from the values for crystalline UQ-10: 183.6 and 184.2 ppm, respectively. Remarkably, the 4-$^{13}C$ resonance could only be observed for temperatures below 255 K, in distinct contrast with the 1-$^{13}C$ resonance, which was observable up to 230 K. In view of the FTIR results discussed above, the slight difference in isotropic chemical shift of the 4-C=O group of $Q_A$ compared to that of UQ-10 *in vitro* is surprising. If indeed the 60 $cm^{-1}$ shift of the 4-carbonyl resonance is related to hydrogen bonding, then this bond should be quite strong, and give rise to a pronounced electronegativity of the carbonyl oxygen. The additional charge on the oxygen would shift the $^{13}C$ NMR line several ppm [32,33], contrary to observation. As mentioned above, FTIR measures changes in bond order, and it is therefore not excluded, and perhaps even likely, that for the neutral quinone, the shift of the 4-C=O stretch is predominantly due to a change in $sp_2$

hybridization, and not to hydrogen bonding. The anisotropy parameter $\eta$ of the two carbonyls differed more than their isotropic shift: For crystalline UQ-10 it was 0.7±0.1 for both carbonyls, whereas for $Q_A$ it was 0.6 and 0.3 (±0.2) for 1- and 4-$^{13}$C=O, respectively. The latter difference in $\eta$ is compatible with the change in bond order of 0.1 inferred from the FTIR results, when it is assumed that it is primarily due to a change in hybridization [34,35].

The marked difference in temperature dependence of the 1- and 4-$^{13}$C resonances was investigated in a MAS-NMR relaxation study [36]. It turned out that neither the longitudinal, nor the transversal relaxation times could be responsible for the difference. The rotational relaxation time $T_1^\rho$, however, was for 1-$^{13}$C practically temperature independent between 250 and 300 K (about 8 ms), whereas for 4-$^{13}$C it decreased from 8 ms at 230 K to 3.8 ms at 50 K. Because $T_1^\rho$ characterizes the depopulation of the energy levels created by the spin-lock field during cross polarization, its marked decrease with rising temperature provides a ready explanation of the loss of 4-$^{13}$C signal above 255 K. Presumably, slow dynamic $Q_A$-protein interactions on a time scale of about 50 kHz, which interfere destructively with the efficiency of cross polarization, are more important around the 4-C=O than at the 1-C=O end of $Q_A$.

## Prospects

In the short survey presented above, the use of site-directed isotope labelling in spectroscopy of photosynthetic RCs has been illustrated, focusing on quinone binding in bacterial RCs. Recently, several other examples of the application of this stratagem on bacterial and plant photosynthetic preparations have been published, for example work by the groups of Babcock, Styring and Barry on the tyrosyl donors in plant photosystem II [37,38], and the MAS-NMR work of De Groot c.s. [39,40] on [4'-$^{13}$C]-tyrosyls in bacterial RCs. There, also site-directed mutagenesis was employed as a powerful technique to assign spectral features, and to explore interactions between the cofactors and the protein matrix. Thus, it appears that the strategy of combining SDIL and SDM with several spectroscopy techniques will become more and more important in biophysical research in photosynthesis.

For the specific topic of this contribution, it is clear that here also SDM can contribute importantly in the search for supportive evidence for the various conclusions and ideas discussed above. For example, the question whether the 4-C=O of $Q_A$ has a bifurcated hydrogen bond to the HisM219 and ThrM222 might be settled by studying RCs in which HisM219 is replaced by a non-hydrogen bonding amino acid. Binding of the 1-C=O group to AlaM260 might similarly be assessed. Complementing these studies with investigations of RCs in which relevant $^{14}$N nuclei are replaced by $^{15}$N will give further information on the hydrogen bonding pattern. Studies along these lines are now in progress in our laboratories.

## Acknowledgements

We thank Drs. R. Brudler, K. Gerwert and H.J.M. de Groot for numerous discussions on the implications of FTIR and NMR for quinone binding to the reaction center protein. The views expressed in this contribution, however, are entirely our own. We are indebted to R.A. Isaacson et al. for sending us a preprint of their work. Work carried out in Leiden was supported by the Netherlands Foundation for Chemical Research (SON), financed by the Netherlands Organization for Scientific Research (NWO). TNK and RIS acknowledge travel grants from NWO and NATO, respectively. PG was a research fellow of the Royal Netherlands Academy of the Arts and Sciences (KNAW).

## References

1. Lugtenburg, J., Mathies, R.A., Griffin, R.G. and Herzfeld, J. (1988) Trends Biochem. Sci. 13, 388-393
2. Breton, J., Burie, J.-R., Berthomieu, C., Berger, G. and Nabedryk, E. (1994) Biochemistry 33, 4953-4965
3. Spoyalov, A.P., Samoilova, R.I., Tyryshkin, A.M., Dikanov, S.A., Liu, B.-L. and Hoff, A.J. (1992) J. Chem. Soc. Perkin Trans. 2, 1519-1524
4. Schelvis, J.P.M., Liu, B.-L., Aartsma, T.J. and Hoff, A.J. Biochim. Biophys. Acta (1992) 1102, 229-236
5. Gunner, M.R. and Dutton, P.L. (1989) J. Am. Chem. Soc. 111, 3400-3412
6. Van Liemt, W.B.S., Steggerda, W.F., Esmeijer, R. and Lugtenburg, J. (1994) Rec. Trav. Chim. Pays-Bas 113, 153-161
7. Paddock, M.L., Rongey, S.H., McPherson, P.H., Juth, A.C., Feher, G. and Okamura, M.Y. (1994) Biophys. J., 66, A1, Abst. Su-AM-3
8. Karplus, M.A. and Fraenkel, G.K. (1961) J. Chem. Phys. 35, 1312-1323
9. Das, M.R. and Fraenkel, G.K. (1965) 1350-1360
10. Samoilova, R.I., Van Liemt, W., Steggerda, W.F., Lugtenburg, J., Hoff, A.J., Spoyalov, A.P., Tyryshkin, A.M., Gritzan, N.P. and Tsvetkov, Yu.D. (1994) J. Chem. Soc. Perkins Trans. 2, 609-614
11. Samoilova, R.I., Gritzan, N.P., Hoff, A.J., Van Liemt, W., Lugtenburg, J., Spoyalov, A.P. and Tsvetkov, Yu.D. (1995) J. Chem. Soc. Perkins Trans., in the press
12. Stone, E.W. and Maki, A.H. (1962) J. Chem. Phys. 36, 1944-1945
13. Gendell, J., Freed, J.H. and Fraenkel, G.K. (1962) J. Chem. Phys. 2832-2841
14. Kropacheva, T.N., Raap, J., Lugtenburg, J. and Hoff, A.J. (1995) Proc. Xth Congress Photosynth., Montpellier, in the press.
15. Kropacheva, T.N., Raap, J., Lugtenburg, J. and Hoff, A.J. (1995) in preparation
16. De Boer, E. and Mackor, E.L. (1963) J. Chem. Phys. 1450-1452

17. Freed, J.H. and Fraenkel, G.K. (1963) J. Chem. Phys. 39, 326-348; (1964) J. Chem. Phys. 40, 1815-1829
18. Prabhananda, B.S., Khakhar, M.P.K. and Das, M.R. (1968) J. Am. Chem. Soc. 90, 5980-5986
19. Meot-Ner, M. (1988) J. Amer. Chem. Soc. 110, 3854-3558
20. Van den Brink, J.S., Spoyalov, A.P., Gast, P., Van Liemt, W.B.S., Raap, J., Lugtenburg, J. and Hoff, A.J. (1994) FEBS Lett. 353, 273-276
21. Wertz, J.E. and Bolton, J.R. (1972) Electron Spin Resonance, McGraw-Hill, New York
22. Isaacson, R.A., Abresch, E.C., Lendzian, F., Boullais, C., Paddock, M.L., Mioskowski, C., Lubitz, W. and Feher, G. (1995) Proc. Feldhafing Conf. (Michel-Beyerle, M.E., ed.) in the press
23. Allen, J.P., Feher, G., Yeates, T.O., Komiya, H. and Rees, D.C. (1988) Proc. Natl. Acad. Sci. USA 85, 8487-8491
24. Chang, C.-H., El-Kabbani, O., Tiede, D., Norris, J.R. and Schiffer, M. (1991) Biochemistry 30, 5352-5360
25. Ermler, U., Fritzsch, U., Buchanan, S. and Michel, H. (1994) Structure 2, 925-936
26. Bosch, M.K., Gast, P., Hoff, A.J., Spoyalov, A.P. and Tsvetkov, Yu. D. (1994) Chem. Phys. Lett. 239, 306-312
27. Mäntele, W. (1993) in The Bacterial Reaction Center (Deisenhofer, J. and Norris, J.R., eds.) Vol. II, pp. 239-283, Acad. Press, San Diego, USA
28. Josien, M.-L., Fuson, L., Lebas, J.-M. and Gregory, T.M. (1953) J. Chem. Phys. 21, 331-340
29. Brudler, R., De Groot, H.J.M., Van Liemt, W.B.S., Steggerda, W.F., Esmeijer, R., Gast, P., Hoff, A.J., Lugtenburg, J. and Gerwert, K. (1994) EMBO J. 13, 5523-5530
30. Breton, J., Boullais, C., Burie, J.-R., Nabedryk, E. and Mioskowski, C. (1994) Biochemistry 33, 14378-14386
31. Brudler, R., De Groot, H.J.M., Van Liemt, W.B.S., Gast, P., Hoff, A.J., Lugtenburg, J. and Gerwert, K. (1995) FEBS Lett. 370, 88-92
32. Naito, A. and McDowell, C.H. (1984) J. Chem. Phys. 81, 4795-4803
33. Sastry, D.L., Takegoshi, K. and McDowell, C.H. (1987) Carbohydr. Res. 165, 161-171
34. Van Liemt, W.B.S., Boender, G.J., Gast, P., Hoff, A.J., Lugtenburg, J. and De Groot, H.J.M. (1995) Biochemistry 34, 10229-10236
35. De Groot, H.J.M. (1995) Proc. Xth Congress Photosynth., Montpellier, in the press
36. Van Rossum, B.-J., Van Liemt, W.B.S., Boender, G.J., Gast, P., Hoff, A.J., Lugtenburg, J. and De Groot, H.J.M. (1995) Proc. Xth Congress Photosynth., Montpellier, in the press
37. Tommos, C., Tang, X.-S., Warncke, K., Hoganson, C.W., Styring, S., McCracken, J., Diner, B. and Babcock, G.T. (1995) J. Am. Chem. Soc., in the press

38. Bernard, M.T., MacDonald, B.M., Nguyen, A.P., Debus, R.J. and Barry, B.A. (1995) J. Biol. Chem. 270, 1589-1594
39. Fischer, M.R., De Groot, H.J.M., Raap, J., Winkel, C., Hoff, A.J. and Lugtenburg, J. (1992) Biochemistry 31, 11038-11049
40. Shochat, S., Gast, P., Hoff, A.J., Boender, G.J., Van Leeuwen, S., Van Liemt, W.B.S., Vijgenboom, E., Raap, J., Lugtenburg, J. and De Groot, H.J.M. (1995) Spectrochim. Acta 51A, 135-144

# Subject Index

# Index of Authors